The Science of Entomology

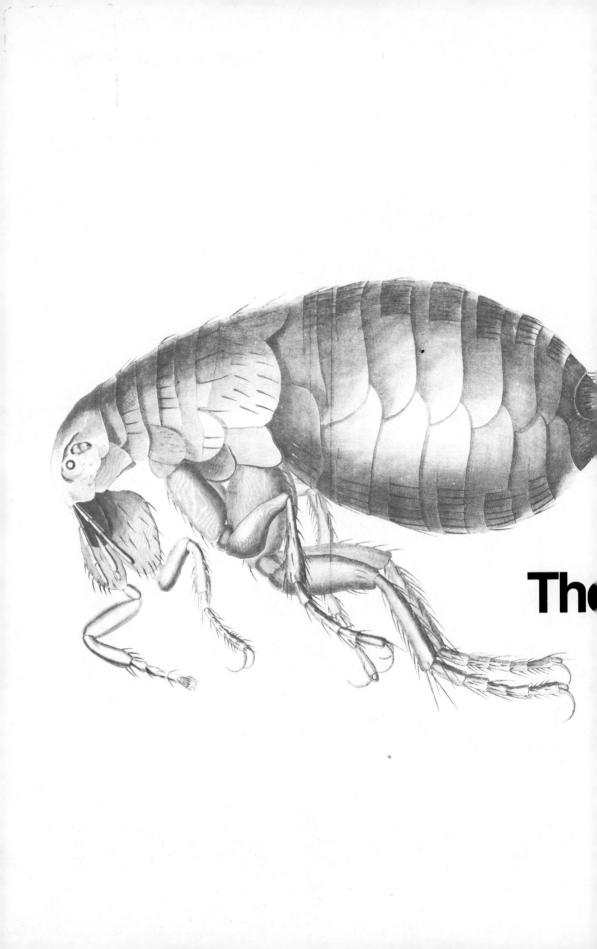

Th

William S. Rom

*Department of Zoology and M
Ohio*

Science of Entomology

SECOND EDITION

Macmillan Publishing Co., Inc.
New York

Collier Macmillan Publishers
London

To Anne and Regan

MACMILLAN PUBLISHING CO., INC.
866 Third Avenue, New York, New York 10022

COLLIER MACMILLAN CANADA, LTD.

Library of Congress Cataloging in Publication Data

Romoser, William S
  The science of entomology.

  Bibliography: p.
  Includes index.
  1. Entomology. I. Title.
QL463.R65   1981   595.7   80-17793
ISBN 0-02-403410-X

Printing: 1 2 3 4 5 6 7 8   Year: 1 2 3 4 5 6 7 8

Acknowledgements

For permission to include materials from the following publications, as indicated in the figure and table captions.

ACADEMIC PRESS, INC.
Gilbert, 1964: Physiology of growth and development: endocrine aspects, by L. I. Gilbert, in *The Physiology of Insecta*, Vol. 1, pp. 149–225, M. Rockstein, ed. Copyright 1964 by Academic Press, Inc., New York

Hoyle, 1965: Neural control of skeletal muscles, by G. Hoyle, in *The Physiology of Insecta*, Vol. 2, pp. 407–49, M. Rockstein, ed. Copyright 1965 by Academic Press, Inc., New York.

Miller, 1964: Respiration—aerial gas transport, by P. L. Miller, in *The Physiology of Insecta*, Vol. 3, pp. 557–615, M. Rockstein, ed. Copyright 1964 by Academic Press, Inc., New York.

Rockstein and Miquel, 1973: Aging in insects, by M. Rockstein and J. Miquel, in *The Physiology of Insecta*, 2nd ed., Vol. I, pp. 371–478, M. Rockstein, ed. Copyright 1973 by Academic Press Inc. (London) Ltd.

Smith and Treherne, 1963: Functional aspects of the organization of the insect nervous system, by D. S. Smith and J. E. Treherne, Adv. Insect Physiol., I:401–84. Used with permission from Advances in Insect Physiology. Copyright by Academic Press Inc. (London) Ltd.

AMERICAN ASSOCIATION FOR THE ADVANCEMENT OF SCIENCE
Locke, 1965: Permeability of the insect cuticle to water and lipids, by M. Locke, Science, 147:295–98. Copyright 1965 by the American Association for the Advancement of Science.

THE AMERICAN ENTOMOLOGICAL SOCIETY
Breland, Eddleman, and Biesele, 1968: Studies of insect spermatozoa I, by O. P. Breland, C. D. Eddleman, and J. J. Biesele, Entomol. News, 79(8):197–216, Philadelphia, 1968.

THE BELKNAP PRESS OF HARVARD UNIVERSITY PRESS
Wilson, 1971: *The Insect Societies*, by E. O. Wilson. © Copyright 1971 by the President and Fellows of Harvard College.

BOREAL LABS
Selected portions of key cards.

AMERICAN SCIENTIST
Harcourt and Leroux, 1967: Population regulation in insects and man, by D. G. Harcourt and E. J. Leroux, Amer. Sci., 55(4):400–15, New Haven, CT, 1967.

Acknowledgments

AMERICAN SOCIETY OF ZOOLOGISTS

Rothenbuhler, 1964: Behavior genetics of nest cleaning in honey bees IV. Response of F_1 and backcross generations to disease-killed brood, by W. C. Rothenbuhler, Amer. Zool., 4:111—23.

ANNUAL REVIEWS, INC.

Gelperin, 1971: Regulation of feeding, by A. Gelperin, Ann. Rev. Entomol., 16:365–78. Reproduced, with permission, from the Annual Review of Entomology Volume 16. © 1971 by Annual Reviews Inc.

Jander, R. 1963. Insect orientation, by R. Jander, Ann. Rev. Entomol., 8:94–114, Palo Alto, CA, 1963.

ASSOCIATED BOOK PUBLISHERS LTD.

Clark et al., 1967: *The Ecology of Insect Populations in Theory and Practice,* by L. R. Clark, P. W. Geier, R. D. Hughes, and R. F. Morris, London, 1967.

Wigglesworth, 1965: *The Principles of Insect Physiology,* 6th ed., by V. B. Wigglesworth, London, 1965.

CAMBRIDGE UNIVERSITY PRESS

Caveny, 1969: Muscle attachment related to cuticle architecture in Apterygota, by S. Caveny, J. Cell Science, 4:541–59, Cambridge.

THE COMPANY OF BIOLOGISTS LTD.

Beament, 1958: The effect of temperature on the waterproofing mechanism of an insect, by J. W. L. Beament, J. Exp. Biol., 35:494–519, Cambridge, 1958.

Wigglesworth, 1930: The formation of the peritrophic membrane in insects with special reference to the larvae of mosquitoes, by V. B. Wigglesworth, Quart. J. Microscop. Sci., 73:593–616, Cambridge, 1930.

Wigglesworth, 1945: Transpiration through the cuticle of insects, by V. B. Wigglesworth, J. Exp. Biol., 21:97–114, Cambridge, 1945.

CORNELL UNIVERSITY PRESS

Comstock, 1918: *The Wings of Insects,* by John Henry Comstock. Copyright 1918 by the Comstock Publishing Company, Inc. Used by permission of Cornell University Press.

Comstock, 1940: *An Introduction to Entomology,* by John Henry Comstock. Copyright 1933, 1936, 1940 by Comstock Publishing Company, Inc. Used by permission of Cornell University Press.

von Frisch, 1950: *Bees: Their Vision, Chemical Senses, and Language,* by Karl von Frisch. Copyright© 1950 by Cornell University. Used by permission of Cornell University Press.

Matheson, 1944: *Handbook of the Mosquitoes of North America,* Second Edition, by R. Matheson. Copyright 1944 by Comstock Publishing Company, Inc. Used by permission of Cornell University Press.

Maynard, 1951: *The Collembola of New York State,* by Elliott A. Maynard. Copyright 1951 by Comstock Publishing Company, Inc. Used by permission of Cornell University Press.

Snodgrass, 1956: *Anatomy of the Honeybee,* by R.E. Snodgrass. © 1956 by Cornell University. Used by permission of Cornell University Press.

West, 1951: *The Housefly: Its Natural History, Medical Importance, and Control,* by Luther S. West. Copyright 1951 by Comstock Publishing Company, Inc. Used by permission of Cornell University Press.

Wigglesworth, 1959a: *The Control of Growth and Form: A Study of the Epidermal Cell in an Insect,* by V. B. Wiggleworth.© 1959a: *The Control of Growth and Form: A Study of the Epidermal Cell in an Insect,* by V. B. Wigglesworth.© 1959 by Cornell University. Used by permission of Cornell University Press.

Snodgrass, 1963b: The anatomy of the honey bee, in *The Hive and the Honey Bee,* pp. 141–90, R. A. Grout, ed., Hamilton, IL, Dadant & Sons, Inc.

ELSEVIER NORTH HOLLAND, INC.

Chapman, 1971: *The Insects—Structure and Function,* 2nd ed., by R. F. Chapman. Copyright 1971 by Elsevier North Holland, Inc. Reprinted by permission of Elsevier North Holland, Inc.

ENTOMOLOGICAL SOCIETY OF AMERICA

Claassen, 1931: *Plecoptera Nymphs of America (North of Mexico),* by P. W. Claassen, The Thomas Say Foundation, Publ. 3.

Ewing, 1940: *The Protura of North America,* by H. E. Ewing, Ann. Entomol. Soc. Amer., 33(3): 495–551.

Froeschner, 1947: Notes and keys to the Neuroptera of Missouri, by R. C. Froeschner, Ann. Entomol. Soc. Amer., 40(1):123–36, College Park, MD., 1947.

Hess, 1917: The chordotonal organs and pleural discs of cerambycid larvae, by W. N. Hess, Ann. Entomol. Soc. Amer., 10:63–78, College Park, MD., 1917.

Wolglum and McGregor, 1958: Observations on the life history and morphology of *Agulla bractea* Carpenter (Neuroptera: Raphidiodea: Raphidiidae), by R. S. Wolglum and F. A. McGregor, Ann. Entomol. Soc. Amer., 51:129–41, College Park, MD, 1958.

Acknowledgments

ENTOMOLOGICAL SOCIETY OF WASHINGTON

Caudell, 1918: Zoraptera not an apterous order, by A. N. Caudell, Proc. Entomol. Soc. Wash., 22:84–97, Washington, DC, 1918.

GUSTAV FISCHER VERLAG

Barth, 1937: Muskulatur and Bewegungsart der Raupen, by R. Barth, Zool. Jahrb., 62:507–66, Jena, Germany, 1937.

Vogel, 1921: Zur Kenntnis des Baues und der Funktion des Stachels und des Vorderdarmes der Kleiderlaus, by R. Vogel, Zool. Jahrb., Anat., 42:229–58, 1921.

HOLT, RINEHART AND WINSTON, INC.

Borror and DeLong, 1971:An Introduction to the Study of Insects, third edition, by Donald J. Borror and Dwight M. DeLong. Copyright © 1964, 1971 by Holt, Rinehart and Winston, Inc. Copyright 1954 by Donald J. Borror and Dwight M. DeLong. Reprinted by permission of Holt, Rinehart and Winston, Inc.

HUTCHINSON AND CO. (PUBLISHER) LTD.

Cornwell, 1968: The Cockroach, Vol. 1, by P. B. Cornwell. © 1968 by P. B. Cornwell; Drawings © 1968 by Rentokil Laboratories Ltd.

LONGMAN GROUP LTD.

Skaife, 1961: Dwellers in Darkness, by S. H. Skaife, London, 1961.

MCGRAW-HILL BOOK COMPANY

Folsom and Wardle, 1934: Entomology, with Special Reference to Its Ecological Aspects,4th ed., by J. W. Folsom and R. A. Wardle, copyright 1934 by McGraw-Hill, Inc. Used with permission of McGraw-Hill Book Company.

Graham and Knight, 1965: Principles of Forest Entomology, 4th ed., by S. A. Graham and F. B. Knight. Copyright 1965 by McGraw-Hill Book Company. Used with permission of McGraw-Hill Book Company.

Johannsen and Butt, 1941: Embryology of Insects and Myriapods, by O. A. Johannsen and F. H. Butt. Copyright 1941 by McGraw-Hill, Inc. Used with permission of McGraw-Hill Book Company.

Metcalf, Flint, and Metcalf, 1962: Destructive and Useful Insects, by C. L. Metcalf, W. P. Flint, and R. L. Metcalf. Copyright 1962 by McGraw-Hill, Inc. Used with permission of McGraw-Hill Book Company.

Snodgrass, 1925: Anatomy and Physiology of the Honey Bee, by R. E. Snodgrass. Copyright 1925 by McGraw-Hill, Inc. Used with permission of McGraw-Hill Book Company.

Snodgrass, 1935: Principles of Insect Morphology, by R. E. Snodgrass. Copyright 1935 by McGraw-Hill, Inc. Used with permission of Mcgraw-Hill Book Company.

Wickler, 1968: Mimicry in Plants and Animals by W. Wickler. Copyright © 1968 by McGraw-Hill, Inc. Used with permission of McGraw-Hill Book Company.

MACMILLAN PUBLISHING CO., INC.

Essig, 1942: College Entomology, by E. O. Essig. Copyright 1942 by Macmillan Publishing Co., Inc., New York.

Essig, 1958: Insects and Mites of Western North America, by E. O. Essig. Copyright 1958 by Macmillan Publishing Co., Inc., New York.

Herms and James, 1961: Medical Entomology, 5th ed. Copyright 1961 by Macmillan Publishing Co., Inc., New York.

Herreid, 1977: Biology, by C. F. Herreid, II. Copyright © 1977, by Macmillan Publishing Co., Inc., New York.

James and Harwood, 1969: Herm's Medical Entomology, by M. T. James and R. F. Harwood. Copyright 1969 by Macmillan Publishing Co., Inc., New York.

Stiles, Hegner, and Boolootian, 1969: College Zoology, 8th ed., by K. A. Stiles, R. W. Hegner, and R. Boolootian. Copyright 1969 by Macmillan Publishing Co., Inc., New York.

THE MARINE BIOLOGICAL LABORATORY

Schneiderman and Williams, 1955: An experimental analysis of the discontinuous respiration of the cecropia silkworm, by H. A. Schneiderman and C. M. Williams, Biol. Bull., 109:123–43, Woods Hole, MA, 1955.

OLIVER & BOYD LTD.

Davey, 1965: Reproduction in the Insects, by K. G. Davey, Edinburgh, 1965

Wigglesworth, 1970: Insect Hormones, by V. B. Wigglesworth, Edinburgh, 1970.

PAUL PAREY

Sturm, 1956: Die Paarung beim Silberfischen, Lepisma saccharina, by H. Sturm, Z. Tierpsychol., 13:1–2, Berlin, 1956.

PERGAMON PRESS, INC.

Engelmann, 1970: The Physiology of Insect Reproduction, by F. Engelmann, New York, 1970. Reprinted with permission.

Acknowledgments

ALVAH PETERSON

Peterson, 1951: *Larvae of Insects*, part II, Coleoptera, Diptera, Neuroptera, Siphonaptera, Mecoptera, Trichoptera, by A. Peterson, Columbus, OH. 1951.

Peterson, A. 1948: *Larvae of Insects*, part I, Lepidoptera and plant infesting Hymenoptera, by A. Peterson, Columbus, OH, 1948.

QUAESTIONES ENTOMOLOGICAE

Rempel, 1975: *The evolution of the insect head: the endless dispute*, by J. G. Rempel, Quaestiones Entomologicae, 11:7–25, 1975.

THE RONALD PRESS COMPANY

Hagen, 1951: *Embryology of the Vivparous Insects*, by H. R. Hagen. Copyright 1951. The Ronald Press Company, New York.

Rolston and McCoy, 1966: *Introduction to Applied Entomology*, by L. H. Rolston and L. E. McCoy. Copyright 1966, The Ronald Press Company, New York.

THE ROYAL ENTOMOLOGICAL SOCIETY

Berridge, 1970: *A Structural Analysis of Intestinal Absorption*, by M. J. Berridge, in *Insect Ultrastructure*, pp. 135–151, A. C. Neville, ed., Symposia of the Royal Entomological Society of London, No. 5. © 1970 by The Royal Entomological Society.

THE ROYAL SOCIETY

Gillett and Wigglesworth, 1932: The climbing organ of an insect, *Rhodnius prolixus* (Hemiptera: Reduviidae). Proc. Roy. Soc. London, Series B, 111:364–76, 1932.

W. B. SAUNDERS COMPANY

Patton, 1963: *Introductory Insect Physiology*, by R. L. Patton, Philadelphia, 1963.

SPRINGER-VERLAG

Baldus, 1926: Experimentelle Untersuchungen über die Entfernungslokalisation der Libellen *(Aeschna cyanea)*, by K. Baldus, Z. vergl. Physiol., 3:375–505, Berlin, 1926.

Fraenkel, 1932: Untersuchungen über die Koordination von Reflexen und automatischnervösen Rhythmen bei Insekten III, by G. Fraenkel, Z. vergl. Physiol., 16:394–460, Berlin, 1932.

Goetsch, 1957: *The Ants*, by W. Goetsch. Copyright © by Springer-Verlag, West Berlin, Germany.

Hertz, 1929: Die Organization des Optischen Feldes bei der Biene, I, by M. Hertz, Z. vergl. Physiol., 8:693–748, Berlin, 1929.

Markl, 1962: Schweresinnesorgane bei Ameisen und anderen Hymenopteren, by H. Markl, Z. verfl. Physiol., 44:475–569, Berlin, 1962.

Nachtigall, 1960:Über kinematik, Dynamik und Energetik des schwimmens einheimischen Dytisciden, by W. Nachtigall, Z. vergl. Physiol., 43:48–118, Berlin, 1960.

Nachtigall, 1963: Zur Lokomotionsmechanik schwimmender Dipterenlarven, by W. Nachtigall, Z. vergl. Physiol., 46:449–66, Berlin, 1963.

UNIVERSITY OF CHICAGO PRESS

Andrewartha and Birch, 1954: *The Distribution and Abundance of Animals*, by H. G. Andrewartha and L. C. Birch. Copyright 1954 by The University of Chicago. All rights reserved. Published 1954.

UNIVERSITY OF MINNESOTA PRESS

Richards, 1951: *The Integument of Arthropods*, by A. Glenn Richards. University of Minnesota Press, Minneapolis, © 1951 University of Minnesota.

VAN NOSTRAND REINHOLD COMPANY

Cummins, Miller, Smith, and Fox, 1965: *Experimental Entomology*, by K. W. Cummins, L. D. Miller, Ned A. Smith, and R. M. Fox, New York, 1965.

DuPorte, 1961: *Manual of Insect Morphology*, by E. M. DuPorte, New York, 1961.

VERLAG CHEMIE GMBH

Schildknecht and Holoubek, 1961: Die Bombardierkäfer und ihre Explosionschemie V. Mitteilung über insekten. Abwehrstoffe, by H. Schildknecht and K. Holoubek, Angew. Chem., 73(1):1–6, Weinheim/Bergstr., Germany, 1961.

GEORGE WEIDENFELD & NICOLSON LTD.

Chauvin, 1967: *The World of an Insect*, by R. Chauvin, London, World University Library.

Wickler, 1968: *Mimicry in Plants and Animals*, by W. Wickler, London, World University Library.

JOHN WILEY & SONS, INC.

Ross, 1965: *A Textbook of Entomology*, 3rd ed., by H. H. Ross, New York, 1965. By permission of John Wiley & Sons, Inc.

Schneirla, 1953: Insect behavior in relation to its setting, by T. C. Schneirla, in *Insect Physiology*, pp. 685–722, K. D. Roeder, ed., New York, 1953.

Preface

The magnitude of the role played by insects in the scheme of life is undisputed. In adaptive diversity and number of species, they are among the most successful of all organisms. The relatively few that are problems for us have taxed our ingenuity to its fullest throughout history, and the battle is far from over. Thus the science of entomology is one of the major areas of basic biology as well as a vital applied science.

My objective in writing this text has been to provide a broad, balanced introduction to the field of entomology. Its major role should be in the one-quarter or one-semester course in general entomology. However, I hope that the professional entomologist or zoologist who is engaged in one of the many specialized areas will find it useful as a reasonably up-to-date review. I have treated entomology as a branch of biology that has applied aspects but is not strictly an applied science. The major portion of the text is developed around two concepts: *structure* and *function* at the various levels of biological organization (Part One) and *unity* and *diversity* as the result of organic evolution (Part Two).

This second edition has been substantially revised. All chapters have been updated and/or expanded; the Selected References have been significantly updated and expanded; the chapter sequence in Part One has been altered; a glossary has been added; new illustrations have been added and few old illustrations omitted.

In Part One, the chapter "Nervous, Glandular, and Muscular Systems" now follows the discussion of the insect skeleton in Chapter 2, thus giving a better feel for skeletal function, and precedes the discussion of the alimentary and other systems to make the discussion of regulation of those systems more meaningful. Because the chapter on reproduction and morphogenesis stresses anatomy and physiology, it now follows the other chapters on anatomy and physiology instead of "Behavior." Thus the chapters dealing with sensory mechanisms, locomotion, behavior, and insect ecology now form a continuous and more logical sequence. As in the first edition, insect ecology is included in Part One because insects not only are functional units but also are parts of higher orders of structure and function—that is, parts of populations and ecosystems.

The chapter on behavior has been reorganized and largely rewritten. In Chapter 10, "Systematics and Evolution," the discussion of systematics is now first and has been condensed; the evolution section has been completely rewritten and expanded. Systematics is still included because I feel that it is important for the student to appreciate how the various groupings of insects have been developed. Chapter 11, "Survey of Class Insecta," has been reorganized somewhat on the basis of recent phylogenetic interpretations, and information on major fossil orders has been added. In this brief overview of the orders of insects I have tried to show how the various orders relate to one another as well as provide information regarding their biology and medical and economic significance.

I have also added several new sections: "Physical properties of cuticle" in Chapter 2; "Glands and endocrine system" in Chapter 3 (this chapter also now contains additional information on nervous function and on the insect brain); "The instar definition controversy" in Chapter 5; "The control of behavior," "Communication," and a subsection on "Origins of social behavior" in Chapter 8; "Adaptations associated with interspecific interactions" (dealing with coevolution) and "Behavior and the fluctuating environment" (dealing with dormancy, dispersal and migration, etc.) in Chapter 9; and "Insect pest management" in Chapter 12.

I have arranged the topics in the sequence I think most appropriate for dealing with the various aspects of entomology. However, each chapter can be read and understood with minimal reference to other chapters. Thus this text should be useful in any organizational framework a given instructor may choose to develop. A bibliography consisting mainly of major review papers, monographs, and specialized textbooks is given at the end of each chapter, and I have included in Chapter 1 a discussion of the literature of entomology with the hope that the student will be encouraged to make full use of the vast information available.

In addition to those persons who contributed to the first edition, I wish to express my sincere appreciation to the following individuals who have played a role in the development of the second edition. Osmond P. Breland, Dean G. Dillery, Elwood R. Hart, John G. Rae, Carl Spirito, and John G. Stoffolano, Jr., reviewed various parts of the manuscript and offered constructive criticism and very useful suggestions. Elliot Rosen, Cathy Walker, and Ned Walker, entomology graduate students, offered encouragement throughout the course of this project. Cathy Walker prepared electron micrographs, which appear in Chapter 3. My wife, Margaret, in addition to providing continuous support, encouragement, and love, prepared the glossary and indexes. David N. Garrison, Gregory W. Payne, and especially Elisabeth Belfer of Macmillan Publishing Co. offered invaluable assistance and advice from the beginning to the end of this project.

W.S.R.

Contents

PART TWO
UNITY AND DIVERSITY

PART THREE
APPLIED ASPECTS OF ENTOMOLOGY

Introduction

Insects are members of the class Insecta in the invertebrate phylum Arthropoda, the largest in the Animal Kingdom. The members of this phylum are characterized by a segmented body that bears a varying number of paired and segmented appendages; bilateral symmetry; an exoskeleton that contains the nitrogenous polysaccharide, chitin; and various internal features, such as an open circulatory system, Malpighian tubules (generally), and in most a system of ventilatory tubules, the tracheae and tracheoles. Present-day arthropods are often divided into two large subphyla, Chelicerata and Mandibulata. The trilobites, a long-extinct group of organisms, comprise a third arthropod subphylum, the Trilobita. The chelicerates bear a pair of appendages called *chelicerae* near the oral opening; the mandibulates are characterized by a pair of grinding structures associated with the mouthparts, the *mandibles*. Both chelicerae and mandibles are subject to considerable variation in structure; consequently, there are rather profound deviations from the "typical" forms. Although both structures usually function as parts of the feeding apparatus, they are not homologous. Spiders, ticks, scorpions, and horseshoe crabs are examples of chelicerates. Insects, together with millipedes, centipedes, and others, make up the subphylum Mandibulata.

Insects can be differentiated from the vast majority of other arthropods by several rather distinct traits. Among these are three well-defined body regions or tagmata: a head, a thorax, and an abdomen; three pairs of legs in the adult stage; commonly one or two pairs of wings; a single pair of segmented antennae on the head; and several less obvious but equally distinctive characteristics that will become apparent as the reader proceeds through this text. The name *Hexapoda* (six legs) is commonly applied to insects. However, the name *Insecta* is preferable since there is some question as to whether all arthropods with six legs in the adult stage actually belong in the same class (Sharov, 1966). Insecta literally means "in-cut," which describes the segmented appearance of the members of this class. The phylum Arthropoda will be discussed in more detail when we consider the evolution of insects.

Significance of Insects

Insects as a group are highly successful organisms. Their significance can be looked upon from two standpoints: (1) their tremendous success relative to organisms other than human beings and (2) their extreme importance from the human point of view.

One useful measure of the success of insects is the number of extant species. Estimates based on the current rates of description of new species of insects run from one to several million. It has been pointed out by a number of entomologists that several large groups of insects have hardly been studied at all. Brues, Melander, and Carpenter (1954) explain that approximately 750,000 species have been described and named and suggest that this figure is possibly only one fifth to one tenth of the insectan species that exist. Insects have been said to outnumber all the other species of animals and all the species of plants combined.

Other important criteria for success include the span of geologic time traversed by a group of organisms and their adaptability to various environmental situations. Insects are thought to have arisen in the Devonian era, approximately 400 million years ago. Mammals as a group are approximately 230 million years old; modern Man has been around for perhaps 1 million years. In this sense, insects have not invaded the human world; we have invaded theirs! The adaptability of the basic insectan plan has been phenomenal. Insects can be found in nearly every conceivable situation. As you proceed through this text, you will come to realize the seemingly unlimited adaptability of insects and gain some insight as to how they have reached their position of success.

From earliest times people have seen certain insect species as arch enemies. Although the pest species are a very small proportion of the total number of insect species, this group has caused and continues to cause astonishing trouble. Insects destroy annually millions of dollars worth of agricultural crops, fruits, shade trees and ornamental plants, stored products of various sorts, household items, and other valuable material goods. They serve as vectors of the causative agents of a sizable number of diseases of humans and domestic animals, and their direct attacks cause irritation, blood loss, and, in some instances, death. However, there are two sides to the picture. Insects have provided over the years, and still provide, many goods and services, so to speak, looked upon very favorably by man. Such insect products as honey and beeswax, silk, shellac, and cochineal are utilized for a variety of applications, ranging from sweetening biscuits to furnishing one of the basic components of many cosmetics. In addition, there are many indirect benefits of insect activities, such as plant pollination. A more detailed consideration of these subjects can be found in Chapter 12.

Although there is much that can be said both for and against insects as they relate to humans, the vast majority of insects are quite neutral, neither bestowing any great benefit nor causing any great harm.

Entomology as a Science

What Is Science?

Before we can intelligently consider entomology as a science, we must first discuss exactly what we mean by science. Science is a body of knowledge obtained in a very special way, which we call the *scientific method*. Scientists assume that entities of the universe interact with one another in a predictable pattern—in other words, follow a very definite set of rules. We might say then that the major objective of the scientist is to discover and seek to understand these rules. From this understanding comes the ability to predict events and in many instances exert controlling or modifying influences on these events. Thus, a knowledge of the life cycle and biology of a given insect enables us to predict the insect's activities at a given time and may allow us to act in such a way that we can either effectively limit the insect or make effective use of its favorable points. For example, understanding of the reproductive processes of screw-worm flies, serious pests of both wild and domestic animals in the southern United States and elsewhere, has enabled investigators to develop the comparatively new, quite successful sterile-male method of insect control; it has apparently eradicated the screw-worm from the southeastern United States. Likewise, extensive knowledge of the behavior of the honeybee has enabled apiarists over the years to harvest greater and greater amounts of honey per hive.

The difference between scientific and other approaches to learning about the universe lies in the rigid application of objective or rational thinking and the scientific method. Basically, this method consists of the following steps:

1. *Observation and description:* Critical observation and accurate description of events are essential prerequisites to the remaining steps.
2. *Hypothesis:* Once an event or series of events has been adequately described, tentative explanations or hypotheses are formulated.
3. *Experiment:* It is the purpose of experimentation to determine which, if any, of the hypotheses is the correct or the "most" correct one. Experimentation may also suggest hypotheses that had not been apparent earlier in the process. An experiment is a manipulation of the entities being investigated in such a way that the influence of changing a single factor or variable can be evaluated. Of course, it is usually not feasible, or, for that matter, possible, to achieve a sufficient degree of constancy of all factors that might vary. It then becomes necessary to introduce a "control" or "controls" into the experiment. A control is carried out concurrently with and is essentially identical in all respects to the experiment except that the variable being evaluated is unchanged. This helps to correct for the effects of variables that may change in an unknown and/or uncontrollable way during the course of the experiment. The introduction of controls is especially important in biological research where one is commonly dealing with a large number of uncontrollable variables.

Hypotheses borne out by repeated experiments over a long period of time often come to be considered laws, but even these well-supported hypotheses may fall in the light of new, more critical or sophisticated experimentation. At some point between hypotheses and laws lie theories, hypotheses that have survived some experimentation but still remain more within the realm of conjecture than do laws. Thus there may be a number of current theories explaining a given natural phenomenon, all of which are supported by a degree of experimental evidence but none of which has as yet been established as the "correct" one.

Every step of the scientific method is not necessarily a part of every scientist's work. A given investigator may be involved only with the observational and descriptive level and for a variety of reasons may not carry on experiments. Another may subject observations and hypotheses to rigorous experimentation. Both these men are scientists. The essential requirement is that they work within the framework of the scientific method and adhere closely to its basic assumptions.

Basic Components of Science

The application of the scientific method to the entities and events of the universe is a progressive and accumulative activity. New information modifies, clarifies, and supplements the old. Old information serves as an ever-increasing foundation for the new. Basically, there are two major components of science: scientists (and their students) and the great body of printed works—the literature. One often wonders how much subtle, unrecorded information acquired through years of professional experience is stored within the minds of scientists and how much of this information is lost with them or is passed on to their students. To learn about the interests, place of employment, and activities of a given contemporary entomologist, one should consult such references as Gupta (1976), Hull and Odland (1979), *American Men and Women of Science,* and *The Naturalist's Directory and Almanac* (International). In addition, the *Bulletin of the Entomological Society of America* periodically includes membership lists. Information on the lives and works of well-known entomologists of the past may be found in the list of selected references. Although probably a significant amount of information is, in fact, passed from scientist-teacher to student, the true long-term memory bank of science is the literature. This topic will be pursued in greater detail later in this chapter.

Acquisition of Scientific Information

We are now in a position to consider how scientific information can be acquired. The most primitive and still one of the most important means is by word of mouth. This the method employed in the classroom, at scientific meetings, conferences, and so on. Perhaps its most valuable aspect is the opportunity for a two-way exchange of information. Ideally, in this situation the lines of communication are wide open and a minimal amount of ambiguity should be the result. Another, equally important—in fact, essential—means of acquiring scientific information is consultation of the literature. The advantage of this method is that one can go back in time as far as one wishes but the two-way exchange of information is impossible if the author of a piece of recorded work is no longer living. Both these means of acquiring information should be considered to be prerequisites for a third means—personal investigation. This method is, of course, the source of new information. All three means are essential to the existence of science, and certainly no one could take precedence over the other two.

In applying the various means of acquiring information, a given scientist may have a variety of motivations. These motivations can be looked upon as placing this scientist somewhere between the two endpoints of a continuum. At one end is the scientist motivated entirely by desire to satisfy a direct human need; at the other end is the scientist motivated purely by human curiosity or academic interest. Individual scientists vary a great deal with regard to position on this continuum.

Where Does Entomology Fit into the Total Framework?

Entomology is, first of all, one of the biological sciences, since it is concerned with "living" systems. The biological sciences can be divided into basic divisions, such as morphology, physiology, genetics, and ecology, and into taxonomic divisions: ornithology, mycology, and so on. Entomology, since it deals with the study of a very specific group of organisms, is, of course, a taxonomic division. This being the case, we can logically approach the science of entomology by considering the "basic" divisions as they apply to insects: insect morphology, physiology, ecology, and so on.

The study of insects has played and will no doubt continue to play a major role in the development of biology. An indication of the extent of this role can be gained by examining a recent general biology text, for example, *Biology Today* (Kirk, 1975), where 8.4% of the total number of pages include at least some mention of insects. The following list provides an idea of the breadth of entomological topics covered in this text: Redi's experiments with maggots and spontaneous generation; coevolution of plants and pollinating insects; the mechanisms of sex determination in insects; mutations in *Drosophila;* linkage groups, sex-linkage, chromosomal mapping, and induced mutations in *Drosophila;* chromosomal puffs in *Drosophila* and their induction by ecdysone; pheromones; behavioral genetics of cricket singing; spatial orientation of the digger wasp; migration of the monarch butterfly; circadian rhythms and adult emergence in *Drosophila pseudoobscura;* several other behavioral examples using insects (including behavior of the honey bee); termite–protozoan mutualism; temperature control in termite mounds and beehives; bees and orchids; ants and acacias; fig wasps and figs; industrial melanism; specialization in the *Drosophila willistoni* complex; pesticide problems . . . .

For an informative introduction to entomology as a science, see Wigglesworth (1976a).

The Literature of Entomology

The literature of entomology consists of a wide variety of publications ranging from rather popularized accounts intended for the layman to highly technical treatises on very specific aspects of the science. The information available on insects is vast and is growing so rapidly that no one person can stay abreast of and learn more than comparatively small portions of it. Associated with this situation is a great deal of specialization, the concentration of effort on a single topic or a group of closely related topics. Thus, among entomologists there are, for example, insect physiologists, ecologists, morphologists, systematists, toxicologists, and economic entomologists. Generally, the specialization goes even further; someone in one of the preceding groups may concentrate on a particular insect species or group of insects, or on a specific topic, or both. For example, there are specialists in the systematics of a particular family of beetles, in mosquito physiology, and so on.

The tremendous number of entomological publications makes a comprehensive review inappropriate within the context of this book. However, a brief discussion of some of the different types of entomological literature will at least broaden the reader's perspective. More extensive treatments of the literature of entomology and zoology in general can be found in Smith and Reid (1972), Chamberlin (1952), and Blackwelder (1967). Arnett (1970) provides insight into the problems involved with the storage and retrieval of entomological information and discusses both current and proposed solutions.

Publications pertinent to entomology (for that matter, the entire field of zoology) can be divided into two basic groups. The first group includes all those publications, irrespective of type, that contain the actual information about animals (insects). This group presents the products of zoological (entomological) research. The second group includes all the publications that attempt to coordinate the vast information contained in the first group, making it more readily available to investigators. Examples of publications from each of these groups are presented below. This method of classification, as with most, is certainly not without exception, as will be shown.

Publications That Contain the Actual Information of Entomology

Textbooks. A textbook is generally designed to give the reader an understanding of the basic principles involved in a given subject. A textbook may be quite general in scope, presenting a survey or overview of an entire field—for example, a general entomology textbook such as the one you are reading or any of several others (see Selected References). Other textbooks deal with a particular area within the field. For example, there are texts devoted to insect physiology—Bursell (1970), Chapman (1971), and Wigglesworth (1972)—or insect ecology—Price (1975)—or insect behavior—Matthews and Matthews (1978). Books of this type treat a subject in more depth than the general text and are more likely to be used in advanced courses.

Monographs.   A monograph is limited in scope, dealing only with a very small area within a science. However, monographs are usually comprehensive in their coverage of pertinent literature and handling of subject matter. An excellent example is Dethier's superb *The Hungry Fly* (1976).

Symposia.   Symposia are published collections of the presentations of several experts in a given area who have met to consider a specific aspect of science. The symposia of the Royal Entomological Society of London—for example, Emden (1972)—fall into this category.

Lectures.   A lecture is an oral discourse presented to an audience or class, particularly for instructional purposes. Good examples of entomological lectures are the talks presented by outstanding scientists to the general sessions usually held several times during each annual meeting of the various entomological societies throughout the world. These talks are commonly published in bulletins issued periodically by these societies.

Essays.   Essays are analytic or interpretative expositions that usually deal with a given topic from a rather personal or limited point of view. A well-known example is the book *Silent Spring* by Rachel Carson.

Reference Works.   Reference books generally attempt to offer comprehensive coverage of a specific area and are designed to be consulted as needed rather than read and digested in their entirety. This category may seem rather arbitrary, since any piece of literature that is consulted can be classified as a reference, and rigorously documented and highly technical entomological texts, such as Richards and Davies (1977, 1978) and *The Insects of Australia* (C.S.I.R.O., 1970), can equally well be viewed as reference works. An especially valuable reference work is Rockstein's six-volume set *The Physiology of Insecta*.

Pamphlets.   Pamphlets are usually rather brief writings on a specific topic, commonly geared to the persons who put many of the findings of entomological research into practice: farmers, exterminators, and so on. Examples of pamphlets are the *Farmer's Bulletins* published by the U. S. Department of Agriculture, which pertain to the biology and control of economically important species of insects.

Reports.   From time to time, groups of experts are called together to investigate or to discuss an issue or problem of particular significance. For example, the World Health Organization (WHO) periodically sponsors meetings of expert committees on various problems pertaining to international health matters, including topics such as malaria and similar diseases with which insects are involved. These committees usually submit reports describing the problems discussed and conclusions reached in the course of the meeting.

Series. Serial publications are issued periodically (sometimes at irregular intervals) and have a certain unity of subject matter; that is, a particular series deals, volume after volume, with more or less the same general subject matter. Series are published by most professional societies and many private and governmental concerns in the form of journals, bulletins, miscellaneous publications, yearbooks, and so on. Hammack (1970) provides a useful and comprehensive description of the serial literature pertinent to entomology. Some examples of entomological journals are *Annals of the Entomological Society of America, Journal of Economic Entomology, Environmental Entomology, Journal of Insect Physiology, Systematic Entomology, Ecological Entomology, Physiological Entomology,* and *Psyche.*

Publications That Attempt to Coordinate the Literature

From the time scientists realized the fantastic rate of growth of the literature of science, many very useful attempts to coordinate and integrate the works in various areas have been made. In this section we want to discuss briefly some of the more common publications in this category. Most of these publications are quite expensive and are seldom purchased by an individual.

Bibliographies. A bibliography, as Smith and Painter (1966) aptly point out, is to the literature of science as an index is to a book, the basic difference being that a bibliography lists only publications of various sorts on a single topic, instead of the subjects, authors, and so on, to be found in the index of a single book. Bibliographies appear in different forms. For example, one form is a list of pertinent references at the end of a chapter of a book or a scientific paper. Another type of bibliography is issued periodically and lists current publications in a particular field. Important examples of this type are *Zoological Record, Bioresearch Index, Bibliography of Agriculture, Cumulated Index Medicus,* and *Current Contents.* The *Zoological Record* has a very broad coverage, both foreign and domestic, and is arranged according to taxonomic groups and by subjects. The section on insects is quite extensive and contains references to papers of interest to most entomologists, not just taxonomists. *Bioresearch Index* is a monthly publication that furnishes bibliographies from various journals and contains citations of research papers. The *Bibliography of Agriculture,* a monthly publication, generally contains a large number of references to entomological papers. The last issue of each year is a cumulative subject index. *Cumulated Index Medicus* is issued four times yearly, covers much of the foreign and domestic medical literature, and may contain references of interest particularly to medical entomologists. *Current Contents* reprints the tables of contents of many journals, several entomologically oriented ones included. It is issued weekly and is probably one of the best ways to keep abreast of the most current literature. This is especially important when one is working in a very active area in which a number of researchers are publishing extensively.

Abstracting Journals. Abstracting journals contain brief descriptions or abstracts of the results reported in the journals that fall within their scope. Abstracts are extremely useful because they give an investigator a better idea of the content of a given reference than does a mere title listing, although they do serve also as bibliographies. This kind of information helps one decide whether or not to consult the actual references and compensates somewhat for the fact that some publications are extremely difficult to obtain or translate from a foreign language. Abstracting journals are usually extensively cross-indexed, which makes them efficient to use. One of the most significant examples of this type of publication is *Biological Abstracts*. This bimonthly publication is quite comprehensive in its coverage of the literature of theoretical and applied biology. In addition to the volumes containing the abstracts, a semimonthly publication, *B.A.S.I.C.*, provides a very elaborate computerized subject index to the issues of *Biological Abstracts*. A cumulative subject index based on all the issues of *B.A.S.I.C.* is published for each year's volume of *Biological Abstracts*. Since January 1970 the abstracts and citations of research papers pertaining to insects and arachnids in *Biological Abstracts* and *Bioresearch Index* have been compiled into a separate publication, *Abstracts of Entomology*. One issue of this publication corresponds to two issues of *Biological Abstracts* and one issue of *Bioresearch Index*.

Other abstracting journals that specialize in covering entomological literature are *Entomology Abstracts* and *Review of Applied Entomology*. Entomology abstracts is published monthly and covers a wide variety of entomological topics. *Review of Applied Entomology* is published in two series: Series A, Agricultural, and Series B, Medical and Veterinary. It is well indexed, both by subject and author, and contains abstracts covering a wide variety of entomological topics. Other periodical publications that are at least partly abstracting journals are *Biologisches Zentralblatt, Physiological Abstracts, Tropical Diseases Bulletin, Apicultural Abstracts,* and several others from various countries.

Another very useful monthly publication is *Dissertation Abstracts*. This contains abstracts of all dissertations by contributing institutions in the United States arranged by the type of subject matter. These abstracts are quite useful since there is commonly a hiatus between the writing of a dissertation and the publication of a research paper or papers based on it. If one decides on the basis of a given abstract that he needs more information, he may readily obtain, for a fee, a microfilm of an entire dissertation.

Review Journals.   Review journals contain papers that discuss the literature on a rather specific topic in a given field. Review papers not only bring together information from the pertinent literature on a given topic but also commonly contain useful syntheses of information that may not occur in any other type of publication. In this sense they may be classed in either of the two rather arbitrary categories we have used to discuss entomological literature. Two very important review journals in the field of entomology are the *Annual Review of Entomology* and *Advances in Insect Physiology*. Other review journals that may contain reviews of entomological interest are the *Annual Review of Ecology and Systematics, Annual Review of Physiology*, and the *Annual Review of Medicine*. In addition, review papers may appear in journals, bulletins, and so on, which contain other types of articles.

Taxonomic Indexes and Catalogs. Taxonomic indexes include literature references to such items as the original description of a given genus or species and revisions of genera. These publications are quite useful for tracing the taxonomic literature pertinent to a given group and for determining the systematic position of a given genus or species. Especially important indexes are *Nomenclator Zoologicus* edited by A. A. Neave, *Zoological Record*, and *Biological Abstracts*. *Nomenclator Zoologicus* lists the names of genera and subgenera of all zoological groups from 1758, the year of the publication of the 10th edition of Carl Linne's (Linnaeus) *Systema Naturae*, to 1950. *Zoological Record* contains the names of all new genera described each year and pertinent literature references from 1864. *Biological Abstracts*, in the section "Systematic Zoology," provides references to the original descriptions of genera and subgenera of animals since 1935. Smith (1958) points out that "Neave's *Nomenclator Zoologicus* and *Biological Abstracts* serve admirably as a complete generic index from 1758 to the present." He further suggests that ". . . for names published since the most recent issues of *Biological Abstracts*, journals in which new genera of the various groups might be expected to occur must be consulted."

For species indexes and catalogs similar to the generic ones just described, one must refer to one or more of several currently available. Sherborn's *Index Animalium* is the only general species index available and covers all the specific names proposed for animals from 1758 through 1800. Smith and Reid (1972), Chamberlin (1952), and Blackwelder (1967) each contain lists of catalogs for various insectan and other groups.

Science Citation Index. *The Science Citation Index* is a recent addition to the literature-coordinating publications. It is published by the organization that publishes *Current Contents* and is composed of two sets of indexes, a citation index and a source index, both of which are cumulative. Its objective is to list and index the current and past research papers that cite a given reference. It enables an investigator to begin with a given reference and find other references that have cited the "starting reference." Since both the "starting reference" and "citing reference" are likely to pertain to the same or very closely related topics, one is able to proceed forward or backward in time, using "citing references" as "starting references" in a cyclical manner and by doing so accumulate a bibliography of the literature on a given topic.

Union List of Serials and New Serial Titles. Most libraries do not have complete sets of all journals useful in entomology or any other science. However, they generally have agreements with other libraries whereby volumes can be borrowed or copies of particular papers can be obtained. *The Union List of Serials in Libraries of the*

U. S. and Canada and *New Serial Titles* are listings of all journals
and of the major libraries that house these journals. Thus, by consulting these lists, one may determine which libraries have the journal he or she is seeking. Both lists are indexed by journal name and
by subject.

Books In and Out of Print.   *Books in Print* is an annual listing
of books currently available on the commercial market. This list is
composed of four volumes: Volumes 1 and 2, titles; and Volumes
3 and 4, authors plus author and title indexes. Out-of-print books
may be found by consulting the *A. B. Bookman's Weekly,* various
companies that specialize in such books, and major libraries. Fac-
similes of out-of-print books are available from University Micro-
films International in Ann Arbor, Michigan.

The "Language" of Entomology

There are are number of publications that are very helpful in
learning the technical terminology that necessarily exists in entomology. Borror's *Dictionary of Word Roots and Combining Forms*
(1960) enables one to determine roughly the word roots of scientific
terms and is thus useful in learning and understanding these terms.
Torre-Bueno (1937) and its supplement (Tulloch, 1960) and the
more recent Leftwich (1976) are also valuable in dealing with complicated and often confusing entomological terms. Recently a preliminary thesaurus of entomology (Foote, 1977) has been developed,
the role of which is explained as follows.

> A carefully built thesaurus is, then, the vocabulary of a specific
> field of activity. It is also the guide required if more than one
> person is to become involved meaningfully in storing and re-
> trieving information. Without specifically defining every term,
> the thesaurus specifies those terms which are to be used for
> specific ''things'' and concepts and interrelates them in a way
> that makes the application clear.

Selected References

GENERAL
   Atkins (1978); Borror, DeLong, and Triplehorn (1976); Chapman
(1971); C.S.I.R.O. (1970); Daly, Doyen, and Ehrlich (1978); Eisner and
Wilson (1953–1977); Elzinga (1978); Fox and Fox (1964); Horn (1976);
Little (1972); Richards and Davies (1977, 1978); Ross (1965); Wiggles-
worth (1964a).

ENTOMOLOGICAL LITERATURE
   Arnett (1970); Blackwelder (1967); Borror (1960); Chamberlin (1952);
Foote (1977); Hammack (1970); Leftwich (1976); Smith and Reid (1972);
Torre-Bueno (1937); Tulloch (1960).

HISTORY OF ENTOMOLOGY
   Cloudsley-Thompson (1976); Cushing (1957); Essig (1931); Howard
(1930); Mallis (1971); Osborn (1937); Richards and Davies (1978); Smith,
Mittler, and Smith (1973); Southwood (1977a, b); Wigglesworth (1976a).

PART ONE

Structure and Function

The Integumentary System

The general body covering, or *integument,* is a complex organ of diverse structure and function. Probably its most obvious role is that of forming a supportive shell, the *skeleton*. The insect skeleton differs from the internal skeleton, or endoskeleton, of vertebrates in being located on the outside of the body, that is, it is an *exoskeleton.* Even the internal integumental processes are continuous with the external shell. A modified form of the cuticular part of the integument lines parts of the following: the alimentary canal, the tracheal system, genital ducts, and the ducts of the various dermal glands.

The insect skeleton affords almost unlimited area for muscle attachment. In many cases the elastic nature of certain parts of the skeleton opposes muscular contraction in a manner similar to the mutual opposition of antagonistic muscles in vertebrates. For example, muscle contraction causes retraction of the pretarsus (terminal segment of tarsus or "foot"), and extension is the result of elasticity of the skeleton. Some elastic regions are involved in the storage and release of energy during jumping and flying. Elasticity may, in some cases, aid in the inspiration of gases during ventilatory movements. The insect skeleton is hardened, or *sclerotized,* to varying degrees and thus serves as a protective armor for the organs encased by it. Other properties of the integument also play a role in the protection of the insect; for example, the integument forms a highly effective barrier against the entry of many pathogens and insecticides. The integument may be relatively impermeable to water, particularly that which tends to escape from the insect. Since the exoskeleton lies between the external environment and the remainder of the insect, external portions of the sensory and exocrine structures are intimately associated with it. Characteristics of the integument usually determine body coloration.

Although some regions of the exoskeleton are more flexible than others, there are definite limits on its expansibility. Increase in size is accomplished by the periodic shedding and renewal of part of the integument, the process of molting.

Extensive information on the insect integument may be found in Ebeling (1974), Hackman (1974), Hepburn (1976), Locke (1974), Neville (1975), and Richards (1951, 1978).

Histology of the Integument

Basic Components

The insect integument (Fig. 2-1A) can be divided into three basic parts: one cellular layer, the *epidermis,* and two noncellular layers, the *basement membrane* (entad of the epidermis) and the *cuticle* (ectad of the epidermis).

The epidermis is typically one cell thick, adjacent cells being joined by *septate desmosomes* (see Fig. 4-3). It secretes the cuticle and may secrete the basement membrane. Since the epidermis secretes the cuticle, it not surprisingly shows greatest activity during molting. Interspersed among the epidermal cells are *dermal glands,* some of which play a part in secreting a portion of the cuticle. As you will see in Chapter 3, other types of single-celled or multicellular dermal (exocrine) glands carry out a variety of functions such as secretion of defensive substances, silk, and pheromones.

The basement membrane is generally 0.5 micrometer or less in

Fig. 2-1
Structure of the integument (diagrammatic). A. Section of generalized integument. B. Daily growth layers and lamellar patterns. C. Generalized epicuticle.

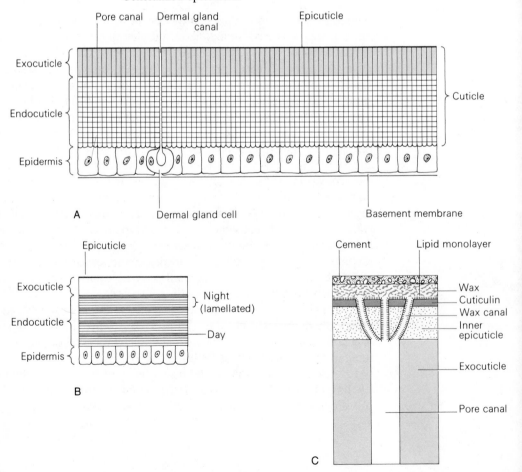

thickness and in electron micrographs appears as a continuous, amorphous granular layer. It is apparently composed of a muco-polysaccharide.

The cuticle comprises up to one-half the dry weight in certain species. The nature of cuticle of necessity varies both spatially and temporally according to the demands of function and growth. The chemical and physical properties of cuticle in a given region at a given time are ultimately a function of the synthetic abilities of the epidermal cells.

With the techniques of light microscopy and various stains, the insect cuticle can be seen to consist of a number of different layers. There are two major layers: the very thin, outer *epicuticle* (0.03–4.0 micrometers thick) and the much thicker (up to 200 micrometers) inner *procuticle*. In hardened regions, the procuticle is typically divisible into at least two layers of variable thickness. The outer layer is called *exocuticle* and the inner layer, *endocuticle*. The endocuticle may or may not be pigmented, the exocuticle is often pigmented, and the epicuticle is unpigmented. Compared with hardened regions, softer areas of cuticle are typically comprised of only endo- and epicuticle.

Under the light microscope, with polarizing filters, the endocuticle can be seen to be composed of successive light and dark layers (Fig. 2-1B), which in all insects studied thus far, except the beetles and Apterygota, correspond to daily growth layers (Neville, 1970). The dark layers are those deposited during the day and the lighter layers during the night. The light layers are further subdivided into alternating layers of light and dark lamellae. Also, typically, striations running vertically through the cuticle can be seen. The use of the electron microscope has clarified somewhat the nature of these lamellae and striations. The lamellar patterns are considered to be the result of the orientation of layers of microfibrils of chitin and probably protein embedded in a protein matrix. These microfibrils are laid down in layered sheets (Fig. 2-2A). Within each sheet, the microfibrils are parallel to one another, but in successive sheets they are aligned at regularly changing angles. This arrangement (helicoidal) is responsible for the parabolic patterning of endocuticle when cut transversely and observed with the electron microscope (Fig. 2-2B). In the dark growth layers deposited during the day, the microfibrils are oriented in successive sheets in a ''preferred'' direction and hence do not have the lamellar appearance of the lighter nighttime layers. The vertical striations have been shown to be tiny tubes, the *pore canals* (Figs. 2-1A, C and 2-2B), which extend from the epidermal layer nearly to the external surface of the epicuticle. The pore canals are usually 1 micrometer or less in diameter, are ribbonlike in appearance, and apparently serve as connecting tubes between the cellular and the cuticular layers. They run a spiral course through the helicoidally arranged layers of endocuticle and may number from several thousand to well over a million per square millimeter of integument.

The exocuticle is a highly stabilized layer of the procuticle that is resistant to the action of molting fluid and is shed as the *exuvia* during *ecdysis*. Exocuticle is typically thin in soft-bodied insects and thicker in hard-bodied insects.

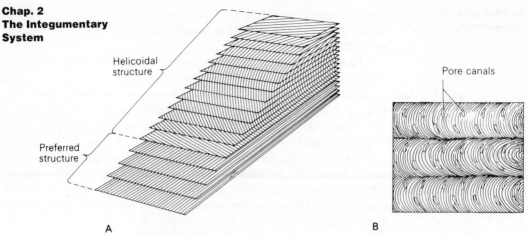

Fig. 2-2
A. Helicoidal and preferred structure of layers of endocuticle. The
parallel lines in each layer represent microfibrils. Note the parabolic
effect in transverse section. B. Transverse section of endocuticle
showing parabolic effect and appearance of pore canals as they pass
through helicoidal layers. [Redrawn with modifications from Neville,
1970.]

In addition to the exo- and endocuticle, two other procuticular
layers have been identified: the *mesocuticle* and a granular *subcuticle*. Mesocuticle lies between exo- and endocuticle and is differentiated on the basis of staining reaction. Subcuticle is probably newly
secreted endocuticle in which the microfibers are not yet oriented.

The epicuticle, despite its comparative thinness, is exceedingly
complex and possesses characteristics that make it an extremely
important layer of the cuticle. Although it appears as a very thin,
sometimes indiscernible, line under the light microscope, the electron microscope has revealed a multilayered structure that is pene-
trated by wax canals 60–130 Ångstroms in diameter containing wax
filaments (Locke, 1974). At least four layers, of varying thickness
depending on the insect species, have been described in the epicuticle
(Fig. 2-1C)

1. The outer *cement layer,* sometimes called tectocuticle (''roof''
   cuticle), less than 0.1 micrometer thick.
2. The *wax layer.*
3. The *cuticulin layer.*
4. The inner *protein epicuticle.*

The cement layer is secreted by dermal glands and may be similar
to shellac, a substance secreted by a particular group of insects in
the suborder Homoptera. This layer, outermost when present, probably determines the surface properties of the cuticle—that is,
whether the cuticle will be water-repellent *(hydrophobic)* or water-
attractant *(hydrophilic)*—and also probably serves as a protective
barrier for the more vulnerable layers beneath. It is apparently absent
in many adult insects that possess cuticular scales. In some insects,
cockroaches, for example, it may form a spongy meshwork in which
wax molecules can move about. In other insects, such as many
Homoptera, a waxy bloom appears on the surface of the cement

layer (Locke, 1974). The wax layer is thought to consist of an ordered monolayer of lipid directly associated with the cuticulin layer and a more ectad, less well-ordered lipid layer. Experimental evidence indicates that this layer is responsible for many of the permeability characteristics of the cuticle. The cuticulin layer has been found in every insect in which the epicuticle has been investigated and covers the entire integumental surface, including the tracheoles and gland ducts. It is thought to be composed of lipoprotein and is 100–200 Ångstroms thick. The innermost layer, the protein epicuticle, is about 1.0 micrometer thick and appears as a homogeneous, dense, refractile layer. It may be the layer that limits the amount of expansion of cuticle possible between molts. The pore canals terminate beneath this inner layer of epicuticle; the dermal gland canals communicate with the surface of the epicuticle.

Tiny *wax filaments* or *wax canals* approximately 100 Ångstroms in diameter are evidently continuous with pore canals, penetrating the inner epicuticle and cuticulin. Since they are continuous with pore canals, these wax canals are evidently involved in transporting wax molecules to the epicuticle (Locke, 1974; Wigglesworth, 1976b).

Chemical Composition of the Cuticle

Quantitatively, the polysaccharide *chitin* and various structural proteins are the major cuticular constituents. Chitin (Fig. 2-3) is a high-molecular-weight polymer of *N*-acetyl-D-glucosamine with the empirical formula $(C_8H_{13}O_5N)_n$. It may make up from one fourth to more than one half of the dry weight of the exo- and endocuticle. Chitin has not been found in the epicuticle. Chitin probably does not exist in a pure state naturally, but is combined with a protein as a glycoprotein. Apparently chitin chains are attached to one another by hydrogen bonds, forming elongate microfibrils. It seems likely that chitin microfibrils and protein chains are in intimate combination with one another and that this complex is impregnated with loosely bound protein (Hackman, 1976).

Cuticular proteins usually make up more than 50% of the dry weight of cuticle. Included in this category are *arthropodins,* a group of soluble proteins; *resilin,* a protein that forms a rubberlike framework and is found sometimes in pure form in skeletal articulations;

Fig. 2-3
Structural formula of chitin.

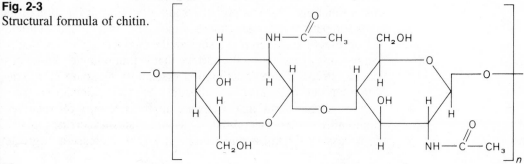

and *sclerotins,* stabilized proteins that are responsible for the hard, horny character of cuticle.

Other constituents of the cuticle include polyhydric phenols and quinones, which play a role in the *sclerotization* (hardening) and *melanization* (darkening) processes; lipids of various sorts associated with the epicuticle; enzymes (nonstructural proteins), which catalyze the many complex biochemical reactions involved in molting and subsequent processes; and very small amounts of inorganic compounds.

Richards (1978) and Anderson (1979) discuss the chemistry of cuticle and cite several useful references.

Sclerotization

Chitin is not the agent responsible for the hardness of the cuticle, although it undoubtedly lends strength to it. In fact, highly sclerotized skeletal regions may contain less chitin than softer membranous areas.

The hardening of insect cuticle is due to the tanning of protein, which follows *eclosion* (emergence of an immature stage from the egg) or *ecdysis* (shedding of the cuticle) and subsequent expansion of an insect. Proteins are rendered hard, dark, and insoluble by the linkage of adjacent polypeptide chains and the blocking of reactive groups by the tanning substances (phenols and quinones). These tanned proteins are the sclerotins mentioned above. The degree of hardness resulting from the tanning process is highly variable. Segmental plates of caterpillars (Lepidoptera larvae) are nearly unsclerotized, whereas the mandibles of certain beetles are capable of biting through metals such as lead, tin, and copper.

In a very few species, for example, in puparial cases of the flies *Musca autumnalis* and *M. fergusoni,* calcium carbonate (lime) is responsible for the hardness of the external shell.

For information on the chemistry of sclerotization, see Hackman (1974), Anderson (1976), and Richards (1978).

Physical Properties of Cuticle

One need only consider the form and various activities of an insect to realize the great importance of the physical characteristics of cuticle. Cuticle in the same insect must be rigid, elastic, impermeable, permeable, flexible, and so on, as appropriate to a given structure and/or function. Physical properties are, of course, a function of the kinds and patterns of organization of molecules that compose cuticle.

The insect exoskeleton is a shell in the engineering sense, that is, a supporting structure on the surface that is thin relative to total size (Locke, 1973). The exoskeleton is essentially made up of a series of tubular, curved, and spherical panels. Physical calculations based on these shapes show that shells are appropriate only for relatively small animals like insects; that is, as an organism increases in size, the necessary thickness and weight of a shell increase at a greater rate than body mass, and eventually the shell becomes prohibitively bulky. Insect cuticle serves nicely as a shell and weight for weight compares favorably with metals (Locke, 1973). For example, the tensile strength of steel is 100 kg/mm^2 while that of cuticle is 10 kg/mm^2, but steel is eight times as heavy as cuticle (Locke, 1973).

The physical characteristics of cuticle in a given region may change periodically. For example, when the bloodsucking bug *Rhodnius prolixus* takes a blood meal, a "plasticizing factor" is secreted, under control of the central nervous system, that changes the pH of cuticle and as a result changes the degree of bonding between cuticular proteins (Bennet-Clark, 1962). Thus the cuticle becomes more flexible at a time when the abdomen will be drastically expanded.

See Hepburn and Joffe (1976) and Locke (1974) for useful discussions of the physical properties of cuticle. Bennet-Clark (1976) discusses physical properties of cuticle in jumping insects.

Coloration

Insect color may be due to various pigments present in the cuticle, scales on the cuticle, epidermal cells, or fat body; to physical characteristics of the cuticle and/or scales; or to a combination of pigments and physical characteristics.

Most insect pigments appear to be metabolic by-products derived from various pigments in food. Coloration is commonly produced by complex mixtures of different pigments, and the same color may be achieved in different ways. Examples of insect pigments are the very common *melanins* (yellow, brown, black), the *carotenoids* (red and yellow) derived from plant tissues, the widely distributed *pterins* (red, yellow, white), and the *ommochromes* (red, yellow, brown). Pterins and ommochromes are commonly found as eye pigments. Other pigments include *anthraquinones* (red, orange), which are common in scale insects (Homoptera); *aphins* (red, orange, yellow), confined to aphids; *chlorophyll derivatives* (greenish); *hemoglobin derivatives* (reddish); *anthoxanthins* (whitish, yellow); *insectoverdins* (green); and *flavines* (greenish yellow).

For discussions of pigments and pigment metabolism, see Fuzeau-Braesch (1972), Needham (1978), and Wigglesworth (1972).

Physical coloration is due to the effects of various structural configurations on light (see Hinton, 1976b). Colors result from the reflection of light in all directions (scattering) by irregular surfaces or granules beneath the surface, from optical interference between light rays reflected from successive thin layers, and from the light-breaking effect of minute, regular, cuticular striations. Whites in several butterflies (and probably most insects) are produced by the light-scattering effect of the irregular surfaces of scales. In several dragonflies, very small, randomly dispersed granules in the epidermal cells produce a blue color, *Tyndall blue*, by scattering light. (The scattering of light by colloidal-sized particles is called the Tyndall effect.) Most iridescent colors of insects are produced by interference (Fig. 2-4), for example, the brilliant metallic coloration of the

Fig. 2-4
Scanning electron micrographs of iridescent scales on a wing of the Buckeye butterfly showing the fine structure responsible for the light-breaking effect. 1, scale; 2, socket. (Scale lines: A, 20 μm; B, 2 μm; C, 1 μm.) [Courtesy of Walter J. Humphreys.]

A

Morpho spp. butterflies and of several beetles. It has recently been shown that some pierid butterflies produce interference colors in the ultraviolet part of the spectrum (Hinton, 1976b).

Some insects are able to change color reversibly. This is accomplished in epidermal cells or in the cuticle (Hinton, 1976b). Some, certain Odonata and Orthoptera among others, change color by movements of pigment granules in epidermal cells. Some beetles in the genus *Dynastes* and several tortoise beetles can change cuticular color. The cuticle of Hercules beetles (*Dynastes hercules*) can appear greenish yellow or black depending on atmospheric moisture. The epicuticle of the elytra is transparent and covers a spongy yellow layer. Beneath the yellow layer the cuticle is black. Under dry conditions (e.g., daytime), the spongy yellow layer contains air and yellow is reflected back through the epicuticle, but when humidity increases (e.g., nighttime), the spongy layer becomes fluid-filled and allows light to pass through, revealing the black layer. These color changes evidently help protect the beetle from predators.

Permeability Characteristics

Insects, being essentially terrestrial animals, are continually faced with the problem of losing water, especially in extremely arid habitats. The small size of insects makes this problem particularly acute, since transpiration (water loss) rate varies inversely with the ratio of surface area to volume. Thus the smaller an organism, the greater the amount of surface area per unit volume and hence the greater the tendency to lose water.

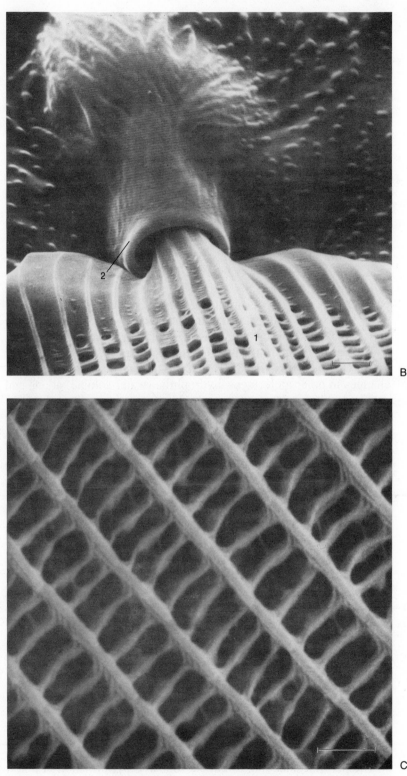

B

1

2

C

Experiments, such as the application of various organic solvents that dissolve a portion of the epicuticle, have demonstrated that the abrasion, adsorption, or dissolution of the epicuticular layers causes an appreciable increase in the rate of transpiration, often resulting in the death of the insect. On the basis of these and other kinds of experiments, it is felt that at least part of the barrier to the exit of water from an insect lies in the epicuticle, particularly the wax layer. Further, the wax layer is absent in many insects and other arthropods found in very moist terrestrial or aquatic habitats.

Transpiration rate in insects varies directly with temperature (Fig. 2-5). However, this relationship does not produce a linear curve throughout its length. Above a certain temperature, depending on the species of insect, within an approximate range between 28 and 60°C, transpiration rate increases abruptly. The temperature at which this occurs has been called the *critical* or *transition temperature*. This abrupt change may be brought about by disruption of the oriented lipid monolayer in the wax layer of the epicuticle (Fig. 2-1C) at the critical temperature, which is somewhat below the melting point of the wax. Permeability of the insect cuticle to substances other than water may be a function of other layers of the epicuticle and of the exo-, meso-, and endocuticle (Richards, 1978).

The permeability of the integument to water may vary at different times (Locke, 1965). This variability may be related to phase changes in polar molecules—hydrophilic (water-soluble) at one end and hydrophobic (nonwater-soluble) at the other end—comprising the wax filaments. These filaments could be liquid crystals composed of lipid and water, which change from the middle phase (Fig. 2-6), in which they impart impermeability, to the reversed middle phase or complex hexagonal phase, in which they allow at least some passage of water. Phase changes may occur in response to changes in temperature and humidity.

Recent work indicates that the epidermis may also play a vital role in integumental permeability (Ebeling, 1976). Injection of most insecticides adversely affects an insect's ability to regulate water loss. This action is presumably due to interference with normal functioning of the epidermal cells. Dead firebrats, *Thermobia domestica*, lose water more rapidly than live ones, suggesting again that an active epidermis is involved in control of water loss.

For further information on cuticular permeability, see Ebeling (1974, 1976) and Wharton and Richards (1978).

Insects conserve water in a variety of ways in addition to the integumental mechanisms mentioned above. These will be mentioned in appropriate sections of other chapters.

Molting

The periodic shedding of cuticle followed by formation of a new cuticle is a mechanism facilitating growth despite a more or less inflexible integument. Molting has been the subject of numerous investigations, and as a result a general description is possible (Fig. 2-7). At the onset of molting, the epidermal cells show much activ-

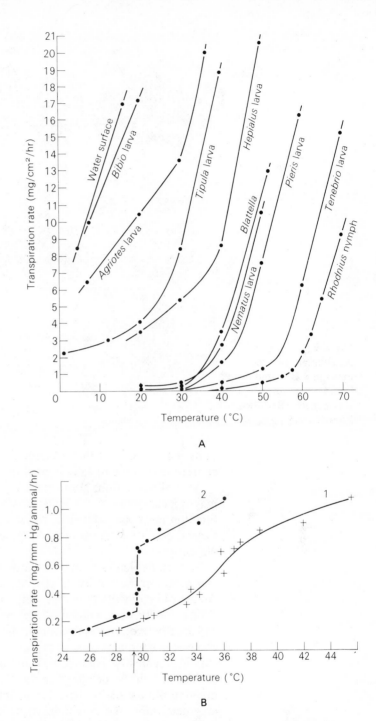

Fig. 2-5
Rate of transpiration in relation to temperature. A. From dead insects. B. From a cockroach nymph at constant saturation deficiency (arrow indicates critical temperature). +, air temperature; ●, cuticle temperature. [A redrawn with slight modifications from Wigglesworth, 1945; B redrawn with slight modifications from Beament, 1958.]

ity. They generally increase in size and often increase in number. During the molting process the epidermal cells separate from the old cuticle and begin to secrete the new. The initial separation of old cuticle and epidermis is called *apolysis* (Jenkin and Hinton, 1966). The epidermal cells then secrete the *molting* or *exuvial fluid*. It contains the chitinase and protease that are capable of digesting endocuticle, which may make up to 80–90% of the old cuticle. The products of this digestion are then absorbed by the epidermal cells.

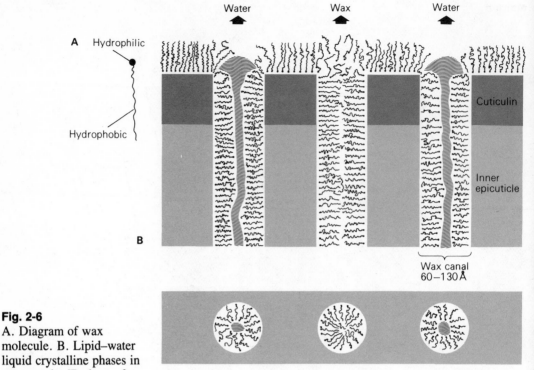

Water   Wax   Water

A   Hydrophilic

Hydrophobic

B

Cuticulin

Inner epicuticle

Wax canal 60–130 Å

Complex hexagonal phase   Middle phase   Reversed middle phase

Fig. 2-6
A. Diagram of wax molecule. B. Lipid–water liquid crystalline phases in wax canals. [Redrawn from Locke, 1965.]

Thus it may be said that a significant portion of the cuticle never entirely leaves the metabolic pool of the insect.

At apolysis, a thin, homogeneous, transparent *exuvial membrane* may appear between the epidermis and old cuticle. Its origin and function are not established, but it is resistant to molting-fluid enzymes, persists throughout molting, and is shed with the undigested old cuticle.

Exo- and epicuticle are resistant to the action of the molting fluid and make up the portion of the integument that is shed at ecdysis. As the old endocuticle is digested, forming the *exuvial space,* new cuticle is deposited, cuticulin layer first, then the protein epicuticle, and finally exo- and endocuticle, the raw materials of which have come at least in part from digestion of the old endocuticle. The new cuticle is typically wrinkled beneath the old, indicative of the greater surface area to be occupied in the "expanded" insect after the remaining old cuticle is shed. Ultimately the protein epicuticle probably determines the maximum to which cuticle can be expanded.

The wax layers are laid down shortly before ecdysis, assuring the waterproofing of the newly emerged insect. The cement layer is the last secreted, forming soon after ecdysis. It is secreted by dermal glands, the canals of which perforate the wax layers and hence allow the cement layer to be formed over the wax layers. Pore canals apparently serve as routes for secretion of the wax layer.

When the secretion of the new cuticle is complete, the insect emerges, the act of *ecdysis*, leaving behind what remains of the old cuticle and the tracheal and gland duct linings, that is, the *exuvia*.

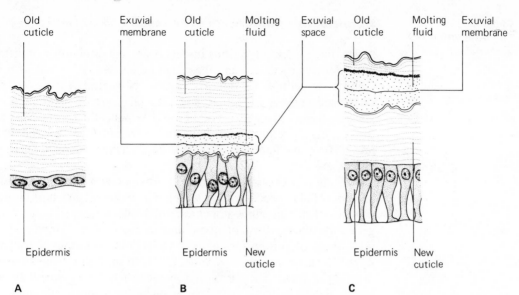

Old cuticle

Old
cuticle

Exuvial
membrane

Old
cuticle

Molting
fluid

Exuvial
space

Old
cuticle

Molting
fluid

Exuvial
membrane

Epidermis

Epidermis

New
cuticle

Epidermis

New
cuticle

A

B

C

Fig. 2-7
Molting process. A. Section of integument prior to molting. B. After
apolysis, digestion of old endocuticle in progress and new cuticle being
deposited. C. Just prior to ecdysis. [Redrawn with slight modifications
from Wigglesworth, 1959.]

This process is facilitated by *ecdysial lines* beneath which only epicuticle and endocuticle are present (Fig. 2-8). Since the endocuticle
is digested during the molting process, a line of weakness develops.
When ready to emerge, the insect may gulp air or water or increase
the hydrostatic pressure of the blood by contracting various body
muscles. These actions exert an internal force on the ecdysial lines,
and subsequently the old cuticle splits wherever they are located.
These lines of weakness are usually located on the dorsum of the
head and thorax with an anterior–posterior orientation. Following
ecdysis an insect may eat the exuvia and hence reclaim nearly all
nutrients that may have been lost during molting. Sclerotization and
melanization occur subsequent to ecdysis.
    Zacharuk (1976) provides a recent review of the structural changes
that occur during molting.

Fig. 2-8
Ecdysis (diagrammatic). A.
Section of integument prior
to molting. B. Section
following digestion of
endocuticle with line of
weakness where
endocuticle was previously
in contact with epicuticle.
C. Splitting of head
capsule along preformed
ecdysial line. [Redrawn
from Snodgrass, 1960.]

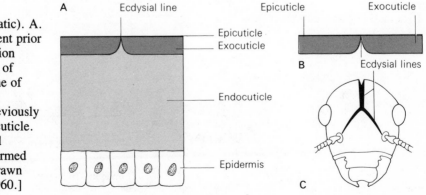

A   Ecdysial line

Epicuticle

Epicuticle
Exocuticle

Exocuticle

B   Ecdysial lines

Endocuticle

Epidermis

C

External Integumentary Processes

The integument of various insects bears a great number of different external processes, and these can be classed (Fig. 2-9) as noncellular or cellular (Snodgrass, 1935). Noncellular processes are composed entirely of cuticle and may take any of several forms, such as spines, ridges, or nodules. Some noncellular processes are in the form of minute fixed hairs (*microtrichia*) that lack the basal articulation characteristic of *setae* (*macrotrichia*), which are cellular processes.

Cellular processes may be further broken down into multicellular and unicellular processes. Multicellular processes are hollow outgrowths of the integument and are lined with epidermal cells. They generally take the form of spines and are found, for example, on the hind tibiae of certain Orthoptera. Unicellular processes are all referred to as *setae* (*macrotrichia*), although they show an extensive diversity of form. They are commonly hairlike, but may be flattened into scales (Fig. 2-4), may bear branches and appear plumose, or may take on other shapes. The setal shaft is formed by a protoplasmic outgrowth of a specialized hair-forming or *trichogen* cell. This projection is surrounded by a setal membrane and lies within a socket. The membrane and the socket are formed by a second cell, the *tormogen* or socket-forming cell. Setae have various functions. For example, many are connected to a sensory nerve cell, some are associated with a specialized gland that may secrete a toxic substance, and the scale form often contains the pigment or possesses the physical characteristics responsible for much of the coloration and patterning.

The Insect Skeleton

The insect skeleton is composed of a series of plates, or *sclerites*, of varying degrees of hardness (Fig. 2-10). These sclerites are separated either by very soft membranous areas, called *conjunctivae*, or by external grooves that indicate internal inflections or very narrow lines of thin, sometimes flexible integument. The external grooves are referred to as *sutures* or *sulci* (sing., *sulcus*). The com-

Fig. 2-9
External integumentary processes.

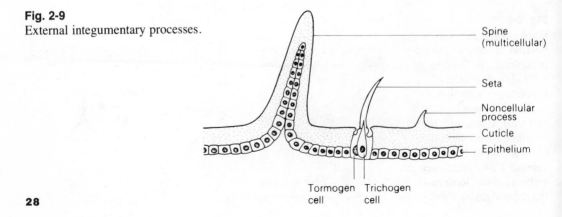

Spine
(multicellular)

Seta

Noncellular
process

Cuticle

Epithelium

Tormogen   Trichogen
cell   cell

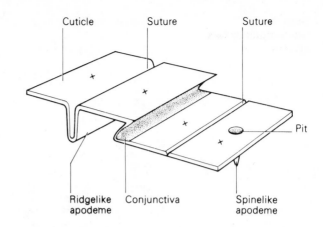

Cuticle   Suture   Suture

Pit

Ridgelike   Conjunctiva   Spinelike
apodeme   apodeme

Fig. 2-10
Sclerites (denoted by ×), sutures, and
internal integumentary processes
(diagrammatic).

bination of hardened plates joined by conjunctivae allows body
movement and articulation of the appendicular joints. The internal
inflections of the integument denoted externally by sutures or pits
are *apodemes* and may be multicellular or unicellular. These inflections may be spinelike or ridgelike and serve to strengthen the exoskeleton or as points for muscle attachment. A spinelike inflection
is called an *apophysis*.

Segmentation

The insect body, like that of other arthropods, is divided into a
series of segments. Examination of an insect will reveal that many
of these segments are rather different from one another. Some bear
appendages, others do not; the appendages of one segment may be
unlike those of another; some segments may be quite movable relative to the others, while some adjacent segments are fused or partially fused; and so on. One aspect becomes immediately apparent—
the obvious functional specialization of the different segments.
These specialized functions will be discussed later, but first we want
to consider briefly the hypothetical explanation of the origin of these
segments.

The primitive arthropod is thought to have been composed of a
series of essentially identical segments called *somites* or *metameres*.
Most of these somites apparently carried a pair of appendages. The
anterior *acron* or *prostomium* and the posterior *telson* or *periproct*,
in which the posterior opening of the alimentary canal, the *anus*,
was located did not bear appendages. The mouth, the anterior opening to the alimentary canal, opened between the acron and the first
postoral segment. During the course of insect evolution these segments fused in different ways, forming the variously divided body
regions of modern arthorpods, and the appendages took on specialized roles appropriate to the body region in which they were located
or disappeared altogether.

The segments in the primitive arthropod were marked externally
by constrictions of the integument (Fig. 2-11A). Internally, these
constricted regions formed folds where the principal longitudinal
muscle bands, the *segmental muscles,* were attached. This arrangement of body units has been called *primary segmentation.* This type

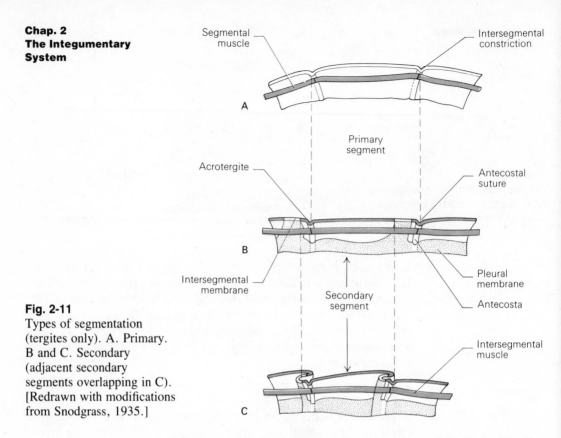

Fig. 2-11
Types of segmentation
(tergites only). A. Primary.
B and C. Secondary
(adjacent secondary
segments overlapping in C).
[Redrawn with modifications
from Snodgrass, 1935.]

of segmentation is found today among the soft-bodied wormlike
larvae of several insects—for example, the larvae of Lepidoptera—
and in all arthropod embryos.

Another type of segmentation (Fig. 2-11B, C), *secondary seg-*
mentation, is found in adult and many larval insects. In this type of
segmentation, the membranous areas between adjacent segments do
not coincide with the points of attachment of longitudinal muscles,
but are slightly anterior to them. Secondary segmentation is considered to be a derivative of primary segmentation; the infolded region
of a primary segment has become sclerotized, and a membranous
area has developed just anterior to it. Secondary segmentation is
well illustrated by a pregenital abdominal segment of a generalized
insect.

The typical abdominal segment consists of a dorsal sclerotized
region, the *tergum* (Fig. 2-11B, C), and a ventral sclerotized region,
the *sternum*. Individual sclerites that form subdivisions of the tergum
or sternum are called *tergites* or *sternites,* respectively. The tergum
and sternum are separated by a lateral membranous area, the *pleural*
membrane. The external groove that corresponds to the infolded
portion forms an internal ridge that, in some cases, may be quite
pronounced. This internal ridge or apodeme is called the *antecosta*.
A small sclerite is demarcated by the antecostal suture and the secondary intersegmental membrane, both in the tergite and in the sternite. The sclerite associated with the tergum is the *acrotergite;* that
with the sternum, the *acrosternite*. There is an advantage to overlapping hardened plates, for these plates can then have a protective

function and also become involved in locomotion (Snodgrass, 1935). Since the arrangement of longitudinal muscles in secondarily segmented insects does not coincide with the definitive segments, these muscles are referred to as *inter-segmental muscles*.

The General Insect Plan

Despite the many features that insects have in common with one another, they display a tremendous diversity in form. This being the case, it is useful to have a hypothetical generalized insect form that subsequently will serve as a conceptual basis for interpreting variations in insect structure, or morphology. For good accounts of general insect morphology and access to recent literature, see Chapman (1971) and Richards and Davies (1977). Matsuda (1976) outlines the major principles of morphology. Richards (1973) discusses the history of insect morphology.

Tagmata

The segments of the insect body are divided into three well-defined regions, or *tagmata:* a head, a thorax, and an abdomen (Fig. 2-12). The head bears the organs of ingestion, or *mouthparts; compound eyes; simple eyes,* or *dorsal ocelli,* which may be lacking; and a pair of appendages called *antennae.* The thorax is composed

Fig. 2-12
Generalized pterygote insect.

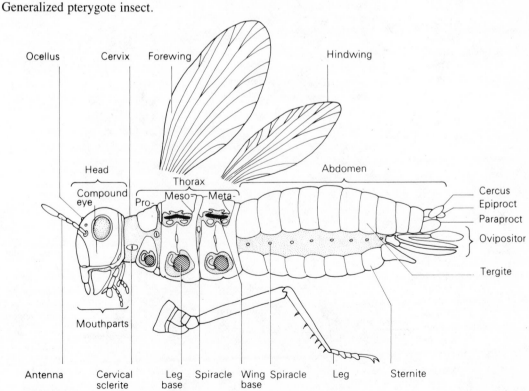

of three basic segments named by their relative positions from anterior to posterior as follows: *prothorax, mesothorax,* and *metathorax*. Each of the three thoracic segments bears a pair of legs in most immature and nearly all adult insects. Wings, if present, are found only on the mesothorax and metathorax. The class Insecta is usually divided into two groups (see Chapter 11) on the basis of the presence or absence of wings: *Apterygota*, primitively wingless insects, and *Pterygota*, winged or secondarily wingless insects. The abdomen consists of a varying number of legless segments, the primitive number being considered on the basis of embryological studies to be 11 plus a terminal segment, the *periproct*, or *telson*, which bears the anus. The external genitalia are borne on one or more of the posterior segments.

Head

The insect head (Fig. 2-13) is a composite structure that has evolved from the fusion of the prostomium with a number of postoral segments and modifications of the appendages of these segments into the organs of ingestion, the mouthparts. There is no general

Fig. 2-13
Generalized insect head. A. Anterior view. B. Posterior view. C. Lateral view. D. Top of head capsule cut away to show tentorium. [Redrawn with slight modifications from Snodgrass, 1935.]

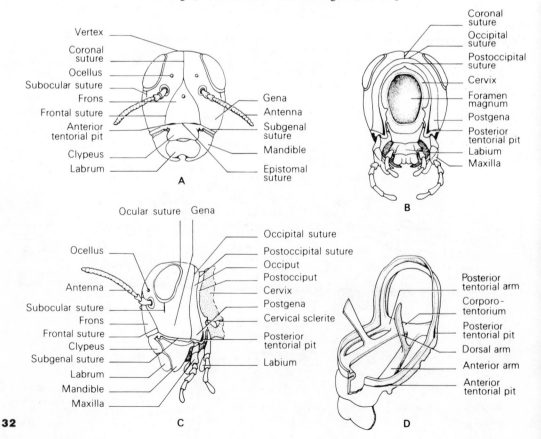

agreement as to the number of primitive segments that contributed to the evolution of the head. Opinions vary from four to seven segments plus the acron (Matsuda, 1965; Rempel, 1975) (see Chapter 10). Basically, the head is composed of a hardened capsule, the *cranium,* which bears the antennae, eyes, and mouthparts. There is little evidence of segmentation in the mature insect head, and some of the lines or sutures that give it a segmented appearance are probably of functional significance rather than relating to any primitive segmentation. However, one of the sutures, the *postoccipital suture* of the cranium, is considered to have persisted, separating the maxillary and labial segments. The head is attached to the thorax by means of a membranous region, the *neck* or *cervix.* The cervical membrane is quite flexible and allows movement of the head. The neck typically bears two pairs of lateral plates, the *cervical sclerites.* Muscles from both the head and thorax attach to these sclerites and are involved with certain movements of the head. The cervix may be derived in part from the prothorax and in part from the labial segment of the head.

Cranial Structure. The cranium is divided into various regions by a series of sutures. These sutures may be quite apparent in some insects and completely lacking in others. Therefore, this description will pertain to a generalized insect cranium and will mention sutures and regions that may or may not be readily discernable in a given insect. The *epicranial suture* is shaped like an inverted Y, the stem forming the dorsal midline of the cranium and the arms diverging ventrally across the anterior portion of the head. The stem of the Y is called the *coronal suture;* the arms form the *frontal sutures.* The frontal sutures are commonly obscure, especially in adults, and often lacking altogether.

The region delimited by the frontal sutures is called the *frons* and the dorsal portion of the cranium bisected by the coronal suture is the *vertex.* These three sutures are lines along which the shed cuticle of the cranium splits during ecdysis (Fig. 2-8C). For this reason it is preferable to call them *ecdysial cleavage lines* (Snodgrass, 1963a). Since the ecdysial cleavage lines are poorly developed in many insects, a better definition of frons is the frontal region of the cranium, which bears the *median ocellus,* if present, and internally bears the origins of the muscles of the anterior mouthpart, the *labrum.* Using this definition, the ecdysial cleavage lines would be located on the frons instead of defining it.

The *occipital suture* forms a line from the posterior termination of the coronal suture to just above the mandibles on either side of the cranium. The *postoccipital suture* lies posterior to, and in the same plane as, the occipital suture. This suture surrounds the posterior opening of the head capsule, the *foramen magnum,* through which the internal organs communicate between the head and thorax. The postoccipital suture forms an internal ridge to which are attached the muscles that move the head. On either side of the cranium immediately above the bases of the mandibles and maxillae are the *subgenal sutures.* Internally, these sutures form ridges which add strength to that portion of the cranium.

Above each subgenal suture and anterior to the occipital suture is a "cheek," or *gena*. A *postgena* lies adjacent to the gena, but posterior to the occipital suture. The postgenae are delimited posteriorly by the postoccipital suture. Dorsally, the region between the occipital and postoccipital sutures is called the *occiput*. The plate posterior to the postoccipital suture, which surrounds the better part of the foramen magnum, is the *postocciput*. This structure generally bears a bilateral pair of processes upon which the cervical sclerites articulate. The subgenal sutures may be connected across the front of the cranium, just beneath the frontal suture, by the *epistomal* or *frontoclypeal suture*. In a fashion similar to the postoccipital and subgenal sutures, the epistomal suture forms an inflected ridge.

A lobelike structure, the *clypeus,* lies immediately ventral to the epistomal suture. In many insects, the clypeus is hinged with the foremost mouthpart, the *labrum*. The compound eyes are commonly surrounded by the *ocular sutures*. Similarly, an *antennal suture* surrounds the base of each antenna. In addition, *subocular sutures* may be present, running vertically beneath the compound eyes.

Tentorium.   At each of four points on the cranium, the cuticle forms an armlike inflection. Internally these inflections join, forming a framework. This internal framework, the *tentorium* (Fig. 2-13D), affords many points for muscle attachment, contributes appreciably to the rigidity of the head capsule, and provides support for the brain and the part of the foregut that passes through the head. Externally the points of inflection form the anterior and posterior *tentorial pits*. The anterior pits are generally contained in the epistomal suture, although in some insects they may be more closely associated with the subgenal sutures. The posterior tentorial pits open bilaterally in the postoccipital suture immediately adjacent to the subgenal sutures. Internally the tentorium is composed of a pair of anterior arms, which arise from the anterior pits and fuse with the posterior arms, which correspondingly arise from the posterior pits. At the point of fusion, a transverse bar, the *tentorial bridge* or *corporotentorium* is formed. In addition, a third pair of arms may arise dorsally from the anterior arms.

Compound Eyes and Ocelli.   On each side of the head is a compound eye, so called because each eye is composed of many individual units. Each eye unit is an *ommatidium*. The surface of each eye is divided into a large number of (usually) hexagonal facets. These facets are the *corneal lenses,* and they are the cuticular parts of the ommatidia. The *dorsal ocelli,* or simple eyes, are commonly three in number and are located on the anterior portion of the cranium, one on either side of the coronal suture and the third between the frontal sutures.

Antennae.   The antennae are paired appendages that articulate with the head capsule and are located on the anterior portion near the compound eyes. Some evidence suggests that they have been derived from appendages associated with a primitive segment (Rempel, 1975). Antennae are usually strictly sensory appendages bearing large numbers of sensilla (see Chapter 6). Typically, they are composed of a series of segments and the generalized form (Fig. 2-14) is filament-like in appearance. There are three basic parts of an antenna in most insects.

1. A basal *scape*.
2. A *pedicel*.
3. A distally located *flagellum,* which is usually long and composed of a number of subsegments.

The pedicel in most insects (except Collembola and Diplura) contains Johnston's organ, a special sensory structure that is discussed in Chapter 6. The scape articulates in an antennal socket, the integument between the antenna and head capsule being membranous and flexible. The rim of the antennal socket associated with the cranium characteristically contains an articular point, the *antennifer.* Antennae are moved about by muscles inserted on the base of the scape. Schneider (1964) provides a review of the research literature on the structure and function of antennae.

Mouthparts. The mouthparts of the generalized *mandibulate* type (Fig. 2-15) are considered to be the primitive form. They typically consist of

1. An anterior "upper lip," or *labrum.*
2. The *hypopharynx.*
3. A pair of *mandibles.*
4. A pair of *maxillae.*
5. A posterior "lower lip," or *labium.*

There is general agreement that the mandibles, maxillae, and labium arose from the appendages of three primitive segments. The labrum may also represent primitive segmental appendages (Rempel, 1975; see Fig. 10-13).

The labrum is suspended from and articulates with the clypeus by a narrow membrane, which allows considerable movement. The integument of the labrum forms a lining that runs dorsally across the inside of the clypeus and terminates at the true mouth. This epipharyngeal wall forms the dorsal lining of the *preoral cavity* formed by the mouthparts and is continuous with the lining of the pharynx.

Fig. 2-14
Generalized insect antenna.

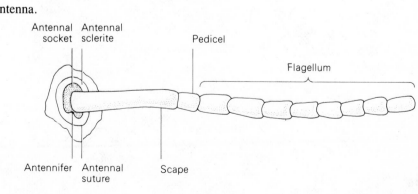

Commonly the epipharyngeal wall bears a lobe referred to as the *epipharynx*. The *hypopharynx* lies in the preoral cavity somewhat in the way a tongue does (Fig. 2-16). Generally the duct from the salivary glands opens between the hypopharynx and the epipharyngeal wall. The portion of the preoral cavity between the hypopharynx and labrum is the *cibarium*. The portion of the preoral cavity between the hypopharynx and labium forms the *salivarium*.

The mandibles (Fig. 2-15) are a pair of highly sclerotized unsegmented jaws, each of which forms two articulations with the cranium, a posterior articulation with the postgena and an anterior secondary articulation near the tentorial pit. Each mandible has a proximal molar or grinding region and a distal incisor or cutting region.

The paired maxillae serve as accessory jaws, aiding in holding and chewing food. They are somewhat more complex than the mandibles. Each is composed of the proximal *cardo,* which bears the *stipes,* which in turn bears two distal lobes, the *galea* and *lacinia*. The galea, a comparatively unsclerotized lobe, is located lateral to the mesal lacinia, which is more sclerotized and contains teeth on its inner edge. The galea commonly covers the lacinia when the mouth parts are "closed." In addition, the stipes also bears a lateral sclerite, the *palpifer,* to which is attached a five-segmented *maxillary palpus*.

Fig. 2-15
Generalized mandibulate mouthparts as illustrated by a cockroach. [Redrawn from James and Harwood, 1969.]

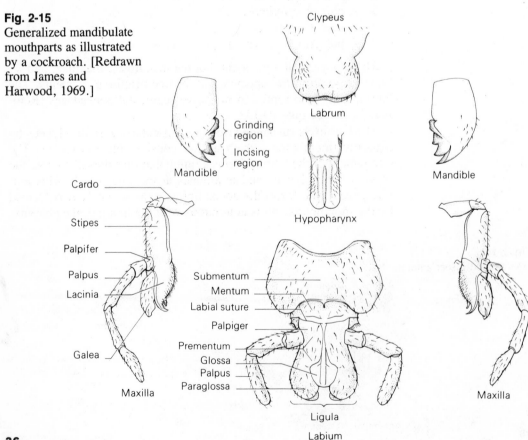

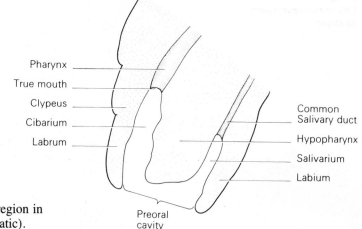

Pharynx

True mouth

Clypeus

Cibarium

Labrum

Common Salivary duct

Hypopharynx

Salivarium

Labium

Preoral cavity

Fig. 2-16
Sagittal section of mouthparts region in
a generalized insect (diagrammatic).

The labium is a composite structure formed from the fusion of
two primitive segmental appendages, and its parts can be seen to
correspond to those of the maxillae. It consists of a basal *postmentum*
attached to the cervix ventral to the foramen magnum and is closely
associated with the postoccipital region near the posterior tentorial
pits. The postmentum is commonly divided transversely into two
portions, a proximal *submentum* and a distal *mentum*. The apical
portion of the labium is the *prementum,* which is hinged to the
postmentum by the *labial suture*. The *prementum* bears laterally a
pair of segmented *labial palpi* and distally four lobes, two inner
lobes, the *glossae,* located between two outer lobes, the *paraglossae*. The labial palpi are attached to lateral sclerites on the prelabium,
the *palpigers*.

The muscles responsible for the movement of the mouthparts are
attached at various points on the head capsule and tentorium. One
needs only to observe a grasshopper or cockroach eating to appreciate the intricate, highly coordinated movements of which the
mouthparts are capable.

Thorax

As mentioned previously, the insect thorax is composed of three
segments: an anterior *prothorax,* a *mesothorax,* and a posterior
metathorax. Functionally, the thorax is the locomotive tagma since
it bears the legs and, if present, the wings. Wings are borne on either
or both the meso- and metathoracic segments. These two segments
are thus often referred to collectively as the *pterothorax* (ptero =
wing). In most winged insects the prothorax is usually separate from,
and somewhat less developed than, the remaining segments. In many
insects at least part of the first abdominal segment has become intimately associated with the thorax, and in many of the Hymenoptera
it has literally become a part of the thorax, being separated from the
rest of the abdomen by a constriction.

Each thoracic segment typically can be divided into four distinct
regions (Fig. 2-17A): a dorsal tergum, or *notum;* a pair of bilateral
pleura (sing., *pleuron*); and a ventral *sternum*. Each of these regions

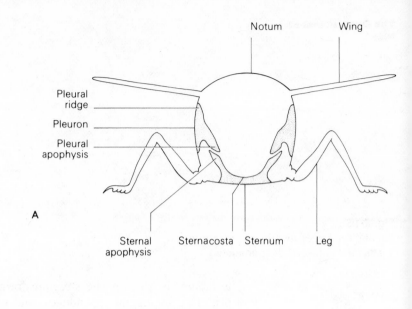

A

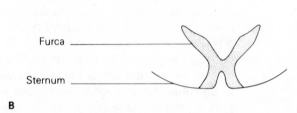

B

Fig. 2-17
A. Cross section of a
generalized thoracic
segment. B. Sternal
apophyses in the form of
furcae (diagrammatic).

is commonly subdivided into two or more sclerites. The legs arise
on the pleura; the wings articulate between the notal and pleural
regions. *Spiracles* (Fig. 2-12), the external openings of the ventilatory system, are usually found, one in each of the pleural regions
between the pro- and mesothorax (mesothoracic spiracles) and one
in each of the pleural regions between the meso- and metathorax
(metathoracic spiracles). Prothoracic spiracles are atypical.

Matsuda (1970) and Manton (1972) should be consulted for rigorous treatments of the morphology of the insect thorax.

Thoracic Terga.　The thoracic terga (nota) of apterygote and
some immature pterygote insects are rather simple when compared
to the more modified terga of the adult winged forms. Each is merely
a single plate associated with the remaining two by secondary segmentation. The dorsal longitudinal muscles of the thorax are attached
to the antecostae as described in the discussion of secondary segmentation in an abdominal segment. The terga of most pterygote
insects are more complex, being divided into smaller sclerites by
various sutures. These divisions of the terga, as well as the other
modifications of the generalized thoracic segment, arose as a result
of the evolution of wings and flight. Typically, the tergum of a wing-
bearing segment is composed of two parts: an *alinotum* and a *postnotum* (Fig. 2-18A and 2-19). The alinotum bears the wings; the
postnotum bears the internally inflected *phragma*, which may be
viewed as the modified antecosta of the next segment posterior. In

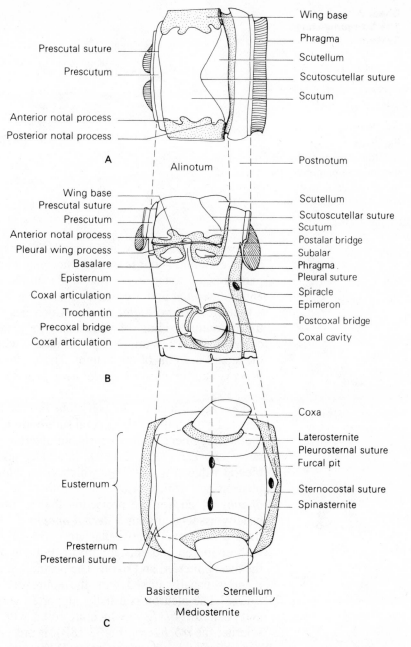

Prescutal suture

Prescutum

Anterior notal process

Posterior notal process

Wing base

Phragma

Scutellum

Scutoscutellar suture

Scutum

A

Alinotum

Postnotum

Wing base
Prescutal suture
Prescutum
Anterior notal process
Pleural wing process
Basalare
Episternum
Coxal articulation
Trochantin
Precoxal bridge
Coxal articulation

Scutellum
Scutoscutellar suture
Scutum
Postalar bridge
Subalar
Phragma
Pleural suture
Spiracle
Epimeron
Postcoxal bridge
Coxal cavity

B

Coxa
Laterosternite
Pleurosternal suture
Furcal pit

Eusternum

Sternocostal suture
Spinasternite

Presternum
Presternal suture

Basisternite   Sternellum

Mediosternite

C

Fig. 2-18
Generalized pterygote
insect thorax. A.
Notum. B. Pleuron. C.
Sternum. [A and B
redrawn from
Snodgrass, 1935, A
with modifications.]

the case of the metathorax this would be the antecosta of the first
abdominal segment. The phragmata are more platelike than ridgelike
and afford a comparatively large surface area for the attachment of
the dorsal longitudinal wing muscles, which are extremely important
in the flight of insects.

According to Snodgrass (1935), the intersegmental membrane
present in the more primitive apterygote insects is reduced or lacking
and the acrotergite of the next segment posterior has become the
postnotum. Correlated with the presence of wings is a strengthening
of the alinotum afforded by various internally inflected ridges. Although there are different sutures associated with these internal
ridges in various insects, one, the *scutoscutellar suture* (Fig. 2-18A),

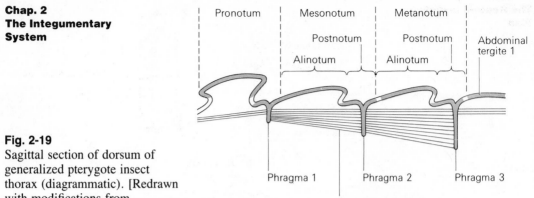

Fig. 2-19
Sagittal section of dorsum of
generalized pterygote insect
thorax (diagrammatic). [Redrawn
with modifications from
Snodgrass, 1935.]

is present in nearly all winged forms. It lies in the posterior portion
of the alinotum and it is shaped somewhat like a V, with the bottom
part directed anteriorly. This suture divides the alinotum into an
anterior *scutum* and a posterior *scutellum.* Many insects also have
a transverse or *prescutal suture,* which lies on the anterior part of
the alinotum dividing off a small anterior plate, the *prescutum.* Processes that serve as articular points for the wings have developed on
the lateral margins of the scutum. These are the *anterior notal wing
process* and the *posterior notal wing process.*

Thoracic Pleura. In pterygote insects an internal inflection,
denoted externally by the pleural suture, divides the pleuron of each
thoracic segment into two parts: an anterior *episternum* and a posterior *epimeron* (Fig. 2-18B). Internally this inflection forms the
pleural ridge (Fig. 2-17A), which gives additional strength to the
pleuron and which bears, in pterothoracic segments, the *pleural
apophysis,* an armlike projection that is directed ventrally and is
usually associated with a *sternal apophysis* (Fig. 2-17A). The pre-
fixes pro-, meso-, and meta- are commonly used in combination
with epimeron and episternum; the epimeron of the mesothorax is
the mesepimeron, and so on. The postnotum of a wing-bearing segment is usually united with the epimeron, forming the *postalar
bridge.* The pleuron is usually supported ventrally by the *precoxal*
and *postcoxal bridges,* which are fused with the sternum. A small
sclerite, the *trochantin* (Fig. 2-18B), is present in many of the more
generalized wing-bearing insects. When present it usually bears one
of the points of articulation of the coxa. A second articular process
is generally located at the ventral extremity of the pleural suture.
Dorsally, the *pleural wing process* serves as a ventral articulation
point for the wing. Two small sclerites, one anterior (*basalar*) and
one posterior (*subalar*) to the pleural wing process, are important in
wing movements (more in Chapter 7).
    The pleural region in the apterygotes is composed of a group of
subcoxal sclerites associated with the *coxa,* the basal segment of the
leg. On the basis of comparative morphological studies it has been
suggested that the primitive subcoxal sclerites were three in number
(Fig. 2-20), a dorsally located *anapleurite,* forming a rather crude

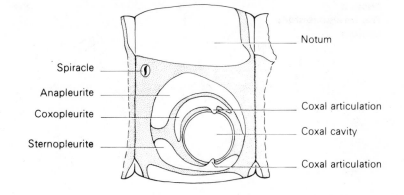

Fig. 2-20
Hypothetical primitive arrangement of subcoxal sclerites. [Redrawn from Snodgrass, 1935.]

semicircle around the coxa; a smaller *coxopleurite,* concentric with and ventral to the anapleurite; and a ventral *sternopleurite.* Presumably the coxopleurite and the sternopleurite each carried a process with which the coxa articulated. With the advent of wings, the anapleurite and coxopleurite apparently fused, forming a sclerotic pleuron, and the ventral sternopleurite joined with the sternum, resulting in the formation of a continuous sclerotic body beneath the notum and the wings (Snodgrass, 1935). A recent interpretation, however, disagrees with this "subcoxal" theory of origin of the pleuron (Manton, 1972).

Thoracic Sterna. In many generalized pterygote forms, the sternum (Fig. 2-18C) of a thoracic segment is composed of a segmental plate, or *eusternum,* and a smaller, posterior plate, apparently derived from an intersegmental plate, the *spinasternum.* The eusternum is commonly divided into three parts: the small, anterior *presternum,* the *basisternum,* and the *sternellum.* The presternum is separated from the rest of the sternum by the *presternal suture.* A pair of internal projections, the *sternal apophyses* (Fig. 2-17A), sternum and are indicated externally by the *furcal pits* (Fig. 2-18C). The furcal pits are often connected by the *sternacostal suture* which is associated with an apodeme, the *sternacosta* (Fig. 2-17A), which connects the sternal apophyses. The furcal pits and sternacostal suture form the division between the basisternum and sternellum (Fig. 2-18C). Lateral plates, or *laterosternites,* are sometimes divided from the sternum by *pleurosternal sutures.* The sternal apophyses are often in the form of a V diverging from the sternacosta (Fig. 2-17B). In this case they are termed *furcae* (sing., *furca*). Furcalike structures also arise from the sternites of some apterygotes. The spinasternum bears an internally inflected median process, the *spina,* that forms an externally apparent *spinal pit* (Fig. 2-18C). The various internal inflections of the sternum provide strength and areas for muscle attachment in the thorax.

Legs

The generalized insect leg consists of six segments (Fig. 2-21).

1. A basal *coxa,* which articulates with the thorax in the pleural region (see Fig. 2-18B).

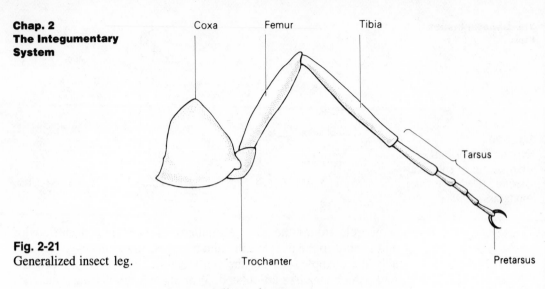

Fig. 2-21
Generalized insect leg.

Trochanter

Pretarsus

2. A small *trochanter*.
3. A *femur*.
4. A *tibia*.
5. A segmented *tarsus*.
6. A *pretarsus*.

The coxa is often divided into two parts, the posterior and usually larger part being called the *meron*. The trochanter articulates with the coxa, but usually forms an immovable attachment with the femur. The femur and tibia are typically the longest leg segments. The tarsus, which is derived from a single segment, is usually divided into individual *tarsomeres*. The pretarsus may consist of a single *claw*, but it is usually composed of a pair of moveable claws and one or more pads or bristles. Legs are usually looked upon as the principal organs of terrestrial locomotion, although, as will be seen, they have undergone many modifications and have been adapted to a wide variety of functions, including swimming, prey capture, and digging.

Wings

The wings arise as outgrowths of the integument between the tergal and pleural sclerites (Fig. 2-17A). They are thus composed of two layers of integument (Fig. 2-22). A series of tracheae grow between these integumentary layers, and when a wing is fully de-

Fig. 2-22
Cross section of wing of dragonfly nymph. [Redrawn from Comstock, 1918.]

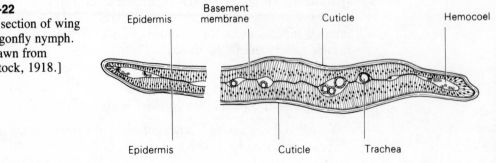

Epidermis   Basement membrane   Cuticle   Hemocoel

Epidermis   Cuticle   Trachea

veloped, they run within the longitudinal and transverse supportive framework, the *wing veins*. The cuticle is often thicker in the region of these veins, lending further rigidity. Since the wings are outgrowths of the integument, the space between the epidermal layers is continuous with the body cavity, or *hemocoel,* of the insect. This space is usually evident only around the veins, since in the other parts of the wing, the *cells,* for example, the two layers of integument become closely appressed to one another. In many instances, blood cells, or *hemocytes,* can be seen circulating in the wing immediately on either side of a vein. Complete circulation of hemolymph in the wing is made possible by the presence of numerous cross veins in many insects. The wings are, of course, the organs of aerial locomotion in most cases, but, like the legs, have undergone extensive adaptive modification.

Wings articulate dorsally with the anterior and posterior notal processes (Fig. 2-18A, B) and ventrally with the pleural wing processes (Fig. 2-18B). Refer to Figures 7-9 and 7-10 for further details on the articulation of wings with the thorax.

Comparative studies of wing venation in many species have led to the development of various hypothetical generalized patterns. One such pattern is given in Figure 2-23. The patterns of wing venation in different insects, interpreted on the basis of the generalized pattern developed by Comstock (1918) and others, are extremely useful in the identification of many insects.

Major *longitudinal veins* (Fig. 2-23) have been identified by their location relative to one another, their form, their association with basal sclerites (see Fig. 7-10), and the presence of tracheae. The study of fossil forms has also helped identify the major longitudinal veins. Longitudinal veins are connected by *cross veins* (Fig. 2-23). The combination of longitudinal and cross veins, or longitudinal veins reaching the wing margin, divides a wing into various *cells.* In the more primitive, generalized insect orders, such as Orthoptera, the wings fold in a pleated fashion so that the longitudinal veins lie on top of crests *(convex veins)* or within troughs *(concave veins).* Concave veins are always concave, and convex veins are always convex; for example, the main stem of the *radius* is always convex. Comstock (1918) and Hamilton (1972) deal in detail with wing venation.

The edges or *margins* of wings are named as follows (Fig. 2-23): (1) the anterior, *costal margin,* (2) the posterior *anal margin,* and (3) the outer *apical margin.*

Fig. 2-23
Hypothetical primitive pattern of wing venation. Longitudinal veins: A, anal; C, costa; Cu, cubitus; M, media; R, radius; Rs, radial sector; Sc, subcosta. Cross veins: h, humeral; m, medial; m-cu, mediocubital; r, radial; r-m, radiomedial; s, sectorial. Costal, apical, and anal margins are indicated. [Redrawn from Comstock, 1918.]

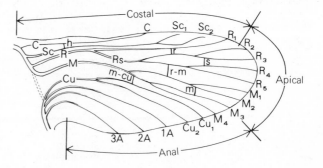

Abdomen

The segmentation of the insect abdomen has already been discussed. As previously mentioned, the primitive number of abdominal segments is considered to have been 11 true metameres plus a
terminal segment, the *periproct* or *telson,* that contained the anus.
The tendency in insectan evolution has been toward a reduction in
the number of segments, and in the generalized insect abdomen (Fig.
2-24) there are 11, the eleventh being reduced and divided into lobes
that surround the anus. This terminal segment may bear a pair of
appendages, the *cerci.* These are considered to be serially homologous with the legs and mouthparts. The plates of the eleventh segment are generally three in number: the *epiproct* dorsal to the anus,
and the two *paraprocts* on either side of the anus. The abdominal
segments are usually numbered from anterior to posterior, number
one being immediately posterior to the metathorax. Spiracles, the
external openings of the ventilatory system, are typically found one
on either side of the first eight abdominal segments.

In the generalized female pterygote insect, modified appendages
of the eighth and ninth abdominal segments form the *ovipositor,* or
egg-laying apparatus, which is composed of two pairs of basal *valvifers.* The valvifers in turn bear the *valvulae,* one pair on the eighth
segmental appendage and two pairs on the ninth segmental appendage. The female *gonopore* (reproductive opening) is usually on or
posterior to the eighth or ninth segment. The male external copulatory apparatus, *penis* or *aedeagus,* is usually borne on the ninth
abdominal segment.

Matsuda (1976) deals specifically with the morphology of the
insect abdomen. Snodgrass (1957), Smith (1969), Tuxen (1970),
and Scudder (1971) deal specifically with genitalia.

Variations on the General Insect Plan

One of the major reasons for the tremendous success of insects
has undoubtedly been the seemingly endless potential of the basic
plan to undergo evolutionary modification. It appears as though no
structure has escaped some modification. Insects thus are excellent

Fig. 2-24
Generalized insect
abdomen. [Redrawn
from DuPorte, 1961.]

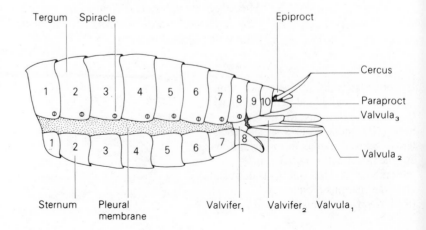

organisms with which to demonstrate the phenomenon of adaptation. Having presented the external structure of a hypothetical, generalized insect as a basis for interpreting modifications, we turn now to some of the more obvious variations in the insect skeleton.

Matsuda (1965, 1970, and 1976), Borror, DeLong, and Triplehorn (1976), and Richards and Davies (1977) provide extensive information on the external anatomy of members of the various insect orders and families.

Patterns of External Integumentary Processes

Setae are commonly arranged in constant patterns that have often been useful in systematics, for example, among members of the suborder Cyclorrhapha, in the order Diptera (true flies). These arrangements and associated nomenclature are referred to as *chaetotaxy.*

Modifications of the Head

Antennae. Although insect antennae vary greatly in length, overall size, size of the individual segments, segmentation, setation, and other aspects, they can usually be described as being of a particular type or combination of types (e.g., capitate–lamellate). In some insects antennae are weakly developed, as in some larval Hymenoptera and some larval Diptera, or absent, as in the Protura. Antennal type is commonly of value in the assignment of a given insect to a family and may, in certain instances, serve to differentiate the sexes, as in most male and female mosquitoes. Male mosquitoes bear distinctly plumose antennae; the female's antennae are less feathery in appearance and bear comparatively few whorls of hairs. This is an example of *sexual dimorphism,* a structural difference between the two sexes. Most modifications of the antennae occur in the flagellum. Antennal structure is closely related to function, for example, the plumose antennae of many moths provide extensive surface area, and the myriads of olfactory sensilla that cover them respond to chemical cues associated with mate location.

Commonly occurring antennal types with examples of insects possessing them are presented in Figure 2-25.

In most insects antennae serve exclusively as sensory structures, but are involved in grasping prey in larval *Chaoborus* spp. (Diptera, Chaoboridae) and serve as claspers by which males of several kinds of insects hold females during copulation.

Compound Eyes and Ocelli. Compound eyes are characteristic of most adult insects and many nymphs, but are absent in a majority of the larvae of insects with complete metamorphosis. In these larvae, structures called stemmata are the light receptors. Compound eyes are also absent in the soldier castes of some termite species, some species of fleas, and certain species of springtails. When present (Fig. 2-26), compound eyes occur in many diverse—sometimes bizarre—forms. Some are exceedingly large and contain many facets, as in adult dragonflies (Fig. 2-26A); others are quite small and have few facets, as in many hemipterous insects (Fig. 2-26B). In

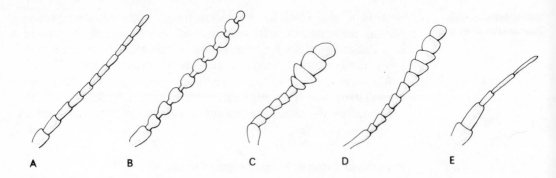

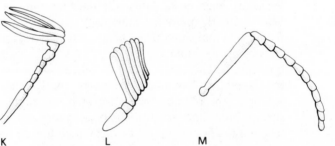

Fig. 2-25
Antennal types. A. Filiform, grasshopper. B. Moniliform, wrinkled bark beetle. C. Capitate, skin beetle. D. Clavate, carrion beetle. E. Setaceous, dragonfly. F. Serrate, click beetle. G. Pectinate, fire-colored beetle. H. Plumose, male mosquito. I. Aristate, flesh fly. J. Stylate, horse fly. K. Lamellate, scarab beetle. L. Flabellate, cedar beetle. M. Geniculate, honey bee.

some insects, the compound eyes are actually divided, appearing as two pairs. Examples are found among the beetles in the families Gyrinidae (Fig. 2-26C) and Cerambycidae, and among mayflies in the genus *Cloeon* (Fig. 2-26D). In the latter the anterior division is borne upon a stalklike outgrowth of the head capsule. The compound eyes of males of the flies in the genera *Bibio* (Fig. 2-26E) and *Simulium* have two rather distinct areas of different-sized facets, giving the appearance of a compound compound eye. The compound eyes, like the antennae, may be involved in sexual dimorphism, as

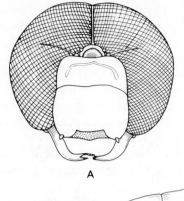

A

B

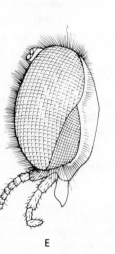

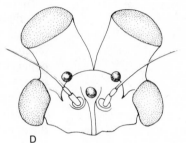

Fig. 2-26
Variations in compound
eyes. A. Dragonfly. B.
Thrips. C. Whirligig
beetle. D. Mayfly,
Cloeon sp. E. March
fly, *Bibio* sp. F. Blow
fly, *Phormia* sp., male.
G. Blow fly, female. [A
redrawn from
Snodgrass, 1954; B
redrawn from Essig,
1958; C redrawn from
Boreal Labs key card; D
redrawn from
Comstock, 1940; E
redrawn from Richards
and Davies, 1957; F and
G redrawn from Folsom
and Wardle, 1934.]

C

D

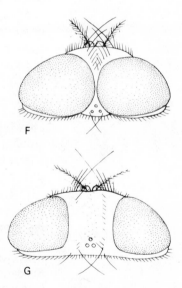

E

F

G

is the case in the fly *Phormia regina*. In the male fly the compound
eyes are large and meet at the dorsal midline of the head; in the
female they are smaller and do not meet at the dorsal midline of the
head. (Fig. 2-26F, G).

In many larval and adult insects, compound eyes are replaced by
stemmata or *lateral ocelli* (Fig. 2-27A). Stemmata are the only eyes
in most larvae of insects with complete metamorphosis. In other
insects (e.g., springtails, silverfish, fleas), stemmata are the only
"eyes" of the imaginal stage; compound eyes never appear. The
number of stemmata is quite variable, some insects having as few
as one on each side of the head (e.g., many fleas) whereas others

have many (e.g., 12 in *Lepisma* and 50 in some adult Strepsiptera).

Many insects possess simple eyes, the *dorsal ocelli,* in addition to compound eyes. When present (Fig. 2-27B, C), these ocelli vary in number from one to three: one is very uncommon; two are found in most Heteroptera (Hemiptera) and in several other insects; and three occur in many members of the suborder Homoptera (Hemiptera) and others. Ocelli may also vary in position. For example, the most common position for three ocelli is one in each of the angles formed by the ecdysial cleavage lines. In members of the genus *Perla* in the order Plecoptera all three lie within the angle of the arms of the Y formed by the coronal and frontal sutures.

Mouthpart Structure. Upon examination of the mouthparts of various insects, it seems that there are nearly as many variations in mouthpart structure and function as there are different feeding situations. The mouthparts discussed previously are considered to represent the primitive condition, for they are found in the more generalized insectan forms and in most cases show greater resemblance to primitive locomotor appendages from which, based on comparative morphological and ontogenetic data, they were derived. Specialized mouthparts can, in most instances, be homologized with the generalized type. In a number of insects, mayfly adults (Ephemeroptera) among others, the mouthparts are reduced and nonfunctional.

Mouthparts can be very broadly classified into two groups: biting, or *mandibulate,* and sucking, or *haustellate* (Fig. 2-28). The mouthparts already discussed represent the mandibulate type. An outstanding characteristic of this type is the presence of a pair of well-developed, usually highly sclerotized, mandibles that articulate at two points with the head capsule and are capable of lateral movement (Fig. 2-28A). Mandibulate mouthparts are generally adapted to chewing activities, the mandibles acting as cutting and grinding structures. However, there are many exceptions (Fig. 2-29). For example, in many predaceous beetles and ants, they are elongate, grasping structures, well adapted for catching and holding prey (Fig. 2-29A). Similarly developed mandibles in the male dobsonfly hold

Fig. 2-27
Ocelli. A. Stemmata (lateral ocelli) of a caterpillar. B. Two dorsal ocelli, stink bug. C. Three dorsal ocelli, cicada. [A redrawn from Snodgrass, 1961; B redrawn with modifications from Snodgrass, 1935; C redrawn from Boreal Labs key card.]

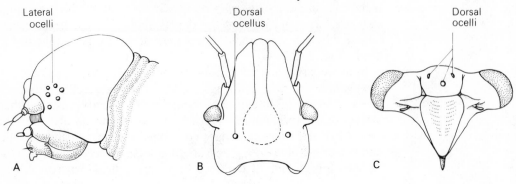

Lateral ocelli

Dorsal ocellus

Dorsal ocelli

A   B   C

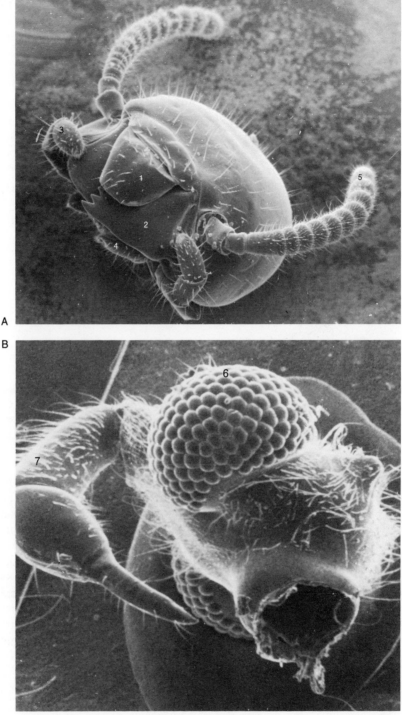

A

B

Fig. 2-28
Scanning electron
micrographs of
mandibulate and
haustellate mouthparts.
A. Head of a termite,
mandibulate type. B.
Head of the spider
bug, *Stenolemus* sp.,
haustellate type. 1,
Labrum; 2, mandible;
3, maxillary palp; 4,
labial palp; 5, antenna;
6, compound eye; 7,
beak (contains stylets).
Highly magnified.
[Courtesy of Walter J.
Humphreys.]

the female during copulatory activities (Fig. 2-29B). The female
dobsonfly does not have such impressively developed mandibles,
and dobsonfly mandibles provide another example of sexual dimorphism. In some insects, such as antlion larvae (Fig. 2-29C), the
maxillae and mandibles are elongate and grasping and together form
a food channel through which the body fluids of prey are sucked.

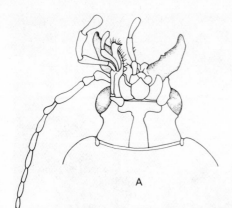

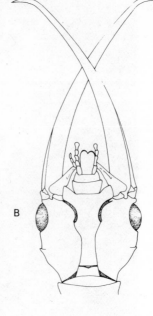

Fig. 2-29
Variations in form of mandibles. A. Ground beetle, *Calosoma* sp. B. Male dobsonfly, *Corydalus cornutus*. C. Antlion larva, *Myrmeleon* sp. [A redrawn from Essig, 1958; B redrawn from Packard, 1898; C redrawn from Peterson, 1951.]

Although these particular mouthparts are functionally sucking, they are obvious modifications of the chewing mandibulate type. In pollen-feeding and dung-feeding beetles, the mandibles are more or less flattened and serve to mold dung or pollen into small pellets or balls.

Haustellate mouthparts (Fig. 2-28B) are generally adapted for sucking activities of various sorts. Many are characterized by the presence of *stylets* (Fig. 2-30), which are swordlike or needlelike modifications of one or more of the generalized mouthpart structures. Stylets may be formed from a combination of one or more of the mouthparts and the hypopharynx. Stylets enable the insects that possess them to pierce or at least abrade plant or animal tissues and subsequently feed on the fluids that exude or are pumped from the host.

However, not all haustellate mouthparts have piercing stylets. For example, the mouthparts found in most butterflies and moths, in the nonbiting muscoid flies, and in many higher hymenopterous insects lack stylets. The mouthparts of each of these groups, lacking stylets, are incapable of penetrating tissues. These insects are thus obliged to feed on exposed fluids or soluble solids of various sorts, such as sugar. In the vast majority of butterflies and moths, an elongate sucking tube or proboscis is formed from the galeae of the maxillae (Fig. 2-31A). The remaining mouthparts are reduced or absent.

These insects feed mainly on flower nectar. When inactive, the proboscis is coiled beneath the head. This type of mouthpart structure
and method of feeding is commonly referred to as *siphoning*. The
nonbiting muscoid flies have a rather peculiar method of feeding,
often referred to as *sponging* (Fig. 2-31B). A basal segment, the
rostrum, which is made up of a part of the clypeus and basal portions

Fig, 2-30
Examples of stylate haustellate mouthparts. A. Sagittal section of head
of a sucking louse. B-D. Mouthparts spread out to show details of
structure: B, mosquito; C, cicada; D, flea. [A after Vogel, 1921; B
redrawn from Snodgrass, 1959; C redrawn from Snodgrass, 1935; D
redrawn with slight modifications from James and Harwood, 1969.]

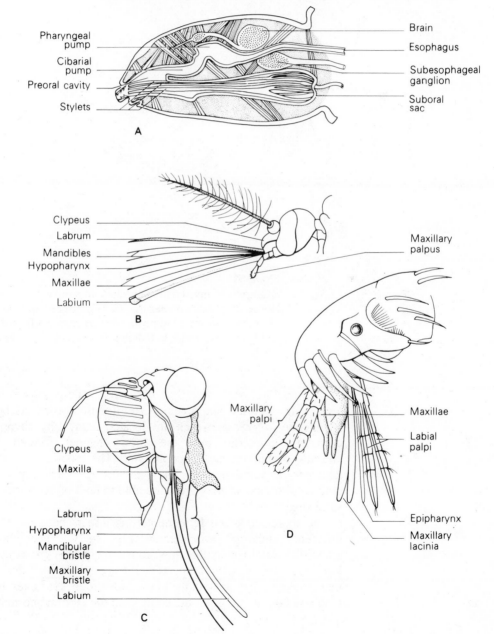

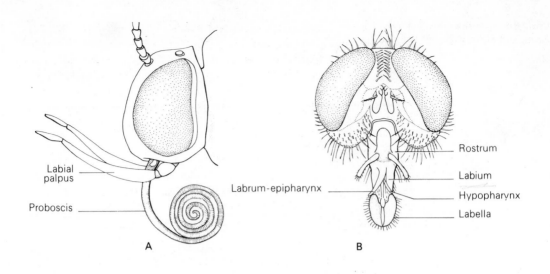

Labial palpus

Proboscis

A

Labrum-epipharynx

Rostrum

Labium

Hypopharynx

Labella

B

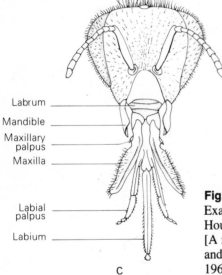

Labrum

Mandible

Maxillary palpus

Maxilla

Labial palpus

Labium

C

Fig. 2-31
Examples of nonstylate haustellate mouthparts. A. Moth. B House fly, *Musca domestica*. C. Honey bee, *Apis mellifera*. [A redrawn from Snodgrass, 1961; B redrawn from James and Harwood, 1969; C redrawn from Herms and James, 1961.]

of the maxillae, bears distally a fleshy retractile proboscis that represents the labium. The apical portion of this proboscis bears the spongelike *labella* in which are located many tiny channels that utimately converge into the food channel formed by the labrum–epipharynx and hypopharynx. The labella are capable of taking up exposed liquids. These insects egest salivary secretions onto solid foods, so they are quite able to feed on such materials as solid sugar.

The more advanced hymenopterous insects have an altogether different "sucking" arrangement (Fig. 2-31C). The labrum and mandibles usually resemble those found in typically chewing insects. For this reason these mouthparts could as easily be included with the mandibulate group as with the haustellate group. However, the maxillae and labium are quite different from the labrum and mandibles,

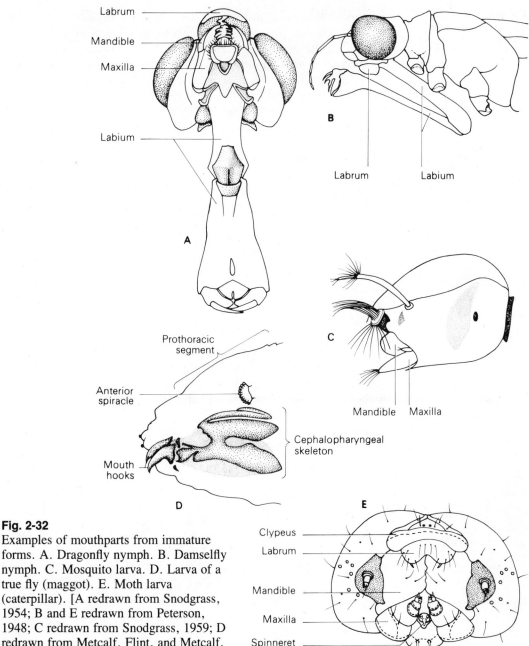

Fig. 2-32
Examples of mouthparts from immature
forms. A. Dragonfly nymph. B. Damselfly
nymph. C. Mosquito larva. D. Larva of a
true fly (maggot). E. Moth larva
(caterpillar). [A redrawn from Snodgrass,
1954; B and E redrawn from Peterson,
1948; C redrawn from Snodgrass, 1959; D
redrawn from Metcalf, Flint, and Metcalf,
1962.]

having become united, as in the honey bee, to form a sort of lapping
structure. The elongate glossae of the labium form the tubular part
of this structure, which is well adapted for thrusting into flower
nectaries. This type of mouthpart structure has been functionally
classified as chewing-lapping.

In most sucking insects, the cibarial region has become modified
and functions as a pump. The pharyngeal region also commonly
forms a pump. An example of both types of food pumps is found in
adult mosquitoes (see Fig. 4-2).

Mouthparts of Immature Forms   In many instances the mouthparts of immature forms are essentially identical to those of the adult. However, in some, the differences are rather profound. For example, the nymphs (aquatic immatures) of dragonflies and damselflies (Fig. 2-32A, B) present an unusual modification of the labium. This structure is quite enlarged, and an "elbow" is found at the junction of the prementum and postmentum. The apical portion of the prementum bears a pair of grasping jaws. When the nymph is at rest or stalking prey, the elbow of the labium is flexed and part of the prementum with its jaws forms a "mask" over the other mouthparts. When prey is within striking range, the labium is thrust out extremely fast, the jaws grasping the prey until it can be brought within reach of the maxillae and mandibles. In the lower Diptera, the mouthparts, although fundamentally chewing, are adapted for filter feeding, as in larval mosquitoes (Fig. 2-32C). In the larvae of higher Diptera, the mouthparts are extremely reduced and have been invaginated into the head, forming the *cephalopharyngeal skeleton,* which anteriorly bears the vertically moving *mouth hooks* (Fig. 2-32D). Lepidopterous larvae (Fig. 2-32E) and others, such as larval trichopterans and hymenopterans, have a rather elaborate apparatus for spinning silk into cocoons. This spinning apparatus, the *spinneret,* is composed of the maxillae, hypopharynx, and labium. Neuropteran larvae, such as antlions and aphidlions, have a grasping-sucking modification of the mandibles and maxillae similar to that in the predaceous diving beetle described earlier.

Position of the Mouthparts   There are basically three positions of the mouthparts relative to the head capsule: *hypognathous, prog-*

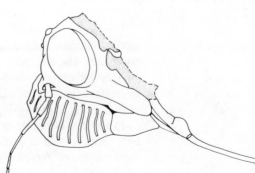

Fig. 2-33
Positions of mouthparts relative to the head capsule. A. Hypognathous. B Prognathous. C. Opisthognathous. [Redrawn from Snodgrass, 1960.]

nathous, and *opisthognathous* (Fig. 2-33A–C). In the hypognathous condition, the mouthparts hang ventrally from the head capsule. This is considered to be the most primitive condition of the three since the mouthparts are apparently modified locomotor appendages and have retained a similar position relative to the insect body. The prognathous condition is characterized by the anteriorly directed position of the mouthparts. Correlated with this modification is an elongation of the genal and postgenal regions and some flattening of the head capsule. Opisthognathous insects are those in which the mouthparts are directed ventroposteriorly relative to the head capsule. This condition is found mainly among the Hemiptera and enables them to place the sucking beak between the legs and out of the way when not feeding.

Head Capsule. The overall shape and structure of the head vary with the position of the mouthparts relative to the head capsule and the necessity for rigidity and muscle attachment associated with the mouthparts. The various sutures and cleavage lines described earlier for the generalized insect may or may not be present and, if present, may be highly modified or reduced. As a result, the cranial regions, which are largely defined by the sutures, also show a great amount of variation. The internal framework, the tentorium, also varies considerably, again correlated with the necessity for muscle attachment and strengthening support.

In many insects with the hypognathous condition, the occipital foramen (foramen magnum) is closed ventrally, separating the postmentum of the labium from the cervical membrane. This closure is due to the fusion of lobes of the posterior or hypostomal region of the subgenae, and the resulting structure is referred to as a *hypostomal bridge* (Fig. 2-34A, B). In other insects, lobes of the postgenae converge mesially on the hypostomal bridge and, in some cases, may fuse to form a *postgenal bridge* (Fig. 2-34C, D).

In some prognathous insects, the parts on the ventral side of the head are essentially equivalent to the posterior parts in the hypognathous forms. However, in some—for example, several species of beetles and neuropterous insects—the hypostomal regions of the postgenae have fused with one another in a fashion similar to the formation of the hypostomal bridge in hypognathous insects and formed a structure called the *gula* (Fig. 2-34E). In many instances, the gula fuses with the submentum, forming a composite structure, the *gulamentum.*

Modifications of the frontoclypeal region of the head are often correlated with the size of the cibarial region and the pharynx or with the development of the cibarial pump. The clypeus may be quite pronounced, as in many Hemiptera (Fig. 2-34F), where it presents a broad surface area for the attachment of the muscles of the sucking pump.

The tentorial structure already described is that of a generalized pterygote insect and would be found, for example, in members of the order Orthoptera. In the entognathous apterygote insects (those in which the mouthparts are not externally apparent), the tentorium is absent. In the ectognathous apterygotes (mouthparts externally

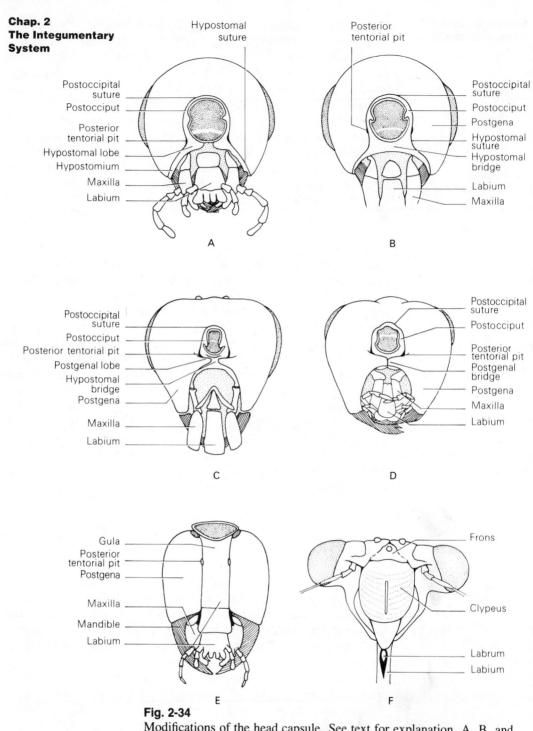

Fig. 2-34
Modifications of the head capsule. See text for explanation. A, B, and
E. Diagrammatic. C. Honey bee. D. Vespid wasp. F. Cicada. [A–D
redrawn from Snodgrass, 1960; E redrawn from Snodgrass, 1959; F
redrawn from Snodgrass, 1935.]

evident) anterior arms arise from the anterior tentorial pits. However,
these arms and the posterior bridge do not connect. In some of the
higher orders there has been a tendency toward a reduction of the

tentorium, and in some the anterior and posterior arms do not
connect.

Modifications of the Thorax

Nota. In wing-bearing segments, the terga or nota vary in the
number, nature, and location of internally inflected ridges present.
These ridges are marked externally by sutures and serve to lend
strength and rigidity to the tergum. The terga of the prothorax of
pterygote insects are quite different from those of the wing-bearing
segments since they are not directly associated with the wing mechanism. Consequently, any sutures associated with internal ridges on

Fig. 2-35
Modifications of the pronotum. A. Grasshopper, *Melanoplus bilituratus.*
B. German cockroach, *Blattella germanica.* C. Treehopper. *Thelia
bimaculata.* D. Hercules beetle, *Dynastes hercules.* [A redrawn from
Ross, 1962; B after U.S. Public Health Service; C redrawn from Borror
and DeLong, 1964; D redrawn from Essig, 1942.]

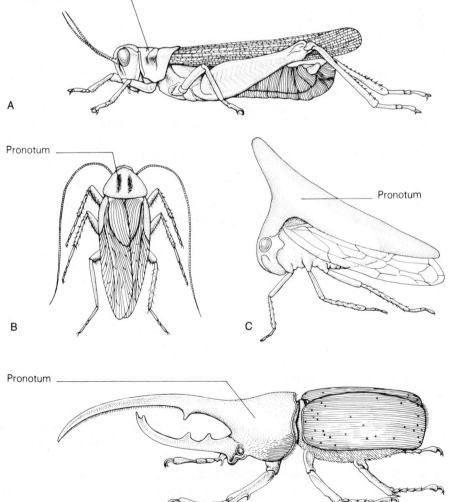

57   D

the pronotum cannot be considered equivalent to those in the pterothorax. The pronotum is commonly quite small and undeveloped compared to the nota of the wing-bearing segments. However, in many insects it is quite pronounced, forming a pronotal shield (Fig. 2-35A, B). In some insect species it has taken on rather bizarre shapes, often mimicking an environmental characteristic such as a thorn (Fig. 2-35C) or an elongate projection over the head (Fig. 2-35D).

Pleura. A pleuron of a winged segment and that of the prothorax are generally similar. However, a propleuron is usually less developed, and there may be secondary modification in a wing-bearing pleuron associated with the wing mechanism. For example, the episternum or epimeron may be subdivided into dorsal and ventral plates, or an anterior plate may be separated from the episternum. The precoxal and postcoxal areas of the pleuron may be separated from the episternum and epimeron. The trochantin is fairly well developed in more generalized pterygotes and somewhat less so in higher orders, being absent in some. In certain dipteran species, the meron of the coxa has actually become quite pronounced, forming a part of the pleuron.

Sterna. In the more generalized pterygote insects the major sternal sclerites are usually present, although they may be reduced, leaving rather large membranous areas between them. However, in the higher pterygote orders, there is considerable modification, such as fusion of various plates and formation of secondary sclerites. It is thus often very difficult to identify homologies between these highly modified sterna and those of the more generalized insects.

Modifications of the Legs

Insect legs, although typically ambulatory in function, have been modified extensively in several directions. Both the immature and adult stages of most insects have thoracic legs. However, there are many examples of apodous (lacking legs) larvae (e.g., fly maggots) and even of apodous adults (e.g., female scale insects). Typically developed insect legs are *cursorial;* that is, they are adapted for walking and running. The cockroach is a good example of an insect with cursorial legs (Fig. 2-36A). In some insects, such as the mole cricket and the nymphs of the periodical cicada, the forelegs are highly modified, bearing heavily sclerotized digging claws (Fig. 2-36B). *Fossorial* is the term commonly used to denote the adaptation for digging. The forelegs of certain other insects (e.g., the praying mantis) (Fig. 2-36C) are *raptorial* or modified for grabbing and holding prey. The forelegs have not been the only ones to undergo modification. For example, the femora of the hindlegs of grasshoppers and katydids are enlarged, accommodating the muscles used in jumping (Fig. 2-36D). Legs adapted for this kind of activity are commonly referred to as *saltatorial.*

The legs of several aquatic insects are modified in such a way that they facilitate swimming, as is the case with adult dytiscid beetles

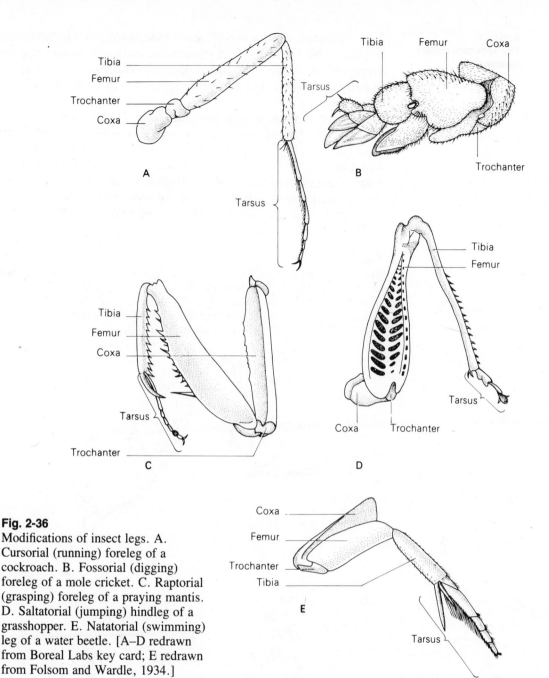

Fig. 2-36
Modifications of insect legs. A. Cursorial (running) foreleg of a cockroach. B. Fossorial (digging) foreleg of a mole cricket. C. Raptorial (grasping) foreleg of a praying mantis. D. Saltatorial (jumping) hindleg of a grasshopper. E. Natatorial (swimming) leg of a water beetle. [A–D redrawn from Boreal Labs key card; E redrawn from Folsom and Wardle, 1934.]

(Fig. 2-36E), which bear two rows of "swimming hairs" on the edges of the flattened tibiae and tarsi of the middle and hindlegs. These hairs are attached to the legs by movable joints. When the legs are thrust posteriorly during the act of swimming, the distal ends of these hairs move out from the legs, greatly expanding the surface area being applied against the water in the paddling action. As the legs are brought anteriorly in the recovery stroke, the hairs become pressed very close to the legs, reducing the surface area applied against the water, much like the feathering of a paddle when paddling a boat in the wind. The term *natatorial* applies to swimming legs. Although similar in their gross morphology to the middle and hindlegs, the forelegs of the Protura are carried in an elevated position anterior to the body. It has been said that these are principally sensory in function and that they suggest a step in the evolution of antennae.

The legs of many insects bear various specialized structures. For example, the legs of honey bees bear structures that are used during their pollen-collecting activities. One of these structures is the *corbiculum* (Fig. 2-37A), or *pollen basket,* composed of two rows of hairs on the outer surface of the hind tibia, where the pollen collected by a foraging worker is stored for transport back to the hive. The forelegs of males of some species of diving beetles bear large suction

Fig. 2-37
Specialized structures borne on the legs. A. Corbiculum or pollen basket on the hind tibia of the honey bee. B. Suction discs on the fore tarsus of a diving beetle. C. Tympanic organs on the fore tibia of a long-horned grasshopper. [A redrawn from Snodgrass, 1956; B redrawn from Folsom and Wardle, 1934; C redrawn from Packard, 1898.]

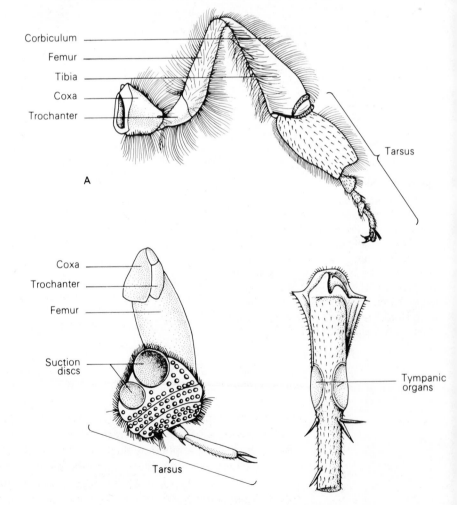

discs on the tarsi (Fig. 2-37B). They are used to hold the female during copulation.

The hind femora of certain species of short-horned grasshoppers have short peglike structures with which they rub the forewings and produce a sound. The legs of different insects may bear sensory structures of various types (see Chapter 6). Several species of flies (e.g., blow flies and house flies) "taste" by means of sensilli on the tarsi of their forelegs. The long-horned grasshoppers and crickets possess oval auditory organs, or *tympana,* at the base of each front tibia (Fig. 2-37C; see also Figs. 6-7 and 11-15).

The tarsal and pretarsal segments are also variously modified. Padlike *pulvilli* may be found on the lower surface of each tarsal segment, as in several members of the order Orthoptera, or in association with each *ungue* or *pretarsal claw,* as in the flies (Fig. 2-38A). A bulbous lobelike structure, the *arolium* (Fig. 2-38B), may be present between the claws. Some insects have a spinelike or lobelike structure, the *empodium* (Fig. 2-38A), which arises from the distal part of the *unguitractor plate* and also is located between the claws.

Modifications of the Wings

Insects may bear a single pair of wings, two pairs, or none at all. Many wingless insects are grouped with the winged forms (Pterygota; see Chapter 11) on the basis of developmental and morphological similarities. Among these insects, the apterous condition is considered to be secondary, having developed from a winged ancestor. On the other hand, some insects (Apterygota; see Chapter 11) and their ancestors never had wings.

There is considerable variation in the structure of insect wings. Examples of some of these variations will be presented in the following paragraphs.

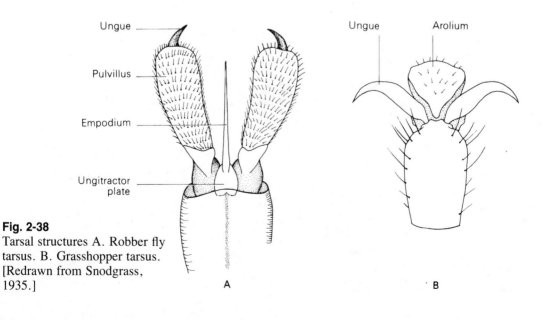

Fig. 2-38
Tarsal structures A. Robber fly tarsus. B. Grasshopper tarsus. [Redrawn from Snodgrass, 1935.]

Size.   The wings may be quite large, as in many of the larger butterflies and moths, or extremely small, as in many of the wasps and flies. The Atlas moth of Australia has been said to reach a wingspan of 14 inches! In many insects, there is a tendency for the hindwings to be smaller than the forewings (Fig. 2-39A).

Venation.   Because of the tremendous variation in wing venation, it has been used as a source of taxonomic characters. Venation ranges from the extensively reduced and simplified system found in many of the wasps (Fig. 2-39A) to the highly complex network in the wings of dragonflies and damselflies (Fig. 2-39B). The regions between principal veins in many of the more primitive insect groups, for example, Odonata (Fig. 2-39B), contain irregular networks of veins. These networks probably represent what remains of the *archedictyon* described from insect fossils. Veins also vary in thickness; for example, those of the periodical cicada are quite thick, whereas those in the scorpionfly are very thin and delicate.

Fig. 2-39
Variations in wing structure. A. Wings of a wasp with reduced venation. B. Wings of a dragonfly with an elaborate network of veins. C. Ladybird beetle with left elytron (forewing) and hindwing extended. D. Hemelytron (forewing) of a true bug. E. Lateral view of the thorax of a true fly. [A, C, and D redrawn from Essig, 1958; B redrawn from Boreal Labs key card; E redrawn from Smart, 1959.]

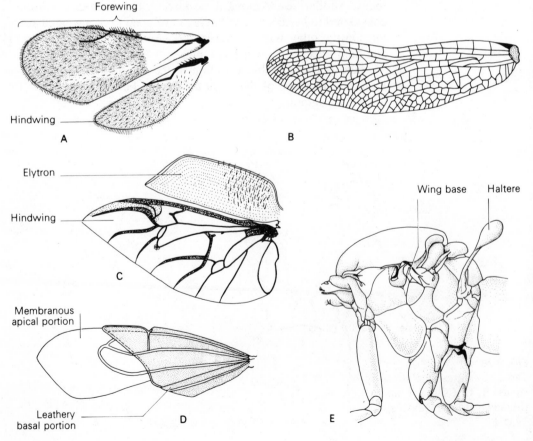

Function and Texture. The most obvious function of wings is, of course, flight. However, in several instances the wings have been modified or are at least used for different purposes. In the beetles, the hindwings are membranous and fold beneath the forewings, which are usually quite hard and form a protective armor for the membranous hindwings when not in use. These modified forewings (Fig. 2-39C) are called *elytra* (sing., *elytron*). Elytra may be sculptured in various ways. Among the Hemiptera (suborder Heteroptera), the forewings are only partly hardened, the distal portions remaining membranous and containing veins. These structures (Fig. 2-39D) are appropriately named *hemelytra* (sing., *hemelytron*), or "half"-elytra. The forewings of orthopterans are parchmentlike and probably afford similar protection to the hindwings. A parchmentlike wing is called a *tegmen* (pl., *tegmina*). In some insects the wings are used for the production of sound. Several examples of insects that produce sound with their wings are to be found among the Orthoptera. The field cricket is well known for this activity. The true flies possess a single pair of well-developed forewings and a pair of highly modified hindwings, the *halteres* (Fig. 2-39E). These club-shaped structures are important in the stability of flight of these insects. In the very hot days of midsummer, honey bees fan their wings in a community effort and thereby reduce the temperature within the hive.

Relationship to One Another. Among the insects with two pairs of wings, the wings may work separately as in the dragonflies, damselflies, mayflies, and Neuroptera. However, in many of the higher pterygote insects, the fore- and hindwings are coupled to one another in various ways, resulting in each pair of wings acting together as a unit. Examples of wing-coupling mechanisms include tiny hooks or *hamuli* found among the Hymenoptera (Fig. 2-40A) and the spinelike *frenulum* (Fig. 2-40B) or lobelike *jugum* (Fig. 2-40C) found in the Lepidoptera. In insects that have wing-coupling mechanisms, the hindwings are usually somewhat smaller than the forewings. The tendency toward a reduction of the hindwings has, of course, reached its maximum in the true flies, which have lost the hindwings as such, the remnants being the previously mentioned halteres.

Resting Position. When not being utilized for flight, the wings are held in various positions relative to the body. Members of the primitive orders Ephemeroptera and Odonata are unable to flex the wings over the abdomen and hence when at rest hold them vertically over the dorsum (mayflies and damselflies) or horizontally (dragonflies). Other insects (bees, wasps, etc.) are able to flex the wings over the abdomen at rest. Many Homoptera and others (e.g., Neuroptera) hold the wings rooflike over the abdomen.

Coloration. Although many insects have clear *(hyaline)* or opaque unpigmented wings, there are groups in which wing coloration is especially well developed. The coloration may be due to pigmentation within the integument itself or to a covering of minute

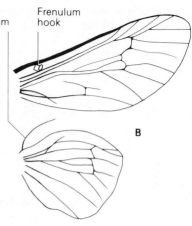

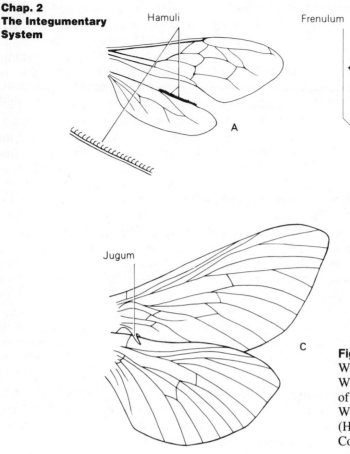

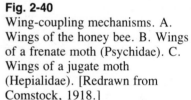

Fig. 2-40
Wing-coupling mechanisms. A.
Wings of the honey bee. B. Wings
of a frenate moth (Psychidae). C.
Wings of a jugate moth
(Hepialidae). [Redrawn from
Comstock, 1918.]

pigmented scales or the covering scales may physically resemble
thin layers or diffraction gratings in their effects upon impinging
light. The bright and decorative colorations on the wings of some
dragonflies are a good example of the first kind of coloration. The
butterflies and moths as a group possess the second type of coloration, colored scales. Variety and complexity of coloration abounds
in the Lepidoptera, and many behavioral patterns are intimately tied
up with coloration. In several instances the coloration matches the
environmental background and in this way affords a degree of protection from predators. The protective value of coloration is discussed in Chapter 9.

Presence of Hairs and Scales. The scales of butterflies and
moths have already been mentioned. These are considered to be
modified setae. The wings of the caddisflies are covered with tiny
hairs, which are also modified setae. Wings of other insects bear
various types of macrotrichia and microtrichia. Although microtrichia are randomly scattered over a wing surface, macrotrichia (true
setae) are not. Macrotrichia tend to be concentrated along the major
veins and branches. Thus distinct rows of macrotrichia, in the absence of a vein, are assumed to indicate the location of a vein that
has been lost. As such, these hairs have been useful in understanding
evolutionary changes in wing veins.

Modifications of the Abdomen

In adult pterygote insects the abdominal segments (except the first) anterior to those that bear the external genitalia are usually quite simple and uniform. Each of these segments consists of a tergum and a sternum separated by a pleural membrane and never bears appendages. As explained earlier, the first abdominal segment in pterygote insects is associated more with the thorax than the abdomen, since the antecostal portion of the tergum furnishes the third phragma, to which the dorsal longitudinal wing muscles are attached. In many of the Hymenoptera the first abdominal segment, the *propodeum* (Fig. 2-41A), is completely associated with the

Fig. 2-41
Modifications of the abdomen. A. Fire ant. B. Parasitic wasp, *Eusemion* sp. C. Gravid termite queen. [A after U.S. Public Health Service; B redrawn from Essig, 1958; C redrawn from Skaife, 1961.]

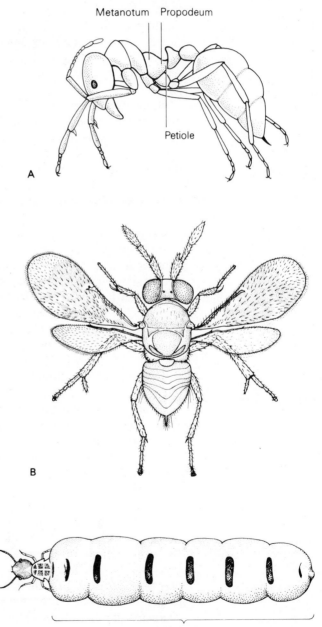

thorax and is separated from the remaining abdominal segments (collectively, the *gaster*) by a construction, the *petiole*.

The abdomen as a whole varies considerably both in size and in the number of segments. As pointed out earlier, the primitive number of segments is considered to have been 12. This condition occurs, for example, among adult Protura and in the embryos of some higher insects. The telson is probably represented in most insects by the membrane surrounding the anus, but in Odonata the anus is surrounded by three small plates that may represent the telson. The tendency has been toward a reduction in the number of abdominal segments. The springtails have 6 segments in their abdomen, the more generalized pterygotes have 11, and the higher pterygotes usually have 10 or less.

The size of the abdomen relative to the remainder of the body ranges from the tiny abdomen characteristic of several parasitic wasps (Fig. 2-41B) to the extremely large abdomen of a gravid termite queen (Fig. 2-41C).

Nongenital Abdominal Appendages. Unlike the pregenital and genital segments in adult pterygotes, the pregenital and genital segments of many larval pterygotes and many apterygote insects do bear appendages of various sorts. For example, the first three abdominal segments of adult proturans bear rather simple, bilateral appendages, the *styli*. Similarly, styli are usually borne on several of the abdominal segments of adult thysanurans (Fig. 2-42A).

Members of the order Collembola are in many ways quite different from other insects. One of the outstanding differences is the presence of three rather unique abdominal structures (Fig. 2-42B). The most anterior structure, the *collophore*, is located on the venter of the first abdominal segment. This structure is roughly cylindrical in appearance when it is protruded by the hydrostatic pressure of the hemolymph. *Collophore* means literally "that which bears glue," based on an early suggestion that it might serve as an organ of adhesion (see Chapter 11). The two posterior structures are the *tenaculum* on the venter of the third segment and the *furcula* on the venter of the fifth segment. The furcula is capable of being moved anteriorly, engaged by the tenaculum, and subsequently released, exerting a force against the substrate and propelling the insect through the air. Hence the name "springtail."

Abdominal appendages of pterygote larvae usually serve either a walking or a ventilatory function. An example of the first is found among the larvae in the order Lepidoptera (Fig. 2-42C), which generally bear bilateral appendages on the first four or five, and commonly the tenth, abdominal segments. These are the *prolegs* and complement the three pairs of thoracic appendages in the locomotion of the insect. Each proleg bears a series of minute hooks or *crochets*. Bilateral abdominal appendages in mayfly nymphs serve as gills (Fig. 2-42D), facilitating the absorption of oxygen from and release of carbon dioxide into the surrounding water. A pair of lobelike projections, *cornicles,* on the posterior dorsum of the abdomen is characteristic of Aphids (Hemiptera, suborder Homoptera).

The structure of the terminal abdominal segments (e.g., the cerci

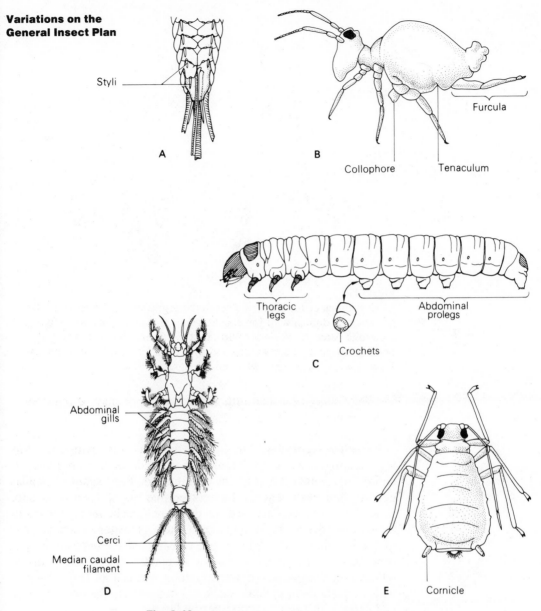

Fig. 2-42
Nongenital abdominal appendages. A. Venter of posterior portion of the
abdomen of a silverfish. B. Springtail. C. Caterpillar. D. Mayfly nymph.
E. Aphid. [A redrawn with modifications from Essig, 1942, after
Oudemans; B after U.S. Public Health Service; C redrawn from
Snodgrass, 1961; D and E redrawn from Boreal Labs key card.]

and the epi- and paraprocts) is variously modified. For example, the
cerci may be forcepslike or clasperlike (Fig. 2-43A), feelerlike (Fig.
2-43B), reduced, or absent. Cerci are characteristic of the more
primitive orders, such as Ephemeroptera and Orthoptera, but are
absent in the hemipteroid orders and most of the higher (holometabolous) orders, except the Mecoptera and possibly some Hymenoptera. Like the cerci, epiprocts and paraprocts may be long and

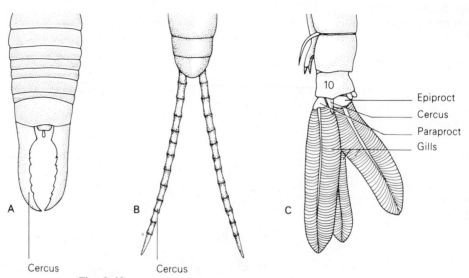

A — Cercus
B — Cercus
C — 10 — Epiproct / Cercus / Paraproct / Gills

Fig. 2-43
Modifications of the terminal abdominal segments. A. Posterior portion of earwig abdomen with forcepslike cerci. B. Stonefly nymph with feelerlike cerci. C. Posterior portion of damselfly abdomen with gill-bearing epi- and paraprocts. [A redrawn from Hebard, 1934; B redrawn from Ross, 1962; C redrawn from Snodgrass, 1954.]

feelerlike, may bear anal gills (Fig. 2-43C), or may be reduced or inconspicuous.

External Genitalia. In insects that have an ovipositor, this structure may show considerable variation, depending upon the situation into which the eggs must be placed. For example, cicadas deposit their eggs beneath the bark on tree twigs. Their ovipositor (Fig. 2-44A) is well adapted for this function, being composed of three rather sharp and rigid blades. Other ovipositors, such as those of the katydids (Fig. 2-44B), are constructed for depositing eggs beneath the surface of the soil or in plant tissue. Some parasitic ichneumon wasps have extremely long ovipositors (Fig. 2-44C), which enable them in some instances to penetrate the bark of a tree and deposit an egg in a wood-boring larva.

The adult female members of the apterygote order Thysanura possess a very primitive ovipositor composed of paired appendages borne on the venter of the eighth and ninth abdominal segments (Fig. 2-44D). Each appendage or *gonopod* is borne on a basal *coxopodite*, which may or may not bear a stylus. Many similarities are apparent between this primitive ovipositor and that found among the pterygotes. Many insects lack an ovipositor altogether and have devised other means of egg deposition. For example, certain members of the orders Thysanoptera, Mecoptera, Lepidoptera, Coleoptera, and Diptera use the abdomen itself as an ovipositor or *oviscapt* (Fig. 2-44E). Some are capable of telescoping the abdomen to great lengths.

Mayflies (Ephemeroptera) and earwigs (Dermaptera) are exceptional in that the female genital opening (gonopore) is just behind the seventh abdominal segment.

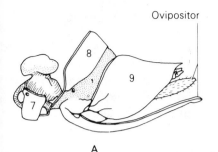

Ovipositor

8

9

7

A

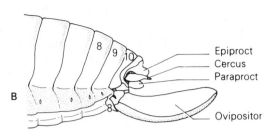

8 9 10

Epiproct
Cercus
Paraproct

8

Ovipositor

B

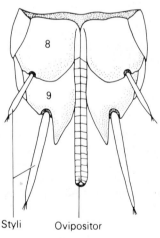

C

Fig. 2-44
Modifications of the ovipositor. A.
Cicada, *Magicicada septendecim*. B.
Katydid, *Scudderia* sp. C. Ichneumon
wasp, *Megarhyssa lunator*. D.
Firebrat, *Thermobia domestica*. E.
Telescoping terminal abdominal
segments serving as an ovipositor in
the house fly, *Musca domestica*.
Numerals indicate abdominal
segments. [A, B, and D redrawn
from Snodgrass, 1935; C redrawn
from Riley, 1888; E. redrawn from
West, 1951.]

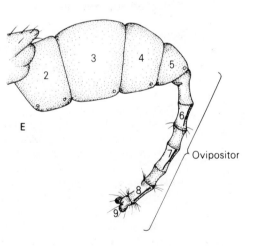

8

9

Styli   Ovipositor

D

2   3   4   5

6

7   Ovipositor

8

9

E

The external genitalia of male insects are commonly extremely
complex structures and probably show more inherent variation than
any other insectan structure. It is for this reason that taxonomists
have made extensive use of these structures in their work. Since
taxonomists usually specialize in a single group, there has been a
proliferation of terminologies applied to male genitalia. This obviously creates a problem for the entomology student and he or she
is best referred to Tuxen (1970). Attempts to homologize the male

genitalia in the diverse insectan groups include Snodgrass (1957) and Smith (1969).

Two exceptional situations deserve mention. Although dragonflies and damselflies (Odonata) are like most other insects in having the gonopore on the ninth abdominal segment, the penis in these flies is located on the venter of the second and third abdominal segments. Male mayflies have paired gonopores and penes.

Modifications of the General Body Form

The shape of many insects is an obvious adaptation to a given environmental situation. An outstanding example of this is found in the fleas. These insects are bilaterally flattened, a characteristic that enables them to move easily between the feathers or hairs of their hosts. One does not realize the efficiency of this adaptation until he attempts to remove one of these insects from a pet cat or dog. An example of an insect flattened dorsoventrally is the bed bug. This body shape enables the bed bug to hide in tiny cracks and crevasses between feedings. Many insects possess rather bizarre shapes and in many instances actually mimic an environmental characteristic, as in the case of protective coloration, which often provides protection from potential predators (see Chapters 8 and 9). Some insects (e.g., walkingsticks, Phasmida) are elongate and tubular in shape. In the sedentary stage, the bodies of scale insects are so modified that they are hardly recognizable as insects.

Selected References

GENERAL

Bursell (1970); Chapman (1971); Rees (1977); Richards and Davies (1977); Rockstein (1978); Roeder (1953); Smith (1968); Snodgrass (1935, 1963a); Wigglesworth (1972).

PHYSIOLOGICAL ASPECTS

Anderson (1979); Ebeling (1974); Fuzeau-Braesch (1972); Hackman (1974); Hepburn (1976); Locke (1974); Needham (1978); Neville (1970, 1975); Richards (1951, 1978); Wharton and Richards (1978).

MORPHOLOGICAL ASPECTS

Borror, DeLong, and Triplehorn (1976); Comstock (1918); DuPorte (1961); Hamilton (1972); Manton (1972); Matsuda (1965, 1970, 1976); Rempel (1975); Schneider (1964); Scudder (1971); Smith (1969); Snodgrass (1935, 1950, 1952, 1957, 1958, 1960, 1963a); Tuxen (1970).

The Nervous, Glandular, and Muscular Systems

The Nervous System

The many, diverse activities of the various sytems of an insect are coordinated in large part by the nervous system. This system is composed of elongated cells, or *neurons,* which carry information in the form of electrical impulses from external and internal *sensilla* (sensory cells) to appropriate *effectors.* The nature and location of the stimulated sensilla determine the nature of the response. The basic effectors are muscles and glands. However, there are other effectors (e.g., the light-producing organ of the firefly). The nervous system and effector organs enable an insect to adjust continually and often very quickly to changes *(stimuli)* in both the internal and external environment and to behave in a manner favorable to the maintenance of life.

Huber (1974) and Treherne (1974) provide good, recent reviews of the structure and function of the insect nervous system.

Structure and Function of the Nervous System

The Neuron. The basic functional unit of the nervous system is the nerve cell or *neuron* (Fig. 3-1A). A neuron may be described as a thin-walled tube that varies from much less than 1.0 mm to more than 1 meter (in larger animals) in length and has a diameter between 1.0 and 500 micrometers. Insect neurons are usually comparatively small in diameter, on the order of 45–50 micrometers for the largest (much less for the smallest). Typically, a neuron consists of a cell body, the *perikaryon* or *soma,* and one or more long, very thin *fibers,* or *axons.* A branch of an axon is called a *collateral.* Associated with the cell body or near it are tiny branching processes, the *dendrites.* Similar branching processes, the *terminal arborizations,* are found at the end of an axon.

Neurons may be *unipolar, bipolar,* or *multipolar* (Fig. 3-1A–C). In unipolar neurons, a single stalk from the cell body connects with the axon and a collateral. In bipolar neurons the cell body bears an axon and a single, branched or unbranched dendrite. Multipolar neurons have an axon and several branched dendrites. The individual neurons are not connected directly to one another, but the finely branching terminal arborizations of an axon come into extremely close association with the dendrites of another neuron. A very small

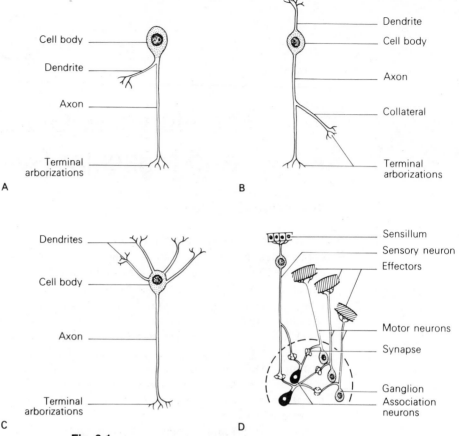

Fig. 3-1
Structural types of neurons. A. Unipolar. B. Bipolar. C. Multipolar. D. Relationship among sensory, motor, and association neurons. [Redrawn from DuPorte, 1961.]

but measurable distance, however, always lies between them. The region of close association between terminal arborizations and dendrites is called a *synapse,* and the space between the arborizations and dendrites is called the *synaptic cleft.*

Most synapses occur within the neuropile of aggregations of neurons called *ganglia* (sing., *ganglion*). The term *ganglion* means a "swelling." Several histological components of a ganglion can be identified (Fig. 3-2; see also Fig. 3-11).

1. An outer connective tissue layer, the sheath or *neural lamella.*
2. The *perineurium,* a layer of cells beneath the neural lamella (the perineurium probably secretes the material that composes the neural lamella).
3. A region containing the neuron cell bodies with associated *glial cells.*
4. A central region consisting of intermingling, synapsing axons encapsulated by processes of glial cells, the *neuropile.*

Distinct *fiber tracts,* groups of axons running parallel to one another, are usually evident in the neuropile. Within a ganglion, extracellular spaces are present between the glial cells surrounding axons. The

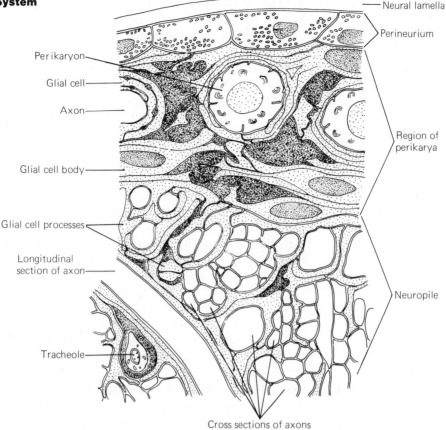

Perikaryon

Glial cell

Axon

Glial cell body

Glial cell processes

Longitudinal
section of axon

Tracheole

Neural lamella

Perineurium

Region of
perikarya

Neuropile

Cross sections of axons

Fig. 3-2
Cross section of part of the caudal ganglion of the cockroach *Periplaneta*
sp. (diagrammatic). Darkly shaded areas indicate extracellular spaces.
[Redrawn from Smith and Treherne, 1963.]

fluid in these spaces contains higher concentrations of sodium and
potassium ions and a lower concentration of chloride ions than the
hemolymph. Maintenance of the proper concentration of this fluid
is critical to neural function. The neural lamella, perineurium, and
glial cells are apparently involved in maintaining the composition of
this fluid as well as providing physical support and transporting and
storing nutrients used by neurons. In addition, glial cell processes
serve to insulate single axons or groups of axons from one another.

Nerves, which are bundles of axons, are invested in a noncellular
layer, the *basal lamina* (similar to neural lamella), and in the perineurium. Individual neurons are surrounded by glial cells. Nerves
provide connections among ganglia and between ganglia and other
parts of the nervous system.

Types of Neurons.   The insect nervous system is composed of
basically three types of neurons (Fig. 3-1D).

1. *Sensory* or *afferent.*
2. *Motor* or *efferent.*
3. *Association* or *internuncial.*

Sensory or afferent neurons are usually bipolar, and cell bodies are located peripherally in the insect. The distal process, or dendrite, is associated with a sensory structure of some type (see Chapter 6); the proximal process usually connects directly with the central nervous system where it may synapse directly with a motor neuron or more commonly with one or more association neurons. Some sensory neurons may be multipolar. Their distal processes may branch, sometimes elaborately, over the inner surface of the integumental wall, through perforations in the integumental wall, or over the alimentary canal, while their axons enter the ganglia of the central nervous system. Some multipolar neurons are involved with stretch receptors (see Chapter 6).

Motor or efferent neurons are unipolar. The cell body lacks dendrites and is located in the periphery of a ganglion. The bundles of axons from the cell bodies form the motor nerves that activate muscles; the collateral of each neuron enters the neuropile, and the terminal arborizations connect with those of association or sensory neurons.

Association or internuncial neurons also have their cell bodies located in the periphery of a ganglion and may synapse with one or more other association neurons, sensory neurons, or motor neurons. Some association neurons are quite large and are connected to "giant axons" that have very large diameters (approximately 45 micrometers) and may run the entire length of the ventral nerve cord (e.g., in *Periplaneta americana,* the American cockroach). These axons have been thought to serve as a rapid conduction system for alarm reactions and are associated with a variety of sensory-motor combinations in different insects. For example, giant axons may be involved in the sudden "jet propulsion" (i.e., the forcible expulsion of water from the rectum) escape response of dragonfly larvae (page 239). It was previously thought that the escape response elicited by a puff of air across the cerci of the American cockroach (*Periplaneta americana*) was mediated via the giant axons. However, stimulation of these axons fails to induce the escape response, and, further, the escape response can be induced in individuals in which the cell bodies of the giant axons have been severed (Parnas and Dagan, 1971, and others cited in Huber, 1974).

Central Nervous System.   The insect central nervous system (Fig. 3-3) is composed of a double chain of ganglia joined by lateral and longitudinal connectives. The anterior ganglion, the *brain,* is very complex and is located dorsal to the foregut in the head. It is usually connected by *circumesophageal connectives* to a ganglion ventral to the foregut. This is the *subesophageal ganglion,* which is also highly complex, being composed of three fused ganglia representing the mandibular, maxillary, and labial segments. The subesophageal ganglion innervates sense organs and muscles associated with the mouthparts, salivary glands, and the neck region. In addition, the subesophageal ganglion in many insects has an excitatory or inhibitory influence on the motor activity of the whole insect. Posterior to the subesophageal ganglion are, typically, three segmental *thoracic ganglia,* each containing the sensory and motor

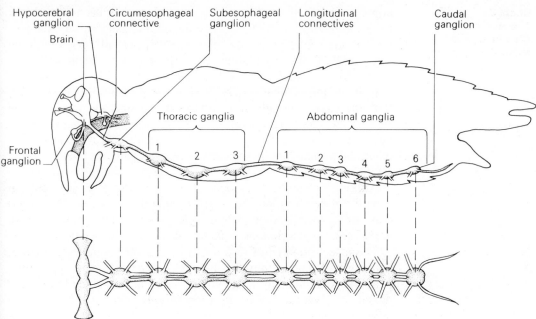

Fig. 3-3
Generalized insect central nervous system.

center for its respective segment. Two pairs of major nerves, one pair supplying the legs and the other the remaining musculature of each segment, arise from each ganglion. In winged insects, the meso- and metathoracic ganglia each give rise to a third pair of nerves associated with the wing musculature. In some insects (e.g., adult Diptera, Hymeoptera, and some Coleoptera; Fig. 3-4), the thoracic ganglia may be more or less fused longitudinally, forming what appears as a single neural mass in the thorax.

In the more primitive insects each of the first several abdominal segments contains a ganglion, the first 8 segments in apterygote insects, 7 in dragonfly and damselfly nymphs, and 5 or 6 in grass-

Fig. 3-4
Variation in the concentration of ventral chain ganglia in two species of beetles. [Redrawn from Packard, 1898.]

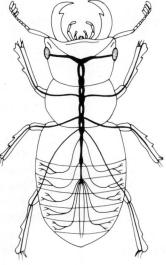

hoppers and their relatives. However, there has been a tendency toward reduction in the number of abdominal ganglia; for example, several adult flies have only one, which is partially fused with the large single thoracic ganglion. Patterns of fusion of ganglia sometimes differ between the larval and adult stages of the same species.

In those insects that possess abdominal ganglia, the most posterior, the *caudal ganglion,* is always compound and furnishes the sensory and motor nerves for the genitalia. This ganglion is therefore intimately involved in the control of copulation and oviposition. The other abdominal ganglia typically give rise to a pair of nerves to the segmental muscles.

Although ganglia are associated with specific body segments, it should not be assumed that a given ganglion provides innervation only for its own segment, for muscles of one segment may receive nerves from a ganglion associated with a different segment.

The insect brain (Figs. 3-5 and 3-6) is a very complex structure, apparently formed from the fusion of three separate, primitive, paired segmental ganglia and composed of at least three lobes from dorsal to ventral, the *protocerebrum, deutocerebrum,* and *tritocerebrum.*

The *protocerebrum* is the most complex part of the brain. Several distinct cell masses and regions of neuropile have been identified.

1. *Optic lobes* (associated with the compound eyes).
2. *Ocellar centers* (associated with the dorsal ocelli).
3. *Central body.*
4. *Protocerebral bridge.*
5. *Pars intercerebralis.*
6. *Corpora pedunculata* ("mushroom bodies").

The optic lobes receive sensory input from the compound eyes and are composed of three neuropiles and associated perikarya. The ocellar centers are associated with the bases of the nerves from the ocelli. A neuropile mass, the central body, connects the two lobes of the protocerebrum and is located in the center of the protocerebrum dorsal to the esophagus. It receives axons from various parts of the brain and may be the source of premotor outflow from the brain. The protocerebral bridge is a mass of neuropile located medially. It is associated with axons from many parts of the brain, save the corpora pedunculata. The pars intercerebralis is located in the dorsal median region above the protocerebral bridge and central body. It contains two groups of *neurosecretory cells* that transport neurosecretory material to the *corpus cardiacum* (to be discussed under glandular system). Each corpus pedunculatum is composed of a central stalk that splits ventrally into α and β lobes and is capped dorsally by a mass of neuropile and associated perikarya, the *calyx.* The corpora pedunculata contain association neurons, which do not extend outside of these bodies, and terminal portions of axons that enter from perikarya located in other parts of the brain. Axons that connect with the calyx and α lobe apparently are mainly sensory; those connecting with the β lobe are premotor axons, which in turn synapse with motor fibers.

In view of the complexity of the protocerebrum, it is not surprising

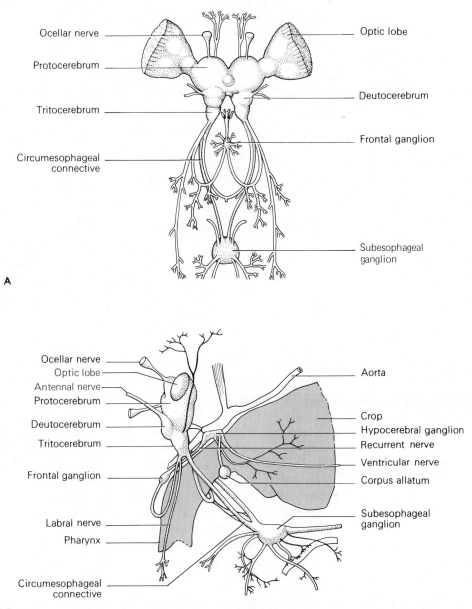

Fig. 3-5
Brain and stomatogastric nervous system of the grasshopper *Dissosteira carolina;* A. Anterior view. B. Lateral view. [Redrawn with slight modifications from Snodgrass, 1935.]

that this part of the brain is considered to be the location of the "higher centers" in the central nervous system, which control the most complex insect behavior. The fact that the corpora pedunculata are comparatively large in the social Hymenoptera (ants, bees, and wasps) and small in less behaviorally sophisticated insects (true bugs, flies, etc.) supports this idea.

The *deutocerebrum* contains the *antennal lobes,* which receive both sensory and motor axons from the antennae. The antennal lobes are connected to one another by a central fiber tract or commissure.

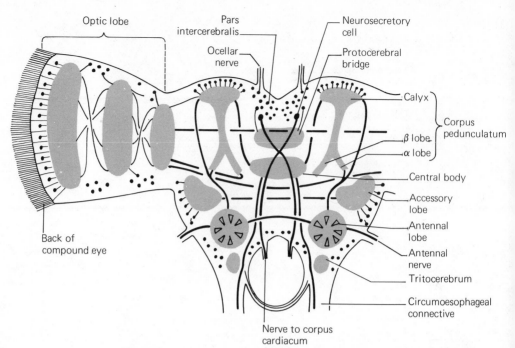

Fig. 3-6
Diagram showing major neuropilar regions (shaded) of the brain and some connections between these regions. Black dots indicate location of perikarya. [Redrawn from Chapman, 1971.]

Tracts of olfactory (''sense of smell'') fibers connect the antennal lobes and corpora pedunculata of the protocerebrum.

The *tritocerebrum* connects the brain to the stomatogastic nervous system via the *frontal ganglion* and to the ventral chain of ganglia (beginning with the subesophageal ganglion) via the *circumesophageal connectives*. The tritocerebrum also receives nerves from the labrum and possibly sensory fibers from the head capsule.

See Strausfeld (1976) for a well-illustrated treatise on the fly brain. Howse (1975) reviews the literature pertinent to brain structure and behavior.

Visceral Nervous System. The visceral nervous system is often referred to as the sympathetic nervous system of insects. It is made up of three separate systems.

1. *Stomatogastric (stomodael),* associated with the brain, aorta, and foregut.
2. *Ventral sympathetic,* associated with the ventral nerve cord.
3. *Caudal sympathetic,* associated with the posterior segments of the abdomen.

The stomatogastric system (Figs. 3-5 and 3-7) arises during embryogeny from the dorsal or the dorsal and lateral walls of the stomodaeum and eventually becomes connected to the brain. Its various components typically include a *frontal ganglion*, which lies on the dorsal midline of the foregut just anterior to the brain. The frontal ganglion connects with the brain by bilateral nerves. The *recurrent*

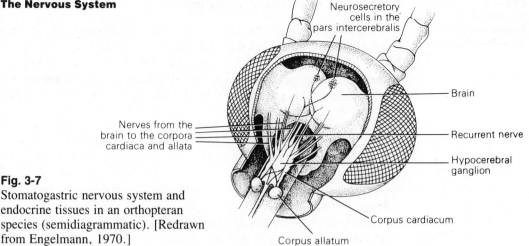

Neurosecretory cells in the pars intercerebralis

Brain

Nerves from the brain to the corpora cardiaca and allata

Recurrent nerve

Hypocerebral ganglion

Corpus cardiacum

Corpus allatum

Fig. 3-7
Stomatogastric nervous system and endocrine tissues in an orthopteran species (semidiagrammatic). [Redrawn from Engelmann, 1970.]

nerve arises medially from the frontal ganglion and extends beneath and posterior to the brain. Beyond this point there is considerable variation in the stomodael system among different kinds of insects. Therefore, the description that follows will pertain only to one of the more common arrangements. The recurrent nerve ends posteriorly in a *hypocerebral ganglion,* which may give rise to one or two *gastric* or *ventricular nerves,* which continue posteriorly and eventually terminate with a *ventricular ganglion.* Also associated with the hypocerebral ganglion are two pairs of endocrine glands, the *corpora cardiaca* (sing., *corpus cardiacum*) and the *corpora allata* (sing., *corpus allatum*). The stomatogastric system apparently exerts some control over the movements of the gut and the heart and possibly, in certain instances, labral muscles, mandibular muscles, and salivary ducts. The corpora cardiaca and corpora allata are involved with hormone secretion and are thus considered as parts of the endocrine system. For more information on the stomatogastric nervous system, see Miller (1975b).

The ventral sympathetic system is associated with the ganglia of the ventral nerve cord. From each segmental ganglion a single median nerve arises and divides into two lateral nerves, which supply the spiracles of the segment in which they are located. These median and lateral nerves may be lacking altogether.

The nerves of the caudal sympathetic system arise from the caudal ganglion of the ventral chain and supply the posterior portions of the hindgut and the internal sexual structures.

Peripheral Nervous System. All the nerves emanating from the ganglia of the central and visceral nervous systems comprise the *peripheral nervous system.* The distal processes, dendrites, of sensory neurons within these nerves are associated with a sensory structure (see Chapter 6), while the axon usually synapses with neurons within a ganglion of the central nervous system. Nerves also typically contain motor fibers, the cell bodies of which are located in ganglia of the central nervous system and fibers of which end in muscles, glands, and other effector structures.

Nervous Conduction and Integration

Information about the external and internal environment is continuously conveyed from sensilla to the central nervous system where it is integrated such that appropriate behavioral and regulatory changes are made. Figure 3-8 provides a simplified outline of the interrelationships among the various parts of the nervous system. In this section we want to take a closer look at the actual operation of the nervous system (additional information may be found in Chapter 8). Although many aspects of insect neurobiology have been studied, only beginnings have been made in understanding the detailed links between neural function and behavior. Biophysically, nervous conduction in insects has not been found to differ significantly from that in other invertebrates.

Input (nervous impulses) from a single sensillum to the central nervous system usually consists of the following events:

1. *Stimulus* (external or internal).
2. *Receptor potential*.
3. *Generator potential* in the perikaryon of the sensory cell.
4. *Action potential* in the axon of the sensory cell.

Fig. 3-8
Major interrelationships of the insect nervous system. [Modified from Cornwell, 1968.]

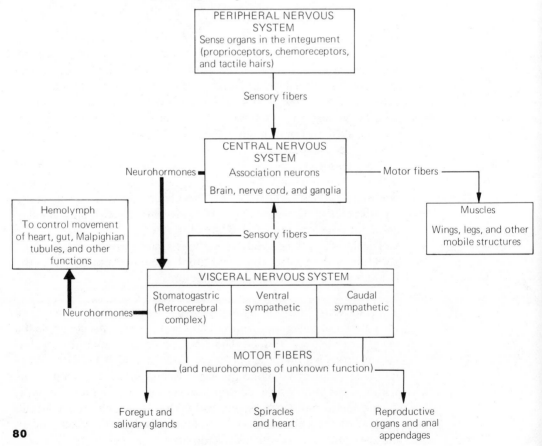

5. Release of a *chemical transmitter substance* at a synapse.
6. Generator potential in the perikaryon of the next neuron (post-synaptic) in line.
7. Action potential, and so on.

Stimuli range from electromagnetic radiation to changes in pH (see Chapter 6). Exactly how the energy of a stimulus is *transduced* (changed) into the receptor potential is not completely understood, but a change in membrane permeability of the dendrite is apparently involved. In any case, a stimulus induces the depolarization or receptor potential in the dendrite which in turn leads to the generator potential (depolarization) in the perikaryon. Both the receptor and generator potentials are graded; that is, their strengths vary with the strength of the stimulus.

The cell membrane of a resting axon—that is, an axon that is not responding with an impulse—is actively maintained in a polarized state. The inside of the axon is negatively charged (approximately 70–80 mV) relative to the outside. This polarity is due to the active pumping of Na^+ out of the cell.

If the generator potential is sufficiently strong—that is, above *threshold,*—there is a sudden increase in the permeability of the axon membrane to Na^+ and a wave of depolarization, action potential ($+85$–100 mV), proceeds along the axon to the terminal arborizations. Whereas the receptor and generator potentials are graded, the action potential is not. It is of constant amplitude and is *all or none* in that if the generator potential is above threshold, a full-strength action potential occurs. The axon returns to the depolarized state as soon as the action potential passes.

The arrival of an action potential at the terminal arborizations induces the release of a chemical transmitter that stimulates the next neuron. *Acetylcholine* has been identified as one among several possible transmitter substances in insects. As soon as acetylcholine is released, the enzyme *acetylcholine esterase* begins to break it down. In some cases there is a tight junction between the terminal arborization of one neuron and the dendrites of the next, and transmission is apparently electrical rather than chemical.

The outcome of a nervous impulse depends on a number of factors. For example, an impulse arriving at a given synapse may induce a subthreshold generator potential and have no effect. However, if a whole series of impulses arrive at a synapse in rapid succession, the resulting generator potentials may reinforce one another sufficiently to exceed threshold and stimulate an action potential (*temporal summation*). When several axons synapse with a single neuron, impulses arriving simultaneously may also summate and induce an action potential (*spatial summation*).

At any given moment, the ganglia, including the brain, receive input from a variety of sensory axons. They integrate the diverse impulses and exert regulatory influences both by stimulation and by inhibition. For example, stimulation of certain parts of the cricket (*Acheta* sp., Gryllidae, Orthoptera) brain stimulates stridulation ("singing"), while stimulation of other parts exerts an inhibitory effect. The copulatory movements of the male mantis increase in vigor when the head, and hence the inhibitory influence, is removed. Under natural circumstances, it is the female mantis that removes the males's head, resulting in more vigorous copulation as well as a highly nutritious meal.

Many axons fire spontaneously without input from a sensory neuron. Such activity is considered to keep the nervous system, or at least a part of it, in a highly sensitive state. Thus a stimulus that would otherwise be subthreshold might induce an action potential by summating with a spontaneous discharge.

McDonald (1975) reviews the literature dealing with the effects of various chemicals on nervous function.

Glands and Endocrine System

Here we consider a diverse group of cells and tissues that secrete a wide variety of substances with an equally wide variety of functions. Secretion of a specific product or products as a major function is what unites this diverse group. All cells secrete to some extent, but secretion is the *main* function of the cells and tissues discussed in this section. There are basically two types of glands: *exocrine* and *endocrine*. Exocrine glands discharge their products via apertures or ducts into the external world or into lumens of various viscera, for example, the reproductive tract or the alimentary canal. Endocrine glands are typically ductless, and their secretions are usually released directly into the *hemolymph*.

Exocrine Glands

Exocrine glands may be single cells or small aggregates of secretory cells (Fig. 3-9). A single secretory cell, for example, one that secretes a toxic substance, may contain an *intracellular ducteole* and may also be associated with another cell, a *ductule cell,* that forms a duct for transfer of secretions from the gland cell to the outside. More complex glands may be formed as invaginated masses composed of large numbers of cells that secrete their products into a common lumen with a single aperture that may in turn open into a common duct (e.g., salivary glands; see Fig. 4-1). Further, complex glands may be associated with a separate, but attached, reservoir in which large amounts of secretion can accumulate (e.g., salivary reservoirs in many insects). Externally, fine hairs may be associated with a gland opening or gland duct opening. Such hairs may aid in rapid dispersal of the secretion product(s), such as one involved in chemical communication between individuals.

Exocrine glands are generally of ectodermal origin and are found widely scattered over the insect body. The specific location of a given gland is often correlated with its function. For example, salivary glands are typically derived from the labial segment of the insect head during embryogenesis, although in the fully formed insect, they are usually located in the thorax on either side of the foregut. Glands associated with the reproductive system are closely connected with the internal genital organs (see Chapter 5). Glands

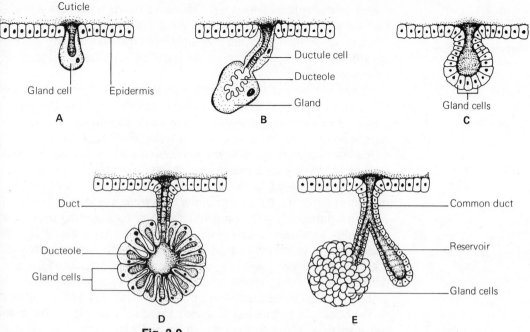

Fig. 3-9
Examples of exocrine glands. A. Simple unicellular gland. B. Unicellular gland with ductule cell. C. Simple aggregation of gland cells. D. Aggregation of gland cells with common duct. E. Complex aggregation of gland cells with common duct and reservoir. [Modified from various sources.]

that secrete their products only into the external world vary a great deal in location. Strictly speaking, the epidermal cells that secrete the cuticle, as well as the secretory cells in the midgut and Malpighian tubules (see Chapter 4), can be viewed as glandular epithelium. From this point of view, the epidermis emerges as the most extensively developed gland in an insect.

Among the many diverse functions of glands are secretion of structural materials, saliva (see Chapter 4), repugnatorial and other defensive secretions, and chemicals involved with intra- and interspecific communication (see Chapter 8).

Examples of structural materials include various waxes, lac, and silk. The wax secreted by epidermal glands located ventrally between the overlapping sternal plates of the fourth through the seventh abdominal segments of honey bees is used to construct honey comb. Members of the Homoptera—scale insects (superfamily Coccoidea), white flies (Aleyrodidae), and so on—secrete waxes in various forms (powdery, filamentlike, etc.) onto the body surface. Certain scale insects, the Lacciferidae and especially *Laccifer lacca,* secrete a resinous material called lac from which commercial shellac is produced (see Chapter 12). Insects in many groups produce silk, which is usually composed of fibrous proteins (Rudall and Kenchington, 1971). Silk is typically produced in the form of fine threads that are then woven into various structures, including the protective pupal cocoons of many Lepidoptera, Hymenoptera, and Siphonaptera. The tunnels of Embioptera (webspinners) are constructed with silk produced in the enlarged foretarsi. Members of the Trichoptera (caddisflies) are aquatic and utilize silk to construct both larval and pupal shelters, often incorporating materials from the environment among the silk threads. Certain predaceous Trichoptera construct silken nets and snares to trap prey.

The production of repugnatorial substances is widespread among the insects (see Chapter 8). Examples include the scent (''stink'') glands, located on the metepisternum or on the dorsum of the abdomen among the true bugs, that secrete a number of different hydrocarbon derivatives. Nasute soldiers found in certain termite species (Isoptera) secrete a sticky defensive material from the enlarged *frontal gland* in the head. Some insects have eversible repugnatorial glands, for example, the *osmeteria* found on the dorsum of the prothorax of swallowtail butterfly larvae and similar structures at various locations among other species of Lepidoptera. *Pygidial glands* at the posterior end of certain beetles also secrete a repugnatorial substance. Glands and associated structures produce and deliver the venoms found in the stinging Hymenoptera. Poison glands are also associated with *urticating* (stinging) *hairs* or *spines* on the bodies of certain Lepidoptera. These are easily detached from the insect's body and are very sharp and can cause considerable discomfort if they come in contact with human skin.

Glands in many insects produce chemicals that serve as signals of various sorts for other members of the same or different species. The various functions of these substances, called *pheromones,* are dealt with in Chapter 8. A few examples will suffice here. In Lepidoptera a variety of glands produce sex pheromones that attract or excite a mate. These glands include the *androconia,* specialized scales and associated glands scattered or clustered among the wing scales, and eversible *hair-pencils.* Both androconia and hair-pencils provide much surface area, facilitating the dispersal of an aphrodisiac secretion. Glands in the mandibular segment of the queen honey bee produce the ''queen substance,'' which inhibits the workers from constructing queen cells. The review by Jacobson (1974) contains information on pheromone-secreting glands.

Except for a few cases of the apparent presence of neurosecretory axons, few dermal glands have been found to be innervated. Other exocrine glands, such as the salivary glands, may (cockroaches) or may not (*Calliphora,* Calliphoridae, Diptera) be innervated (Noiret and Quennedy, 1974).

Endocrine Glands

Endocrine glands commonly function in close coordination with one another, in contrast to exocrine glands, which often function independently. The endocrine system is involved with the secretion of substances called *hormones.* These diverse substances complement the coordinating activities of the nervous system. They are generally secreted into the hemolymph from a rather well-defined tissue and circulate throughout the insect body. However, each exerts its influence quite specifically upon a target organ or tissue. Like the nervous system, hormones help the insect to adjust to external and internal environmental changes, but physiological and behavioral responses mediated by hormones occur more slowly than those mediated directly by the nervous system. It might be said that hormones are involved with long-term adjustments and nerves with short-term adjustments to changes in the internal and/or external environment.

The best identified endocrine structures whose functions are at least partially understood are the *corpora cardiaca,* the *corpora allata,* and the *thoracic glands* (see Fig. 5-26). In addition, there are groups of *neurosecretory cells* in the protocerebrum (Fig. 3-10) as well as in other parts of the brain, in the subesophageal and other ventral chain ganglia, and in the corpus cardiacum. Neurosecretory cells (Rowell, 1976) are modified neurons that secrete hormones. They are usually identified by the use of specific stains; for example, they stain deeply with paraldehyde fuchsin (Fig. 3-11). They are also characterized by the presence of electron-dense granules when examined with the electron microscope.

The locations of the corpora cardiaca and corpora allata have already been described, and specific functions will be discussed in later chapters. In addition to containing intrinsic secretory cells, the corpora cardiaca receive axons from the neurosecretory cells in the brain and serve as storage and release sites for their secretions.

Fig. 3-10
Neurosecretory cells in the brain of the mosquito *Aedes taeniorhynchus.* Cells exposed by retraction of posterior part of head capsule. 1, Neurosecretory cells; 2, brain; 3, pharyngeal muscles; 4, retracted portion of head capsule; 5, compound eye; 6, anterior portion of the thorax. [Courtesy of A. O. Lea.]

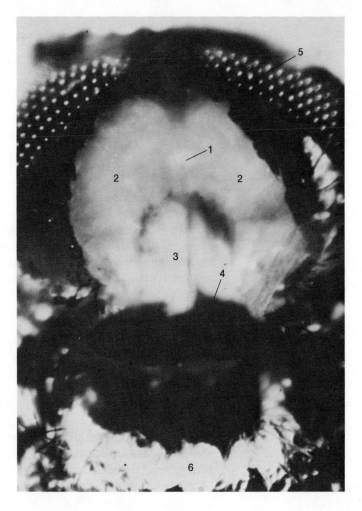

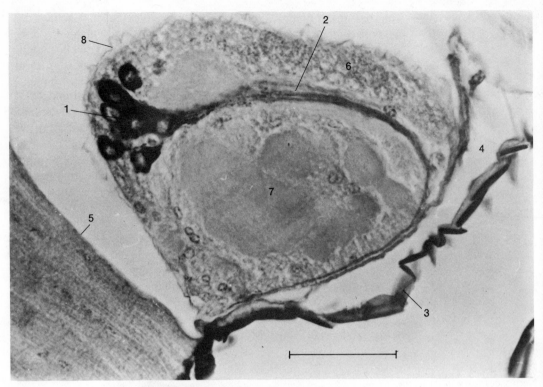

Fig. 3-11
Photomicrograph of neurosecretory cells exposed by staining with paraldehyde fuchsin (PF). Sagittal section of the brain of a mosquito, showing neurosecretory cells and axons running anteriorly and then posteriorly. The axons eventually reach the corpora cardiaca (see Fig. 3-7). 1, Neurosecretory cells; 2, axons of neurosecretory cells; 3, cuticle of pharyngeal pump; 4, recurrent nerve; 5, pharyngeal pump muscle; 6, cell bodies of neurons; 7, neuropile; 8, neural lamella. (Scale line = 50 μm.) [Courtesy of S. M. Meola.]

Structures with a storage and release capacity are called *neurohaemal organs*. The thoracic glands are irregular masses of tissue of ectodermal origin that are usually intimately associated with tracheae. They may or may not be innervated. They have been found to secrete the hormone that initiates the molting process. As will be discussed in more detail in Chapter 5, the secretory activity of the thoracic gland is stimulated by hormones from neurosecretory cells in the brain.

In larvae (maggots) of the higher Diptera (true flies, suborder Cyclorrhapha), there is a small ring of tissue, supported by tracheae, called the *ring gland* or *Weisman's ring*. The different cells that compose it are considered to be homologous with the corpora allata, corpora cardiaca, and the thoracic glands.

Several endocrine functions have been identified in insects. Examples include influencing color changes; regulation of the excretion of water; control of heartbeat rate and amplitude; regulation of metabolic activities (Steele, 1976), such as the maintenance of carbohydrate level in the hemolymph, synthesis of proteins, and metabolism of lipids; control of sclerotization and melanization of the cuticle;

control of growth and metamorphosis; (possibly) control of circadian rhythms; determination of readiness to mate; regulation of dormancy; and regulation of migratory behavior.

The Muscular System

The conversion of chemically stored energy into mechanical energy occurs within muscle tissue. This conversion is associated with the muscle shortening in length, or *contraction*. Muscle contraction in turn produces a variety of results, depending upon the location, attachments, and degree of stimulation of the contracting muscles. Muscle contraction is responsible for most forms of locomotion, for the maintenance of posture, and for movements of the viscera, such as the peristaltic propulsion of food along the alimentary canal. Quantitatively speaking, muscle is probably the most abundant tissue in the higher animals.

Muscles are commonly classified as striated, those containing definite transverse lines or striations, and smooth, those lacking these striations. Insect muscles are all of the striated variety, although in some types of muscle, the striations are extremely difficult to discern. They are typically colorless or grayish in appearance, but may be tinged with yellow, orange, or brown, as is wing muscle in many insects. In view of the small size of insects, one might assume that they have comparatively few muscles. Actually, the reverse is true; larger insects possess perhaps two or three times as many individual muscles as, for example, humans do.

Based on location, insect muscles can be divided into two groups: *skeletal* and *visceral*. Skeletal muscles (Fig. 3-12) are attached at both ends to regions of the integument and are those associated with the maintenance of posture and the various movements of the skeleton. Visceral muscles are involved with movements of various internal organs, such as the alimentary canal, the ovaries, and the Malpighian tubules. Visceral muscles commonly attach to other visceral muscles.

Skeletal Muscles

Skeletal muscle in most cases has a somewhat stationary attachment area, the *origin,* and a movable attachment area, the *insertion*. Areas of attachment may be either directly on the body wall (Fig. 3-13A–D) or on the inner surface of an apodeme (Fig. 3-13E). A number of different means exist by which muscle is attached to integument, but in all cases where it is present, the outer membrane of a muscle fiber, the *sarcolemma,* is continuous with the basement membrane of the epidermis. Typically, fibrillar structures, the *tonofibrillae,* appear to extend through the epidermal cells and into the cuticle (Fig. 3-13B–D). Ultrastructural studies have revealed that tonofibrillae are bundles of microtubules that extend from desmosomes, which link them to muscle cells, to hemidesmosomes at the junction between the epidermis and the cuticle (Fig. 3-14). Additional fibers run from the hemidesmosomes through pore canals to

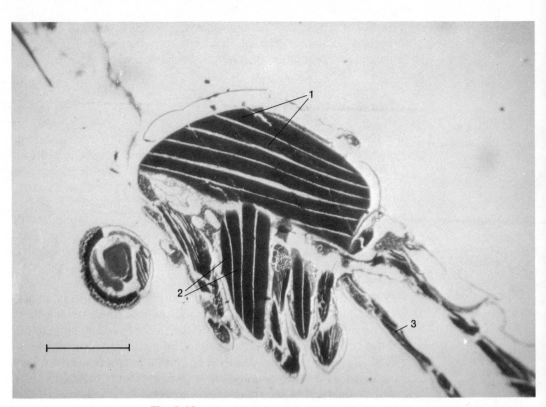

Fig. 3-12
Photomicrograph of a sagittal section of a mosquito showing the indirect flight muscles in the thorax and intersegmental muscles in the abdomen. 1, Longitudinal flight muscles; 2, tergosternal flight muscles; 3, intersegmental muscle. (Scale line = 0.5 millimeter.) [Tissue prepared by S. M. Meola.]

the epicuticle (Caveney, 1969). Desmosomes are local points of attachment between cells. Hemidesmosomes (''half'' desmosomes) and the additional fibers apparently serve as local points of attachment between epidermal cells and cuticle. In some cases, muscle fibers may attach directly to unmodified epidermal cells (Fig. 3-13A). The attachments between the cuticle and epidermal cells are resistant to molting fluid, and hence muscles remain functionally attached to the cuticle between apolysis and ecdysis.

Groups of Skeletal Muscles   The skeletal muscles of the head effect three movements: of the entire head, of mouthparts, and of antennae. Muscles that originate on the anterior part of the prothorax and insert on the tentorium and various parts of the head capsule are responsible for head movement. Mouthparts and antennae are moved both by *extrinsic muscles,* which originate on the inner surface of the head capsule or tentorium and insert within a given appendage, and by *intrinsic muscles,* whose origins and insertions are entirely within a given appendage (Fig. 3-15A).

Thoracic skeletal muscles are involved mainly with the legs and wings. Legs (Fig. 3-15B), like mouthparts and antennae, are operated by both extrinsic and intrinsic muscles; the extrinsic muscles

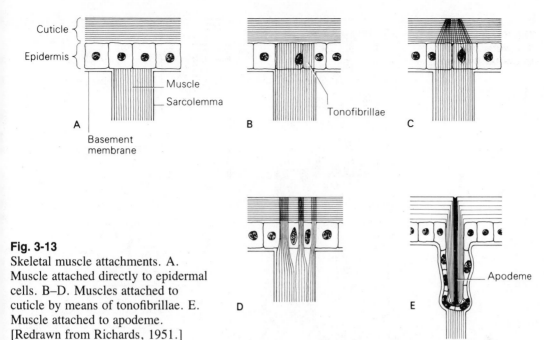

Fig. 3-13
Skeletal muscle attachments. A.
Muscle attached directly to epidermal
cells. B–D. Muscles attached to
cuticle by means of tonofibrillae. E.
Muscle attached to apodeme.
[Redrawn from Richards, 1951.]

are responsible for leg movement and the intrinsic muscles for movement of the individual segments. The origins of the various extrinsic
leg muscles are on the terga, pleura, and sterna of each thoracic
segment.

In the typical wing-bearing segment, well-developed longitudinal
muscles run between the phragmata of the pterothorax (Figs. 3-16
and 3-12; see also Fig. 7-9), and dorsoventral muscles attach to the
tergum and sternum of each segment (see Fig. 7-9). Contraction of
the longitudinal muscles causes arching of the tergum, which, owing
to the construction of the thorax, causes the depression of the wings.
The action of the dorsoventral muscles is antagonistic to that of the
longitudinals, contraction resulting in the depression of the tergum
and consequent elevation of the wings. Since the action of these two
sets of muscles on the wings is indirect, they are referred to as
indirect wing muscles. In addition to indirect wing muscles, there
are muscles that attach directly to the bases of the wings. Direct
muscles are responsible for various wing movements during flight
and for the flexing of the wings over the abdomen in those insects
that possess this ability. In some insects direct muscles provide the
main force of propulsion. Their action will be considered in more
detail in Chapter 7.

In addition to the muscles associated directly or indirectly with
leg and wing movements, there are muscles that run between the
pleura and terga or sterna and lateral intersegmental muscles. Also,
since there are spiracles located in the thorax, muscles associated
with their closure mechanisms are present.

The skeletal musculature of the abdomen is somewhat simpler
than that of the head and thorax. The most prominent abdominal
muscles are the longitudinal intersegmental or segmental ones (Fig.
3-17). These run between successive antecostae of both the terga

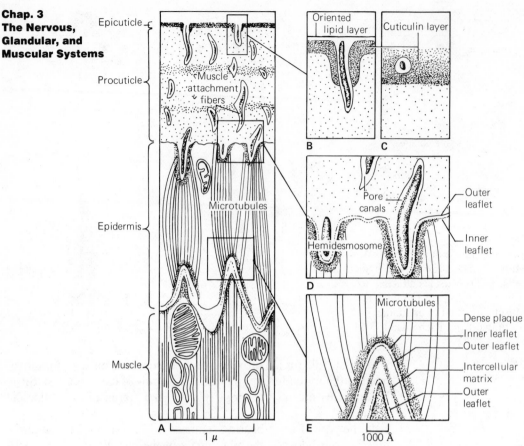

Fig. 3-14.
Diagram showing ultrastructural details of muscle attachment to cuticle (in Apterygota). B–E represent specific regions in A. The parts of a desmosome are identified in E. [Redrawn from Caveney, 1969.]

and the sterna. Also present are lateral abdominal muscles, which may be oblique in orientation, the *oblique sternals,* but typically are dorsoventral in orientation, the *tergosternals.* These may be inter- or intrasegmental. Spiracular muscles are also present. In addition, special muscles are involved in the various movements of the copulatory structures, ovipositor, and cerci.

For more detailed information on the musculature of insects, consult Chapman (1971), Richards and Davies (1977), and Snodgrass (1935, 1952).

Structure of Skeletal Muscles. The basic structural unit of a muscle is the *muscle fiber* (Fig. 3-18). Typically, it is composed of an outer membranous layer, the *sarcolemma* (cell membrane), and the inner *sarcoplasm* (cytoplasm). Within the sarcoplasm is a bundle of tiny fibers, the *myofibrils,* each of which has a cross-sectional diameter of about 1 micrometer. Numerous cell nuclei and mitochondria and sarcoplasmic (endoplasmic) reticulum are also located within the sarcoplasm. Deep invaginations of the cell membrane form the T-system (transverse tubular system). The myofibril is the contractile element. By weight a muscle fiber is approximately 20%

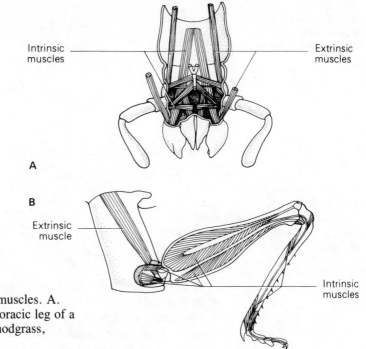

Intrinsic muscles — Extrinsic muscles

A

B

Extrinsic muscle — Intrinsic muscles

Fig. 3-15
Extrinsic and intrinsic skeletal muscles. A.
Labium of a cricket. B. Metathoracic leg of a
grasshopper. [Redrawn from Snodgrass,
1935.]

protein; the remaining 80% is primarily water, plus small quantities
of salts and metabolic substances (Huxley, 1965).

Ultrastructural studies of insect muscle fibers reveal regular arrays
of thick and thin filaments organized in the same fashion as vertebrate muscles (Figs. 3-18 and 3-19). As in vertebrate muscle, the
thick filaments are considered to be composed of the protein myosin,
the thin filaments of the protein actin. The actin filaments are more
numerous and in cross section can be seen to be arranged in regular
orbits around myosin filaments. For example, in flight muscle each
myosin filament is orbited by 6 actin filaments; in leg muscle orbits

Fig. 3-16
Cross section of the mesothorax showing the action of the indirect flight
muscles. A. Notum depressed, wings elevated. B. Notum arched and
elevated, wings depressed. Arrows indicate direction of movement.
[Redrawn from Snodgrass, 1963b.]

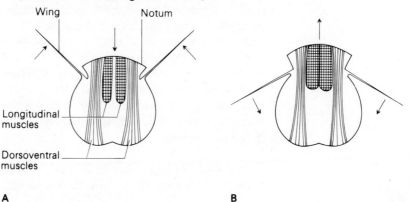

Wing   Notum

Longitudinal muscles

Dorsoventral muscles

A   B

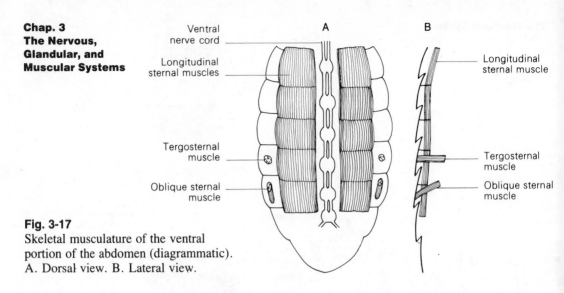

Fig. 3-17
Skeletal musculature of the ventral
portion of the abdomen (diagrammatic).
A. Dorsal view. B. Lateral view.

of 9–12 actin filaments have been observed. In longitudinal section actin filaments are seen to be anchored in a region called the Z-line. A sarcomere is a longitudinal division of a fibril from one Z-line to the next. Sarcomere length in insects ranges from about 2.5 to more than 9.0 micrometers (Hoyle, 1974). The myosin filaments are located centrally and do not reach the Z-line; the actin filaments, on the other hand, are attached at the Z-lines, but do not reach the center of a sarcomere. Regularly occurring branches extend from the myosin filaments to the actin filaments, forming cross-bridges. The organization of the thick and thin filaments is responsible for the transverse striations mentioned earlier. There are only actin filaments in the I-bands and only myosin filaments in the H-band. Hence these bands do not stain as deeply as do the A-bands, the regions where both actin and myosin filaments overlap. Patterns of transverse bands are also evident when fresh unstained muscle fibers are viewed under an ordinary light microscope with the condenser stopped down or under a polarizing light microscope. The banding pattern changes when a muscle is induced to contract. Smith (1972) should be consulted for a survey of insect muscle ultrastructure.

Individual insect skeletal muscle fibers are organized into discrete morphological entities, the *muscle units* (see Fig. 3-21). The fibers, typically 10–20, composing each unit are enveloped by a tracheolated membrane. Muscle units, in turn, compose the various skeletal muscles. Larger muscles are composed of a few to several muscle units, while a very small muscle may consist of a single unit (Hoyle, 1974, 1975). This is in distinct contrast to vertebrate muscles, which are typically composed of very large numbers of muscle units.

Oxygen is carried to the muscle fibers by tracheoles (see Chapter 4) that are typically in very close contact with each fiber. In the flight muscles of most insects these tracheoles actually fit into indentations in the sarcolemma.

Types of Skeletal Muscle Fibers.   Skeletal muscle fibers vary widely in diameter (less than 10 micrometers to greater than 1.0

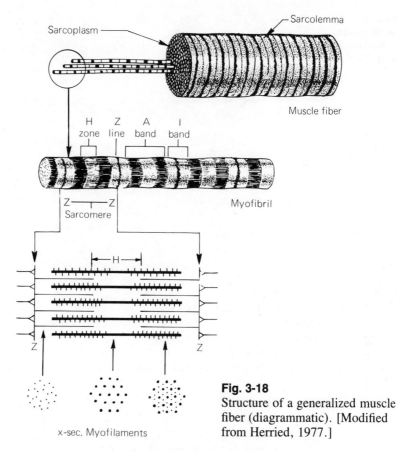

Fig. 3-18
Structure of a generalized muscle fiber (diagrammatic). [Modified from Herried, 1977.]

millimeter), location of nuclei, location of mitochondria, organization of myofibrils, and so on.

Based on studies at the light microscope level, three broad categories of flight muscles (Fig. 3-20) have been described (Elder, 1975): *tubular*, *close-packed*, and *fibrillar*. Tubular and close-packed are also found in skeletal muscles other than those associated with flight. Tubular muscle fibers (Fig. 3-19A), 10–25 micrometers in diameter, are viewed as primitive types, being found in cockroaches, dragonflies, and damselflies. The nuclei are arranged in a central column with the myofibrils disposed radially. The mitochondria are slablike and located among the fibrils. Close-packed fibers (Fig. 3-20B), 10 –100 micrometers in diameter, are found in the Orthoptera, Trichoptera, and Lepidoptera. In contrast with tubular fibers, the nuclei are flattened and located peripherally. The mitochondria are large and located among the relatively small myofibrils. Fibrillar muscle fibers (Fig. 3-20C), 100 micrometers to 1.0 millimeter in diameter, are characteristic of the flight muscles of Coleoptera, Diptera, Hymenoptera, and most Hemiptera. The nuclei are found peripherally in longitudinal rows, while the large mitochondria are found in longitudinal columns in the spaces between the fibrils.

A muscle fiber found in the Apterygota, some larvae, and legs of some adult insects provides an additional example of skeletal muscle

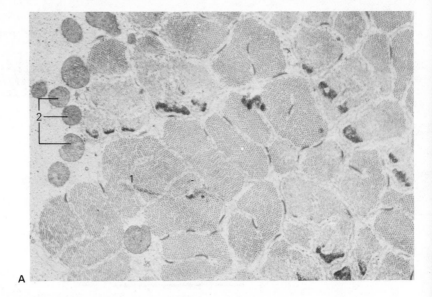

A

Fig. 3-19
Ultrastructure of skeletal muscle in the mosquito *Aedes aegypti*. A. Cross section (9600 ×): 1, center of a single myofibril; 2, mitochondria. B. Longitudinal section (25,400 ×): 1–2 and 3–4, I-bands; 5, mitochondrion; the dark, thick (myosin) and thin (actin) parallel, horizontal lines between 2 and 3 are myofilaments. C. Portions of cross sections of myofibrils showing arrays of thick (myosin) and thin (actin) myofilaments (73,000 ×): 1, parts of T-system. [Courtesy of M. Catherine Walker.]

(Fig. 3-20D). The nuclei are located peripherally and myofibrils centrally, and more of the fiber is occupied by sarcoplasm than in muscle fibers described above.

Muscle Contraction. As in the vertebrate muscle fibers, contraction of insect muscle fibers is thought to involve the sliding of the myosin and actin filaments past one another with the consequent shortening of the sarcomeres. Although the mechanism of shortening is not understood, it is thought to involve the formation and breakage of temporary linkages between the two kinds of filaments via the cross-bridges, as if the myosin filaments "walked" along the actin filaments. Muscle contraction in insects might also involve the coiling of myosin filaments (Chapman, 1971).

Contraction of a fiber is induced by the arrival of a nerve impulse, which leads to release of a transmitter substance, possibly L-glutamate (Pichon, 1974). The membrane of a resting muscle fiber is maintained in a polarized state (i.e., the inside is on the average − 60 mV relative to the outside). Arrival of the transmitter substance apparently causes a change in permeability that results in an influx of sodium ions and consequent depolarization, sometimes to the point where the inside is positive relative to the outside. The potential difference across the muscle membrane returns to its original level with the movement of potassium ions out of the cell. The change in membrane potential induced by nervous impulse is called the postsynaptic potential. As the postsynaptic potential spreads from the neuromuscular junction, it decreases rapidly and hence its effect is localized. Evidently the T-system helps carry the postsynaptic po-

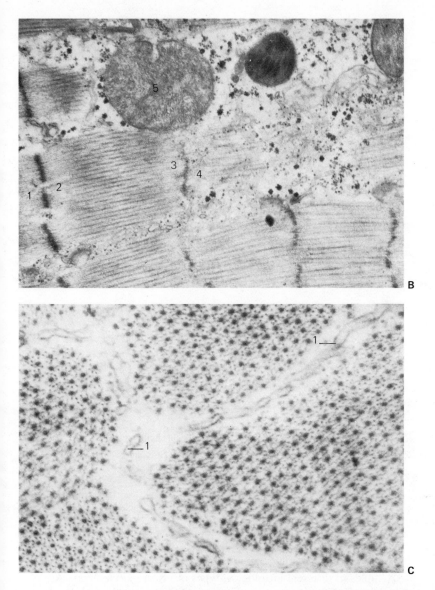

B

C

tential deep into muscle cells. The actual contraction of muscle fibers involves the release of calcium ions, probably from vesicles in the sarcoplasmic reticulum, within the muscle fiber. The energy for muscle contraction comes from the hydrolysis of ATP to ADP, and the enzyme (ATPase) that catalyzes the breakdown of ATP is thought to be localized in the cross-bridges. Calcium ions presumably stimulate the action of ATPase and hence muscle contraction. Muscle relaxation, then, appears to involve the sequestration of calcium ions and consequent termination of ATPase activity.

Innervation and Contraction of Skeletal Muscle. Insect skeletal muscle units commonly receive more than one motor axon. This situation is called *polyneural innervation*. In contrast with the single end-plate characteristic of vertebrate neuromuscular junctions, insect motor axons branch out, forming terminals at several regularly spaced points along a fiber (Fig. 3-21). This is called *multiterminal*

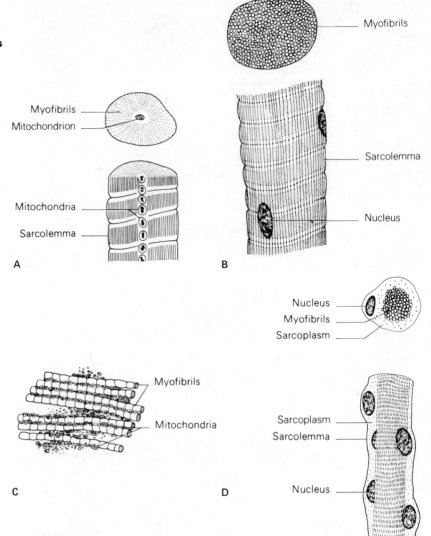

Fig. 3-20
Examples of different histological types of insect muscle fibers. A. Tubular. B. Close-packed. C. Fibrillar. D. Nonflight skeletal. [Redrawn with modifications from Snodgrass, 1925.]

innervation. Since the postsynaptic potential decays rapidly and influences only a portion of a fiber, the occurrence of many terminals promotes stimulation of a whole fiber.

In the majority of cases the number of axons that innervate a given muscle unit is two. However, a third has been identified in the jumping hindlegs of *Romalea* and other locusts. In a typical muscle unit (Fig. 3-21) one axon is fast and the other slow. An individual fiber may receive branches from one or both kinds of axons. The designations "fast" and "slow" refer to the kind of contraction produced. Stimulation of the fast axon characteristically produces a postsynaptic potential of constant size and a rapid, powerful twitch. A rapid train of impulses along a fast axon produces a sustained strong contraction *(tetanus)*. Stimulation of the slow axon has a different effect.

A single impulse produces a single, weak postsynaptic potential and a single, weak twitch. A train of closely spaced impulses induces successively greater postsynaptic potentials and contractions, that is, a graded response. This graded responsiveness of muscles to impulses arriving along the slow axon affords precision in control of muscle contraction. This method of control of contraction is associated with the small size of insect muscles as compared with those of vertebrates. Vertebrates effect slow, precise movements by varying the number of different muscle units stimulated within a given muscle.

Fast axons are involved in activities like jumping that require sudden strong muscle contraction. Slow axons are involved with muscle tonus and mediate actions that require slow, smooth, and precisely controlled contractions. The term *tonus* refers to the continuous tension applied by a muscle as opposed to alternate contraction and relaxation. Muscles in the tonic state support an insect in a given stance. In soft-bodied larvae, such as caterpillars, muscles in the tonic state maintain the rigidity of the exoskeleton. The third motor axon mentioned above, when present, may function in an inhibitory fashion. Although fast and slow axons often function independently, they also commonly function such that the action of one reinforces the action of the other.

Chapman (1971), Usherwood (1975), Hoyle (1974), Maruyama (1974), and Usherwood (1974) all contain information on innervation and contraction of insect muscles.

Physiological Types of Insect Muscle. Insect muscles can be classified as *resonating (asynchronous)* and *nonresonating (synchronous)*. Resonating muscle in insects is the fibrillar type described earlier. In appropriate mechanical situations it is capable of undergoing several successive contractions when stimulated with a single nervous impulse. This type of muscle is almost exclusively

Fig. 3-21
Innervation of a typical muscle unit showing fast and slow axons and multiterminal endings. [Redrawn from Hoyle, 1965.]

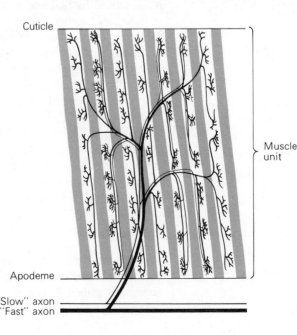

Cuticle

Muscle unit

Apodeme

"Slow" axon
"Fast" axon

associated with flight, but it is associated also with the halteres of the true flies and with the sound-production mechanism of the cicada. Nonresonating muscles include the other types of muscle fibers described earlier and are involved with movement other than that produced by fibrillar muscle. The differences between the resonating and nonresonating types of muscle will become more clear when we discuss the flight of insects.

Energy for Muscle Contraction. The biochemical processes that result in the production of adenosine triphosphate (ATP) occur in the cytosol and mitochondria of muscle. The phosphate that replenishes the ATP utilized in muscle contraction is furnished by the phosphogen, arginine phosphate, which has been identified in a number of insects (in vertebrate muscle the phosphogen is creatine phosphate). Carbohydrates are commonly the primary fuel for muscle contraction, although many insects utilize fat and occasionally protein or amino acids. For example, the amino acid proline serves as the primary flight fuel in the tsetse fly. For further information on this topic see the reviews by Bailey (1975), Crabtree and Newsholme (1975), Gilmour (1965), Rees (1977), Rockstein (1978), and Sacktor (1970, 1974, 1975).

Muscle Power. Great feats of strength have been ascribed to insects. For example, it has been said that if a flea were the size of a human being, he would be able to leap over tall buildings. This is a common fallacy. It is true that some insects can leap comparatively great distances and that some are capable of lifting or moving objects many times heavier than themselves, but it is *not* true that if these insects were to increase in size, their strength would increase correspondingly. The force a muscle is able to exert varies directly with its cross-sectional area, while the volume or mass of a body varies cubically. As an insect body would increase in size, its volume and mass would increase at a greater rate than a cross-sectional area of muscle, and hence the muscles would become relatively weaker. Thus we can say that insect muscles are relatively stronger than those of larger animals only because of the comparatively small size of insects. Actually, when the absolute muscle strength (maximum force a muscle can apply per square centimeter) of vertebrates and insects is compared, it is found that there are no great differences (Wigglesworth, 1972).

Although the forces exerted by insect and vertebrate muscles do not differ much, the work per unit time (i.e., power output) capability of insect flight muscles is comparatively large. In fact, active insect flight muscles display the highest metabolic rates of any known tissue (Chapman, 1971).

Visceral Muscle

As mentioned earlier, visceral muscles are associated with the various internal organs, including the muscles of the dorsal vessel, accessory pulsatile organs, alary muscles, dorsal and ventral diaphragm, the alimentary canal, Malpighian tubules, and reproductive organs. Visceral muscles may be *extrinsic* or *intrinsic*. Extrinsic visceral muscles originate on the integument and insert on organs; for example, the muscles responsible for the dilation of the cibarial or pharyngeal pumps (see Chapter 4). Intrinsic visceral muscles occur in regular meshes of circular and longitudinal strands or in irregular networks surrounding a given organ.

Some visceral muscles, like the cibarial and pharyngeal dilators, are indistinguishable histologically from skeletal muscle. However, most visceral muscles are different from skeletal muscles. The contractile material in visceral muscles is not grouped into distinct myofibrils, but simply fills all or part of a fiber. However, the contractile material is composed of thick and thin filaments, apparently myosin and actin. Whereas insect flight muscle typically has six thin filaments orbiting around a single thick filament, visceral muscle tends to have a higher thin filament/thick filament ratio: 10 or 11 or 12 thin to one thick filament. The T-system may be regular or irregular, or it may be only weakly developed. Visceral fibers tend to have longer sarcomere lengths than do flight muscles.

Much remains to be done to elucidate the innervation and control of visceral muscles. Many are innervated by and respond to electrical stimulation, but the neurotransmitter substance is unknown. Some apparently do not receive nerves and may be under neurosecretory control. Contractions range from irregular to slow and rhythmic. The specific functions of visceral muscles are dealt with as appropriate in other chapters.

Miller (1975a) provides a good review of insect visceral muscle structure and function.

Selected References

GENERAL

Bursell (1970); Chapman (1971); Rees (1977); Richards and Davies (1977); Rockstein (1978); Roeder (1953); Smith (1968); Snodgrass (1935, 1952); Wigglesworth (1972).

NERVOUS SYSTEM

Howse (1975); Huber (1974); Lunt (1975); Miller (1975b); Pichon (1974); Strausfeld (1976); Treherne (1974).

GLANDULAR SYSTEM (EXOCRINE)

Jacobson (1974); Noiret and Quennedy (1974); Rudall and Kenchington (1971).

GLANDULAR SYSTEM (ENDOCRINE)

Gilbert and King (1973); Highnam and Hill (1977); Novák (1975); Rowell (1976); Steele (1976); Wigglesworth (1970).

MUSCULAR SYSTEM

Bailey (1975); Davey (1964); Hoyle (1974); Huxley (1965); McDonald (1975); Maruyama (1974); Miller (1975 a&b); Rees (1977); Rockstein (1978); Sacktor (1970, 1974, 1975); Smith (1972); Usherwood (1974, 1975).

Alimentary, Circulatory, Ventilatory, and Excretory Systems

Every cell in the insect body, regardless of its function, requires a source of energy, a source of oxygen, and the raw materials with which to carry out its own maintenance and synthesizing activities. As a result of carrying out these processes, a cell produces carbon dioxide and other waste products of various sorts. Those systems directly involved in the transport of nutrients and oxygen to the individual cells and the removal of accumulated waste materials and carbon dioxide are discussed in this chapter.

The Alimentary System

The alimentary system is involved in the initial steps in the transport of nutrients to individual cells. Ingestion, trituration (chewing), digestion, absorption into the hemolymph, and egestion are all associated with this system. Insects possess a tube (often coiled), the alimentary canal, which extends from the anterior oral opening, the *mouth,* to the posterior *anus.* The gut is formed by a one-cell-thick layer of epithelial cells, and a non-cellular basement membrane is present on the hemocoel side of this layer of cells.

The insect alimentary canal (Fig. 4-1) is divided into three distinct regions, the anterior *foregut* or *stomodaeum,* the *midgut* or *mesenteron,* and the posterior *hindgut* or *proctodaeum.* The fore- and hindguts are lined with a chitinous *intima,* which is continuous with the cuticle of the integument. The fore- and hindguts arise as invaginations of the ectoderm, which also gives rise to the integument, and the gut epithelial cells are continuous with the epidermal cells. The midgut is generally believed to be of endodermal origin. Longitudinal and circular (intrinsic) muscles are usually associated with each of the three regions and by means of rhythmical *peristaltic contractions* move food along the alimentary canal. Some muscles originate on the integument and attach to certain parts of the alimentary canal (i.e., extrinsic muscles), for example, the muscles that dilate the pharyngeal pump. The anterior alimentary canal muscles are innervated by the stomatogastric system, and the posterior muscles by nerves from the posterior ganglion of the ventral chain (see Chapter

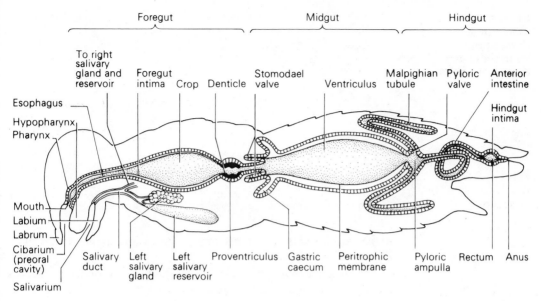

Foregut   Midgut   Hindgut

Fig. 4-1
Generalized insect alimentary system.

3). In addition to extrinsic muscles, tracheae and tracheoles (discussed later in this chapter) provide support for the gut.

The alimentary canal tends to be shorter in species that exist on high-protein diets and longer in those with high-carbohydrate diets, but there are many exceptions.

Foregut

Although there is considerable variation in the foregut, several morphological regions can usually be recognized. The *true mouth* lies at the base of hypopharynx within the preoral cavity *(cibarium)* formed by the mouthparts. The true mouth communicates directly with the *pharynx,* a structure that varies greatly among different insects. The cibarium or pharynx or both may be highly modified, forming pumps with well-developed extrinsic visceral musculature. These pumps, cibarial and/or pharyngeal, are most highly developed in sucking insects, such as Lepidoptera, Hemiptera, and many Diptera (Fig. 4-2). Next to the pharynx is the *esophagus,* which is commonly enlarged posteriorly to form the *crop* (Fig. 4-1). In some insects the posterior enlargement may be in the form of one or more blind sacs, or *diverticula* (Fig. 4-2). Immediately posterior to the crop is the *proventriculus* (Fig. 4-1). The luminal side of this structure often bears sclerotized denticles (teeth) or spines. The proventriculus typically communicates with the midgut by means of an intussusception, which consists of both fore- and midgut tissue. This structure will be called the *stomodael valve* here. Midgut tissue *(cardial epithelium)* surrounds the foregut portion of the stomodael valve. The true flies (Diptera) lack a proventriculus in the sense described. In these insects, the stomodael valve lies between the esophagus—or diverticulum(a), if present—and the midgut posterior

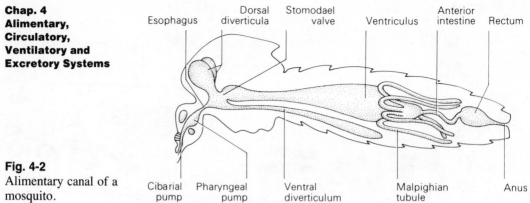

Esophagus | Dorsal diverticula | Stomodael valve | Ventriculus | Anterior intestine | Rectum

Cibarial pump | Pharyngeal pump | Ventral diverticulum | Malpighian tubule | Anus

Fig. 4-2
Alimentary canal of a
mosquito.

to the cardia and is commonly referred to by investigators who study
Diptera as the proventriculus.

The foregut with its various morphological divisions serves
mainly as a conducting tube, carrying food from the cibarial cavity
to the midgut. The enlarged crop functions, at least in part, as a site
of temporary food storage and partial digestion in some cases.

The proventriculus, when armed with denticles, may serve as a
grinding structure in addition to the mouthparts. Spines in the proventricular region may act together as a food sieve or filter, in honey
bees, for example, proventricular spines allow the movement of
pollen into the midgut without admitting ingested flower nectar. The
stomodael valve is developed to varying degrees in different insects.
Whether this structure actually acts as a valve in most insects is not
known.

Although the foregut is not the major digestive region of the alimentary canal, some digestion occurs in the crop by the action of
salivary enzymes and enzymes regurgitated from the midgut, for
example, in the Orthoptera. Except for the possible passage of small
amounts of lipid, the foregut intima has been found to be impermeable. So the foregut probably plays no major role in the absorption
of materials into the hemolymph.

Midgut

The midgut (Fig. 4-1) typically begins with the cardial epithelium
(*cardia*) associated with the stomodael valve. Immediately posterior
to the cardia there is commonly a group of diverticula, the *gastric
caeca*. The number of these caeca varies in different species, and
similar pouches may be present in other sections of the midgut. The
remainder of the midgut, the *ventriculus,* is usually a somewhat
enlarged sac and serves as the insect's stomach. In some insects, the
midgut is divided into distinct regions; two, three, and four regions
have been identified among various true bugs (Hemiptera).

Although a number of different cell types have been described,
midgut cells (Fig. 4-3) are typically columnar with a striated border
formed by *microvilli* on the luminal side. On the hemocoel side, the
basal plasma membrane is characteristically infolded, and mitochondria are associated with these folds. These cells usually contain

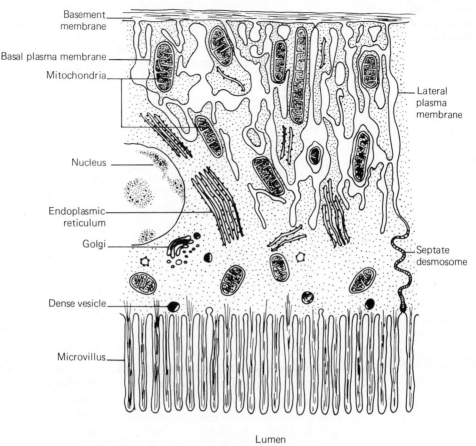

Hemolymph

Basement membrane

Basal plasma membrane

Mitochondria

Lateral plasma membrane

Nucleus

Endoplasmic reticulum

Golgi

Septate desmosome

Dense vesicle

Microvillus

Lumen

Fig. 4-3
Generalized midgut epithelial cell. [Redrawn from Berridge, 1970.]

extensive rough endoplasmic reticulum, much of which is probably involved in the synthesis of digestive enzymes. The microvilli and folded basal plasma membrane provide the extensive surface area one would expect to find in actively absorbing and secreting cells.

At the bases of the midgut epithelial cells are small *regenerative* or *replacement cells*. These cells replace the actively functioning gut cells which die or which degenerate as a result of *holocrine* secretion. In holocrine secretion a cell breaks apart completely when it releases its products. Regenerative cells may be dispersed individually among the epithelial cells (e.g., in caterpillars and true flies; Fig. 4-4A) or be concentrated in discrete groups as *nidi* (e.g., in grasshoppers and relatives, dragonflies, and damselflies) or *crypts* (e.g., in many beetles). Nidi ("nests") are groups of regenerative cells located at the bases of the active epithelial cells but are within the confines of the muscle layers (Fig. 4-4B). Crypts are packets of cells that project through the muscle layers of the gut (Fig. 4-4C).

The midgut is the principal site of digestion and absorption. It may be divided into different regions, which absorb different components of the food materials.

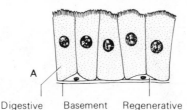

A

Digestive cell | Basement membrane | Regenerative cell

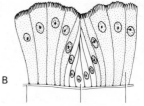

B

Basement membrane | Nidus of regenerative cells | Digestive cell

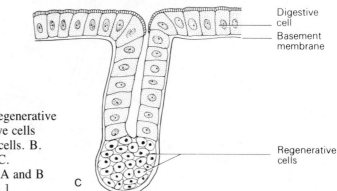

Digestive cell

Basement membrane

Regenerative cells

C

Fig. 4-4
Location and arrangement of regenerative cells. A. Individual regenerative cells scattered at bases of digestive cells. B. Regenerative cells in a nidus. C. Regenerative cells in a crypt. [A and B redrawn from Snodgrass, 1935.]

A chitinous *peritrophic membrane* (Fig. 4-1) is secreted by either the cardial epithelium or the general ventricular epithelium, or by both. This membrane (peri = around; trophic = food) surrounds the food bolus and may protect the epithelial cells from possible abrasion by food particles. That the peritrophic membrane acts in other ways is suggested by its presence in insects that do not ingest solid, potentially abrasive food particles (i.e., fluid-feeders of various types) and its absence in some insects that do. In some beetle larvae, just before pupation, the peritrophic membrane continues to be produced after feeding has ceased and the gut emptied of undigested material. This membrane is extruded from the anus and collapses and dries, forming a thread that is used like silk to form a cocoon (Kenchington, 1976).

Richards and Richards (1977) review the recent literature on peritrophic membranes in general, and Peters (1976) reviews the literature dealing specifically with the Diptera.

Hindgut

The hindgut (Fig. 4-1) is composed of cuboidal epithelial cells. It commences with the *pylorus,* which is associated with a variable number of typically slender, elongate excretory structures, the *Malpighian tubules,* and which usually contains a valvular structure, the *pyloric valve.* The Malpighian tubules can be used as landmarks indicating the beginning of the hindgut. The hindgut is divisible into a tubular *anterior intestine,* just posterior to the Malpighian tubules, and a highly musculated, enlarged *rectum (posterior intestine),* which terminates with the *anus.* The anterior intestine may be differentiated into an anterior *ileum* and posterior *colon*. The rectum usually contains a number of pads or *papillae* (usually six) that project into the lumen. These structures receive a good supply of tracheae and are metabolically very active. They play an especially important role in the excretory system.

Little or no digestion occurs in the hindgut. It serves primarily to carry undigested food material away from the midgut, ultimately egesting it from the insect. Unlike the foregut intima, the hindgut intima is permeable and allows the passage of at least relatively small molecules. Before egestion, the hindgut absorbs, to varying degrees, water, salts, and amino acids previously removed from the hemolymph by the Malpighian tubules. Thus it plays a major role in the water and salt balance of an insect.

Digestion

Although the initial stages of breakdown of ingested food may occur in the foregut, most digestion occurs in the midgut. Digestion occurs by a series of progressive enzymatically catalyzed steps, each producing a simpler substance until molecules of absorbable size or nature are produced. For example, polysaccharides are broken down into small chains, disaccharides, and finally into simple, absorbable monosaccharides (e.g., glucose); proteins, to peptones, small polypeptides, dipeptides, and finally amino acids, which are absorbable.

There is some correlation between the kinds of food material eaten by and the kinds of enzymes present in a given insect. Thus, one would expect an insect, such as a cockroach, that eats many different kinds of food to have more enzymes than an insect, such as a tsetse fly, that feeds primarily on blood. In addition, different enzymes may be secreted by different parts of the midgut epithelium. It is also likely that in some cases enzymes that act in the midgut actually originate in the salivary glands.

Several proteases, lipases, and carbohydrases have been found in insects, some of which are rather unusual. For example, cellulase has been found in several wood-boring insects. In some, microbes may furnish this enzyme, but others such as *Ctenolepisma lineata,* a thysanuran, are able to digest cellulose in the absence of intestinal microflora. Other enzymes in the unusual category include chitinase, lichenase, lignocellulase, hemicellulase, and hyaluronidase. Certain insects are able to digest ordinarily stable substances. For example, chewing lice and a few other insects are able to break down keratin, a protein that occurs in wood, hair, and feathers. Larvae of the wax moth, *Galleria mellonella,* are able to digest beeswax.

Midgut pH (typically pH 6–8; Dadd, 1970), buffering capacity, and oxidation-reduction potential are important factors in the digestive process. These factors vary from species to species and may also vary from one region of the midgut to another within the same insect.

Absorption

Although there is relatively little information available, it appears that simple sugars (monosaccharides), amino acids, and lipids diffuse down concentration gradients between the midgut lumen and hemolymph. The diffusion of simple sugars like glucose and fructose is enhanced by the rapid conversion of these sugars to trehalose in the hemolymph, a process (called *facilitated diffusion*) that maintains a concentration gradient across the gut epithelium. Although it has been generally assumed that carbohydrates and proteins must be broken down to monosaccharides and amino acids in order to be absorbed, there is some evidence that disaccharides and peptide fragments might be absorbed intact. The midgut caeca appear to be particularly active in lipid absorption.

The active transport of sodium may play a key role in the passive absorption (diffusion) of other molecules (Berridge, 1970). The plasma membranes on the hemocoel side of the midgut epithelium are considered to be less permeable than those on the luminal side. So when sodium molecules are pumped (active transport) from the midgut cells across the plasma membrane and into the hemocoel, they are replaced by sodium molecules diffusing into the midgut cells from the lumen. The movement of sodium molecules across the cells would tend to produce a water gradient between the lumen and the cells, that is, to concetrate water in the lumen. Hence, water would diffuse into the cells. Movement of water from the lumen would, in turn, tend to concentrate other molecules that would then diffuse down gradients into the cells. In short, the work necessary to produce the gradients for diffusion (a passive process) of water and other absorbable molecules would be the active transport of sodium ions.

Some cells in the midgut, especially in the posterior region, may actually secrete fluid into the lumen. This possibility, combined with evidence that the anterior midgut is a major region of absorption, suggests that there is a cycle of water movement from the hemocoel into the posterior midgut lumen (more specifically into the space between the peritrophic membrane and epithelium) and then to the anterior midgut region where water is reabsorbed. Such a cycle of water movement would allow digestion to occur along the entire length of the alimentary canal while the absorbable products of digestion are swept along in the peritrophic space to the major anterior sites of absorption (Berridge, 1970).

Absorption in the hindgut is discussed later in this chapter.

Regulation of the Alimentary System

Regulation of the alimentary system in insects involves control of food movement, control of enzyme secretion, and control of absorption. The alimentary canal is regulated in part through the action of the stomatogastric nervous system (see Chapter 3). The control of feeding is dealt with in Chapter 8. Dethier (1976) deals extensively with the regulation of the alimentary system in the blow fly, *Phormia regina*. Gelperin (1971) reviews the literature on regulation of feeding, including consideration of gut emptying.

Food is ingested by the actions of the mouthparts, cibarium, and pharynx and is typically stored in the crop. It is then released gradually, via the stomodael valve, into the midgut, where digestion and absorption occur. In most insects that have been studied, stretch receptors associated with the crop provide information to the brain (via the frontal ganglion) regarding crop distension and help prevent overfilling of this organ. In some insects, stretch receptors in the abdominal wall have a similar role.

The destination of ingested food may vary with the kind of food. For example, in female mosquitoes, sugar meals (flower nectar) are directed to the diverticula and blood meals go to the midgut. Sensilla in the roof of the cibarial pump, acting via the frontal ganglion, are thought to be involved in this so-called "switch mechanism." In other blood-feeding insects, such as tsetse flies, ingested blood goes to the crop first.

Control of passage of food from the crop to the midgut (rate of crop emptying) has been studied mainly in the cockroach, *Periplaneta americana* (Davey and Treherne, 1963), and the blow fly, *Phormia regina* (Gelperin, 1966). Passage of food from the cockroach crop is inversely related to the osmotic pressure of the food—that is, the higher the concentration of food, the slower the passage. In the blow fly, crop emptying appears to be controlled by the osmotic pressure of the hemolymph—the higher the osmotic pressure of the hemolymph, the slower the rate of crop emptying. Osmotic receptors have been identified in the wall of the cockroach pharynx, but the receptors in the blow fly have not been found.

Three mechanisms for the control of enzyme secretion in the insect gut have been suggested:

1. *Secretogogue:* a substance in the ingested material may stimulate enzyme secretion.
2. *Neural:* enzyme secretion is induced by nervous impulses initiated at some point along the gut or as a result of distension of the abdomen. This is considered unlikely since the midgut is weakly innervated and there is a definite time lag between feeding and increase in enzyme levels (Dadd, 1970).
3. *Hormonal:* stimuli inducing enzyme secretion are hormones that act via the hemolymph. There is increasing evidence for this mechanism (House, 1974b).

Absorption appears to be controlled by the availability of absorbable molecules, release of food material from the crop being so regulated that digestion and subsequent absorption occur at an optimal rate for a given circumstance.

Many insects ingest foods with a very high water content. Some of these insects (e.g., butterflies and many true flies) store the dilute food in the impermeable crop and pass it gradually to the midgut. In others (e.g., many blood-feeding insects) food may go to the midgut where excess water is rapidly excreted via the Malpighian tubules. Both mechanisms probably prevent extensive dilution of the hemolymph, and removal of water concentrates solid food, increasing the efficiency of digestion. In aphids, cicadas, and other Homoptera, the more posterior regions of the alimentary canal (posterior midgut or anterior hindgut) are closely associated with the more anterior regions (posterior foregut or anterior midgut). Such associations are called *filter chambers*. Although there is much variation in structural detail, these filter chambers apparently all facilitate the passage of water directly across the closely associated epithelial layers, bypassing most of the midgut. This action concentrates solid food in the esophagus or anterior midgut from which the food then passes into the region of digestion and absorption.

Movements of the alimentary canal that complement the actions of the digestive enzymes and aid absorption have been reported to be under neural or neurosecretory control in some insects (Miller, 1975b). In others, no neural connections can be found, and gut movements in these cases are assumed to be myogenic. Hormonal stimuli may also have a great deal to do with the rate of gut movements. In cockroaches, a hormone is released that stimulates peristalsis and also increases the heartbeat rate. Both of these effects probably aid in absorption by increasing the flow of hemolymph over the midgut cells.

Salivary Glands

Although there may be glands associated with the mandibles and maxillae, salivary glands are typically associated with the labial segment. The *salivary* or *labial glands* (Fig. 4-1), two in number, lie ventral to the foregut in the head and thorax and occasionally extend posteriorly into the abdomen. They vary in size and shape and are often described as being *acinar* (i.e., resembling a cluster of grapes). Each "grape," or *acinus,* bears a tiny duct that communicates with other similar ducts, eventually forming a *lateral salivary duct*. Lateral salivary ducts run anteriorly and merge as the *common salivary duct,* which empties between the base of the hypopharynx and the base of the labium. This region is called the *salivarium* and in some sucking insects forms a salivary syringe that "injects" saliva into whatever is being pierced. The lateral salivary ducts may communicate with *salivary reservoirs,* as in the cockroaches. Depending on stage and species salivary glands may be a single cluster of cells or be composed of two or more lobes. In some species, salivary glands are absent.

The secretory products of the salivary glands are generally clear fluids that serve a variety of functions in different insects.

1. They moisten the mouthparts.
2. They act as a food solvent.
3. They serve as a medium for digestive enzymes and various anticoagulins and agglutinins.
4. They secrete silk in larval Lepidoptera (caterpillars) and Hymenoptera (bees, wasps, and relatives).
5. They are used to "glue" puparial cases to the substrate in certain flies.

The most common enzymes found in insect saliva are amylase and invertase, although lipase and protease also have been found. Aphids secrete a pectinase that aids their mouthparts in the penetration of plant tissues. The spreading factor, hyaluronidase, which attacks a constituent of the intercellular matrix of many animals, has been found in the assassin bug. Anticoagulins and agglutinins have been found in the saliva of bloodsucking insects such as mosquitoes and tsetse flies. Anticoagulins apparently facilitate the pumping of blood by blocking the clotting mechanism of the host; however, they have not been found in all bloodsucking insects examined. Gooding (1972) reviews digestive processes in bloodsucking (*hematophagous*) insects and includes a discussion of saliva.

Since salivary enzymes are commonly injected into the tissues of host plants or animals or are applied to the surface of food materials, a certain amount of digestion often occurs external to the alimentary canal. Some predatory insects (e.g., larval dytiscid beetles) actually regurgitate digestive juices from the gut and inject them into the prey through their grooved mandibles. In this situation a large portion of digestion is extraintestinal, the tissues of the prey becoming a soupy material that is ingested through the mandibles.

Microbiota and Digestion

Under normal circumstances, all insects possess intestinal symbionts that may or may not contribute to their host's well-being. Among the intestinal microflora and microfauna are found bacteria, protozoa, and fungi of various species. In some cases, as will be discussed later, these microbes may provide substances of nutritive value (e.g., vitamins) for their hosts. In other cases these symbionts may synthesize an enzyme that enables the insect to digest substances it would otherwise be unable to digest. Leafhoppers (order Homoptera, family Cicadellidae) harbor yeasts capable of digesting starch and sucrose. Bacteria that ferment cellulose are probably always present in the alimentary tracts of wood-feeding insects. A considerable amount of cellulose fermentation probably occurs in the wood itself, but undoubtedly much also occurs in the guts of certain insect species. In some insects, such as the lamellicorn beetles (stag, bess, and scarab beetles), a specialized region of the gut houses these cellulose-fermenting bacteria. In the lamellicorn beetles a portion of the hindgut is enlarged, forming a "fermentation chamber." Wood-feeding termites similarly possess an enlarged sac in the hindgut, but in this case the sac harbors mainly cellulase-secreting flagellate protozoans required for the survival of the termites on a cellulose diet. Certain microbes may aid significantly in the processing of the food; for example, in the larvae of the blow fly, *Lucilia* sp., microbes produce an alkaline state that aids in the liquefaction of ingested animal tissues. Fungi may aid certain wood-feeding insects in the breakdown of cellulose.

Insect Nutrition

Given a source of chemical energy and the basic raw materials, insects, like other heterotrophic animals, are able to synthesize many, if not most, of the more complex molecules necessary for the maintenance of life. However, certain molecules they cannot synthesize, which are necessary in one way or another for survival and reproduction, also are part of the nutritional requirements of insects.

Not only are specific molecules required, but the quantities of each are of great importance. Too little of one may result in the impairment of some vital function of the insect. Too much of a given molecule in the diet results in excretion of the amount in excess, and since excretion is an energy-utilizing process, this is metabolically inefficient. High-energy-yielding molecules such as carbohydrates and lipids are needed in comparatively large amounts (i.e., grams per kilogram of body weight per day). In adult insects, amino acids, purines, and certain lipids are required in somewhat smaller, but still fairly large, amounts (milligrams per kilogram of body weight per day). Immature insects probably require larger amounts of amino acids since they are more actively synthesizing structural protein. Other molecules, in particular vitamins and minerals, are required in comparatively minute amounts (micrograms per kilogram of body weight per day).

Dadd (1973) and House (1974a) provide recent reviews of the literature on insect nutrition.

Amino Acids.   Amino acids are the building blocks of protein. Different insects have different requirements, depending upon which amino acids they are capable of synthesizing. Those which they cannot synthesize fall into the required category. Required amino acids may be present in the food material in a free state or more commonly in the form of protein. The qualitative and quantitative amino acid composition of a given protein determines its nutritive value.

Carbohydrates.   Carbohydrates are not considered to be essential nutritive substances for most insects, but they are probably the most common source of chemical energy utilized by insects. However, many insects (e.g., many Lepidoptera) do, in fact, need them if growth and development are to occur normally.

Lipids.   Lipids or fats, like carbohydrates, are good sources of chemical energy and are also important in the formation of membranes and synthesis of hormones. It is considered that most insects can probably synthesize from carbohydrate and protein most of the necessary fatty acids that make up the larger lipid molecules. However, some insect species do require certain fatty acids and other lipids in their diets. For example, certain Lepidoptera require linoleic acid for normal larval development. Apparently all insects require a dietary sterol such as cholesterol or ergosterol for growth and development.

Vitamins.   Vitamins include a diverse group of compounds that are required in very small amounts for the normal functioning of any animal. They are not used for energy, nor do they form part of the structural framework of the insect tissues. Vitamin A is required for the normal functioning of the compound eye of the mosquito *Aedes aegypti* (Brammer and White, 1969). Vitamin A is a fat-soluble vitamin. However, vitamins required by insects are principally of the water-soluble type (e.g., B complex vitamins and ascorbic acid). These compounds are coenzymes, working in conjunction with enzymes in specific metabolic reactions.

Minerals.   Like vitamins, various minerals are required by insects, in very small amounts, for normal growth and development. Minerals required by insects include potassium, phosphorus, magnesium, sodium, calcium, manganese, copper, and zinc. Some insects (e.g., the aquatic larvae of mosquitoes, which possess very thin-walled anal papillae; see Fig. 4-17) are able to absorb mineral ions from the water through the cuticle of these structures.

Purines and Pyrimidines.   The nucleic acids, DNA and RNA, are the molecules that carry and mediate the genetic code. Although insects are able to synthesize nucleic acids, dietary nucleic acids have been shown to have an influence on growth. For example, RNA exerts a positive effect on the growth of certain fly larvae.

Water.   All insects require water, whether it be from food, by drinking, absorption through the cuticle, or a by-product of metabolism. Insects vary greatly with respect to amounts of water needed. Some, like the mealworm, *Tenebrio molitor,* can survive and reproduce on essentially dry food. Others, for example, honey bees and muscid flies, require large amounts of water for survival. The excrement of the mealworm is hard and dry, with almost all the water having been reclaimed by the insect, while the excrement of bees and muscid flies contains large amounts of water.

Microbiota and Nutrition

Experiments involving the treatment of eggs to destroy microbes, for example, those in insect's food, have shown that some insects require the presence of certain of these microbes for normal growth and development, and in some instances, for survival. In many cases necessary microbes are passed from generation to generation. In some insects in which this occurs, the microbes are housed in specialized cells, *mycetocytes,* and the tissues composed of these cells, *mycetomes,* are associated with the gut, fat body, or, appropriately, the gonads. The latter would ensure the infection of any eggs produced, thus furnishing one of the possible means for bridging the gap between generations. Whether "hereditary" or not, microbes are commonly found in the alimentary canals of insects, often in the various diverticula of the midgut. The kinds of microbes reported to be likely contributors to their host's nutrition include yeasts, fungi, bacteria, and protzoans. These microbes probably benefit the insect in various ways. Suggestions include the possible fixation of atmospheric nitrogen; synthesis of protein from nitrogenous waste materials; and the provision of vitamins (particularly those of the B group), sterols, and amino acids.

For more information on microbiota and nutrition (and digestion) see Brooks (1963), Buchner (1965), and Koch (1967).

The Circulatory System

The circulatory system of insects is different from those of vertebrates and many other invertebrates in that the major portion of the blood or hemolymph is not found within the confines of a closed system of conducting vessels. Instead, insects have an *open circulatory system*, so called because the blood bathes the internal organs directly in the body cavity or *hemocoel*. This "blood cavity" is not a true *coelom;* that is, it is not lined entirely with mesodermal tissue. Although coelomic sacs do occur in the embryo, almost the entire hemocoel is formed from the epineural sinus and lined with ecto- and endodermal tissue (see Chapter 5). The only conducting tube is the *dorsal vessel,* which is a pulsatile structure that generally extends the length of the insect from the posterior part of the abdomen to just beneath the brain in the head. The blood of insects is commonly called *hemolymph*, a term which implies that it carries out the functions of both blood and lymph, which are distinctly different fluids in vertebrates. The epithelia of most, if not all, organs are separated from the hemocoel by a basement membrane. Thus, hemolymph probably does not bathe individual cells directly, the immediate environment of a cell being determined by the nature of the basement membrane.

Jones (1977) provides a recent, useful monograph on the insect circulatory system.

General Characteristics of the Hemolymph

Hemolymph is a clear fluid that is usually colorless or, because of certain pigments, slightly green or yellow. Outstanding exceptions are the red hemolymphs of some midge (Diptera, Chironomidae) larvae, certain species of backswimmers (Hemiptera, Notonectidae), and the horse bot fly (Diptera, Gasterophilidae), all of which contain the pigment hemoglobin. The hemolymph makes up approximately 5–40% of the total body weight of an insect, depending on the species. However, the hemocoel is not usually filled with blood, and its volume varies with the physiological state of the insect. Blood pH is usually slightly acid (between pH 6 and 7) and may vary slightly within a species. In a few insects the pH may be slightly alkaline, pH 7–7.5.

Insect hemolymph is slightly more dense than water. The specific gravity of hemolymph typically lies between 1.015 and 1.060 and is subject to increase during periods of molting (Patton, 1963). The total molecular concentration in the hemolymph is fairly high, a fact that accounts for the osmotic pressure being somewhat higher than that of mammalian blood. In the more advanced insect orders, free amino acids, organic acids, and other organic molecules play significant roles as osmolar effectors, whereas inorganic anions and cations are largely responsible for the osmotic pressure of vertebrate blood. As with vertebrates, insect blood can be separated into a fluid portion, or *plasma,* and a cellular portion, the *hemocytes.* In addition to the two basic fractions, several nonhemocytic elements may be found, including muscle fragments, free fat body cells, oenocytes, free crystals, spermatozoa, various parasitic organisms (e.g., bacteria, protozoans, and nematodes), tumor cells, and so on (Jones, 1977).

Chemical Composition of the Hemolymph

The chemical composition of the hemolymph shows considerable variation, both qualitatively and quantitatively, among the various insectan groups and is subject to variation in the same species depending upon the physiological state, age, sex, food, and so on, of the organism. Florkin and Jeuniaux (1974) provide extensive information on this topic. A brief listing of some of the major constituents that have been identified follows.

Water.   As with other organisms, water is the major component of the internal body fluid of insects. The typical water content of hemolymph, 84–92% (Buck, 1953), is comparable to that found in human blood plasma. In certain insects, water content may be much lower, perhaps less than 50%.

Inorganic Constituents.   Sodium, potassium, calcium, sulfur, magnesium, chloride, phosphorus, and carbonate comprise the major inorganic materials that have been identified in insect hemolymph. Chloride content is comparatively low relative to that in mammals, whereas phosphate, calcium, and magnesium commonly occur in somewhat higher concentration in many insects than in mammals.

In phytophagous insects like Lepidoptera and certain Hymenoptera and Coleoptera, magnesium and/or potassium replace sodium as major cations. This is moderately correlated with diet in that plants contain relatively high concentrations of these elements. Thus zoophagous (carnivorous) insects, which commonly take in large amounts of dietary sodium, do not show high levels of magnesium and potassium. Copper, iron, aluminum, zinc, manganese, and other metallic elements have been found in very small amounts in insect hemolymph.

Nitrogenous Waste Materials.   The breakdown of proteins and amino acids produces nitrogenous waste materials. In insects, these wastes are usually in the form of uric acid, which occurs in a very high concentration in the hemolymph. It is produced mainly in the fat body and is usually excreted by the Malpighian tubules. Other nitrogenous waste materials produced by insects include urea, allantoin, allantoic acid, and ammonia, the last being formed mainly in aquatic insects.

Organic Acids.   Succinate, malate, fumarate, citrate, lactate, pyruvate, α-ketoglutarate, and several organic phosphate compounds are the major organic acids found in insect blood. The first seven are formed during the Krebs cycle. It has been suggested that these acids are important in balancing the cations in the blood.

Carbohydrates. Much of the carbohydrate found in insect blood is in combination with protein, that is, as glycoprotein. A rather surprising discovery has been that α-trehalose, a disaccharide composed of two glucose molecules, is the major blood sugar in most insects. Prior to this discovery, trehalose was known only from certain plants and from the cocoons of a species of beetle. However, trehalose is not the major blood sugar in every insect; glucose, fructose, ribose, and others have also been found. Only very small amounts of glycogen have been found in the hemolymph. Gycerol has been found in quite high concentrations in certain insects exposed to cold seasons. In some cases it lowers the freezing point of the hemolymph, thus serving as a kind of insect antifreeze. Salt (1961) and Baust and Morrissey (1977) discuss this topic at length.

Lipids. Lipids occur in insect blood either as minute fat particles or as lipoprotein (i.e., lipids in combination with protein).

Amino Acids. The presence of free amino acids in the hemolymph in concentrations higher than that reported in any other animal group is one of the outstanding characteristics of most groups of insects (especially Lepidoptera, Hymenoptera, and many Coleoptera). These free amino acids may be either dietary or synthesized by the insect. They may represent an excess, derived from the diet, which is stored in the hemolymph prior to excretion, or they may serve as a reservoir of raw materials for the synthesis of protein needed in the construction of new cells during periods of growth and metamorphosis. Whatever their function(s), amino acids are major osmotic effectors, replacing in large part inorganic ions.

Proteins. The concentration of proteins in insect blood is similar to that in the blood of vertebrates and somewhat higher than that in the internal fluids of other invertebrates. Several protein fractions have been isolated, many of which are combined with other constituents (e.g., as lipoprotein and glycoprotein). A number of enzymes, among them lipase, protease, sucrase, and amylase, have been found in the hemolymph and are most active during metamorphosis. The protein pool in the hemolymph may also serve as a reservoir of raw materials and one of the sources of the free amino acids.

Pigments. Several pigments have been identified in insect blood. Among them are hemoglobin (a conjugated protein), already mentioned; kathemoglobin, derived from the blood meals taken by the bloodsucking bug, *Rhodnius*; carotene and xanthophyll in plant-eating (phytophagous) insects; riboflavin and fluoroscyanine. In insects with thin transparent cuticle, blood pigments may determine body coloration.

Gases. Both oxygen and carbon dioxide may occur in the hemolymph, usually in very low concentrations since the major route of gaseous exchange is via the tracheal system. Exceptions to this are found, of course, in those species that contain the oxygen-carrying pigment hemoglobin in their blood.

In summary, the composition of the insect hemolymph is in some respects similar to that of the internal body fluids in other animals. However, there are many blood characteristics that are rather distinctive in insects. Among the more significant of these are the high aminoacidemia (amino acids in the hemolymph), the functioning of amino acids and other organic acids as important osmotic effectors, the presence of the unique blood sugar trehalose, the fact that oxygen is usually not carried over great distances by the blood, and the relatively high concentration of uric acid in the blood.

Hemocytes

A number of morphologically distinct cells have been identified in the hemolymph of insects (Fig. 4-5). Some of these hemocytes are found in all groups of insects; others are less common, and some are quite rare, found only in a few or even a single species. Jones (1977) describes nine hemocyte types (based on light microscopy) some of which are considered to give rise to other forms. Of the nine types listed, the first seven are the most commonly found, and the first three have been found in every insect studied.

Fig. 4-5
Examples of hemocyte types. A. Prohemocyte. B. Plasmatocytes. C. Granular hemocyte. D. Spherule cell. E. Cystocyte. F and G. Oenocytoid cells. H. Adipohemocyte. I. Podocyte. J. Vermiform cell.

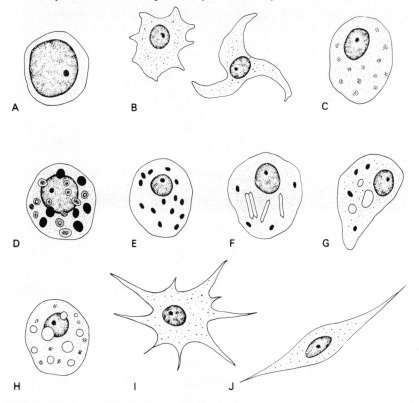

Gupta (1969) provides a listing of the synonomies of hemocyte terms used by different investigators. Information on the ultrastructure of hemocytes can be found in Arnold (1974), Crossley (1975), and Jones (1977).

A listing of the hemocytes described by Jones follows.

Prohemocytes. Prohemocytes are the smallest blood cells and contain a large round nucleus and a small amount of cytoplasmic material, which gives a basic (basophilic) stain reaction. They have been identified in all groups of insects and are commonly observed undergoing mitotic division.

Plasmatocytes. Blood cells of the plasmatocyte type are widely diverse in form (pleiomorphic); some are round, some spindle-shaped, and some have other shapes. They usually possess a single large nucleus located in the center and have a granular cytoplasm, which, as in the prohemocytes, gives a basic stain reaction. These are the dominant hemocytes in many species.

Granular Hemocytes. Like the plasmatocytes, granular hemocytes contain a large centrally located nucleus. However, this type contains a large number of inclusions of uniform size that give an acid stain reaction (acidophilic). They are thought to be derived principally from prohemocytes, although some may arise from plasmatocytes.

Spherule Cells. Spherule cells contain spherical or sometimes elliptical cytoplasmic inclusions, which commonly obscure the nucleus. The inclusions, like those in granular hemocytes, are acidophilic.

Cystocytes. Cystocytes are also sometimes called *coagulocytes* on the basis of their involvement in coagulation and plasma precipitation. Blood cells of this type have a single small nucleus with a weakly basophilic or hyaline (colorless; neither basophilic nor acidophilic) cytoplasm with distinct acidophilic inclusions.

Oenocytoids. Oenocytoids are large cells that may have one or two eccentric (not centrally located) nuclei and a weakly acidophilic cytoplasm, which may be filled with a variety of inclusions, depending on the species. They occur in various shapes.

Adipohemocytes. Adipohemocytes are round or ovoid cells that generally have distinct droplets of fat or fatlike substance in the cytoplasm. The cytoplasm also has granular inclusions.

Podocytes. A podocyte is a rare type of hemocyte that is quite large and flattened and bears a varying number of cytoplasmic extensions.

Vermiform Cells. As is well described by the term *vermiform* (in the form of a worm), vermiform cells are long and threadlike, 100–300 micrometers in diameter. They are uncommon.

Origin of Hemocytes

Embryologically, all the cells in the circulatory system are of mesodermal origin. Most postembryonic hemocyte multiplication is apparently by division of circulating cells and is a more or less continuous process (Arnold, 1974).

Prohemocytes and plasmatocytes are the cells most often observed dividing, but mitosis has also been observed in granular hemocytes and spherule cells. Jones (1977) provides a table of mitotic rates of circulating hemocytes. Many consider the prohemocytes to be the basic stem cells from which others are derived (Jones, 1964; Arnold, 1974).

In addition to the transformation of hemocytes in the blood as described above, there have been identified in several orders of insects potential *hemocytopoietic organs* (blood-cell-producing organs) in which hemocytes may multiply and/or differentiate. These organs are apparently most active during periods of metamorphosis. Jones (1977) points out the lack of experimental evidence and suggests that many so-called hemocytopoietic organs are probably artifacts.

Number of Hemocytes

Counts of the total number of hemocytes per unit volume of insect hemolymph have proved to be quite variable, depending upon species, developmental stage, various physiological states, the technique applied to obtain the count, and so on.

The number of hemocytes tends to increase during the larval instars, decline in the pupal stage, and increase initially and then decline in the adult stage. Jones (1977) provides a table listing hemocyte counts in more than 100 species of insects. Among the species listed, the number of cells per microliter ranges from 10 to 167,000 and averages about 20,000.

There is evidence that hormones may be involved in the regulation of the amount of hemolymph and also in cyclic changes in the number of circulating cells.

Functions of Hemolymph

The hemolymph is a complex mixture of a variety of materials and different cell types. This complexity is reflected in the many diverse functions of the components of the hemolymph. A brief description of some of the major functions of hemolymph follows. See Arnold (1974), Crossley (1975), and Jones (1977) for further information.

Lubricant   Hemolymph serves as a lubricant, allowing easy movement of the internal structures relative to one another.

Hydraulic Medium.   Like the hydraulic fluid in an automobile braking system, insect blood is incompressible. Thus, forces that tend to reduce the blood volume in one portion of the insect's body (e.g., compression of the abdomen) are transferred hydrostatically ("hemostatically") via the hemolymph to other parts of the body and may effect a change there. A good example of this is found among the Diptera (suborder Cyclorrhapha). These flies, when ready to emerge, literally pop the preformed lid from the end of the puparium (i.e., the hardened case in which they pass the pupal stage). This is accomplished by means of the hydrostatic extrusion of the bladderlike *ptilinum* from the anterior portion of the head.

As mentioned in Chapter 2, hydrostatic pressure may oppose muscular contraction in the extension of an appendage. In addition, ecdysis involves hydrostatic pressure, and newly emerged adult insects expand the wings hydrostatically.

Transport and Storage. Like the body fluids of other animals, insect hemolymph transports various substances from one tissue or organ to another. Substances transported include nutrients (amino acids, sugars, fats, etc.) absorbed through the gut wall or released from cells that store such materials; metabolic wastes (nitrogenous materials) and foreign materials to be excreted or taken in and stored by certain cells; possibly hormones; and, usually over very short distances, oxygen and carbon dioxide. Any of these may be carried in the plasma and, in some instances, may be carried by certain hemocytes. *Pinocytosis,* the intake of fluids or tiny particles (e.g., colloids) by minute infoldings of the cell wall, has been observed in insect hemocytes (Jones, 1977).

In addition to transport, the hemolymph serves as an important storage pool for the raw materials used in the construction of new cells.

Protection. Following are ways by which the hemolymph has been found to provide protection against invading parasites, inanimate particles, and cell fragments.

1. *Phagocytosis.* Certain hemocytes, in particular plasmatocytes and granular hemocytes, actively ingest foreign particles of various sorts, bacteria, and cellular debris. This is important in the destruction of invading Protozoa, viruses, bacteria, and so on. Phagocytic hemocytes are also important during periods of molting and metamorphosis, when many tissues are in a state of disintegration and fragments of cells freely invade the hemolymph.
2. *Encapsulation.* In certain situations, large numbers of hemocytes may actually become layered around an invading entity, such as a parasitic worm, that is too large to be phagocytized. Such a process would be significant in inhibiting the activities of a parasite, possibly by interfering with its feeding or oxygen supply. In some instances, a capsule may melanize (darken).
3. *Detoxification.* Some hemocytes are capable of rendering toxic metabolites and certain insecticidal materials nontoxic.
4. *Hemostasis, coagulation,* and *plasma precipitation.* Hemocytes are involved with inhibiting hemolymph loss at a wound site by a mechanical plugging action and/or promotion of coagulation or plasma precipitation. Mechanical plugging may

be a nonspecific function of hemocytes that are normally circulating and that settle out at a wound site. On the other hand, coagulation and plasma precipitation are usually functions of particular hemocytes, the coagulocytes (cystocytes). Coagulocytes have been observed ejecting material that induces the rapid formation of a fine granular precipitate or of veil- or threadlike networks that enmesh the cells around them. Grégoire (1974) examines coagulation and related processes in detail.

5. *Wound healing.* Hemocytes of various types (e.g., plasmatocytes and possibly spherule cells) tend to accumulate at sites of injury where they may be phagocytic, promote coagulation (coagulocytes), and form protective sheets (e.g., plasmatocytes), all of which aid healing. Plasmatocytes have been observed undergoing mitotic division in association with a wound.

6. *Noncellular protective factors.* There is evidence that implicates factors other than blood corpuscles in the protection of an insect via the hemolymph. For example, injection of toxoids or killed or attenuated microorganisms can induce immunity to subsequent infection (*acquired active immunity*). Injection of plasma from an immune insect into a nonimmune individual may protect the nonimmune (*passive immunity*). However, globulin antibodies like those in vertebrates have not been found in insects. Nonspecific protective functions have been found in the hemolymph. For example, lysozyme which kills many different bacteria has been found in members of several insect orders.

See Whitcomb, Shapiro, and Granados (1974) for an especially useful review of defense mechanisms in insect hemolymph.

Formation of Other Tissues.   Certain hemocytes may be capable of forming other tissues; however, there is little evidence available at present.

Heat Transfer.   In the sphinx moth, *Manduca sexta,* and probably most insects, the hemolymph is involved in the transfer of heat from one body region to another (see "Behavior, Temperature, and Humidity" in Chapter 8).

The Dorsal Vessel and Accessory Pumping Structures

The dorsal vessel (Fig. 4-6) is the principal organ responsible for blood circulation. It lies along the dorsal midline of the insect body, extending from the posterior region of the abdomen to just behind or beneath the brain in the head. It may be a simple straight tube or have bulbar thickenings along its length or be a complex, branching structure. It is largely composed of circular muscle, but it may also have semicircular, oblique, helical, or longitudinal fibers and may have a connective tissue sheath around the outside. Fine elastic fibers, which arise from the dorsal integument, alimentary canal, so-

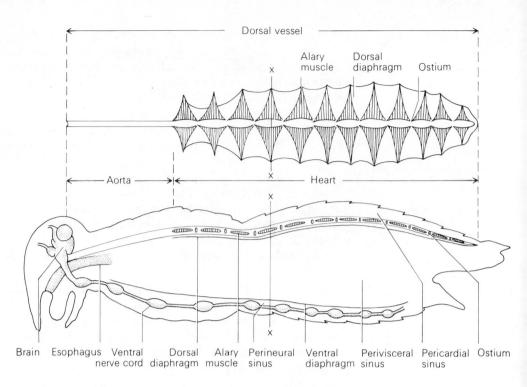

Brain Esophagus Ventral Dorsal Alary Perineural Ventral Perivisceral Pericardial Ostium
nerve cord diaphragm muscle sinus diaphragm sinus sinus

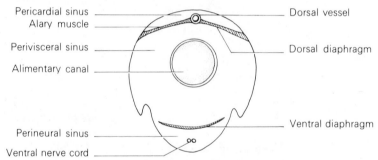

Pericardial sinus _____ Dorsal vessel
Alary muscle _____

Perivisceral sinus _____ Dorsal diaphragm

Alimentary canal _____

_____ Ventral diaphragm

Perineural sinus _____
Ventral nerve cord _____

Fig. 4-6
Generalized insect circulatory system. x—x's in top two diagrams
indicate cross-sectional plane of bottom diagram.

matic muscles, and other structures, serve as suspensors for the
dorsal vessel. The dorsal vessel may or may not be extensively
tracheated.

It is commonly possible to identify two major divisions of the
dorsal vessel (Figs. 4-6 and 4-7): a posterior *heart* and anterior *aorta.*
The heart is usually closed at its posterior end and bears a number
of valvular openings, or *ostia,* which allow hemolymph to enter
(*incurrent ostia*) or exit from (*excurrent ostia*) the heart. Ostia usually occur in laterally oriented pairs but may be dorsally or ventrally
located. The heart may be constricted between successive ostia,
giving it a chambered appearance, but its lumen is in nearly every
instance continuous throughout. On either side of the heart, in a
segmental arrangement, are the "wing" or *alary muscles,* so named
because when viewed from either a dorsal or ventral aspect, they
resemble wings. These fibromuscular structures are attached lat-

erally to the body wall and vary in number from 1 to 13 pairs, depending on the species of insect. Lateral segmental vessels associated with the heart have been identified in several insects, for example, many cockroach species. The heart may extend into the thorax but is generally confined to the abdomen. The aorta extends anteriorly from the heart and opens behind or beneath the brain. It usually lacks any valvular openings and in some insects may be variously thrown into one or two vertical loops or arches, lateral kinks, or coils. Also it may have dilated portions along its length.

The dorsal vessel may or may not be innervated or may receive nerves from the stomatogastric system and/or the ventral nerve cord and associated ganglia. For example, the dorsal vessels of crane fly (Diptera, Tipulidae) larvae, *Anopheles* sp. mosquito (Diptera, Culicidae) larvae and pupae, and cecropia moth (Lepidoptera, Saturniidae) adults do not receive any nerves. Cockroach hearts (especially *Periplaneta americana*, the American cockroach) are extensively innervated. *Lateral cardiac nerves* arise from the stomatogastric nervous system and run along either side of the dorsal vessel. These nerves contain both neurosecretory and nonneurosecretory fibers. Segmental nerves arise from the ventral nerve cord and dorsal branches supply the alary muscles *(alary nerves)* and the heart *(segmental cardiac nerves)*. Some insects fall in between extremes. For instance, the dorsal vessels of several receive segmental nerves, but receive no innervation from the stomatogastric system.

In addition to the pumping activities of the heart, various accessory pulsatile structures (Fig. 4-7) aid the movement of hemolymph. They have been found at the bases of the antennae, at the bases of and within the legs and wings, and within the meso- and metathorax.

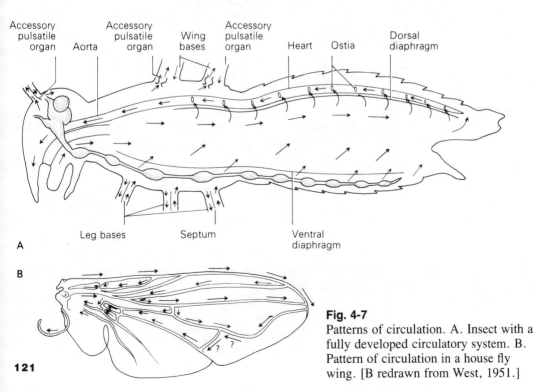

Accessory pulsatile organ · Aorta · Accessory pulsatile organ · Wing bases · Accessory pulsatile organ · Heart · Ostia · Dorsal diaphragm · Leg bases · Septum · Ventral diaphragm

A

B

Fig. 4-7
Patterns of circulation. A. Insect with a fully developed circulatory system. B. Pattern of circulation in a house fly wing. [B redrawn from West, 1951.]

Sinuses and Diaphragms

The hemocoel, or body cavity, particularly in the abdomen, is usually separated into two and sometimes three cavities or *sinuses*. This compartmentation is produced by the presence of one or two fibromuscular septa. The one that is generally present is the *dorsal diaphragm* or *pericardial septum*. It typically consists of two layers, which enclose the alary muscles associated with the heart. It may be imperforate, but is usually not, containing lateral *fenestrae* or "windows" through which hemolymph can readily pass. This diaphragm divides the dorsal *pericardial* ("around the heart") *sinus* from the *perivisceral* ("around the gut") *sinus*. In many insects, a second septum, the *ventral diaphragm,* is present. It also contains muscle fibers and is located ventral to the alimentary tract but dorsal to the ventral nerve cord. Like the dorsal septum, it is usually fenestrated around its periphery. The ventral diaphragm separates the perivisceral from the *perineural* ("around the nerve cord") *sinus*. Undulation of both the dorsal and the ventral diaphragm may aid appreciably in the circulation of blood.

Vertically oriented septa, attached dorsally and ventrally, have been identified in a few insects.

Circulation

The general pattern of circulation in insects (Fig. 4-7) can be described as follows. Blood enters the heart through the ostia. It is then directed anteriorly by a wave of peristaltic contraction that passes along the dorsal vessel in the direction of the head. The direction of propagation of the wave of peristaltic contraction has been observed to reverse on occasion. Blood usually returns to general circulation in the head. With the aid of undulatory movements of the diaphragms, action of the accessory pulsatile structures, and visceral and body movements, blood circulates throughout the general body cavity and appendages. Blood returns to the pericardial sinus through the openings in the dorsal diaphragm and enters the heart.

The alternate phases in the heartbeat cycle, contraction *(systole)* and relaxation *(diastole),* can be measured mechanically at a single point along the heart (Fig. 4-8). Local decrease in heart volume (movement of the curve upward) results from contraction, and increase in heart volume (movement of the curve downward) results from relaxation. A period of rest or *diastasis* occurs between successive beats. Toward the end of this rest period, a small but definite additional expansion may occur. This is represented by the *presystolic notch* and may be the result of contraction of the alary muscles or blood being forced from behind into the region of measurement.

Experiments with denervated dorsal vessels or fragments of vessels have demonstrated that rhythmic contraction usually continues despite the absence of exogenous stimuli (i.e., these dorsal vessels are *myogenic*). Since the waves of contraction are commonly from posterior to anterior, there must commonly be a *pacemaker* (i.e., a

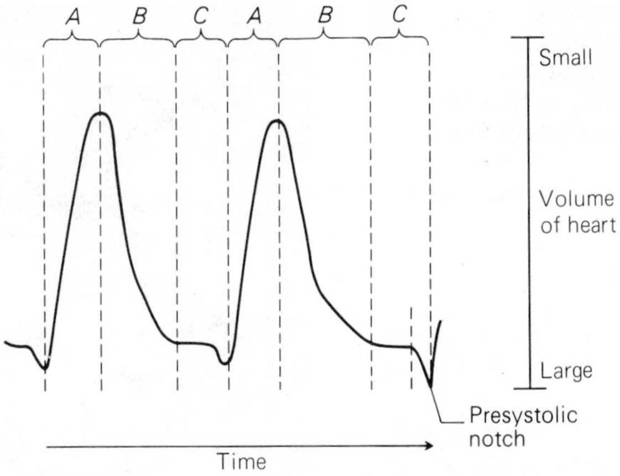

Fig. 4-8
Mechanical recording of two heartbeats (cardiac cycles) in an isolated cockroach, *Periplaneta americana*, heart. A, Systolic phase; B, diastolic phase; C, relaxation phase with presystolic notch. [Redrawn with modifications from Yeager, 1938.]

region from which waves of contraction propagate) at the posterior end of the dorsal vessel. However, since the direction of contraction waves is known to reverse and may even vary from one region to another along the same dorsal vessel, there are probably local pacemakers distributed along the vessels of some insects. In insects with innervated dorsal vessels, nervous input has been found to regulate (accelerate or decelerate as physiologically appropriate) the heartbeat rate, but these hearts are also myogenic. This situation is called *innervated myogenic*. An example is found in the cockroaches in which the lateral nerves are thought to synchronize contractions of the various heart regions and segmental nerves are thought to accelerate and possibly decelerate heartbeat rate. It is possible that there are insects in which the dorsal vessel is controlled entirely by neural input, but so far all studied have proved to be fundamentally myogenic with or without nervous input.

Evidence derived mostly from semi-isolated dorsal vessels (i.e., still in place, but completely exposed by dissecting away other viscera) suggests involvement of hormones in regulation. Secretions of both the corpora cardiaca and corpora allata have been shown to influence heartbeat rate.

For a recent review of the electrophysiology of the insect heart, a topic beyond the scope of this text, see Miller (1974).

The rate and amplitude of the heartbeat are affected by a variety of factors including intensity of activity, ambient temperature, metabolic rate, developmental stage, and the presence of biologically active chemicals, insecticides, drugs, and so on. The accessory pulsatile structures are independent of the dorsal vessel, but are influenced by the same factors.

Hemolymph pressure is a function of the amount of fluid in the hemocoel and the forces exerted by the body muscles. In hard-bodied insects, average pressure is generally quite low, although by contraction of regions of the body wall, considerable hydraulic pressure can be brought to bear locally. In soft-bodied insects (e.g., caterpillars), tonic contractions of the body musculature acting in conjunction with the hemolymph maintain body shape hydrostatically ("hydrostatic skeleton").

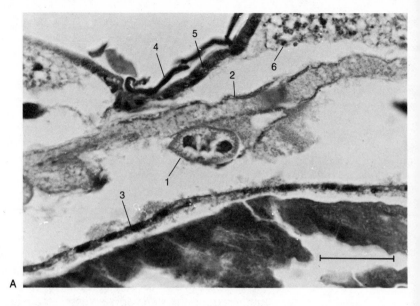

Fig. 4-9
Photomicrographs of tissues associated with the circulatory system in adult mosquito, *Aedes triseriatus*. A. Pericardial cell. B. Fat body. C. Oenocytes. 1, Pericardial cell; 2, heart; 3, midgut epithelium; 4, cuticle; 5, muscle; 6, fat body; 7, oenocytes. (Scale lines = 50 μm.)

Other Tissues Associated with the Circulatory System

Athrocytes.　*Athrocytes* are a diverse group of widely scattered cells, of mesodermal origin, in the homocoel. They are unified by the fact that they all take in particles of colloidal size from the hemolymph. They have been compared with the reticuloendothelial system of vertebrates. Some athrocytes are stationary; others circulate. *Pericardial cells* are comparatively large, sessile, usually multinucleate athrocytes found in association with the heart, alary muscles, and surrounding connective tissue (Fig. 4-9A). *Disseminated nephrocytes* are similar to pericardial cells, but are located in the perivisceral sinus.

Athrocytes are generally considered to be excretory, but it is possible that they are involved with the regulation of hemolymph composition.

Fat Body.　The fat body (Fig. 4-9B) is variously distributed in the insect hemocoel depending on the species. It has been described as a loose meshwork of lobes, composed of cells *(adipocytes)* of mesodermal origin and invested in connective tissue strands. Fat body is an unfortunate term since these cells synthesize and store proteins and carbohydrates (including glycogen and trehalose) as well as lipid.

The fat body plays a central role in the regulation of metabolism, synthesizing and releasing its products at physiologically appropriate times (Price, 1972). Especially large reserves accumulate during the larval stages in insects that later undergo complete metamorphosis. These reserves are probably important in the provision of nutrients and raw materials during the nonfeeding pupal stage. A primary function of the fat body in adult female insects is the synthesis of

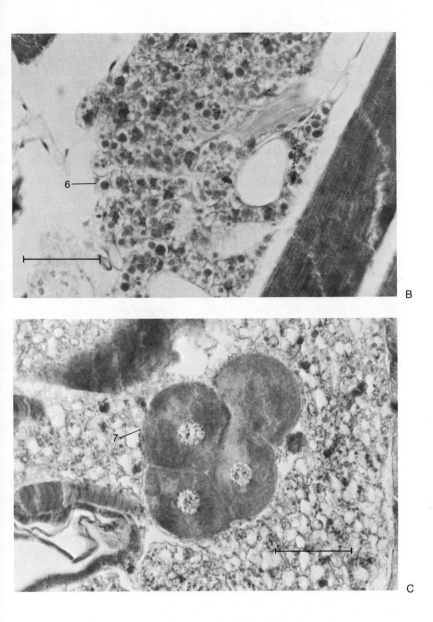

B

C

proteins *(vitellogenins* taken up by developing oocytes in the ovaries (Chen, 1978; Hagedorn and Kunkel, 1979).

The fat body is also intimately involved with the regulation of hormone activity. For example, it influences the level of juvenile hormone by releasing degradative enzymes (Nowock and Gilbert, 1976). In turn, the synthetic activities and release of products by the fat body are influenced by hormones. For instance, ecdysone induces the formation of vitellogenin in female mosquitoes (Fallon and Hagedorn, 1972).

The fat body may or may not be innervated. Fat body cells in some larval insects have been observed undergoing mitosis.

Oenocytes. These are highly specialized cells of ectodermal origin found in nearly all insects except certain Thysanura (Fig. 4-9C). They may be arranged segmentally or be dispersed randomly.

The function of the oenocytes has not been completely elucidated, although there is evidence that they secrete the lipoprotein that forms the cuticulin layer of the epicuticle (Wigglesworth, 1976b). Other functions have been postulated, but have not been supported by conclusive evidence.

The Ventilatory System

The ventilatory system *(tracheal system)* is involved with the transport of oxygen to and carbon dioxide from the various tissues of the insect body. Use of the term *ventilation* to denote this gaseous exchange seems preferable to *respiration,* which in the strict sense is an intracellular process involving the oxidative breakdown of carbohydrate to carbon dioxide and water. Miller (1974a) reviews several aspects of the ventilatory system, including ultrastructure and electrophysiological aspects. Ventilation by insect eggs will be discussed in Chapter 5. Whitten (1972) discusses the comparative anatomy of the tracheal system.

Structure of the Ventilatory System

Tracheae.   In the vast majority of insects, gaseous exchange is accomplished by a system of branching tubules or *tracheae* (Fig. 4-10). In most terrestrial and many aquatic insects these structures communicate with the outside by means of comparatively small openings, the *spiracles.* Internally the tracheae divide and subdivide, their diameters becoming successively smaller and smaller as they probe deeply into the tissues of the insect.

Tracheae are of ectodermal origin and are continuous with the integument. In the embryo they originate in each segment independently of one another. However, with the exception of a few apterygote insects, in postembryonic immature and adult stages they do not retain this early developmental arrangement, but become

Fig. 4-10
Cross section of insect thorax showing the major tracheal branches (diagrammatic). [Redrawn from Essig, 1942.]

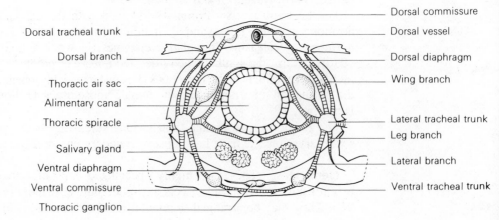

Dorsal tracheal trunk

Dorsal branch

Thoracic air sac

Alimentary canal

Thoracic spiracle

Salivary gland

Ventral diaphragm

Ventral commissure

Thoracic ganglion

Dorsal commissure

Dorsal vessel

Dorsal diaphragm

Wing branch

Lateral tracheal trunk

Leg branch

Lateral branch

Ventral tracheal trunk

joined by longitudinal trunks so that all parts of the tracheal system are in communication.

From each original developing trachea, branches are given off dorsally to the body wall, heart and aorta, various muscles, and dorsally located tissues; mesially to the alimentary tract and associated structures; and ventrally to the nerve cord, body wall, various muscles, and so on. The tracheae are circular or somewhat elliptical in cross section. Histologically (Fig. 4-11A, B) they are similar to the integument being composed of a layer of epithelial cells that secrete a cuticular layer, the *intima*, on the luminal side (i.e., the side opposite from the hemocoel). The intima is shed along with the old integument with each molt. Near the spiracles, the cuticle (which is continuous with the epicuticle of the integument) is composed of a cuticulin layer with a chitoproteinous layer beneath and possibly

Fig. 4-11
Tracheal structure. A. Portion of trachea. B. Fine structure of trachea. C. Tracheoles in close contact with muscle. D. Air sacs in the honey bee. [A and D redrawn from Snodgrass, 1963b; B redrawn from Miller, 1964; C redrawn from Wigglesworth, 1930.]

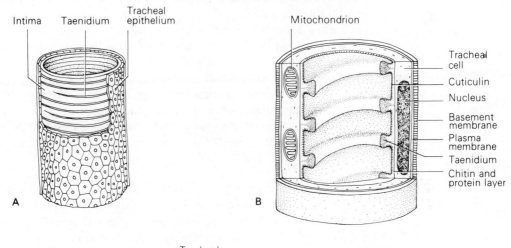

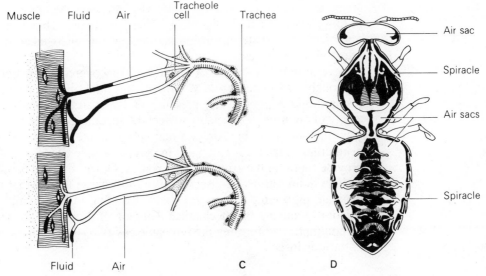

a wax layer on the luminal side. Chitin is lacking in the smaller tracheal branches. The cuticular lining of the tracheae is thrown into a series of folds that usually run spirally around the lumen. These folds are called *taenidia* and lend a measure of support to the tracheae, protecting against collapse with changes in pressure. In the smaller tracheae the folds of the intima may be annular. Although tracheae are resistant to compression in a transverse direction, they can be stretched longitudinally to some extent. These properties are important in insects in which the abdomen becomes greatly distended with food, for example, in several bloodsucking forms.

Tracheoles. The tracheoles are the smallest branches of the tracheal system, ranging in size from 0.2 to 1.0 micrometer in diameter. When examined with the electron microscope, they are seen to have very tiny *taenidia*. The lining of the tracheoles, unlike that of the tracheae, is not shed at molting. A trachea typically ends with a *tracheal end cell,* which gives rise to several tracheoles, which are all a part of this cell (Fig. 4-11C). Some tracheae merge gradually into a tracheole, and in other cases tracheoles arise directly and abruptly from the side of a trachea. Tracheoles may be involved with other tissues in a variety of ways. For example, they may intertwine and form a network around certain organs, such as ovaries and testes, or they may merely lie on the surface of some tissues, such as those of the alimentary tract. Tracheoles may also be interstitial—that is, located between cells—and may actually penetrate some muscle cells. These smallest branches of the tracheal system nearly always end blindly at a diameter of about 0.2 micrometer.

Collectively the tracheoles provide a tremendous surface area for gaseous exchange. Consider that there are 1.5 million tracheoles in a fifth instar silkworm larva!

Air Sacs. Air sacs (Fig. 4-11D) are tracheal dilations of varying size, number, and distribution found mainly in flying insects. They are typically oval in cross section, their cuticular layers lack chitin, and the walls of the cuticular lining lack oriented taenidia. Air sacs are quite distensible and collapsible. They are especially pronounced in those insects that are the most active ventilators. Thus one may observe the alternate distension and collapse as air is taken in and released by the insect.

Several functions of air sacs have been discovered; the major one is probably to increase the volume of the tidal air (the air that is inspired and expired). The presence of large air sacs in the body cavity of an insect appreciably lowers the specific gravity and in this way may aid in flight. Air sacs may also provide room for growth of internal organs; for example, in certain female flies the air sacs provide room for the growth of the ovaries. Other observed functions include aiding in heat conservation in those large insects that must generate high temperatures for flight, assisting in hemolymph circulation, reducing the mechanical damping of flight muscles by the hemolymph, and forming the tympanic cavity of the hearing organs of various insects.

Spiracles. Spiracles are the external openings of tracheae. They
occur in two basic types: simple and atriate. The simple type (Fig.
4-12A) of spiracle is merely an opening to the tracheal system. The
atriate type is formed as a result of the entad migration of the primitive (simple) spiracular opening. Thus, in the fully developed atriate
spiracle (Fig. 4-12B, C), the opening to the trachea, the tracheal
orifice, lies at the bottom of a spiracular chamber or *atrium*. In this
type the opening to the outside of an insect is referred to as the *atrial
aperture* or *orifice*.

In addition to being permeable to oxygen and carbon dioxide, the
tracheal system also readily allows the passage of water. This presents a problem for insects, since they would naturally tend to lose
water very rapidly through the integument or tracheae by evaporation, a phenomenon called *transpiration*. The development of various types of spiracular closing mechanisms has been one of the
evolutionary solutions to this problem. These closing mechanisms
are found in most atriate spiracles. Two principal types of closing
mechanisms may be found (Snodgrass, 1935): the lip type (Fig. 4-
12B), in which folds of the integument form opposing lips that can
be pulled together and effect closure of the atrial aperture, and the
valvular type (Fig. 4-12C), which lies at the inner end of the atrium
and regulates the size of the tracheal orifice.

Whatever the type of mechanism, closure is ultimately accomplished by contraction of the associated muscles. Atriate spiracles
with the second type of closing mechanism may be lined with tiny
hairs, which form a *felt chamber*, or may bear other structures, such
as *sieve plates*, which are porous covers over the atrial aperture that
probably serve to retard water loss and to prevent airborne particles
from entering the tracheae. In addition to these closing mechanisms
and accessory structures, glandular tissue is commonly associated
with spiracles. The secretions of these glands may serve as lubricants
for the movable parts of the spiracular closure mechanism, may
prevent water from entering the tracheae, and may improve the seal
of the closure mechanism. In some insects spiracular glands may
secrete a repugnant substance.

Types of Ventilatory Systems

Spiracles, tracheae, air sacs, and tracheoles compose the venti-

Fig. 4-12
Types of spiracles. A.
Simple, nonatriate
type. B. Atriate
spiracle with lip
closure mechanism. C.
Atriate spiracle with
filter apparatus and
valve closure
mechanism. [Redrawn
from Snodgrass,
1935.]

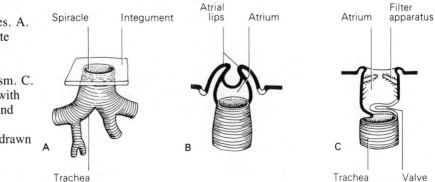

latory systems of most insects. A few, such as most Collembola, many Protura, and certain endoparasitic wasp larvae, lack tracheae, and all gaseous exchange occurs via the integument. In the remainder, tracheae and tracheoles are always present, but air sacs may be absent and spiracles nonfunctional or absent.

The organization of tracheae may be comparatively simple, as in some springtails (Collembola, Sminthuridae) and some Protura in which tracheae arise from each spiracle, but do not connect to any other tracheae. However, tracheal organization in most insects is more complex (Figs. 4-10 and 4-13). Typically there is a pair of *lateral longitudinal trunks* into which the spiracles open, a similar pair of *dorsal longitudinal trunks,* and often a pair of *ventral longitudinal trunks*. The dorsal, lateral, and ventral trunks are connected by more or less dorsoventrally oriented tracheae, and the longitudinal trunks on either side are connected by *transverse tracheal commissures*.

Although the basic pattern of tracheation is genetically determined, new tracheae and tracheoles can be induced to develop if an insect is reared in an atmosphere with a very low oxygen content. New tracheae and tracheoles do not develop between successive molts, but changes in the distribution can occur at the time of molting if there is a demand. For example, if the tracheae and tracheoles in one portion of an insect are destroyed, this region will receive tracheation from an adjacent portion in the next developmental stage.

Major tracheal branching patterns are constant within a given species and are often very similar among members of a given family or order. For this reason, tracheal patterns are sometimes of value in assessing phylogenetic relationships (Whitten, 1972).

Based on the presence or absence and functional or nonfunctional nature of spiracles, there are principally two types of ventilatory systems, open and closed, with a variety of modifications within each type.

The open ventilatory system (Fig. 4-13A, B) is found among most terrestrial insects and many aquatic forms and is characterized by the presence of one or more pairs of functional spiracles. In the more generalized insects, there are two lateral rows of ten spiracles each; the first two spiracles on each side are located on the meso- and metathorax and the following eight on the first eight abdominal segments. This is the *holopneustic* arrangement. During the evolution of the various groups of insects there has been a tendency toward reduction in the number of spiracles *(hemipneustic* arrangements), in some to the point where there is only a single pair, for example, those located at the posterior extremity of the abdomen of mosquito larvae (Fig. 4-13B).

The closed type of ventilatory system, referred to as *apneustic* (Fig. 4-13C), is found in many aquatic and endoparasitic insect larvae. There are no functional spiracles in this type of system; gaseous exchange between the atmosphere and the tracheal system occurs directly through the integument.

The Ventilatory Process

Passive Ventilation. Calculations based on measurements of

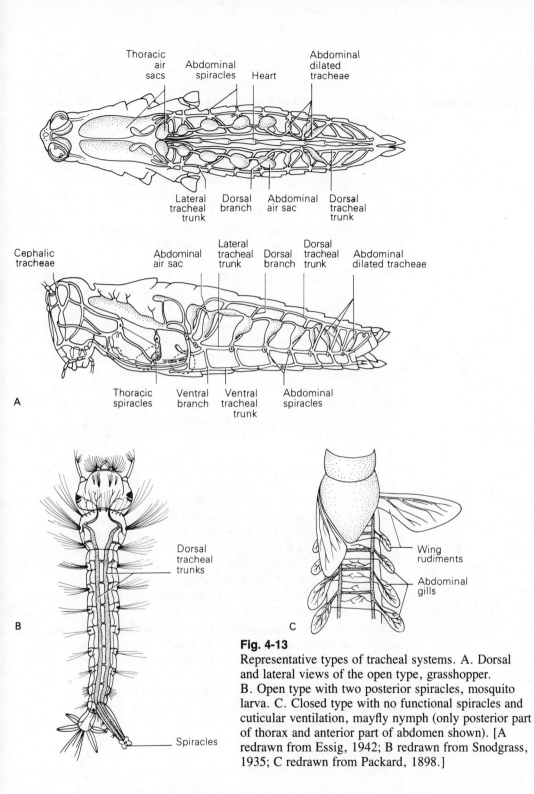

Thoracic
air
sacs

Abdominal
spiracles

Heart

Abdominal
dilated
tracheae

Lateral
tracheal
trunk

Dorsal
branch

Abdominal
air sac

Dorsal
tracheal
trunk

Cephalic
tracheae

Abdominal
air sac

Lateral
tracheal
trunk

Dorsal
branch

Dorsal
tracheal
trunk

Abdominal
dilated tracheae

Thoracic
spiracles

Ventral
branch

Ventral
tracheal
trunk

Abdominal
spiracles

A

Dorsal
tracheal
trunks

Spiracles

B

Wing
rudiments

Abdominal
gills

C

Fig. 4-13
Representative types of tracheal systems. A. Dorsal
and lateral views of the open type, grasshopper.
B. Open type with two posterior spiracles, mosquito
larva. C. Closed type with no functional spiracles and
cuticular ventilation, mayfly nymph (only posterior part
of thorax and anterior part of abdomen shown). [A
redrawn from Essig, 1942; B redrawn from Snodgrass,
1935; C redrawn from Packard, 1898.]

tracheae and on the physical properties of oxygen have shown that
simple diffusion from the outside of smaller insects and from well-
ventilated air sacs in the larger ones can supply sufficient oxygen to
the body tissues to maintain life. Simple diffusion is a passive form
of ventilation, that is, one in which no pumping or other movements

aid the passage of gases in the tracheae and tracheoles. Use of the phrase *passive ventilation,* however, does not imply that insects ventilating exclusively in this manner do not control diffusion. Diffusion control is accomplished by the opening and closing of the spiracles. Thus, in addition to their function in the prevention of excessive water loss from the tissues, the spiracles are able to regulate to some degree the entrance and exit of gases. The spiracles respond to decreased oxygen in the air by remaining open for longer periods of time. Increase in the carbon dioxide content of air produces a similar effect. Spiracular muscles receive nerves from the segmental ganglia of the ventral cord, and there is evidence that opening and closing are under both nervous and hormonal control. Humidity and water balance may also be involved in spiracular control.

Bulk Flow and Active Ventilation. As with hemolymph circulation, general movements of the body and viscera no doubt aid incidentally in the movement of gases (bulk flow) in the tracheal system. In larger insects passive ventilation and general body movements may be inadequate to supply the needs for oxygen when demand is high, for example, during flight. In these insects, air sacs, if present, and larger tracheae are often ventilated by rhythmical pumping movements of the body, *active ventilation.* Movements that are known to cause the inspiration and expiration of gases include peristaltic waves over the abdomen, telescoping or dorsoventral flattening of the abdomen, and, in some, movements in the thorax or even protraction and retraction of the head. The elasticity of the cuticle is also thought to play a part, especially in expiration. In addition, muscular movements other than those already mentioned, such as heartbeat and movements of the gut, may assist in ventilation by pressing against adjacent tracheae. Whatever type they may be, these pumping movements acting via the incompressible hemolymph renew the air in the tracheae and air sacs. Tracheae that are circular in cross section do not respond to compressive forces but, as mentioned, are quite extensible. Both tracheae that are oval in cross section and air sacs are collapsible and hence can serve to increase the volume of tidal air.

In certain insects, ventilatory movements and opening and closing of the spiracular valves are coordinated, sometimes producing a unidirectional flow of gases through the body, for example, into the thoracic spiracles and out the abdominal ones (Fig. 4-14). This coordination mechanism has been shown to be under neural control. As in simple spiracular opening and closing, oxygen deficiency and carbon dioxide excess can both serve as regulatory stimuli of pumping movements.

Passive Suction Ventilation and Elimination of Carbon Dioxide. The rate of diffusion of carbon dioxide through air is not too much different from that of oxygen, but in tissues carbon dioxide diffuses about 35 times more rapidly than oxygen. This being the case, carbon dioxide is much more likely to be eliminated from the body through the tracheal linings and integument than is oxygen to be absorbed along the same routes. Thus, although most of the

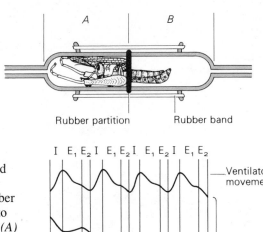

Rubber partition   Rubber band

Fig. 4-14
Apparatus used to
demonstrate directed
ventilation in a
grasshopper. A rubber
partition was used to
isolate the thoracic *(A)*
from the abdominal *(B)*
spiracles. I, inspiration,
thoracic spiracles open,
abdominal spiracles
closed; E^1, initial part of
expiration, all spiracles
closed; E^2, final part of
expiration, abdominal
spiracles open. [Redrawn
from Fraenkel, 1932;
apparatus drawn from
photograph.]

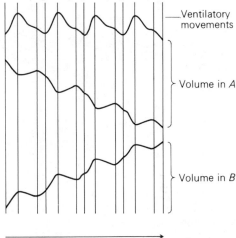

$I \quad E_1 \, E_2 \, I \quad E_1 \, E_2 \, I \quad E_1 \, E_2 \, I \quad E_1 \, E_2$

Ventilatory movements

Volume in *A*

Volume in *B*

Direction of air flow

carbon dioxide produced by respiration is eliminated via the tracheae
and tracheoles, some of it may escape through the general body
surface of soft-bodied insects and the intersegmental membranes of
those with hard bodies.

Many insects have been found to eliminate carbon dioxide through
the spiracles in regular "bursts," while oxygen consumption remains constant (Fig. 4-15). Between these bursts the spiracles remain closed (but apparently not sealed) or "flutter" between slightly
open and closed. The spiracles open completely during a carbon
dioxide burst. Presumably, as oxygen is removed from the tracheoles
and tracheae by respiration, at least a portion of the carbon dioxide
produced goes into solution as bicarbonate in the hemolymph. This
would cause the development of negative pressure in the tracheae
and tracheoles. As a result, air would be sucked in (bulk flow)
through closed (but not sealed) or fluttering spiracles, a process
called *passive suction ventilation*. A carbon dioxide burst probably
indicates the previous buildup of carbon dioxide (in the hemolymph
and tracheae) to a threshold, above which complete spiracular opening occurs. Spiracular opening would then allow equilibrium among
the atmosphere, the tracheae, and the hemolymph to be reached.

Passive suction ventilation has been observed particularly in insects that are inactive owing to a low ambient temperature or to
being in a dormant developmental stage. This phenomenon has also
been induced experimentally in insects in which it does not normally
occur by lowering the temperature and by injuring certain parts of
the brain. The ability to release carbon dioxide periodically in this

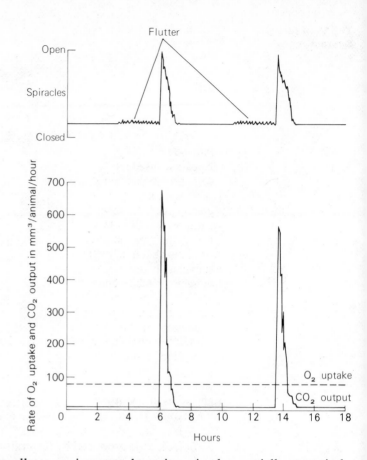

Fig. 4-15
Graph of oxygen uptake and
carbon dioxide output in a
diapausing cecropia moth,
Hyalophora cecropia, pupa at
25°C. [Based on Schneiderman
and Williams, 1955.]

manner allows an insect to keep its spiracles partially or entirely
closed most of the time and hence is thought to be an adaptation that
favors the conservation of water by diminishing the rate of transpiration. In addition, the bulk flow of air into the tracheae would tend
to retard the outward flow of water vapor.

Passive suction has also been observed in mosquito larvae and
pupae and enables these aquatic insects to inspire air rapidly with
only brief contact with the surface.

Oxygen Transport in the Hemolymph.   Except over the very
short distances between most tracheolar endings and the cells they
oxygenate, the hemolymph does not generally function as an oxygen
carrier. However, as mentioned earlier, certain chironomid (Diptera,
Chironomidae) larvae (bloodworms), the endoparasitic bot fly larva
(*Gasterophilus,*) and the backswimmers *Anisops* and *Buenoa* (Hemiptera, Notonectidae) have the oxygen-carrying pigment hemoglobin in the hemolymph. Under conditions of high oxygen tension this
hemoglobin is saturated with oxygen and thus does not serve as a
carrier, but under conditions of low oxygen tension, a common occurrence in the bloodworm's natural habitat (stagnant water), the
hemoglobin is unsaturated and hence available to carry oxygen.
Horse bot fly larvae ventilate periodically when they come into contact with an air bubble in the intestine of their host. Hemoglobin
permits these insects to take in a larger supply of oxygen than they

could otherwise. Through increasing the amount of oxygen that can be stored, hemoglobin enables the aquatic backswimmers to remain submerged for long periods of time.

Ventilation in Aquatic Insects. Many insects spend all or part of their lives in an aquatic environment. These insects must either be able to utilize oxygen in solution or have some means of tapping a source of undissolved oxygen whether it be at an air–water interface (at the water's surface or in the form of submerged bubbles) or from aquatic vegetation. A wide variety of structural and physiological adaptations associated with ventilation in an aquatic environment have evolved. Mill (1974) considers aquatic insect respiration in detail.

Aquatic insects with closed ventilatory systems (which are found only in immature forms) depend entirely upon the diffusion of dissolved oxygen through some region of the integument. However, even in many of these insects the tracheal system is involved. These insects obtain oxygen in a variety of ways. Many possess *tracheal gills* (Fig. 4-16A), which are integumental evaginations covered by a very thin cuticle and are well supplied with tracheae and tracheoles. Such gills may be found anywhere on the body, but are commonly found along the abdomen. In some insects they are located at the posterior end of the abdomen (Fig. 4-16B). Thin areas of the integument may be well supplied with tracheae and tracheoles and function in a similar fashion. Other aquatic insects with closed ventilatory systems possess *spiracular* or *cuticular gills* (Fig. 4-16C; see Hinton, 1968). These are filamentous outgrowths, consisting mostly of very thin cuticle (about 1 micrometer thick), that open directly into the tracheae.

Many aquatic insects with closed ventilatory systems lack any specialized gill structures and depend upon diffusion of oxygen across the general body surface, a process called *cutaneous ventilation*. In fact, all forms with or without gills probably depend, to a greater or lesser extent, on cutaneous diffusion of oxygen.

Most aquatic insects have open tracheal systems and usually obtain oxygen at the water's surface. They are of two general types: those that must surface periodically and depend at least in part on atmospheric oxygen and those that may remain submerged for an indefinite period of time and are somewhat independent of atmospheric oxygen. These two groups are rather indefinite, with intermediates between them. In addition, cutaneous ventilation may under certain circumstances play a role in some of these insects.

Most of the members of the group of aquatic insects that must surface (e.g., most mosquito larvae; Fig. 4-13B) possess hydrofuge structures. These structures are generally associated with particular spiracles and are highly variable from insect to insect. However, all have essentially the same function, the breaking of the surface film of the water and by this action exposing associated spiracles to the atmosphere. They also serve to keep water out of the tracheae when the insect is submerged. Hydrofuge structures are usually made up of hairs and are resistant to "wetting" by water. Thus, when an

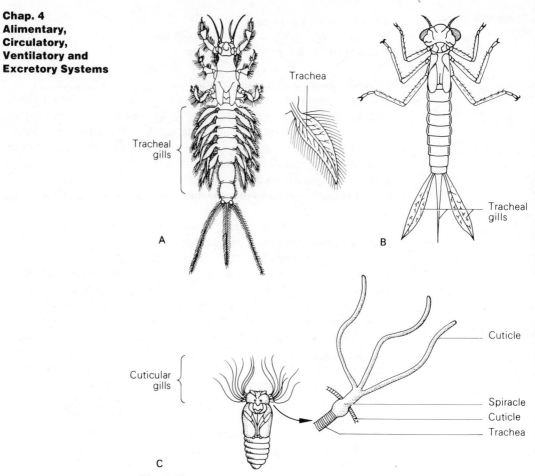

Fig. 4-16
Ventilatory structures in aquatic insects. A. Lateral abdominal tracheal gills in a mayfly nymph. B. Terminal abdominal tracheal gills in a damselfly nymph. C. Cuticular gills on the thorax of a black fly pupa. [A and B redrawn from Boreal Labs key card; C redrawn from Packard, 1898.]

insect approaches the surface, the cohesive properties of water cause it to be drawn away from the hydrofuge areas.

Many insects with open tracheal systems—aquatic Hemiptera and adult Coleoptera, for example—carry air stores in the form of bubbles or films into which spiracles open. These air stores are often held in place by a pile of erect hydrofuge hairs, but in some insects the body is so shaped that it forms a storage area without the use of hydrofuge hairs. Films or bubbles of air would obviously be a temporary source of oxygen if the insect were forced to remain submerged. How long an insect could survive beneath the surface by utilizing stored oxygen depends on a number of factors, but unless it has some way of replenishing the store from oxygen dissolved in the water, its survival time is not likely to be very long. The mechanisms involved in replenishing air stores vary. Water scorpions (Hemiptera, Nepidae), for example, use a caudal siphon, a long hollow tube extending from the rear of the body.

Many aquatic insects that carry stores of air are able to replenish the oxygen without surfacing. This is accomplished by the air store acting as a "physical gill." As the oxygen in reserve is used up, a point is reached where the partial pressure of oxygen is less in the air store than it is in the surrounding water. At this point oxygen diffuses from the water into the air store. Nitrogen in the air store does not readily diffuse into the water and hence tends to keep the air store from collapsing.

Air stores usually make an insect positively buoyant and may play a role in hydrostatic balance. By decreasing specific gravity air stores may also reduce the amount of energy expended in locomotion.

Insects that are able to remain submerged indefinitely usually possess a structure known as a *plastron* or obtain their oxygen from submerged vegetation. A plastron is a very thin layer of gas held firmly in place by tiny hydrofuge hairs or other very fine cuticular networks. The latter are typically associated with spiracular gills. Hair plastrons are found on adults of certain aquatic beetles (e.g., *Elmis*), nymphs and adults of the aquatic bug *Aphelocheirus aestivalis,* and adult females of the wingless lepidopteran *Acentropus niveus*. Plastrons composed of cuticular networks are found in the larval and/or pupal stages of certain beetles and flies. Unlike typical air stores, the gas layer held by a plastron cannot be displaced by water. The spiracles open into the plastron, and it functions in a manner similar to a physical gill except that it does not require repletion by a visit to the surface.

Insects that obtain oxygen from submerged vegetation may do so in a variety of ways. Many are able to capture bubbles on the surface of plants by means of hydrofuge structures. Others penetrate the tissues of submerged plants by biting into them or by inserting a specialized ventilatory structure into the intercellular air spaces. Examples include certain mosquito larvae (Fig. 4-17), other flies, and some larval beetles.

Fig. 4-17
Ventilatory apparatus adapted for penetration of aquatic plants in the mosquito *Mansonia perturbans*. The anal papillae are involved in osmoregulation. [Redrawn from Matheson, 1944.]

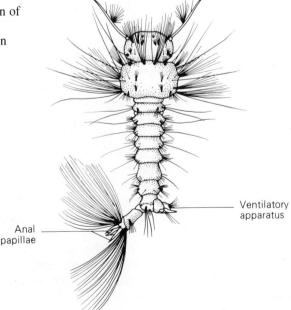

Ventilatory
apparatus

Anal
papillae

Aquatic insects may facilitate gaseous exchange by various movements such as rhythmically undulating the tracheal gills and/or undulating the body. Such movements bring oxygen-fresh water to sites of absorption.

Ventilation in Endoparasitic Insects. Endoparasitic insects are those that invade the tissues of their host, as opposed to external parasites. The great majority of the insects in this group are endoparasitic only in the immature stages. The environment of these insects presents problems similar to those of the aquatic environment in that the insects must either obtain oxygen in solution or from the atmosphere or both. In several, the tracheal system is nonfunctional and ventilation is cutaneous, gaseous exchange occurring directly between the tissues of the parasite and body fluids of the host. Some endoparasitic insects have tracheal gills similar to those found in insects inhabiting aquatic environments. Others depend, at least partly, on atmospheric air, obtaining it either by tubes or other structures that communicate with the tracheal system and that extend out of the host to the atmosphere. As with most aquatic forms, cutaneous ventilation plays a greater or lesser role in most endoparasites.

The Excretory System

The function of the excretory system is the maintenance of a more or less constant internal environment. Since the hemolymph bathes the tissues and organs of the insect body, it largely determines the nature of this internal environment. Thus the excretory system is responsible for the maintenance of the uniformity of the hemolymph. It accomplishes this by the elimination of metabolic wastes and excesses, particularly nitrogenous, and the regulation of salt and water (Maddrell, 1971; Stobbart and Shaw, 1974). The Malpighian tubules are the major organs of excretion, although other tissues may play a role in some instances.

Malpighian Tubules

Malpighian tubules (Figs. 4-1 and 4-18) are usually long slender tubes closed at their distal ends and found in association with the posterior portion of the alimentary canal. They are named after their discoverer, Marcello Malpighi, a seventeenth-century Italian scientist. These tubules are commonly convoluted and vary in number, depending on species, from 2 to 250 or more, usually occurring in multiples of two. They are apparently lacking only in members of the order Collembola (springtails) and the family Aphididae (aphids; suborder Homoptera). Malpighian tubules lie at the junction between the mid- and hindgut. However, their embryonic origin is a matter of controversy, and they may open directly into the midgut or hindgut or more commonly into a dilated ampullar structure.

The Malpighian tubules are usually free in the body cavity and are bathed directly by the hemolymph. At least some are always in close proximity to the fat body and parts of the alimentary canal,

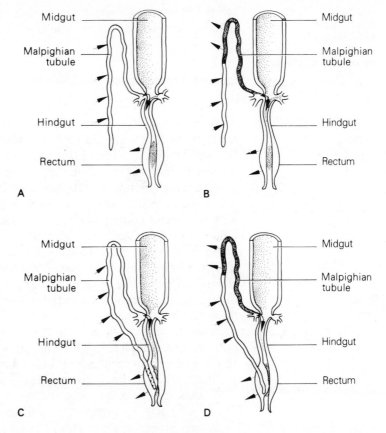

Fig. 4-18
Major types of Malpighian
tubule–hindgut systems. A.
Orthopteran type. B.
Hemipteran type. C.
Coleopteran type. D.
Lepidopteran type. Arrows
indicate directions of
movement of substances in
and out of the tubule
lumen. [Redrawn from
Patton, 1963.]

other than where they open into it. In some groups of insects, for
example, most Lepidoptera (butterflies and moths) and Coleoptera
(beetles), the distal ends of the tubules are embedded in the tissues
surrounding the rectum (Fig. 4-18C, D). This is the *cryptonephridial*
(''hidden kidney'') arrangement. In some insects, the distal ends of
two tubules anastomose, forming a closed loop. In addition, there
may be anatomical differences between the tubules within the same
insect.

In many insects the Malpighian tubules are capable of a variety
of movements by contraction of variously oriented muscles associated with the tubules. Apparently only members of the orders Thysanura (silverfish and relatives), Dermaptera (earwigs), and
Thysanoptera (thrips) do not have muscles associated with the Malpighian tubules. The function of the tubule movement is currently
a matter of conjecture. Suggestions have included such functions as
propulsion of the contents of the lumen toward the opening into the
alimentary canal, mixing of the luminal contents, and exposure of
the tubules to more blood. The tubules are usually well tracheolated
and are one cell thick. A basement membrane lies beneath the muscles and surrounds the tubule cells. The cytoplasm of these cells
varies in appearance. It is usually colorless but may have a faint
green or yellow appearance. It is generally filled with various refractile or pigmented inclusions and sometimes contains needlelike
crystals but may be nearly clear. The tubule cells communicate with
the lumen by means of a finely striated border. The electron micro-

scope has revealed a very large number of mitrochondria (consistent with the occurrence of active transport) in the cytoplasm of the tubule cells.

Salt and Water Balance

Different environmental situations pose different salt and water problems for insects. Terrestrial forms are constantly faced with the tendency to lose water through transpiration and are generally dependent on ingested food for needed water and salt. Depending on the water content of their diet, the fecal material may be quite watery, as from plant-feeding insects that take in an excess of water, or a dry powdery pellet, as from those insects that feed on materials of very low water content. Freshwater insects must excrete the large amounts of water absorbed through the integument and by the gut along with ingested food and at the same time must conserve the inorganic ions. Saltwater forms, like terrestrial insects, must constantly overcome the tendency to lose water, in this case due to a difference in concentration between the insect's internal fluid and the surrounding medium.

In the majority of insects the regulatory problems outlined in the preceding paragraph are, at least in part, solved through the activities of the Malpighian tubules and the rectum in the hindgut. The Malpighian tubules are freely permeable to most small molecules. These molecules may be passed into the lumen of a tubule by simple diffusion or by active transport of potassium ions, which generates fluid flow. Secretion of water and ions is not a selective process. However, substances that are required are reabsorbed into the hemolymph either in the proximal portions of the tubules in some insects and/or in the rectum. These reabsorption processes may also be active transport or simple diffusion. It has been suggested that it is more energy-efficient for organisms to develop, in the evolutionary sense, ways of reabsorbing needed substances than to develop ways of excreting every possible type of unneeded substance.

Absorption of materials in the rectum is considered to be a function of the anal papillae described earlier in this chapter (alimentary system). Ultrastructural and histochemical studies of these papillae have led to the development of a hypothetical explanation of the mechanism of absorption (Berridge, 1970). Without going into unnecessary detail here, the mechanism proposed involves the creation, by the active transport of ions, of gradients down which water, ions, and other molecules can passively diffuse. The proposed mechanism also, interestingly, provides a means for some molecules to move into the hemolymph against a concentration gradient.

Active transport is probably not always involved in rectal absorption. For example, in *Dysdercus* (Hemiptera, Pyrrhocoridae), absorption is entirely passive and occurs only when the rectal fluid is hypotonic relative to the hemolymph.

Evidence is accumulating that excretion is influenced by diuretic hormones. For example, the release of a diuretic hormone is induced by abdominal stretch receptors responding to the intake of a blood meal by the bug *Rhodnius prolixus* (Hemiptera, Reduviidae). Diuretic hormones have been extracted from various ventral chain ganglia, from neurosecretory cells in the pars intercerebralis of the brain, and from the corpus cardiacum.

A number of Malpighian tubule–rectal cycling systems have been described. In the simplest situation, found in members of the order Orthoptera (grasshoppers and relatives), the tubules are composed of a single cell type and have only fluid in the lumen (Fig. 4-18A). This fluid passes down the lumen and mixes with the gut contents, and as it passes down the hindgut, particularly in the rectum, needed water and ions are reabsorbed into the hemolymph. A more complex cycling system is found in most Hemiptera (Fig. 4-18B). In this case movement of materials into the lumen of a tubule occurs in the distal portion; reabsorption of needed water and ions into the hemolymph occurs in the proximal portion and in the rectum. A third system is that typically found in beetles (Fig. 4-18C), in which the distal portion of a Malpighian tubule is embedded in the tissues surrounding the rectum. This is the cryptonephridial arrangement mentioned earlier. Secretion into the lumen of a tubule apparently occurs in its exposed portion, whereas reabsorption of needed materials likely takes place through the portion embedded in the rectum. It has been suggested that cryptonephridiae increase the efficiency of the cycling process.

A cryptonephridial arrangement is also typical of members of the order Lepidoptera (Fig. 4-18D) except that in these reabsorption of needed materials into the hemolymph occurs in the proximal portions of the tubules in addition to in the rectum. Other situations have been described, but those mentioned here should give a general idea of the processes involved. It should be pointed out that factors such as spiracular control, integument permeability, food selection, and habitat selection are also involved with the regulation of salt and water in insects. Also, in certain aquatic insects there exist mechanisms for the uptake of ions other than via the gut. For example, chloride ions were taken into the hemolymph of mosquito larvae by way of four papillae, which surround the anus (Fig. 4-17). This is an active process, occurring against a rather severe concentration gradient. In addition, these papillae are responsible for sodium, potassium, and water uptake.

Nitrogenous Excretion

Nitrogenous products of various types tend to accumulate in the hemolymph as a result of protein, amino acid, and nucleic acid metabolism. These materials are generally of no use to an insect and may be toxic. This being the case, they must either be excreted or stored in an inert situation until they can be excreted. Uric acid is the major nitrogenous waste product formed and excreted, making up 80% or more of the nitrogenous end products found in the urine of most terrestrial and many aquatic insects. It does not require a large amount of water for its elimination and is thus quite appropriate in insects in which water conservation is a problem. On the other hand, ammonia is the major nitrogenous waste formed by many aquatic insects and by blow fly larvae. However, it is highly toxic and requires large amounts of water for its elimination. Other nitrogenous products that have been identified in insect urine are allantoin, allantoic acid, and urea. These arise from the breakdown of uric acid and are usually present in very small amounts. Amino acids are sometimes found but usually in very small quantities.

The Malpighian tubules are the major organs involved in the excretion of nitrogenous materials, but other tissues may be involved to a greater or lesser extent depending upon the species concerned. As mentioned earlier, the nitrogenous wastes may be stored by certain hemocytes until they can be eliminated. This occurs most commonly in the fat body in the *urate cells,* but also in certain other tissues.

Insect Urine

The nature and composition of insect urine is, not surprisingly, highly variable. Its physical appearance varies from the dry powdery material egested by terrestrial insects that inhabit dry environments to a clear fluid in those in which water conservation is not a problem (e.g., plant-feeders and freshwater forms). Its chemical composition is dependent upon the dietary substances that are in excess of an insect's needs or not usable by the insect, the nitrogenous wastes formed as a result of protein and amino acid metabolism, and the excesses of salts and water that may occur.

Selected References

GENERAL

Bursell (1970); Chapman (1971); Rees (1977); Richards and Davies (1977); Rockstein (1978); Roeder (1953); Smith (1968); Snodgrass (1935); Wigglesworth (1972).

ALIMENTARY SYSTEM AND NUTRITION

Berridge (1970); Brues (1946); Dadd (1970, 1973); Dethier (1976); Gelperin (1971); Gilbert (1967); Gooding (1972); House (1974a b); Miles (1972); Miller (1975a, b); Richards and Richards (1977); Peters (1976); Treherne (1967); Waterhouse (1957).

CIRCULATORY SYSTEM AND FAT BODY

Arnold (1974); Crossley (1975); Florkin and Jeuniaux (1974); Grégoire (1974); Hagedorn and Kunkel (1979); Jones (1964, 1977); Miller (1974); Price (1972); Whitcomb et al. (1974); McCann (1970).

VENTILATORY SYSTEM

Hinton (1968); Mill (1974); Miller (1974a); Whitten (1972).

EXCRETORY SYSTEM

Berridge (1970); Cochran (1975); Maddrell (1971); Stobbart and Shaw (1974).

5

Reproduction and Morphogenesis

In this chapter we consider the systems and processes involved in reproduction and then discuss the major events that begin with the newly fertilized egg (*zygote*) and terminate with the death of the insect. The behavioral aspects of reproduction are discussed in Chapter 8.

Reproductive System and Gametogenesis

Reproduction in most insects is bisexual. The male reproductive system functions in the production, storage, and delivery of spermatozoa; the female system produces and stores eggs, stores spermatoza, is the site of fertilization, and deposits eggs or larvae when appropriate. As will be seen, the male and female systems are similar in many respects.

General reviews of this topic include Adiyodi and Adiyodi (1974), Davey (1965), de Wilde and de Loof (1973 a, b), Engelmann (1970), and Highnam (1964). Matsuda (1976) contains descriptions of both internal and external aspects of male and female reproductive systems in all orders of insects.

Male Reproductive System, Spermatogenesis, and Spermatozoa

The male reproductive system (Fig. 5-1) is located in the posterior portion of the abdomen and typically consists of paired gonads (*testes*) connected by various ducts, which ultimately open into the intromittent organ, the *penis* or *aedeagus. Accessory glands* of various sorts are usually associated with the ducts. Although fundamentally similar, this system varies among the different species of insects. The following description is of a generalized male reproductive system with comments on variations.

Testes. Although the testes are usually bilateral, paired structures, they have undergone a medial fusion in some insects (e.g., Lepidoptera). Basically each testis is composed of a varying number of *testicular follicles* (sperm tubes), which are usually encased by a layer of connective tissue. Each follicle is in turn enclosed by a layer of epithelial cells, which are thought to serve a trophic (nutrient) function, absorbing nutrients from the hemolymph and making them available to the germ cells within.

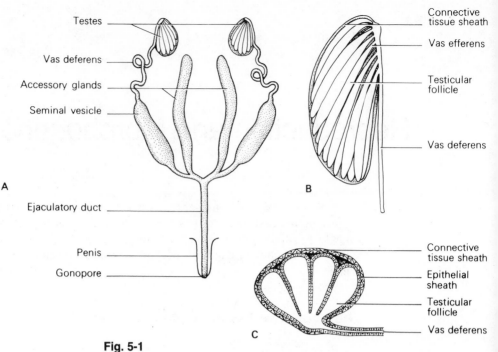

Fig. 5-1
Generalized male reproductive system. A. Principal male organs. B.
Detailed structure of a testis. C. Section of a testis. [Redrawn from
Snodgrass, 1935.]

Ducts. A tiny duct, the *vas efferens,* lead from each follicle to
a common lateral duct, the *vas deferens,* and finally the vasa deferentia from each testis join to form the *ejaculatory duct,* which ends
in the penis or aedeagus at the *gonopore.* The vasa deferentia are at
least partly of mesodermal origin. In some insects (Protura, Ephemeroptera, and some Dermaptera) each vas deferens has its own opening to the exterior. Both the vas deferens and ejaculatory duct are
invested with a layer of muscles, which are involved in the propulsion of semen. Each vas deferens may be formed into a series of
convolutions forming an *epididymis* or have a dilated portion (i.e.,
seminal vesicle) in which the spermatozoa are stored in a quiescent
state.

Accessory Glands. Various glandular structures are typically
associated with the vasa deferentia or with the ejaculatory duct,
although some insects (e.g., Apterygota and some Diptera) lack
accessory glands. These accessory glands and their ducts are either
of mesodermal origin (i.e., evaginations of the vasa deferentia) or
of ectodermal origin (i.e., evaginations of the ejaculatory duct).
They usually occur as a single pair, but in some insects there may
be several in a cluster (e.g., the mushroom body in male cockroaches). In some insects portions of the vasa deferentia or ejaculatory duct may also have glandular functions.

Among the known functions of male accessory glands is the secretion of seminal fluid, activation of spermatozoa, and production
of spermatophores. In addition, accessory gland secretions may in-

fluence an inseminated female in a variety of ways including: stimulation of oviposition; acceleration of oocyte maturation; stimulation of contractions of the genital ducts that aid in sperm movement; and inhibition of subsequent insemination by formation of vaginal plugs or by exerting an effect on the female's behavior. Hinton (1974) and Leopold (1976) discuss functions of male accessory gland secretions at length.

There is evidence for hormonal control of seminal fluid production and growth of the accessory glands. Removal of the corpora allata from young males of several species retards accessory gland growth.

Spermatogenesis. The testicular follicles contain the *germ cells* and are hence the sites of the meiotic cell divisions that give rise to *spermatozoa*, the entire process being referred to as *spermatogenesis*. This process usually occurs during the last larval instar or pupal stage and in some species continues in the adult stage. Each follicle contains a large apical cell or complex of cells that apparently serve a trophic function, providing nutrients for the developing *spermatogonia*. Each follicle is divided apically to basally into zones that represent the different stages of spermatogenesis (Figs. 5-2 and 5-3). Apically the *germarium* or *zone of spermatogonia* is comprised of the germ cells (spermatogonia) and somatic mesodermal cells. In the next region, the *zone of growth* or *zone of spermatocytes,* the spermatogonia underto several mitotic divisions, forming *primary spermatocytes* that become encysted in somatic cells. The primary spermatocytes (diploid cells) undergo meiosis and produce haploid daughter cells in the next region, the *zone of maturation and reduction*. With the first and second meiotic divisions the primary spermatocytes become *secondary spermatocytes* and *spermatids,* respectively. In the basal *zone of transformation*, the secondary spermatids become transformed into flagellated spermatozoa. When the cysts in which they have been encased throughout spermatogenesis

Fig. 5-2
Section of a testicular follicle (diagrammatic). [Redrawn from Snodgrass, 1935.]

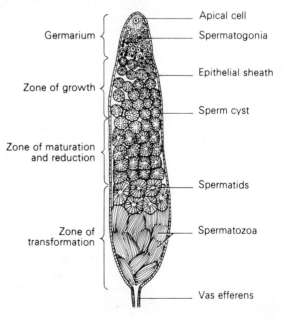

Germarium

Zone of growth

Zone of maturation and reduction

Zone of transformation

Apical cell

Spermatogonia

Epithelial sheath

Sperm cyst

Spermatids

Spermatozoa

Vas efferens

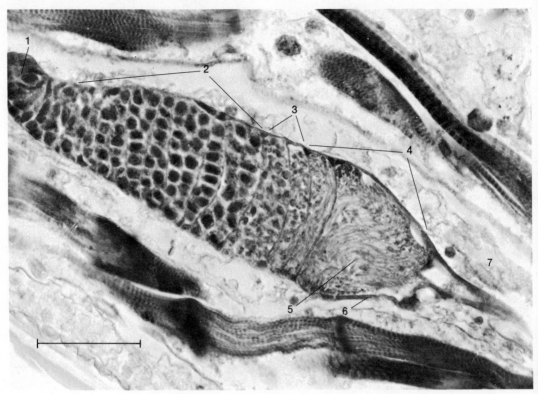

Fig. 5-3
Photomicrograph of a sagittal section of a testis in a male mosquito, *Aedes triseriatus*. 1, Germarium; 2, zone of growth; 3, zone of maturation and reduction; 4, zone of transformation; 5, spermatozoa; 6, epithelial sheath; 7, vas efferens. (Scale line = 50 μm.)

rupture, the spermatozoa enter the vas efferens and vas deferens and lodge in the seminal vesicles. Commonly, when the spermatozoa are released into the ducts, they remain in bundles (*spermatodesms*) held together by gelatinous material. The spermatozoa of most insects studied are filamentous with poorly developed "head" regions (Fig. 5-4).

The movement of spermatozoa within the male reproductive system is due not to their inherent motility but to contractions of the muscles associated with each vas deferens and the ejaculatory duct.

Further information on insect spermatozoa and spermatogenesis can be found in Baccetti (1972), King (1974), and Phillips (1970).

Control of Spermatogenesis.   In cell culture experiments a humoral factor has been implicated in the differentiation of spermatocytes in giant silkworm moths (Lepidoptera, Saturniidae). Ecdysone may play a role in influencing the permeability of the testis walls to this humoral factor.

Spermatophores.   In some insects the spermatozoa are produced and transferred to the female in specialized packets held together by proteinous secretions of the accessory glands. These packets are called *spermatophores* and often assume rather distinct

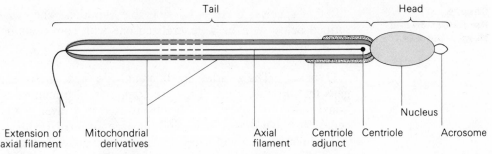

Tail | Head

Extension of axial filament | Mitochondrial derivatives | Axial filament | Centriole adjunct | Centriole | Nucleus | Acrosome

Fig. 5-4
Generalized insect spermatozoon. [Redrawn from Breland, Eddleman, and Biesele, 1968.]

forms (Fig. 5-5). Spermatophores are common in the lower orders (e.g., the apterygotes) and rare or absent in some of the higher orders, such as Hymenoptera. Once spermatophores are emptied of spermatozoa, they may be digested and absorbed within the genital ducts if placed there during copulation or be eaten by the female in those species in which the male places the spermatophore on the substrate.

Female Reproductive System, Oogenesis, and Ova

Like the male reproductive system, the female system (Fig. 5-6) is located in the posterior part of the abdomen. It typically consists of paired gonads (*ovaries*) connected by a series of tubes to the *vagina,* which opens to the exterior and receives the penis during copulation. There are also a variety of *accessory glands* present. Although the female reproductive system is basically similar among

Fig. 5-5
Spermatophore from a male silverfish, *Lepisma saccharina* (see also Fig. 8-7). [Redrawn from Sturm, 1956.]

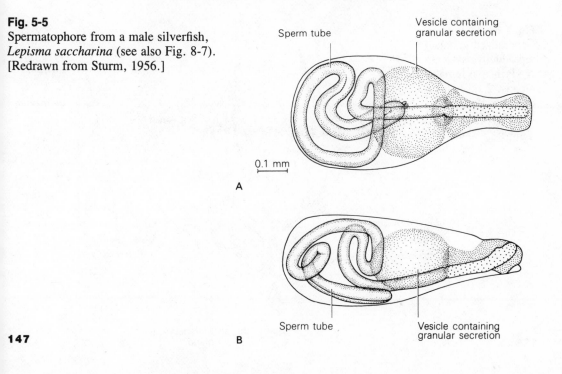

Sperm tube | Vesicle containing granular secretion

0.1 mm

A

Sperm tube | Vesicle containing granular secretion

B

different species, there is considerable variation. The system described below will be a generalized one with comments on some variations.

Ovaries. The ovaries are bilaterally located, mesodermal organs that produce eggs. They are composed of a number of functional units, *ovarioles,* which are invested in a layer of epithelial cells. This layer has muscles and tracheae associated with it.

The number of ovarioles per ovary varies greatly, from 1 in the tsetse flies, *Glossina* spp., and some aphids to over 2000 in the queens of certain termite species. In most insects a terminal thread from the cephalad portion of each ovariole joins those of its neighbors, forming a *terminal filament,* which attaches to the dorsal diaphragm.

Ducts. At the base of each ovariole is a small duct or *pedicel,* which joins those of the other ovarioles in a bulbous *calyx;* this in turn opens into the *lateral oviduct.* The lateral oviducts, which are, like the ovaries, of mesodermal origin, join to form the *common oviduct.* The common oviduct serves as a communicating tube between the lateral oviducts and the *bursa copulatrix* or *vagina,* which opens to the outside. The bursa copulatrix, when it occurs (Fig. 5-7), is a saclike expansion of the vaginal region. The common oviduct and bursa or vagina are of ectodermal origin and, like the fore- and hindgut and tracheae, are lined with a modified form of cuticle. There is usally a single outpocketing from the bursa, vagina, or common oviduct in which spermatozoa are stored prior to fertilization. This outpocketing is called the *spermatheca,* and in some insects (e.g., certain flies) it is a paired structure.

Fig. 5-6
Generalized female
reproductive system.
A. Principal female
organs. B. Single
ovariole. [Redrawn
from Snodgrass,
1935.]

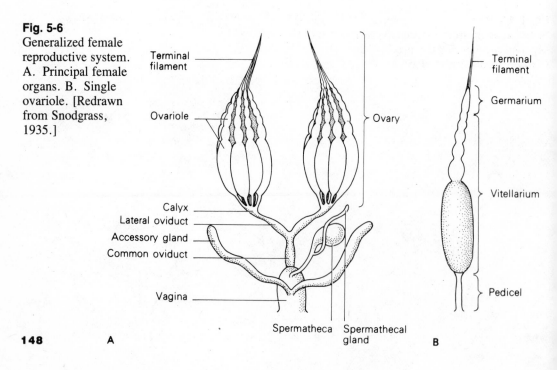

Terminal filament

Ovariole

Ovary

Calyx
Lateral oviduct
Accessory gland
Common oviduct

Vagina

Spermatheca   Spermathecal
gland

Terminal filament

Germarium

Vitellarium

Pedicel

B

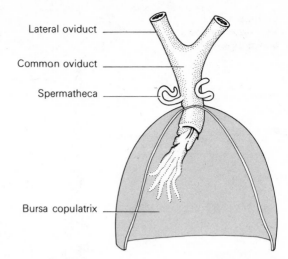

Lateral oviduct

Common oviduct

Spermatheca

Bursa copulatrix

Fig. 5-7
Ducts of a female triatomid bug, *Rhodnius* sp.
(Hemiptera; Reduviidae), showing opened
bursa copulatrix and entrance to the common
oviduct (ventral view). [Redrawn from Davey,
1965.]

Accessory Glands.   There are generally one or two pairs of accessory glands, which usually open into the apical portion of the
bursa copulatrix, vagina, or unpaired oviduct. These glands vary in
structure and function. They are commonly involved in the secretion
of adhesive materials (in which case they are called *colleterial
glands*) and serve to cement eggs to the substratum or hold them
together in masses. In cockroaches, secretions of the accessory
glands form a capsule or *ootheca* around eggs that have accumulated
in the bursa copulatrix. In many aquatic insects, mayflies, stoneflies,
and so on, the gelatinous masses that surround the eggs are accessory
gland secretions.

Oogenesis.   Under the heading oogenesis we include all those
processes that ultimately lead to the development of a mature ovum,
capable of being fertilized, and development of the nutritive capacity
to support embryonic development. Oogenesis may be completed
prior to or during the imaginal stage.

Each ovariole is divided into zones that contain germ cells or
oocytes in various stages of development and maturation (Figs. 5-
6B and 5-8). There are two broad zones, the apical *germarium* and
the basal *vitellarium* both of which are invested in an outer layer of
cells, the *ovariole sheath*. The germarium contains the primary female germ cells, the *oogonia,* which divide mitotically and eventually become *primary oocytes*. The germarium also contains
prefollicular tissue, which comes to form the *follicular epithelium*
in the vitellarium. The vitellarium in an active ovariole is comprised
of oocytes that are undergoing the deposition of nutrients, a process
referred to as *vitellogenesis* (yolk deposition; Fig. 5-9). Yolk is composed of proteins (as *proteid bodies*), neutral lipids, carbohydrates
(e.g., glycogen), and RNA.

Each developing oocyte is surrounded by a follicular epithelium,
and the oocyte and its associated epithelial layer comprise a *follicle*.
The oocytes in the vitellarium have progressively more yolk in an
apical to basal sequence, the most mature basal oocyte being separated from the lumen of the pedicel by an epithelial plug, which
ruptures when the oocyte is ready to leave the ovariole and proceed

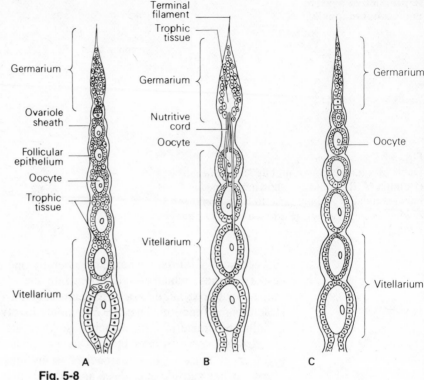

Fig. 5-8
Ovariole types. A. Polytrophic. B. Telotrophic. C. Panoistic.

into the lateral oviduct. This process of exiting from the ovariole is called *ovulation*. Following ovulation, the follicular epithelial cells remain behind and eventually degenerate.

There are three major types of ovarioles, based on the method by which yolk deposition occurs (Fig. 5-8): *polytrophic, telotrophic,* and *panoistic*. Bonhag (1958), Mahowald (1972), and Telfer (1975) review ovarian structure and vitellogenesis.

Polytrophic ovarioles have nutritive cells, *nurse cells* or *trophocytes,* associated with each developing oocyte. Trophocytes are within the follicular epithelium, which surrounds each oocyte. Characteristically, the oocyte and trophocytes all originate from a single oogonium that undergoes a series of mitoses with incomplete cytokinesis. This results in the presence of cytoplasmic canals interconnecting the trophocytes and oocyte. For further information, see Telfer (1975). A follicular plug exists between each oocyte with its accompanying trophocytes. Polytrophic ovarioles have been described in Neuroptera, Lepidoptera, some Coleoptera, Diptera, and Hymenoptera.

Telotrophic ovarioles differ from polytrophic ones in that the trophocytes are not directly associated with each developing oocyte but are all located in the apical region of the ovariole and are connected to the various oocytes by means of "nutritive cords." Otherwise, the two types are similar, each oocyte being surrounded by a follicular epithelium and being separated from the others by follicular plugs. Telotrophic ovarioles are characteristic of Hemiptera and some Coleoptera.

Polytrophic and telotrophic ovarioles are sometimes collectively referred to as *meroistic ovarioles*.

The third type of ovariole listed, panoistic, has no trophocytes. However, each developing oocyte is surrounded by the follicular epithelium, and follicular plugs exist between adjacent oocytes. Panoistic ovarioles are found in the apterygotes, Orthoptera, Isoptera, Odonata, Plecoptera (stoneflies), Siphonaptera, and some Coleoptera.

In polytrophic and telotrophic ovarioles, the trophocytes furnish all or nearly all of the RNA contained in the mature egg. In the panoistic type, the oocyte nucleus produces all of the RNA in the mature egg. Much of the yolk protein originates in the fat body (as *vitellogenin*) and is transferred to the ovaries via the hemolymph (Hagedorn and Kunkel, 1979). Yolk protein is called *vitellin*.

There are a few exceptional ovariole structures that do not fit into the above classification. For example, in ovarioles from the beetle *Steraspis speciosa,* the germarium and vitellarium are very short, there are no nutritive cords or trophocytes, and there is a long region of glandular tissue proximal to the vitellarium (Engelmann, 1970).

In the vast majority of insects oogenesis occurs in the last larval instar, the pupa, and the adult stages. However, in some species immature stages are capable of producing mature oocytes that may commence and even complete embryogenesis. This phenomenon, *pedogenesis,* has been described in a number of insect species, among which are *Micromalthus debilis* (Coleoptera) and cecidomyiid flies in the genera *Miastor* and *Oligarces*.

See King (1974) for more information on oogenesis.

Eggs. Mature insect eggs are typically elongate and oval in longitudinal section (Fig. 5-10), although some assume other forms (Fig. 5-11). In most instances the largest portion of an egg is filled with *yolk* or *deutoplasm,* and the cytoplasm and nucleus occupy only a small portion. The yolk contains carbohydrates, protein, and lipid bodies, the protein bodies being the most abundant. Cytoplasm is located around the nucleus (*nuclear cytoplasm*) and around the periphery of the yolk (*periplasm* or *cortical cytoplasm*). The egg may be encased in two membranes: the *vitelline membrane,* which apparently represents the cell membrane of the egg, and the *chorion,* or eggshell, which is secreted by the follicle cells. Several species lack a chorion (e.g., some viviparous species). When a chorion is present, it is nonchitinous, may be composed of two to several layers, varies considerably in thickness in different species, and may be smooth or sculptured in a variety of ways.

Since insect eggs are usually deposited outside the parent female, they are subject to a drying atmosphere, attacks by parasites and predators, and other dangers. The chorion serves the same functions as the cuticle. It serves as a protective coating (against physical damage, attack by parasites, etc.) and a barrier against water loss and is important relative to ventilation of the egg. Coatings secreted by the accessory glands that cover eggs may aid in water conservation in addition to serving to hold eggs together or glue them to the substratum. Some eggs laid in moist situations are capable of absorbing water from their surroundings. In insects in which the

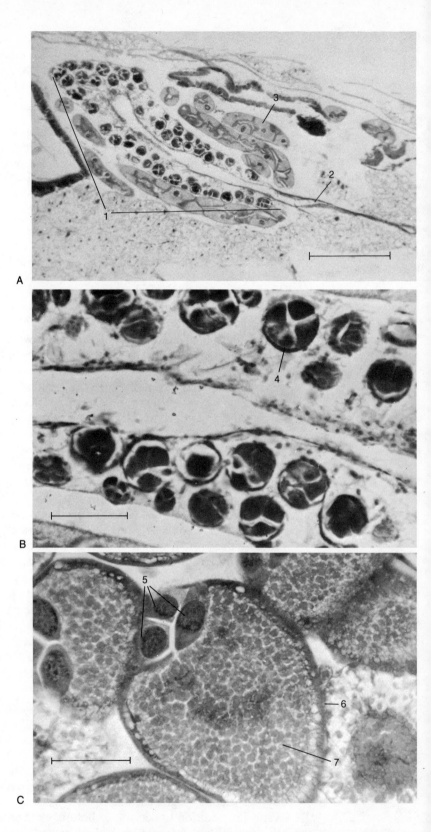

A

B

C

chorion is thin, ventilation may occur over the entire surface. In
others the chorion may be lined with a porous, gas-filled layer that
communicates with the outside of the egg by means of channels or

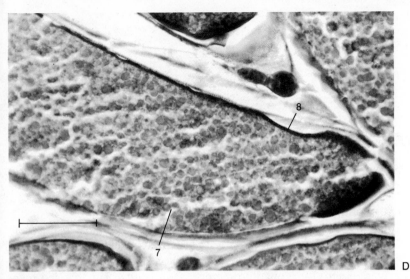

Fig. 5-9
Photomicrographs of vitellogenesis in the mosquito *Culex nigripalpus.*
A. Sagittal section of ovary in an unfed female. B. Same as A, higher
magnification. C. Oocyte, 48 hours following a blood meal. D. Oocyte,
96 hours following a blood meal. 1, Ovary; 2, lateral oviduct; 3,
Malpighian tubule; 4, follicle; 5, nurse cells; 6, follicular epithelium; 7,
yolk; 8, chorion. (Scale lines: A, 250 μm; B–D, 50 μm.) [Tissues
prepared by Elaine Cody.]

aeropyles. Some eggs possess specialized structures that act as physical gills or plastrons, allowing them to obtain oxygen from water.
Hinton (1969) reviews and synthesizes the major literature pertinent
to respiratory systems of insect eggs. Miller (1974a) also provides
information on this topic. Spermatozoa gain entrance to an egg by
means of one to several special channels or *micropyles,* which are
perforations of the chorion and are located at various places on the
eggs of different species.

Hinton (1979) provides a comprehensive three-volume treatise on
insect eggs. Furneaux and Mackay (1976) consider the structure,
formation, and composition of the chorion and vitelline membrane.

Control of Oogenesis. It has been established for a number of
species that the development of eggs—vitellogenesis, in particular—
is controlled by a secretion of the corpora allata (or in the cyclorrhaphous Diptera, the region of the ring gland that is homologous
with the corpora allata). Some of the kinds of experiments that have
enabled investigators to arrive at this conclusion are

1. Histological observation of a correlation between ovarian activity and activity in the corpora allata.
2. Ovarian activity being attenuated by the microsurgical removal
   of the corpora allata (allatectomy) and restored by implantation.
3. Allatectomy, attenuating ovarian activity and hemolymph
   transfusion from a donor with active corpora allata restoring
   this activity.

The corpus allatum hormone may affect egg maturation by stimulating the incorporation of yolk into the oocyte and simultaneously

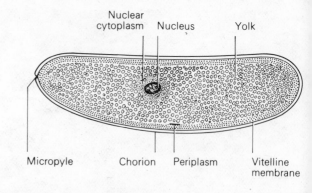

Nuclear
cytoplasm   Nucleus   Yolk

Fig. 5-10
Sagittal section of a tpyical insect egg.
[Redrawn with modifications from Hagen,
1951.]

Micropyle   Chorion   Periplasm   Vitelline
membrane

regulating metabolism, particularly of proteins. In certain species (e.g., *Rhodnius* sp.) corpora allata from adults have been transplanted into larvae, and their secretions have had the same effect as the juvenile hormone. Conversely, larval corpora allata stimulate ovarian activity (gonadotropic effect) in adults. This has led many to hold the view that juvenile hormone and the gonadotropic hormone may be one and the same.

Recently, certain substances (e.g., farnesol and farnesyl methyl ether) have been found that mimic the effect of the corpus allatum hormone, acting as "juvenile" hormone and having a gonadotropic action (i.e., stimulating the ovaries). In some species of insects allatectomy does not prevent egg maturation, but there is evidence that the corpus allatum hormone may be present in the hemolymph from earlier stages. In most species studied, secretions from the median neurosecretory cells in the pars intercerebralis in the brain are necessary for oogenesis. These secretions have been found to activate the corpora allata or stimulate protein synthesis (necessary for yolk formation) or produce a gonadotropic hormone. Recall that the corpora cardiaca serve as neurohaemal organs (see Chapter 3), storing and releasing the products of the neurosecretory cells in the brain. This allows the accumulation of secretory products in the corpora cardiaca for some time prior to their release from this organ. Thus, the action of the median neurosecretory cells is a function of the release of their products from the corpora cardiaca. Such release of material is probably under neural control.

Other factors, mediated hormonally or via the nervous system, that have been found to influence the activity of the corpora allata in different species include mating, light (photoperiod), chemical stimulation (pheromones from males), various nutritional factors, and the presence or absence of eggs in the brood chamber. These factors may have a stimulatory or inhibitory effect on the corpora allata.

In addition to secretions from the median neurosecretory cells and the corpora allata, ecdysone has been found to be involved in the control of oogenesis in female mosquitoes (see references in Hagedorn and Kunkel, 1979). In response to ingestion of a blood meal, secretions from the median neurosecretory cells (i.e., *egg development neurosecretory hormone*) induce the ovaries to release ecdysone, which in turn triggers the synthesis of vitellogenin in the fat body (Fig. 5-12). Juvenile hormone secreted by the corpora allata "activates" the fat body and ovaries.

Several factors probably interact in the control of oogenesis in any given insect species, and the present level of understanding does not permit broad generalizations. The reviews of Davey (1965), de Wilde and de Loof (1973b), Engelmann (1968), Riddiford and Truman (1978), and Telfer (1965) should be consulted for detailed information.

Seminal Transfer, Fertilization, and Sex Determination

Seminal Transfer

Internal fertilization, which prevents exposure of gametes to the drying atmosphere, is looked upon as one of the several prerequisites

Fig. 5-11
Representative insect eggs.
A. California green lacewing, *Chrysopa californica*. B. Egg cases (oothecae) and eggs of a praying mantid. C. Eggs and egg masses of the mayfly, *Ephemerella rotunda*. D. Eggs of the convergent ladybird bettle. E. Egg raft of the mosquito *Culex pipiens*. F. Eggs and egg masses of the stink bug, *Pentatoma lignata*. G. Ootheca of the German cockroach, *Blattella germanica*. [Redrawn from Boreal Labs key card.]

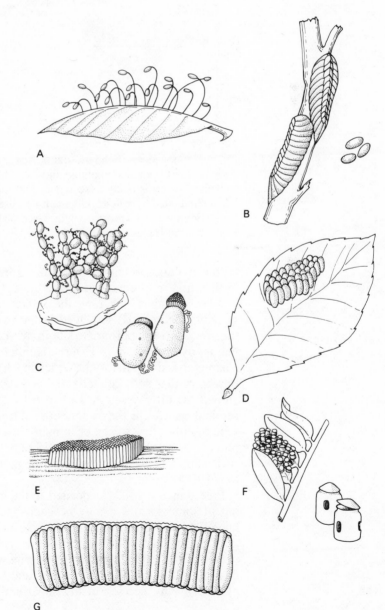

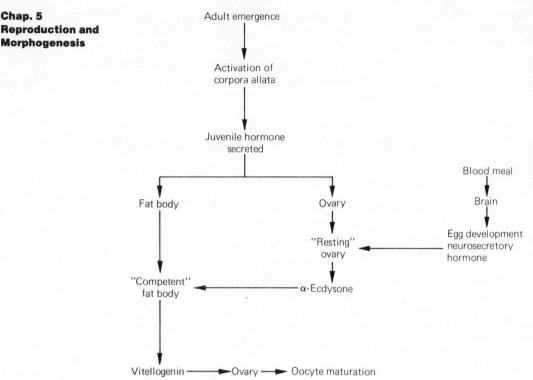

Fig. 5-12

Control of oogenesis in the mosquito, *Aedes aegypti*. A "competent" fat body is able to produce vitellogenin when stimulated by α-ecdysone; a "resting" ovary is ready to secrete α-ecdysone and undergo maturation when stimulated by egg development neurosecretory hormone. (Modified from Riddiford and Truman, 1978; based on Hagedorn, 1974, and Flanagan and Hagedorn, 1977).

for the evolution of animal life in the terrestrial environment. Internal fertilization in insects is generally brought about by the act of copulation, during which time the *semen* (spermatozoa plus various glandular secretions) produced in the male reproductive system is transferred to an appropriate site in the female reproductive system, that is, by *insemination*. External genital structures are discussed in Chapter 2 and copulatory behavior in Chapter 8. Suffice it to say at this point that both genitalia and copulatory behavior vary greatly among the different groups of insects, but their common "goal," seminal transfer, is the same. Seminal transfer may involve the passage of either free semen or in many insects one or more spermatophores from the male to the female. The involvement of a spermatophore is considered to be the more primitive situation (Hinton, 1964).

Free semen is usally deposited in the bursa copulatrix or vagina, but in some species it may be deposited in the common oviduct, lateral oviducts, or even directly into the spermatheca. Male dragonflies and damselflies deposit semen in a specialized organ on the venter of the second abdominal sternite. A portion of that organ is then placed in the female's vagina and seminal transfer is accomplished. Many species in the superfamily Cimicoidea (e.g., bed bugs; family Cimicidae) have a rather bizarre method of seminal transfer. The males of these species inseminate the females by perforating the integument at a specialized site in their abdomen and ejaculating semen directly into the hemocoel. This method of semen introduction has been called *hemocoelic* or *traumatic insemination*. The seminal fluid and many of the spermatozoa are digested by the female, while some of the spermatozoa reach the ovaries. The adaptive value of this method of insemination is thought to be that it provides the female with additional nutritive substance (Hinton, 1964).

Spermatophores are usually deposited by the male somewhere in the female reproductive system: the bursa copulatrix, vagina, or, rarely, the spermatheca. However, in the apterygotes (e.g., Thysanura; see Fig. 8-7), the male deposits a spermatophore on the substrate and the female picks it up and deposits it within herself. This method of indirect sperm transfer without copulation also occurs among the Diplura, Collembola, and in several noninsect arthropod groups (Schaller, 1971). Gerber (1970) considers the methods of spermatophore formation in pterygotes.

Once a spermatophore is in the female reproductive system, various mechanisms may account for the release of the spermatozoa. In most insects the spermatozoa are either forced out by pressure applied by the female or by the mechanical performation of the spermatophore. In the house cricket an osmotic-pressure mechanism is involved where there is a significant difference between the osmotic pressure of a block of protein or pressure body within the spermatophore and the osmotic pressure of the evacuating fluid, which surrounds the spermatophore at the time of ejaculation. The evacuating fluid is absorbed by the protein body, which swells up and forces the exit of the spermatozoa (Hinton, 1964). In some insects the spermatophore may be digested away, causing the release of the spermatozoa. After the spermatozoa have been released, the spermatophore is in most cases probably digested and absorbed by the female and thus may have some nutritional significance.

After release from the spermatophore or deposition in the form of free semen, the spermatozoa move to the spermatheca. This is apparently due to muscular contractions of the female ducts but may also be dependent upon sperm motility. In the honey bee the spermathecal duct has a "pumping" structure, which seemingly transports the sperm to the receptacle following copulation and subsequently releases them when it is time for an egg to be fertilized.

Fertilization

The processes involved in fertilization may be divided into three parts.

1. Release of spermatozoa from the spermatheca.
2. Entry of the egg by spermatozoa.
3. Formation and fusion of the male and female pronuclei.

As mentioned earlier, spermatozoa are stored in the spermatheca of the female until it is time for fertilization. The females of many insect species (e.g., *Apis,* honey bee; *Glossina,* tsetse fly; and *Rhodnius* and *Triatoma,* conenose bugs) mate only at a single time during their lives with one or more males, and the spermatozoa introduced at that time are stored and used to fertilize their eggs for the rest of the reproductive period. Spermatozoa stored in this fashion may survive for several months or years. Other insect species are inseminated periodically throughout their reproductive lives, and in these species the storage of spermatozoa in the spermatheca may only be for a short period of time.

The mechanisms for release of spermatozoa from the spermatheca are not clearly understood. Stimulation of sensory hairs by the passage of eggs in the oviducts, hemocoelic pressure forcing the exit of the sperm, and the inherent motility of spermatozoa that have been "activated" by some secretion have all been advanced as possible mechanisms in various species.

Following ovulation, the egg is oriented in the reproductive passage in such a way that the micropylar region is in rough proximity to the site of sperm release. The sperm migrate to the micropylar region of the egg, possibly responding chemotactically, and one or more enter the egg through the micropyle. In the vast majority of insects more than one spermatozoon enters the egg, but usually only one fuses with the egg pronucleus. Excess sperm usually degenerate without disrupting the development of the zygote.

Shortly following the entry of sperm into the egg, the egg nucleus undergoes meiotic division, forming the female pronucleus (see Fig. 5-14). The spermatozoon that will fuse with this female pronucleus loses its tail, becoming the male pronucleus. The fusion of the two pronuclei forms the *zygote* and signals the commencement of morphogenesis.

Sex Determination and Parthenogenesis

Nearly all insects are bisexual; that is, male sex organs occur in one individual, and the female sex organs are in another individual. Several species in different groups are capable of reproducing *parthenogenetically* (i.e., they are able to produce individuals from unfertilized eggs). Insects in which both male and female sex organs occur in the same individual are said to be *hermaphroditic.* The cottony cushion scale (*Icerya purchasi*) and one or two relatives are the only insects in which true hermaphroditism has been established. Reproduction in these hermophroditic insects usually occurs by self-fertilization.

Sex determination in bisexual insects is considered to depend upon a balance between genes for maleness and genes for femaleness. This balance is in most forms tipped in the direction of one sex or the other by a sex-chromosome mechanism in which one sex possesses two X (sex) chromosomes (i.e., the *homogametic sex,* XX) and the other a single X chromosome (X0) or a single X chromosome plus a smaller Y chromosome (XY). Individuals possessing the X0 or XY configurations comprise the *heterogametic sex.* In most insectan groups the males are heterogametic and the females homogametic. However, the reverse is true in the Lepidoptera and Trichoptera. In some insects (e.g., Hymenoptera, Thysanoptera, and certain Homopterous insects) males develop from unfertilized eggs, females from fertilized eggs.

White (1964) and Bergerard (1972) discuss sex determination in insects.

In some instances environmental factors have been shown to exert an influence on sex determination. For example, in the butterfly *Talaeporia* sp., more males than females are produced at high temperatures; the converse is true at low temperatures (Davey, 1965).

Since extirpation of gonads or implantation of gonads into a previously castrated individual produces no effects, it is generally assumed that secondary sexual characters are not determined by humoral secretions associated with the gonads. However, in the beetle *Lampyris noctiluca*, there is evidence that special secretory tissue associated with the testis rudiment controls the development of male structures, and implants of this tissue masculinize females. The appearance of this secretory tissue is determined by the presence of active neuroendocrine cells in the brain (Naisse, 1966, 1969).

Apparently every cell in the insect body is involved in sex determination, as is evident in the occasional occurrence of *gynandromorphs*. These individuals are literally sexual mosaics, some parts of the body possessing typically male traits and other parts typically female traits (Fig. 5-13). This phenomenon is explained by differences in the sex chromosomes in the cells comprising the various tissues (i.e., some cells are "male" and others are "female"). These differences are known to occur by a number of mechanisms. One such mechanism is the loss of one X chromosome in the cleavage cells of a female (XX becoming X0, e.g., in *Drosophila*, vinegar or fruit flies). These cells thereafter give rise to male traits in whatever tissue they happen to form. Another mechanism occurs in the honey bee, in which the fusion nucleus and an extra sperm that has entered the egg both undergo cleavage. The cells that result from cleavage of the fusion nucleus (diploid) give rise to female traits, and those from the sperm nucleus (haploid) produce male traits.

Embryogenesis

Embryogenesis includes those developmental events that occur between the formation of the zygote and the exit of the fully developed individual from the egg (*eclosion*). The changes that occur following hatching from the egg will be included in the discussion of postembryonic morphogenesis. *Morphogenesis* comprises all developmental events that occur between the formation of the zygote and the emergence of a sexually mature adult. Although there is much variation in detail from group to group, a general account of insect embryogenesis is possible.

Anderson (1973, 1979), Counce and Waddington (1972, 1973), Hagen (1951), Johannsen and Butt (1941), Sharov (1966), Snodgrass (1935), and Weygoldt (1979) are recommended for further readings in the area of insect embryogenesis.

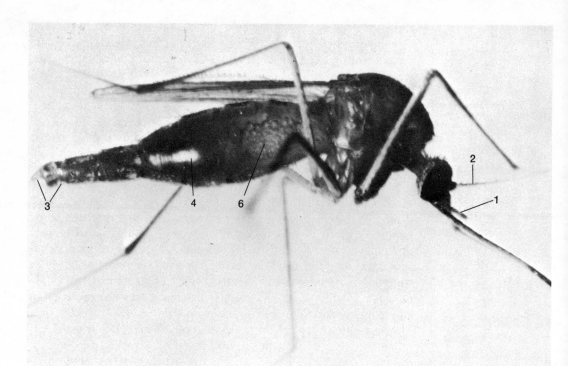

A

Fig. 5-13
Example of gynandromorphy in a mosquito, *Culex salinarius*. A. Wild-caught gynandromorph; the antennae and palpi are characteristic of a female; the terminalia are characteristic of a male. B. Normal female. Both specimens have taken a blood meal, a female trait, 48 hours earlier. 1, Palp; 2, antenna; 3, terminalia; 4, blood in midgut; 5, developing ovary; 6, bubbles in ventral diverticulum (see Fig. 4-2). [Courtesy of J. D. Edman.]

B

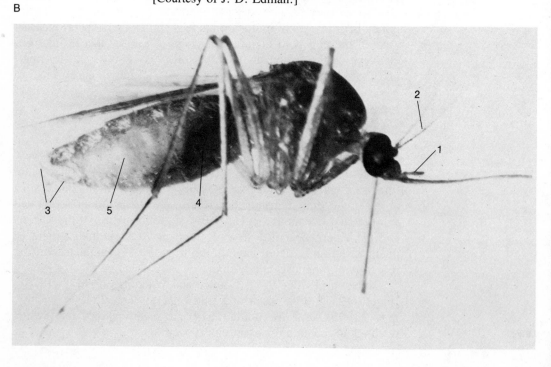

Formation of the Blastoderm and Germ Cells

The first distinct layer of cells to form in the embryogenesis of any metazoan animal is the *blastoderm,* which is composed of a single layer of cells, the *blastomeres.* The means by which this layer of cells arises varies in different kinds of animals and is correlated with the quantity of yolk material initially present in the egg. In animals with little yolk material the zygote divides into two equal parts; each of these daughter cells divides into two equal parts; and so on. This is the process of *cleavage* and eventually leads to a ball of cells, the *morula,* which subsequently develops an internal cavity, the *blastocoel,* surrounded by the blastoderm. Since in this method of blastoderm development, the entire zygote divides, the term *holoblastic* (holo = whole; blast = bud or sprout) cleavage is applied. However, in the vast majority of insects—Collembola (springtails) and certain parasitic Hymenoptera being the most outstanding exceptions—the eggs have large quantities of yolk. In most insects the fusion nucleus, with associated cytoplasm, behaves as though it were an individual cell and proliferates mitotically (Fig. 5-14B). The daughter (cleavage) nuclei eventually migrate to the periphery of the egg and form the *blastoderm* (Fig. 5-14C, D). During the course of this process, the individual cleavage cells develop cell membranes. This form of cleavage is referred to as *meroblastic* (mero = part). The eggs of some arthropods and a few insects (e.g., Collembola) undergo total cleavage initially and subsequently peripheral cleavage in the formation of the blastoderm. This is called *combination cleavage.*

During meroblastic cleavage, some of the cleavage cells remain in the yolk or return to it after reaching the periphery of the egg. These are the *vitellophages* (yolk cells; Fig. 5-14C, D) and are considered to be responsible for the initial digestion of the yolk, making it more readily assimilable by the other embryonic cells.

At the same time the blastoderm forms, some of the cleavage cells differentiate into *germ cells* (Fig. 5-14C–E), which will give rise to gametes or reproductive cells in the late larval, pupal, and adult stages. In many embryos the differentiation of germ cells is correlated with the passage of cleavage cells through a specialized region of the egg called the *oosome.*

Formation of the Germ Band and Extraembryonic Membranes

Following the completion of the blastoderm, the cells on one side of the egg become columnar along the longitudinal midline of the egg (Fig. 5-14E, F). In a lateral direction from this midline, the cells become successively less columnar, finally merging with the remaining cells of the blastoderm, which tend to become squamous. This thickened area of columnar cells of the blastoderm is the *germ band,* which subsequently elongates and develops into the embryo, while the remaining cells take part in the formation of the extraembryonic membranes. In most insects folds from the area outside the germ band grow over the germ band (i.e., overgrowth; Fig. 5-15A), eventually meeting along the longitudinal midline. The inner and

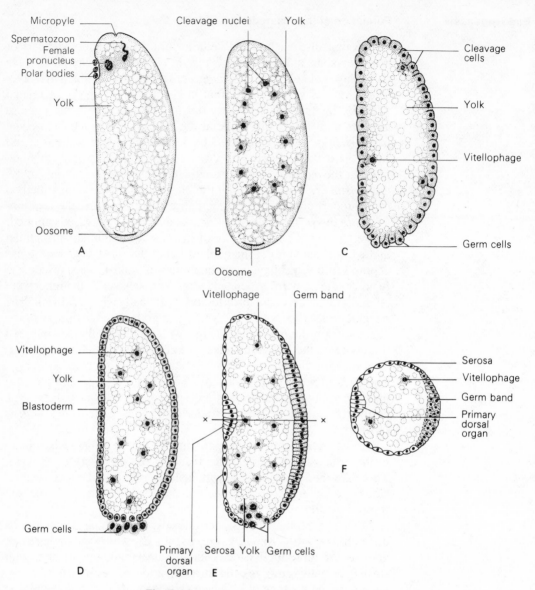

Fig. 5-14
Fertilization—germ band formation (chorion omitted; A–E, sagittal sections). A. Just prior to fertilization (i.e., fusion of spermatozoon and female pronucleus). B. Cleavage. C. Blastoderm forming and germ cells differentiating. D. Blastoderm formation complete. E. Germ band and primary dorsal organ formation. F. Cross section at same stage as E. The location of the section is indicated in E by the line x—x. [Redrawn from Johannsen and Butt, 1941.]

outer layers of one fold then merge with the respective layers of the other, forming an inner *amnion* around the developing embryo and an outer *serosa* around the yolk, amnion, and embryo. In some insects the extraembryonic membranes may form by invagination or involution of the embryo instead of overgrowth (Fig. 5-15B, C). The first method is found in the Apterygota; the second occurs among the orders Odonata, some Orthoptera, and Homoptera. As the extraembryonic membranes are forming, the germ cells become located at what will be the posterior end of the embryo.

As the germ band forms in some insects, particularly in Apterygota, but also in some Pterygota, a cluster of cells form the primary dorsal organ (Fig. 5-14E, F). This structure disappears at dorsal closure and may be glandular in function.

Differentiation of the Germ Layers

In animals with little yolk material and in which a blastocoel develops, an invagination eventually occurs that subsequently differentiates into well-defined mesodermal and endodermal layers. These three germ layers (*ectoderm, mesoderm,* and *endoderm*) then differentiate into the various tissues and organs of the fully developed organism. The formation of the mesoderm and endoderm is referred to as *gastrulation*. Unfortunately, this process in insects is not straightforward and clear. Unequivocal interpretation of insect morphogenesis in terms of this *germ-layer theory* has proved to be extremely difficult; insect embryologists hold widely divergent opinions. Fox and Fox (1964) point out that there is much to be said for not

Fig. 5-15
Formation of extraembryonic membranes (diagrammatic.)

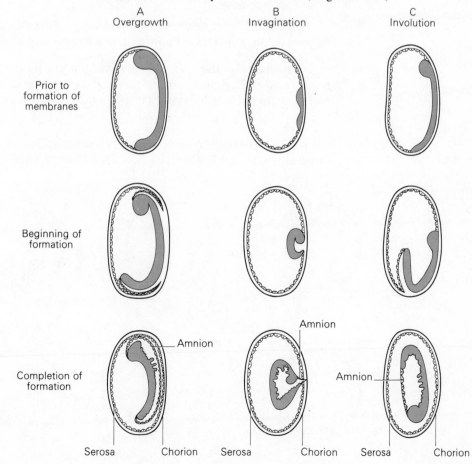

attempting to distinguish between mesoderm and endoderm in the insect embryo but for simply recognizing an outer ectoderm and inner layers. For convenience of discussion, we shall label those cells that give rise to the midgut as endodermal in origin, keeping the aforementioned qualifications in mind.

Gastrulation in most insects occurs as the amnion and serosa are forming. It begins, in most species, as a longitudinal, furrowlike invagination (Fig. 5-16A–C), which runs most of the length of the venter of the germ band. Eventually the invagination flattens out and the outer edges come together and fuse, forming a longitudinal band (*inner layer* or *mesentoderm*) of cells surrounded by an outer layer, which at this point is appropriately referred to as *ectoderm*. Another type of inner layer formation (Fig. 5-16D, E) consists of a ventral longitudinal band of cells in the germ band sinking into the yolk and being overgrown by the remaining cells of the germ band. In a third and less common method of inner layer formation, a longitudinal band of cells proliferates from the germ band (Fig. 5-16F). Whatever the method of formation, the end result is essentially the same (an outer or ectodermal layer and an inner or mesentodermal layer). Eventually the inner layer differentiates into two lateral longitudinal bands (*mesoderm*) and a median strand with cell masses located at its anterior and posterior ends (Fig. 5-17). For our purposes the median strand with its anterior and posterior cell masses will comprise the *endoderm*.

Segmentation, Appendage Formation, and Blastokinesis

Segmentation of the embryo begins very soon after the germ band has formed and originates in the mesoderm. Segmentation later becomes quite evident in nearly all the organs of mesodermal origin

Fig. 5-16
Germ-layer differentiation (yolk and chorion omitted). A–C. Invagination. D. and E. Overgrowth. F. Delamination. [Redrawn with modifications from Snodgrass, 1935.]

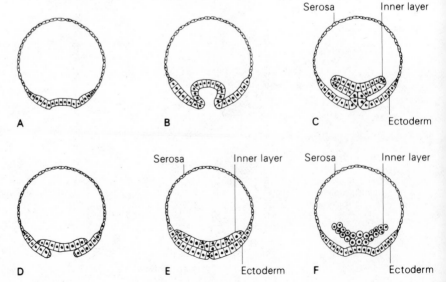

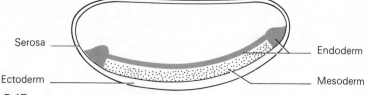

Fig. 5-17
Sagittal section of an embryo (diagrammatic). [Redrawn from
Snodgrass, 1935.]

(dorsal vessel, muscles, etc.) as well as those of ectodermal origin
(nervous system, tracheal system, etc.). The endoderm and the portions of the ectoderm that give rise to the fore- and hindgut are
unaffected by the segmentation process.

As segmentation proceeds, transverse furrows become externally
evident in the ectoderm (Fig. 5-18). Soon after segmentation begins,
bilateral evaginations of the ectoderm, which contain mesodermal
tissue, begin to appear. These will form the various body appendages. Initially the germ band (Fig. 5-18A) is comprised only of the
protocephalon, which is bilaterally expanded, and a narrow lobe
extending posteriorly, from which the remaining body segments will
develop. When the segmentation of the embryo is essentially complete (Fig. 5-18B–D) and all appendage rudiments have formed, the
portions of the embryo that will form the three tagmata of the insect
body can be discerned.

The protocephalon, the antennal and intercalary segments, and
the following three segments with their paired appendages will form
the definitive head. The antennal and intercalary segments usually
cannot be clearly separated, but the next three segments, the appendages of which will form the mandibles, maxillae, and labium, respectively, are easily outlined. These last three segments comprise
the *gnathal segments.* The region of the developing head anterior to
the gnathal segments has been interpreted in several different ways
regarding the true number of primitive segments involved. Rempel
(1975) discusses the various interpretations and makes a strong case
for the insect head having arisen from the amalgamation of an anterior body cap (*acron* or *prostomium*) plus three segments (*labral,
antennal,* and *intercalary*) plus three gnathal segments. This topic
is considered further in Chapter 10.

The three segments posterior to the gnathal segments will form
the definitive thorax, and their appendages will form the thoracic
legs. The remaining segments, which never number more than
twelve, including the posterior telson, will compose the definitive
abdomen. Pairs of limb buds appear on the abdominal segments, but
the first seven pairs and the tenth pair are resorbed. Those on segments eight and nine form the external genitalia, and those on the
eleventh segment form the cerci. As mentioned in Chapter 2, the
tendency in insect evolution has been toward a reduction in the
number of abdominal segments, the most primitive contemporary
forms having only eleven, with the eleventh containing the anus,
and the more advanced forms often having fewer than this number.

Blastokinesis is a term used to denote "all displacements, rota-

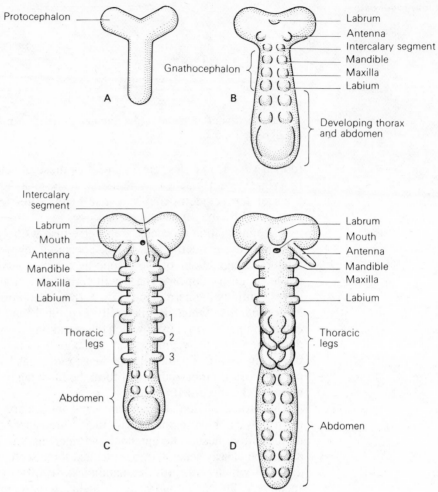

Fig. 5-18
Segmentation and appendage formation. A. Prior to segmentation. B and C. Successive stages. D. Segmentation complete.

tions, or revolutions of the embryo within the egg'' (Johannsen and Butt, 1941). These movements occur in a variety of predictable ways characteristic for each species. They may represent an adaptation to the large quantity of yolk characteristic of the vast majority of insect eggs and enable the embryo to make the most efficient use of the yolk material. These movements probably also result in protection from desiccation by causing the embryo to be surrounded by yolk.

Organogenesis

Subsequent to the formation of the three germ layers, each undergoes further differentiation, eventually forming the various tissues and organs of the fully developed embryo.

Fate of the Mesoderm. As segmentation occurs, the intersegmental portions of the mesoderm either become very thin or separate altogether. Eventually each segmental mesodermal layer develops two lateral lumens (Fig. 5-19A, B), the *coelomic sacs,* and finally the mesoderm differentiates segmentally into the following parts (Fig. 5-19C).

1. The *splanchnic* or *visceral* layer.
2. The *somatic layer.*
3. The fat body.
4. *Cardioblasts.*
5. The *genital ridge.*

As development progresses, the splanchnic layer comes to form the visceral muscles, and the somatic layer, the skeletal muscles. At dorsal closure, to be discussed briefly later, the cardioblasts form the heart portion of the dorsal vessel. The aorta arises from the median walls of the antennal coelomic sacs (Johannsen and Butt, 1941). The genital ridges are later suppressed in all but the eighth and ninth abdominal segments. In these segments, the germ cells, which, you will recall, differentiated while the blastoderm was form-

Fig. 5-19
Cross section of an insect embryo at successive stages of differentiation (chorion and most of yolk omitted; serosa omitted in B and C). [Redrawn from Johannsen and Butt, 1941.]

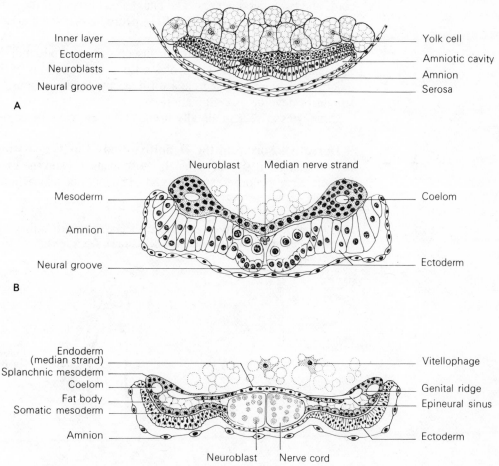

A

B

C

ing, migrate into the genital ridges and together with these meso-dermal cells form the *anlage* (precursors) of the gonads.

Fate of the Ecto- and Endoderm.　The alimentary canal begins its development as two invaginations of the ectoderm, one in the cephalic region (*stomodael*) and one in the last abdominal segment (*proctodael;* Fig. 5-20A). These two invaginations subsequently form the fore- (*stomodaeum*) and hindgut (*proctodaeum*), respectively (Fig. 5-20B). The midgut (*mesenteron*) arises from the proliferation of the cell masses at each end of the median strand of endoderm (Fig. 5-17). Some of the cells freed from the middle strand may also contribute to midgut formation in some species. The endodermal cells eventually envelop the remaining yolk material (Fig. 5-20B). The alimentary canal is completed when the membranes between the stomodaeum, mesenteron, and proctodaeum perforate. The Malpighian tubules develop as evaginations of the proctodaeum immediately posterior to the mesenteron.

The brain, subesophageal ganglion, segmental ganglia, and associated paired nerve cords are of ectodermal origin. The earliest evidence of their development is the presence of a longitudinal neural groove in the ectoderm and the differentiation of *neuroblasts* (Fig. 5-19). The neuroblasts subsequently give rise to the ganglia, nerve cord, and associated nerves. The ganglia and nerves of the stomatogastric nervous system develop from neuroblasts, which arise from the stomodaeum.

All the ventilatory structures of insects are of ectodermal origin. Spiracles, tracheae, and tracheoles arise as invaginations of the ectoderm and eventually secrete a form of cuticle; the various ''gill'' structures develop as evaginations.

Oenocytes arise segmentally in the abdomen from the ectoderm.

Dorsal Closure and the Definitive Body Cavity.　As the embryo develops, it spreads over the yolk material, and the extraembryonic area becomes smaller and smaller. The extraembryonic

Fig. 5-20
Sagittal sections of an insect embryo at successive stages in the development of the alimentary canal. [Redrawn from Johannsen and Butt, 1941.]

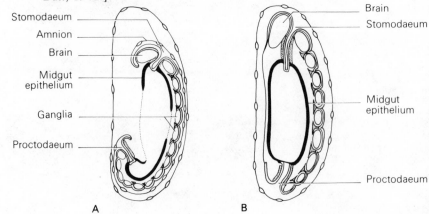

Stomodaeum
Amnion
Brain
Midgut epithelium
Ganglia
Proctodaeum

Brain
Stomodaeum
Midgut epithelium
Proctodaeum

A　　　　B

membranes, the amnion and serosa, usually disappear before the embryo has completed its development. Generally, they fuse at some point ventrally and a longitudinal cleft forms. The resultant folds, which consist of amnion fused to serosa, are drawn back over the embryo and the serosa condenses, forming the *secondary dorsal organ*. Eventually the dorsal organ sinks into the remaining yolk material, and the amnion provides a temporary dorsal closure. Subsequently, the ectoderm grows over the dorsal organ, which is detached and absorbed into the yolk. The ectoderm effects the final dorsal closure.

As should be evident from the preceding discussion, the definitive body cavity of insects is not a true coelom but develops from the blastocoel, which is invaded by mesodermal tissue (Snodgrass, 1935). Thus, as explained in Chapter 3, the body cavity is correctly referred to as a hemocoel or blood cavity.

Polyembryony. Among several groups of parasitic Hymenoptera (e.g., Chalcidoidea, Proctotrupidae, Vespoidea, Braconidae, Ichneumonidae) and members of the Strepsiptera, a developing egg may divide mitotically and produce several embryos. This asexual multiplication of individuals is called *polyembryony*. This phenomenon in the Hymenoptera is considered to be an adaptation associated with parasitism, while in the Strepsiptera, it is associated with ovoviviparity (i.e., "live birth"; offspring "born" as larvae instead of eggs). In the Strepsiptera, the eggs develop in the mother's body cavity.

The eggs and embryogenesis in polyembryonic insects differ from those of other insects as follows.

1. The eggs are extremely small.
2. There is no yolk.
3. The chorion, if present, is very thin and permeable.
4. Cleavage is holoblastic.

These differences are not surprising when one compares the "environments" of polyembryonic and most other insect eggs.

Ivanova-Kasas (1972) reviews the literature on polyembryony in insects.

Control of Embryogenesis

Morphogenesis may be looked upon as a process in which primordial cells become increasingly differentiated in a stepwise fashion, finally becoming the specialized parts that comprise the total functional organism. Events that occur during this process are ultimately under the control of the nucleus in the newly forming egg. However, in the completed egg, the activities of the cleavage nuclei are controlled by an interaction of the nuclei themselves and factors within the cytoplasm, and the tissues these nuclei eventually form are determined largely by their position within the egg.

Experiments that have elucidated the control of embryogenesis have been based on the removal, destruction, disruption, or separation of various parts of the egg at different times during its development. Techniques utilized in these experiments include the use of microsurgery, ultraviolet light, Xray, cautery, and ligation. Recent reviews pertaining to control of embryogenesis are those of Agrell and Lundquist (1973), Chen (1971), Counce (1973), and Sander (1976). *Milestones in Developmental Physiology of Insects* (Bodenstein, 1971) contains reprints of several classic papers in the control of embryogenesis.

Cleavage and migration of the nuclei and their accompanying islands of cytoplasm begin at the *cleavage center,* which is located in the region where the future head will form. The earliest effect of the cytoplasm on cleavage cells occurs at the *activation center,* which is initially located near the posterior pole of the egg. The interaction of the cleavage cells and the activation center results in a change that initiates further growth and development. The effects of this change then proceed anteriorly. Blocking the effects of the cleavage center by removal or exposure to ultraviolet light results in a failure of germ-band formation. Further developmental direction comes from the *differentiation center,* located near the middle of the presumptive germ band, at which point the prothorax will eventually form. From this center all subsequent changes are induced, both in an anterior and posterior direction, and differentiation not only begins here but is always at an advanced stage relative to the rest of the developing insect. It has been shown by ligation experiments that the differentiation center does not release material out into the rest of the egg, but is the point of initiation of a contraction of the yolk from the chorion, an action that creates a space into which the cells forming the blastoderm can move. (It must be mentioned that the activation-center and differentiation-center descriptions are based on the study of comparatively few species of insects. Although there are probably similar centers in the embryos of most species, one should be cautious in applying these ideas to all species.)

Earlier in this section mention was made of the specialized region near the posterior pole of the egg. This is quite similar to the "centers" described above in that it determines which cleavage cells (those which migrate through it) will form the germ cells.

Insect eggs are quite variable as to when the cleavage cells become "determined" (i.e., when they become irreversibly destined to form a specific tissue). One extreme, found, for example, in *Musca* and *Drosophila,* is the "mosaic" egg, in which determination is completed before the egg is deposited and the elimination of any part of it will result in the formation of an embryo missing whatever structure the excised or destroyed cells would have formed. In eggs at the other extreme, determination is not complete until well after deposition, the cells having retained the "potency" to form any tissue. Such eggs are referred to as "regulation," this term alluding to the fact that the capacity for a change in developmental fate (regulation) persists through a much more advanced state than in the mosaic type.

Platycnemis (Odonata, Zygoptera) is an example of an insect with a regulation type of egg, and if a recently deposited egg is ligated in the middle, a "dwarf" embryo forms, demonstrating its well-developed capacity for regulation (Wigglesworth, 1972). Many gradations between the extremes of mosaic and regulation have been described and seem to correlate with phylogenetic level, the mosaic condition occurring in such groups as Diptera and the capacity for regulation being found to an increasing extent among the "lower" orders, such as Odonata. A correlation also exists between the regulative capacity and the amount of cytoplasm; the greater the cytoplasmic volume, the less the regulative capacity (Agrell and Lundquist, 1973).

The control of later development in the embryo is not as well understood as in the earlier stages, but there is evidence that the ectoderm is almost completely autodifferentiating, while the mesoderm depends on the ectoderm for the "induction" of further differentiation.

Adult characters are somewhat determined during embryogenesis, but apparently later than and independent of the earlier determinative change, which affects larval characters. In a manner analogous to the malformations produced in the larva by the removal or destruction of determined cells, similar removal or destruction following the determinative change that affects adult characters cause malformations in the adult.

Oviparity and Viviparity

Most insects are *oviparous;* that is, when they are released from the parent, they are surrounded by the eggshell or chorion. Depending on the species involved, the chorion ranges from very thin and delicate to very hard and thick, and the developmental stage of the insect within it varies from being in early cleavage to being ready to hatch as an active, free-living individual.

On the other hand, many insects are *viviparous* (i.e., they give "birth" to individuals having no chorionic covering). The terms *larviparous, nymphiparous,* and *pupiparous* are commonly used to refer to viviparous larvae, nymphs, ,and pupae, respectively. Although a group of Diptera are referred to as Pupipara, they are, in fact, larviparous; their name is really inappropriate. Members of the genus *Glossina* probably represent an extreme of viviparity in insects. One larva develops at a time in the highly specialized "uterus" and receives nutriment by means of specialized *uterine* (accessory) *glands*. When released from the parent insect, the tsetse larva is ready to pupate and does so within a few hours. While in the uterus the larvae ventilate by protruding their posterior spiracles through the parent's genital opening.

The term *ovoviviparity* is sometimes used and refers to instances where an egg with a chorion develops, but the egg hatches within the parent before it is deposited (Hagen, 1951). In some insects the embryo develops in the hemocoel of its parent (e.g., members of the Strepsiptera, as explained in the section on polyembryony).

Oviposition

The maternal parent insect typically deposits eggs in situations favorable for the survival of the offspring. Oviposition behavior is discussed in Chapter 8.

In oviparous insects the eggs are propelled down the oviducts by peristaltic waves and may be deposited either in groups or singly, depending on the species. The ovipositing structure may be a well-developed appendicular ovipositor (see Fig. 2-44A–D), or the abdomen may be modified in such a way that it can telescope into a relatively long tube and hence function effectively as an ovipositor. The latter structure is commonly referred to as an *ovitubus* and is found among the Thysanoptera, Diptera (see Fig. 2-44E), and others. Ovipositors are reduced or lacking in the following orders: Odonata, Plecoptera, Mallophaga, Anoplura, Coleoptera, and the panorpoid orders.

Eggs (see Fig. 5-11) are deposited in a variety of ways, some merely being dropped passively (e.g., walkingsticks, Phasmida) and others being "glued" singly or in masses to some substratum. Lacewings (Neuroptera, Chrysopidae) deposit their eggs at the tips of stiff stalks. Several orthopterous insects (e.g., locusts, mantids, and cockroaches) deposit their eggs in packets, or *oothecae*. The locusts secrete a frothy material that encases an egg mass, which is deposited in the ground. Mantids deposit their eggs on twigs in a foamy secretion, that subsequently hardens to produce the ootheca. Cockroaches have oothecae with a cuticlelike surface; some species carry these around with them, while others glue them to a substratum. In most species the secretions involved with forming, covering, and gluing the egg masses to the substrate come from the accessory glands. Like the cuticles of the integument and chorion, oothecae of various sorts probably permit gaseous exchange without undue water loss. Insects such as parasitic Hymenoptera use their ovipositors to inject eggs deep into a host insect. Aquatic insects commonly surround their eggs with a gelatinous secretion. Insects that parasitize vertebrates often attach their eggs to hair or feathers.

Oviposition is evidently under neural and hormonal control and may in some instances be induced by insemination (de Wilde and de Loof, 1973b).

Eclosion

Eclosion is used here to denote the process of hatching or exiting from the egg (it is sometimes used to mean emergence of the adult insect). Although the details of this process vary from group to group, eclosion generally first involves the swallowing of amniotic fluid by the larva with the attendant diffusion of air into the egg. In addition to the amniotic fluid, some of this air may also be swallowed. The problems faced at eclosion are the rupture of the chorion and other embryonic layers and escape from the confines of the egg. This rupture may occur irregularly about the surface of the egg or along preformed lines of weakness. In some instances the embryonic membranes may be weakened by the action of digestive enzymes. The actual force involved in the perforation of the chorion and embryonic membranes may be applied via various structures, including spines or eversible bladders or may simply involve the forced expansion of one region of the body by contraction of another, a process aided by the previous swallowing of amniotic fluid and, in some instances, air. Some insects (e.g., Lepidoptera) chew their way out of the eggshell. In one instance [*Glossina* (the tsetse fly)] in which the egg is retained and hatches within the "uterus" of the parent, the larva splits the chorion, but a parental structure removes it from the larva.

Postembryonic Morphogenesis

Here we shall include those events that occur between eclosion (or from the completion of embryogenesis in viviparous forms) and emergence of the adult insect.

Growth

Associated with the evolution of an external, relatively inexpansible exoskeleton has been the concurrent evolution of a mechanism that allows for increase in size, that is, the process of *molting*. Molting involves the periodic digestion of most of the old cuticle, secretion of new cuticle (usually with increased surface area), and shedding of undigested old cuticle. The last-mentioned step, shedding of undigested old cuticle (the *exuvium)*, is commonly referred to as *ecdysis*. The term *molt* is also sometimes used synonymously with ecdysis.

As a typical insect progresses from the newly hatched immature form, it goes through a series of molts, generally increasing in size with each one. Each developmental stage of the insect itself is called an *instar,* and the interval of time passed in that instar is sometimes referred to as a *stadium*. For now, we will view an instar as the individual between successive ecdyses.

In many insects, especially those with a small number of instars (e.g., mosquitoes), it is possible to determine exactly the instar of a given individual larva by characteristic morphological traits. In others, there may be little or no change between instars other than growth, which may vary considerably with the availability of food and other environmental factors. The final instar, during which sexual maturity and functional wings (if they are going to develop at all) are realized, is the adult, or *imago*. A few insects (e.g., thysanurans) continue to molt after they reach the adult stage; however, it is unlikely that there is any growth associated with these molts. In members of the order Ephemeroptera (mayflies) there are two winged instars, the first being the *subimago* and the second the adult. Other than mayflies, adult winged insects have never been known to molt under natural conditions.

The number of instars varies among insect species, the majority having between 2 and 20. In some insects the number of instars is constant; in others it may be variable, in response to environmental factors (e.g., availability of food and temperature). In some species the number of male instars may be different from the female number. The more specialized insects tend to have fewer instars.

Growth in insects occurs as the result of an increase in the number of cells by mitotic division and/or an increase in cell size. The increase in weight between a newly hatched immature and a fully grown immature is usually quite pronounced. For example, a fully grown larva of *Cossus cossus* (Lepidoptera; Cossidae, the carpenter moths) weighs 72,000 times its first instar weight, and it takes three years to accomplish this growth (Richards and Davies, 1977). In many other insects the magnitude of growth is somewhat smaller, varying from about 1000 times on up. Increase in weight may approximate an S-shaped curve, although its continuity may be disrupted by the swallowing of water by aquatic insects, or the ingestion of food, as in blood-feeding insects.

Growth in length and surface area appears to be discontinuous, owing to the relative inflexibility of the cuticle. From observations, in several species, of a degree of regularity in the extent of linear growth in various structures, certain "growth laws" have been induced. The most important of these is *Dyar's rule*, which is based on the assumption that growth as reflected in various linear dimensions follows a geometric progression.

According to this rule, the ratio between a given dimension in one instar and the same dimension in the next is constant throughout all instars and constant for each species. Dyar (1890) showed, for example, that in 28 species of lepidopteran larvae, the head capsule increases by a ratio of 1.4 at each molt. Thus if we were to plot the logarithm of the measurement of some linear dimension at each instar of a given species against the number of the instar, we would expect to obtain a straight line (Fig. 5-21). Where this rule applies, it is useful in determining the number of instars of an insect, since if the dimensions of the final instar and the ratio of a given dimension between any two successive instars are known, the dimensions of all the other instars can be deduced by interpolation and their number found by counting the number of points on the generated growth curve. However, in instances where the number of molts varies in response to certain environmental or other factors, Dyar's rule will not allow the determination of the number of instars.

An extension of Dyar's rule is *Przibram's rule*, based on the assumption that weight doubles during each instar, which says that

Fig. 5-21
Graph depicting Dyar's rule as it might apply to three hypothetical species of insects. A, B, and C are curves generated by plotting the log of a linear measurement against the number of a given instar. Given the information for the known instars in curve C, one can determine the total number of instars and the magnitude of the linear dimension pertinent to each one.

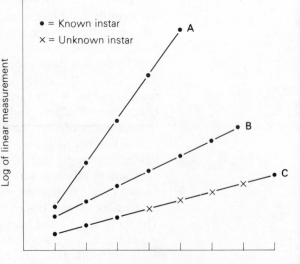

all linear dimensions at each molt are increased by the ratio of 1.26 or the cube root of 2 (Wigglesworth, 1972). Although this rule applied in the species studied when it was devised and in certain other species, it fails to apply in many other instances, for example, where weight more than doubles during each molt or where growth is attained by increase in cell size instead of cell number, as, for example, in the larvae of certain true flies. In addition, Przibram's rule assumes *harmonic* or *isogonic* growth (i.e., that all parts of the insect and the body as a whole increase by the same ratio during each molt). However, in the majority of insects growth during each molt is *heterogonic* or *allometric* (i.e., some parts of the insect body develop at different rates than other parts).

Metamorphosis

Most insects at eclosion are morphologically different from the adult. The degree of difference varies from relatively slight to extreme, with many intermediates. The developmental process by which a first-instar immature stage is transformed into the adult is called *metamorphosis,* which means literally "change in form." This process may take place gradually, with the immature being in general appearance comparatively similar to the adult, or it may be quite abrupt, the immature instars being drastically different from the adult with the transformation from the immature to the adult form occuring in a single stage.

The class Insecta can be divided into groups based on the type of metamorphosis, if it is present. Members of the Apterygota do not undergo any change in form, the immature instars differing from the adults only in size and the development of the gonads and external genitalia (Fig. 5-22A). These insects are sometimes grouped as Ametabola and said to undergo *ametabolous* development.

Members of the Pterygota can be divided into two groups relative to the degree of metamorphosis that occurs: *hemimetabola* and *holometabola.* Among the hemimetabolous insects the immatures resemble the adults in many respects, including the presence of compound eyes, but they lack wings, gonads, and external genitalia. During the course of development (Fig. 5-22B, C) the wings become externally apparent as *wing pads.* Thus the orders that fall into this group are those that are classified as *exopterygotes* (in reference to wings developing externally). The hemimetabolous form of development is often called simple, direct, or incomplete metamorphosis. The immature instars in this group of insects are commonly known as *nymphs,* although they may also be correctly referred to as *larvae.*

In the past insects with hemimetabolous development were subdivided further. Those insects (Odonata, Ephemeroptera, and Plecoptera) that pass the immature instars in an aquatic environment and the adult instar in a terrestrial/aerial environment, and the immatures of which at least superficially appear to be quite different from the adult stage (e.g., immature stoneflies, with highly specialized ventilatory gills, versus adults, which lack these structures and have well-developed wings) were classified as being *hemimetabolous,* or undergoing an incomplete metamorphosis (Fig. 5-22B).

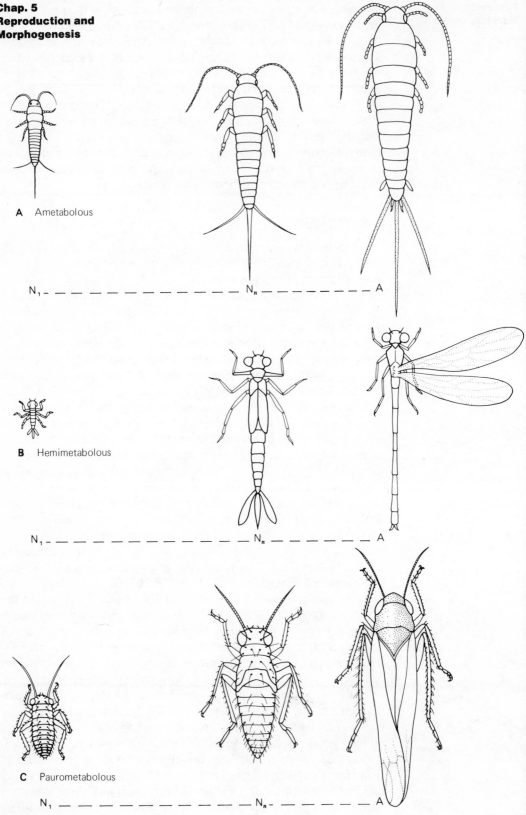

A Ametabolous

N_1 — — — — — — — — — — N_n — — — — — A

B Hemimetabolous

N_1 — — — — — — — — — N_n — — — — — A

C Paurometabolous

N_1 — — — — — — — — — — N_n — — — — — A

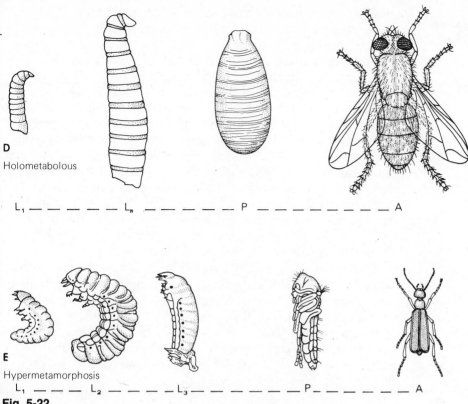

D
Holometabolous

L_1 — — — — — L_n — — — — — P — — — — — — — A

E
Hypermetamorphosis

L_1 — — — L_2 — — — — L_3 — — — — — — P — — — — — A

Fig. 5-22
Types of insect metamorphosis. A. Ametabolous, silverfish. B. Hemimetabolous, damselfly. C. Paurometabolous, leafhopper. D. Holometabolous, house fly. E. Hypermetamorphosis, beetle, *Epicauta cinerea* (Coleoptera; Meloidae). A, adult; L_1, first instar larva; L_2, second instar larva; L_3, third instar larva; Ln, nth instar larva; N_1, first instar nymph; N_n, nth instar nymph; P, pupa. [C redrawn from Essig, 1958; D, E redrawn from Packard, 1898.]

Those insects passing both the immature and adult instars in essentially the same environment were classified as being *paurometabolous,* or undergoing gradual metamorphosis (Fig. 5-22C). The immature instars of hemimetabolous insects were called *naiads;* those with paurometabolous development were referred to as nymphs. Morphological studies have indicated that although Odonata and Ephemeroptera show some affinity with each other, they are not closely related to Plecoptera, and hence these three orders hardly form a cohesive phylogenetic group. The paurometabolous and hemimetabolous groupings have therefore been abandoned and the two merged under the heading "hemimetabolous."

The remaining group of orders, those classified as *endopterygotes* (in reference to wings developing inside the body of the larva), all undergo what is referred to as *holometabolous* development or complete metamorphosis (Fig. 5-22D). In these insects the immature instars are quite dissimilar to the adults and generally are adapted to different environmental situations. Immature instars in this group of insects are called *larvae*. Individuals in the larval stage typically lack compound eyes and usually have mandibulate mouthparts, whether or not they are mandibulate in the adult instar. They may or may not have thoracic or abdominal legs. Most of the changes in the transformation from the last instar larva to the adult are compressed into a single intervening instar, the *pupa*. The pupa is typically a resting stage protected in some way (by a silken cocoon, hidden in leaf litter, in a puparial case, etc.), but the pupae of some insects are quite active. For instance, the aquatic pupae of mosquitoes (''tumblers'') are very active and dive in response to potentially threatening stimuli. In most endopterygotes the larval instars resemble one another except for a few minor morphological details, which are useful in distinguishing one instar from another. However, some holometabolous insects pass through one or more larval instars that are distinctly different from the others (Fig. 5-22E). This phenomenon is called *hypermetamorphosis* and has been described in certain species in the orders Neuroptera, Coleoptera, Diptera, and Hymenoptera, and in all species in the order Strepsiptera.

The grouping of insects based on metamorphosis as described above is not without exceptions. For example, members of some orders that are included in Hemimetabola are secondarily wingless in the adult stage, and their development more closely resembles that of the apterygotes in that it is essentially ametabolous. More important, certain of the *hemimetabolous* insects in fact have a pupal stage. For example, thrips (Thysanoptera) and certain members of the Homoptera (whiteflies, and certain scale insects, e.g., *Pseudococcus* spp.) have a distinct resting ''pupal'' stage between the last ''larval'' and the adult instars (Hinton, 1948). Even members of the ''primitive'' order Odonata undergo considerable morphological change during the transformation from the aquatic nymph to the terrestrial/aerial adult. These changes include loss of rectal gills, restructuring of the labium, modifications in the head and abdomen, complete reconstruction of the alimentary canal, and nearly complete replacement of the abdominal musculature.

It is difficult not to look upon the development of these insects as being ''holometabolous,'' indicating that holometabolism has probably evolved on separate, independent occasions. However, Hinton (1963b) recognizes a distinct difference between the holometabolous exopterygotes and the endopterygotes. He points out that since the wings develop internally in endopterygotes, there is insufficient room in the larval thorax for the development of both the wings and the wing muscles. Therefore, a molt is required to evaginate the wings (the larval–pupal molt) and provide space for the development of wing muscles. A second molt (pupal–adult) is then required to release the adult from the confines of the pupal cuticle. A pupal stage, in this sense, is unnecessary in exopterygotes since the wings develop externally. Hinton accounts for the ''pupae'' in certain exopterygote insects as follows.

In some exopterygotes such as the Aleyrodidae and male Thysanoptera and Coccoidea, the general structure of the feeding larval instars has departed very widely from that of the adult. The structural differences between the feeding larval instars and the adult are normally bridged in the last larval instars. As the differences between the two stages became greater and involved a greater degree of re-organization of the internal tissues, it would seem that the last or last two larval instars became more and more quiescent and eventually ceased to feed. It may be noted here that the structural re-organization required to bridge the gap between the feeding larval stages and the adult of some exopterygotes is greater than in the primitive endopterygotes, e.g. some Megaloptera. No difficulty necessarily arises if these quiescent or semi-quiescent stages of exopterygotes are called pupae provided that it is recognized that their origin is quite independent from that of the endopterygote pupa and their initial functional significance is different.

Wigglesworth (1972) and others confine usage of the term "metamorphosis" to those changes that occur when an insect becomes an adult, regardless of whether the adult stage is reached by a single molt of the last immature instar or in two molts with a pupal stage in between. Wigglesworth (1954) points out that metamorphosis is commonly looked upon as a renewal of embryonic development. However, he regards the larvae and adults of holometabolous insects as essentially two organisms, which are latent within the genome of the embryo and which are expressed in sequence. He views this as a sort of temporal polymorphism. The evolution of a pupal stage in which comparatively drastic changes can occur has evidently enabled the divergent evolution of larvae and adults, which are usually adapted to radically different environmental modes of existence. As will be explained later, the same basic endocrine mechanism probably governs development in all insects.

The immature stages of insects take on a wide variety of forms. In most instances the nymphs of hemimetabolous species closely resemble the adults, but in the holometabolous species, the larvae are often drastically different from the adults. Although there is considerable variation in the appearance of larvae of the different holometabolous groups, there are sufficient similarities to allow recognition of distinct "larval types." Some types of larvae (Fig. 5-23) and terms commonly used to describe them are as follows.

1. *Campodeiform* larvae resemble diplurans in the genus *Campodea,* having flattened bodies, long legs, and usually long antennae and cerci (e.g., several beetles, Neuroptera, and Trichoptera).
2. *Carabiform* larvae resemble the larvae of carabid beetles, which are similar to the campodeiform type but have shorter legs and cerci (e.g., several beetles).
3. *Elateriform* larvae resemble click beetle larvae (Coleoptera; Elateridae) and have cylindrical bodies with a distinct head, short legs, and a smooth, hard cuticle (e.g., certain beetles).
4. *Eruciform* larvae are the typical Lepidopteran "caterpillars" or the caterpillarlike larvae of certain Hymenoptera and Mecoptera and have a cylindrical body with a well-developed head, short antennae, and short thoracic and abdominal legs (*prolegs*).
5. *Platyform* larvae have flattened bodies with or without short thoracic legs (e.g., certain Lepidoptera, Diptera, and Coleoptera).

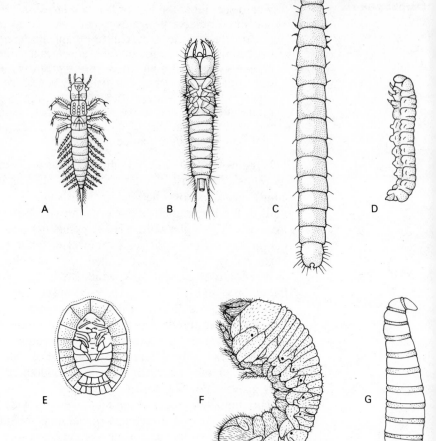

Fig. 5-23
Larval types. A. Campodeiform, alderfly, *Sialis* sp. (Neuroptera;
Sialidae). B. Carabiform, ground beetle, *Harpalus* sp. (Coleoptera;
Carabidae). C. Elateriform, click beetle (Coleoptera; Elateridae). D.
Eruciform, clear-winged moth (Lepidoptera; Aegeriidae). E. Platyform,
aquatic beetle, *Eubrianax edwardsi* (Coleoptera; Dascillidae). F.
Scarabaeiform, branch and twig borer, *Polycaon confertus* (Coleoptera;
Bostrichidae). G. Vermiform, flesh fly, *Sarchophaga* sp. (Diptera;
Sarcophagidae). [A, B, D, G redrawn from Packard, 1898; C, E, F
redrawn from Essig, 1958.]

6. *Scarabaeiform* larvae are the ''grubs'' and have a cylindrical
   body typically curled into a C shape, a well-developed head,
   and thoracic legs (e.g., several beetles).
7. *Vermiform* larvae have wormlike bodies with no legs and may
   or may not have a well-developed head (e.g., several Diptera,
   Coleoptera, Hymenoptera, Siphonaptera, and Lepidoptera).

It must be emphasized that this classification of larval types is a
pragmatic one reflecting only gross similarities and not phylogenetic
affinities. Obviously one should expect to find numerous examples
of larvae that display a mixture of characteristics of two or more of
these types.

Berlese classified insect larvae based on the assumption that they are essentially free-living embryos that are released to the external environment in different stages of development in different species (Richards and Davies, 1977). He recognized three such stages: *protopod, polypod,* and *oligopod.* Protopod larvae are very uncommon and represent a very early state of development in which comparatively little segmentation has occurred. These larvae are found, for example, among certain parasitic Hymenoptera that larviposit in the hemocoel of other insects, placing the "embryo" in the only kind of environment possible for survival—a protected one in which it is surrounded by nutriment. Polypod larvae resemble caterpillars, and hence the eruciform type described above would fall readily into this group. Oligopod larvae lack the abdominal prolegs and may or may not have well-developed cerci. The campodeiform, carabiform, elateriform, platyform, and scarabaeiform types listed above fall nicely into this category. The vermiform (apodous) type of larva is considered to have developed secondarily from the polypod or oligopod type, depending on the kind of insect, and hence is not recognized as a separate group.

A subscriber to Berlese's view of holometabolous larvae must recognize a fundamental difference between these larvae and the nymphs of hemimetabolous insects, which in this context represent postoligopod stages of development. However, one can as easily look upon holometabolous larvae as specialized nymphs with a concurrent concentration of the changes to the adult form in a single stage, the pupa. From this point of view there are no fundamental differences between larvae and nymphs, and the terms could appropriately be used as synonyms. As will be explained subsequently, the results of comparatively recent investigations of the physiological mechanisms that control metamorphosis have supported this latter conception of the relationship between hemimetabolism and holometabolism.

As with larvae, pupae have been grouped according to similarities. Hinton (1964) classified pupae based on whether or not they have articulated mandibles that are used in escaping from a cocoon or pupal cell. Those pupae having such mandibles he described as being *decticous*. Examples of decticous pupae include members of the orders Neuroptera, Mecoptera, Trichoptera, and certain lepidopterous families. Pupae without functional mandibles used in escape from cocoon or pupal cell are termed *adecticous* and are found in all or some members of the orders Strepsiptera, Coleoptera, Hymenoptera, Diptera, and Siphonaptera. Another approach to grouping pupal types (Fig. 5-24) is based on whether the appendages are free or adherent to the body. *Exarate* pupae (Fig. 5-24A) have the appendages free and are usually not covered by a cocoon. In *obtect* pupae (Fig. 5-24B) the appendages adhere closely to the body and are commonly covered by a cocoon. All decticous pupae are exarate, whereas the adecticous types may be obtect, exarate, or *coarctate,* a third type (Fig. 5-24C). A coarctate pupa is encased in the hardened cuticle of the next-to-last (penultimate) larval instar, the *puparium*. However, the pupa itself is of the adecticous exarate form.

Many histological changes occur during the transition from im-

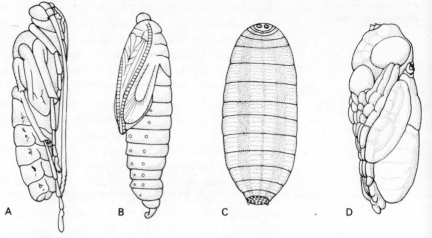

A   B   C   D

Fig. 5-24
Pupal types. A. Exarate, ichneumon wasp. B. Obtect, moth. C.
Coarctate, house fly. D. Puparial case removed, exposing exarate house
fly pupa. [A redrawn from Boreal Labs key card; B redrawn from
Packard, 1898; C, D redrawn from Wigglesworth, 1970.]

mature to adult (Whitten, 1968). In hemimetabolous insects these
changes are comparatively gradual, being spread throughout the
nymphal instars, although there are generally more changes during
the last instar than during earlier ones. In holometabolous insects,
these changes occur mostly in the pupal stage. The extent of change
during the pupal period is variable; in some species a large amount
of tissue change occurs. These changes are accomplished by means
of tissue breakdown (histolysis) and tissue reorientation, growth,
and differentiation (histogenesis). Histolyzed tissues may simply
dissolve in the hemolymph or may be ingested by phagocytic
hemocytes.

Many adult tissues are formed from masses of cells that have
persisted in an undifferentiated state throughout the larval instars
even though they increase in number by mitotic divisions. These
masses of embryonic cells are generally referred to as *imaginal buds,*
or *discs* (Figs. 5-25) and are involved in the formation of structures
such as mouthparts, antennae, wings and legs. Imaginal discs, especially from *Drosophila melanogaster,* have proved to be extremely
valuable in studies of cellular differentiation (Gehring and Nöthiger,
1973; Ursprung and Nöthiger, 1972). In addition to imaginal discs,
rings of embryonic cells are generally found at the posterior extremity of the foregut (*anterior imaginal ring*) and the anterior extremity
of the hindgut (*posterior imaginal ring*). These undergo considerable
growth and differentiation during the pupal stage and contribute
significantly to the formation of the adult alimentary canal. For example, the ventral diverticulum in mosquitoes (see Fig. 4-2) develops as the result of mitotic division of cells in the anterior imaginal
ring. In some instances, certain larval tissues persist into the adult
stage, and these vary with the different groups of insects.

Also of interest in the development of insects are the changes and
control of changes in integumental pattern (distribution of bristles,

etc.) that occur during growth and metamorphosis. Lawrence (1970, 1973) and Waddington (1973) review the literature on the development of spatial patterns in the integument.

Although some pupae (e.g., mosquitoes) are active and can evade potential predators or adverse environmental conditions, most are not and hence are quite vulnerable. A number of mechanisms have evolved that decrease this vulnerability. Probably the most common of these is the use of silk in constructing a cocoon of some sort (e.g., many Lepidoptera, Hymenoptera, Neuroptera, Trichoptera, and Siphonaptera). This cocoon may be composed solely of silk, or it may be basically silk but with bits of environmental debris incorporated into it. Recall the beetle, mentioned in Chapter 4, that constructs a cocoon from collapsed peritrophic membrane. Many insects (e.g., Coleoptera and Lepidoptera) construct cells beneath the surface of the soil and pupate within them. Another example of a pupal protection mechanism is the puparium mentioned above. Obviously the pupae of endoparasitic forms derive protection from being within the body of the host.

Concurrent with the evolution of pupal protection mechanisms was the development of means for escaping at the time of adult emergence. Decticous pupae chew their way out of their cocoon or cell and may be aided in this process by posteriorly directed spines. Escape methods for adecticous pupae include a variety of spines or other hard and sharp protuberances and an eversible bladder, the *ptilinum*, in the head that is used to force open the "cap" of a puparial case in the cyclorrhaphous Diptera. The labial glands of recently emerged adult saturniid and bombyliid moths secrete a solution containing the enzyme *cocoonase*, which digests the silken cocoon and allows easy exit (Kafatos et al., 1967; Kafatos, 1972).

When an adult is nearly ready to emerge, the pupal cuticle splits in various places, particularly along the dorsal midline of the body. At emergence hemostatic pressure is exerted on the pupal cuticle by the contraction of body muscles (likely aided by the swallowing of

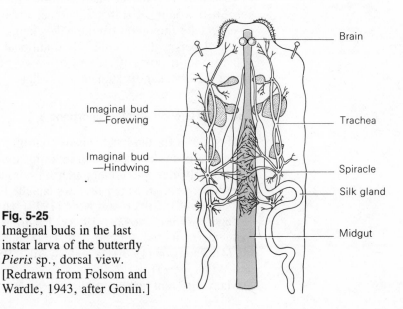

Brain

Imaginal bud
—Forewing

Trachea

Imaginal bud
—Hindwing

Spiracle

Silk gland

Midgut

Fig. 5-25
Imaginal buds in the last
instar larva of the butterfly
Pieris sp., dorsal view.
[Redrawn from Folsom and
Wardle, 1943, after Gonin.]

air or water). The newly emerged adult is in a very vulnerable state until the cuticle hardens and the wings expand and harden. The term *teneral* is commonly used to refer to such a newly emerged, pale, soft-bodied individual regardless of stage.

In addition to morphological, histological, and cellular changes during growth and development, there are biochemical changes as well. Major reviews on this topic include Agrell and Lundquist (1973), Chen (1971), L'Helias (1970), and Wyatt (1968).

The Instar Definition Controversy

A controversy exists over the precise definition of an instar (Hinton, 1946b, 1958, 1971, 1973, 1976a; Whitten, 1976; Wigglesworth, 1973a). Traditionally, since Linneaus, an instar has been defined as the individual between successive ecdyses. However, Hinton (1958, 1971) argues that an instar should be defined relative to the cuticle to which the epidermis is actually attached. Thus he suggests that a new instar actually begins with each *apolysis* (the time of separation between the epidermis and old cuticle prior to secretion of new cuticle; Jenkin and Hinton, 1966). He denotes the period of time between apolysis and the beginning of new cuticle secretion as the *exuvial phase,* and the period of time between the first appearance of new cuticle beneath the old and ecdysis as the *cuticular phase.* Between apolysis and ecdysis, the insect is still within the confines of the old cuticle. Hinton refers to an insect in this state as being *pharate.*

By the traditional terminology the pupal instar, for example, would be an individual between larval–pupal ecdysis and pupal–adult ecdysis. By Hinton's terminology, on the other hand, the pupal instar actually begins with larval–pupal apolysis within the confines of the fully grown larva's cuticle and remains a pharate pupa until larval–pupal ecdysis, after which it is called a pupa, until pupal–adult apolysis. At pupal–adult apolysis, it becomes a pharate adult.

The controversy over the definition of an instar has not been resolved. It would seem that adherence to Hinton's definition would be especially important from the developmental, histological, physiological perspective, while the traditional definition would be of value from the point of view of behavior, ecology, and systematics where the organism is studied as a whole.

Control of Growth and Metamorphosis

The search for the mechanisms controlling growth and metamorphosis has occupied the time of investigators for decades. As a result a reasonably clear picture of patterns of tissue and hormone involvement has emerged. Major reviews include Doane (1973); Etkin and Gilbert (1968); Gilbert and King (1973); Highnam and Hill (1977); Kroeger (1968); Novák (1975); Riddiford and Truman (1978); Schneiderman and Gilbert (1964); Sláma, Romanuk and Sorm (1974); Thomson (1976), Wigglesworth (1954, 1959, 1964b, 1970); and Willis (1974).

Many different techniques have been employed in the elucidation of the mechanisms controlling growth and metamorphosis in insects. Early workers utilized ligation and decapitation extensively, and these techniques still find application today. Later techniques employed have included microsurgical removal of various tissues (see Fig. 3-10), implantation of various tissues, hemolymph transfusion, and joining the hemolymph circulation of one individual with that of one or more other individuals (*parabiosis*). Recently, sophisticated biochemical techniques have made possible the extraction— and, in the case of ecdysone, purification—of the various hormones, and cytological techniques have brought a degree of understanding of the nature and modes of action of these hormones (Gilbert, 1976; Menn and Beroza, 1972; Morgan and Poole, 1976; Rees, 1977; Riddiford and Truman, 1978).

Among the structures and tissues involved in the control of growth and metamorphosis (Fig. 5-26) are

1. The *median neurosecretory cells* in the *pars intercerebralis* region of the brain.

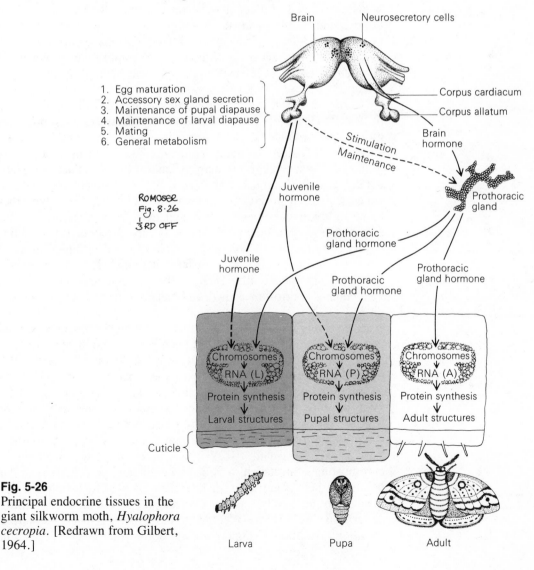

Fig. 5-26
Principal endocrine tissues in the giant silkworm moth, *Hyalophora cecropia*. [Redrawn from Gilbert, 1964.]

2. The *corpora cardiaca*.
3. The *prothoracic glands* located in the prothorax in close association with tracheae.
4. The *corpora allata*.

The neurosecretory cells in the brain secrete the *brain hormone* (also called *prothoraciotropic hormone* and *ecdysiotropin*) and *eclosion hormone*, both of which accumulate in the corpora cardiaca and are subsequently released into the hemolymph. Brain hormone stimulates the secretory activity of the prothoracic gland. The stimuli that cause the secretion of brain hormone vary. In the blood-sucking bug *Rhodnius* (Hemiptera, Reduviidae), abdominal distension resulting from feeding leads to brain hormone secretion. In the grasshopper *Locusta* (Orthoptera, Acrididae), stretch receptors in the wall of the pharynx are involved. Eclosion hormone plays a major role in pupal–adult ecdysis. The prothoracic glands secrete *ecdysone* (also called *prothoracic gland hormone* and *molting hormone*), which initiates the growth and molting activities of cells. Ecdysone is also produced in other tissues, for example, the ovaries of mosquitoes and locusts, but these sources are not considered to have bearing on molting. The corpora allata secrete *juvenile hormone,* which promotes larval development and inhibits development of adult characteristics. It has been described as having a "status quo" effect.

During the larval instars both ecdysone and juvenile hormone are produced. However, during the last immature instar, juvenile hormone is not produced or decreases below some threshold; hence, the expression of adult characters is not inhibited, and metamorphosis to the adult stage occurs. In hemimetabolous insects there is a gradual decrease in the concentration of juvenile hormone produced and hence a gradual progression toward the adult stage. On the other hand, in holometabolous forms the changes from immature to adult are compressed into the pupal stage, but the principle of discontinuation or reduction below some threshold of secretion of juvenile hormone during the last larval instar still applies. Thus there appears to be no significant difference in the fundamental physiological mechanisms controlling growth and metamorphosis in either of these groups.

Ecdysone is actually a generic term for a number of similar steroid compounds, some of which have even been found in certain plants (*phytoecdysones;* Williams, 1970) (more on this in Chapter 9). Ecdysone apparently acts directly on genes, determining which ones are brought into action at a given time, and as a result influences the kinds of proteins (both enzymes and structural proteins) synthesized. Support for this idea of the action of ecdysone has been found in studies of *chromosomal puffs* in giant chromosomes (Ashburner, 1970, 1972; Beerman, 1972). These puffs, localized swellings in chromosomes, are taken to indicate heightened gene activity at the points of their occurrence. It has been shown, for example, that injection of pure ecdysone into *Chironomus* larvae produces puffing patterns in chromosomes identical to those patterns that occur during pupation. Additional support for the notion that ecdysone acts on genes comes from its involvement in sclerotization of the puparium in higher Diptera. This hormone is thought to activate a gene that directs the synthesis of a key enzyme in the sclerotization process (Karlson and Sekeris, 1976). For detailed information on the extraction and analysis of ecdysone, see Morgan and Poole (1976).

Little is known about the chemical nature of brain hormone, but its mode of action is considered to be its influence on cyclic AMP in the cells in the prothoracic gland.

The chemical structure of juvenile hormone has been elucidated (Gilbert, 1976). Actually, three different hormones have been found. They are terpenoids. The mode of action of the juvenile hormones is not known for certain, but they may influence RNA translation (Ilan, Ilan, and Patel, 1972; Williams and Kafatos, 1972). As with ecdysone, juvenile hormone mimics have also been found in plants (Williams, 1970). Menn and Beroza (1972) deal with several aspects of juvenile hormone.

At least two other hormones are involved in regulating processes associated with the growth and development of some insects, *eclosion hormone,* mentioned earlier, and *bursicon.*

Eclosion hormone (Truman, 1971–1973; Riddiford and Truman, 1978), a protein, has been found in several Lepidoptera. It is synthesized in the median neurosecretory cells of the pars intercerebralis in the brain and stored in the corpora cardiaca. Release is controlled by a circadian clock. This hormone influences several aspects of pupal–adult ecdysis (''eclosion''), including the behavior associated with ecdysis, and the subsequent degeneration of the abdominal intersegmental muscles used in the act of ecdysis. Eclosion hormone is also apparently present in the Diptera, but has not been found in hemimetabolous insects. That it has been found only in holometabolous forms and only associated with pupal–adult ecdysis suggests that it may be restricted to Holometabola and used solely to regulate the changes in the central nervous system that occur at adult emergence (pupal–adult ecdysis).

Bursicon (*tanning hormone;* Riddiford and Truman, 1978; Cottrell, 1964; Fraenkel and Hsaio, 1965) is a neurosecretory hormone synthesized and/or released from a variety of different sites according to species. Commonly, it is found in *neurohaemal organs* (similar to the corpora cardiaca) associated with the ventral chain ganglia. Bursicon stimulates sclerotization of the cuticle following ecdysis. It was initially discovered in blow flies (*Calliphora* spp., Diptera) and has subsequently been found in several different insects and at different life stages.

Polymorphism

The term *polymorphism* means literally ''many forms.'' From this definition it follows that any number of phenomena could be cited as examples of polymorphism: differences between the sexes (*sexual dimorphism*), differences between individuals caused by the external environment, differences between developmental stages (i.e., the ''temporal polymorphism'' referred to earlier), and so on. However, the more restricted definition of Richards (1961) seems more useful, although there is still disagreement regarding the meaning of polymorphism. Richards' definition is as follows: Polymorphism exists when

one or both sexes of a species occur in two or more forms which are sufficiently sharply distinct to be recognizable without a morphometric analysis; the occurrence is regular or recurrent; the rarer of the two forms makes up a reasonable proportion of the population (say, at least 5 per cent) or, as in some social species, the rarest type is at any rate essential to the survival of the species.

Actually, since form and function are inseparable, functional as well as morphological differences should be included in one's definition and concept of polymorphism.

Polymorphism is expressed in a variety of traits in different insects. Examples of these traits include coloration and patterns of color, as, for example, in many butterflies; the presence, absence, or attenuation of wings, as found in aphids and other insects; chromosomal differences; the relative size of different structures; and various integumental structures, including horns, spines, and other protuberances. Species in several orders exhibit one or more types of polymorphism, but polymorphism has not been found in the following orders (Richards, 1961): Thysanura, Diplura, Protura, Ephemeroptera, Grylloblattoidea (included with the Orthoptera in this text), Embioptera, Mallophaga, Thysanoptera, Mecoptera, and Trichoptera.

Following Richards' definition, there are basically two types of polymorphism, both of which are controlled by genes.

In one form of polymorphism (i.e., *balanced polymorphism*) the relative abundance of the different forms depends on variation in the selection pressures acting on members of the same species. Hence different populations may show significant differences in the relative abundance of the different forms. An example is found in the variations in relative numbers of different forms of the mimetic butterflies (*Papilio dardanus*) in different regions depending on the abundance of the species that they mimic (Sheppard, 1961).

In the other type of polymorphism, expression of form is determined by various environmental factors, often influencing the endocrine system and ultimately the action of different genes, all individuals in a population having essentially the same genotype. A number of environmental influences have been shown to play a role in polymorphic expression in different species.

Nutrition, both in terms of quantity and dietary balance, plays a significant role in polymorphism in honey bees. Female larvae (from fertilized eggs) all have the potential to develop into either workers or queens, depending on the period of time they are fed royal jelly, a highly nutritious substance produced by the salivary glands of workers. If larvae are fed royal jelly for only two or three days, they develop into worker adults, but if they are fed royal jelly throughout the larval stadia, they develop into queens. By experimentally varying the length of time larvae are fed royal jelly, it has been possible to produce intermediates between workers and queens.

Variations in the quantity of food available can produce polymorphism in insects that grow heterogonically (see page 175) by inducing early or late pupation. For example, the mandibles in stag beetles (Lucanidae, Coleoptera) develop at a greater rate than other parts of the body. So a larger individual that had an abundance of food as a larva will have disproportionately larger mandibles in comparison with a smaller individual that had less food.

Invasion by certain parasites is known to influence morphologic expression. For instance, mermithid nematodes cause worker or soldier ants to show female traits. Parasitization of certain insects by strepsipterans may also cause such changes in form.

Several environmental factors are known to influence polymorphism in aphids (Hemiptera, Homoptera). Aphids may exist as *winged (alate)* or *wingless (apterous) parthenogenetic* forms or as *wingless* or *brachypterous* (very short wings) *sexual forms*. These forms vary with the environmental situation, and the life cycle of a given species may be complex, including generations of different forms at different times. A variety of environmental factors have been found to influence polymorphism in different species, for example, photoperiod, temperature, changes in host plant (e.g., wilting and nutritive changes), and crowding. The changes in form induced by environmental changes are adaptive. For example, there may be several parthenogenetic generations during long photoperiods when host plants are abundant, and with the shortening photoperiods associated with autumn, sexual forms appear, mate, and produce eggs that will overwinter. Alate forms may be produced in response to crowding, facilitating dispersal. As with termites, the endocrine system (especially juvenile hormone and ecdysone) is thought to be involved in the expression of polymorphism.

Crowding results in "phase changes" in some Orthoptera and Lepidoptera. For example, in the migratory locust, *Schistocerca gregaria,* there are changes in color pattern and behavior between solitary individuals and the same individuals when they enter the migratory (gregarious) phase in response to the constant agitation brought about by crowding.

Pheromones (chemicals produced by one individual that influence another individual of the same species; see Chapter 8) may influence the expression of particular forms and functions, for example, among the lower termites. These insects typically have an elaborate caste system consisting of winged adults (kings and queens), wingless supplementary reproductive adults, larvae and nymphs that serve as workers and as a pool for transformation into other castes, and soldiers. This caste system is very flexible. Following hatching from the egg, there are several instars called larvae followed by several instars called nymphs. Older larvae have the capacity to develop into nymphs, soldiers, or supplementary reproductives. Nymphs can develop into winged reproductives, supplementary reproductives, or soldiers and can even regress, becoming larvae again. The caste system is at least partly mediated by substances transferred from one individual to another by *proctodael feeding* (i.e., one individual ingests material from the anus of another individual). These substances regulate the numbers of a given caste in a colony (Lüscher, 1961a). For example, in many termite species supplementary reproductive castes do not appear as long as the original king and queen are present. The king and/or queen produce a substance that circulates throughout the colony and inhibits the formation of supplementary reproductives. Death of the royal pair would then result in the loss of this inhibitory substance, and hence the supplementary reproductives would appear, assuring the continued survival of the colony. Soldiers may also produce a substance that passes from individual to individual throughout the colony and inhibits development of more soldiers. The substances that inhibit development in larvae and nymphs apparently exert their effect on the endocrine system, in particular on the balance between juvenile hormone and ecdysone (Lüscher and Springhetti, 1963). For example, the implantation of additional corpora allata can induce development of larva or nymph into a soldier. Juvenile hormone itself may be passed from individual to individual.

The Royal Entomological Society's symposium, *Insect Polymorphism* (Kennedy, 1961), and Lüscher (1976) should be consulted for further information.

Regeneration

Most, if not all, insects are probably capable of at least some regeneration, particularly with regard to wound healing. This process often involves hemolymph clotting and hemocytes, but the epidermal cells play the major role. Dead or injured epidermal cells in *Rhodnius prolixus* (Hemiptera, Reduviidae) apparently produce a substance that has an attractive effect on the surrounding epidermal cells, which migrate to a wound and lay down new cuticle (Wigglesworth, 1972). Concurrently, mitoses in the regions from which the epidermal cells migrate restore the original density of cells in those regions. Some insects (e.g., the walkingstick, *Carausius*, Phasmida) can be decapitated and the head replaced with the result that the epidermis and gut, but not neural tissue, grow together again. However, this is not the case in other insects that have been studied (e.g., *Cimex* and *Rhodnius*, Hemiptera) in which only the continuity of the integument (cells and cuticle) is reestablished.

If appendages of developing larvae are removed, they may be regenerated during later instars, indicating that at least some of the cells surrounding an appendage are sufficiently undifferentiated to retain the ability to reform that appendage. Some insects (e.g., walkingsticks and certain other Orthoptera) exhibit the ability to spontaneously amputate a leg, generally between the trochanter and femur. This phenomenon is called *autotomy* and has the obvious adaptive value of allowing escape from the grasp of a predator. Many immature insects (e.g., walkingsticks and mantids) can completely regenerate a lost appendage, and this regeneration usually requires a molt. Evidently the capacity for regeneration of the external form of an appendage is wholly within the epidermal cells, since removal of associated ganglia has failed to block regeneration of form, at least in the species studied.

In some instances abnormal regeneration may occur. Abnormalities include the duplication or triplication of an appendage at its tip and *heteromorphous* regeneration, where the regenerated appendage is like another appendage on the body but unlike the one it replaces. An example of the latter would be the growth of a leg where an antenna had originally been.

Aging

Brief consideration of those processes that ultimately bring about termination of the life of an insect is an appropriate way to complete a discussion of postembryonic morphogenesis. The reviews by Clark and Rockstein (1964) and Rockstein and Miquel (1973) cover aging at length.

Aging includes all those changes in structure and function that occur from the beginning of life to its termination. These changes are predictable and reproducible. *Sensescence* refers to all those changes in structure and function that decrease an individual's capacity for survival and ultimately lead to the death of the individual.

Theories on the causes of senescence are based in two areas, heredity and environment. Probably in the majority of situations, both heredity and environment act together to produce the observed senescence. Hereditary factors are as follows (Rockstein and Miquel, 1973).

1. Aging as the result of programmed retardation or cessation of growth.
2. Programmed retardation and/or failure of some substance necessary for maintenance of the nonsenescing state.
3. Depletion of DNA, RNA, enzymes, coenzymes, and other substances essential for cell function.
4. Scheduled production or accumulation of an aging substance.
5. Scheduled accumulation of material(s) that may become harmful to an organism in time.

Environmental factors are as follows (Rockstein and Miquel, 1973).

1. Cumulative radiation (ionizing, infrared, etc.) effect.
2. Cumulative pathological effects from parasitic invasions.
3. Cumulative effects of physical insults such as extremes in temperature or mechanical injury.

Where information is available, species have a constant distribution of individual life spans under defined (genetically and environmentally) conditions. Among species, life spans are quite variable, and the overall range is probably fairly accurately reflected by 1 day for adults of certain mayflies to more than 25 years for certain termite species. Differences in mean life span have been found between different genetic strains, between populations with different diets, and between populations exposed to different temperature regimes. Some examples of mean adult life span in different insects are given in Table 5-1.

Table 5-1
Mean Adult Life Span (in Days) of Insects[a]

Insect	Male	Female	Reference[b]
DIPTERA			
Drosophila melanogaster	38.1	40.1	Gonzalez (1923)
Wild (line 107)			
Vestigial mutant	15.0	21.0	
D. subobscura	40.0	36.4	Maynard Smith (1959)
9 inbred lines			
(average)			
4 outbred populations	56.8	60.0	
(average)			
Musca domestica	17.5	29.0	Rockstein and Lieberman
			(1959)
M. vicina	20.8	23.3	Feldman-Muhsam and
			Muhsam (1945)
Calliphora	35.2	24.2	
erythrocephala			
Aedes aegypti		15	Kershaw et al. (1953)
LEPIDOPTERA			
Acrobasis caryae	6.5	7.3	Pearl and Miner (1936)
Bombyx mori (unmated)	11.9	11.9	Alpatov and Gordeenko
			(1932)
B. mori (mated)	15.2	14.2	
Fumea crassiorella		5.5	Matthes (1951)
(unmated)			
F. crassiorella (mated)		2.3	
Samia cecropia	10.4	10.1	MacArthur and Baillie
			(1932)
S. californica	8.7	8.8	
Tropea luna	5.9	6.0	
Philosamia cynthia	5.9	7.1	
Telea polyphemus	8.1	10.0	
Collasamia promethea	4.6	7.0	
Pyrausta nubilalis	13.0	17.4	
P. penitalis	6.8	7.7	
Carpocapsa pomonella	9.4	10.2	
HYMENOPTERA			
Apis mellifera			
Summer bees		35	Ribbands (1952)
Winter bees		350	Maurizio (1959)
Habrobracon juglandis	24	29	Georgiana (1949)
Wild type			
Small wings, white	20	24	
eyes, mutant			
H. serinopae	62	92	Clark and Rubin (1961)
ORTHOPTERA			
Blatta orientalis	40.2	43.5	Rau (1924)
Periplaneta americana	200	225	Griffiths and Tauber (1942)
Schistocerca gregaria	75	75	Bodenheimer (1938)

Table 5-1 (continued)

Insect	Male	Female	Reference[b]
	COLEOPTERA		
Tribolium confusum	178	195	Park (1945)
T. madens	199	242	
Procrustes	374	338	Labitte (1916)
Carabus	323	386	
Necrophorus	232	291	
Dytiscus	854	740	
Hydrophilus	164	374	
Melolontha vulgaris	19	27	
Cetonia aurata	57	88	
Lucanus cervus	19	32	
Dorcus	327	375	
Ateuchus	338	467	
Sisyphus	198	266	
Copris	497	623	
Geotrupes	700	642	
Oryctes	37	55	
Blaps mortisaga	848	914	
B. gigas	700	728	
B. magica	700	728	
B. edmondi	700	728	
Akis	854	951	
Pimelia	669	714	
Timarcha	135	182	

[a]From Clark and Rockstein (1964).
[b]Full citations can be found in Clark and Rockstein (1964).

Selected References

GENERAL
   Chapman (1971); Rees (1977); Richards and Davies (1977); Roeder (1953); Snodgrass (1935); Wigglesworth (1972).

REPRODUCTION
   General: Adiyodi and Adiyodi (1974); Davey (1965); de Wilde and de Loof (1973a, b); Engelmann (1970); Highman (1964); Matsuda (1976).
   Spermatozoa, Spermatophores, and Spermatogenesis: Baccetti (1972); Gerber (1970); King (1974); Phillips (1970); Schaller (1971).
   Eggs, Oogenesis, and Control of Oogenesis: Bonhag (1958); Davey (1965); de Wilde and de Loof (1973a, b), Hagedorn and Kunkel (1979); Hinton (1979); King (1974); Mahowald (1972); Riddiford and Truman (1978); Telfer (1965, 1975); Telfer and Smith (1970).
   Sex Determination: Bergerard (1972); White (1964).

EMBRYOGENESIS
   General: Anderson (1973; 1979); Counce (1961); Counce and Waddington (1972, 1973); Hagen (1951); Johannsen and Butt (1941); Sharov (1966); Snodgrass (1935); Weygoldt (1979).
   Polyembryony: Ivanova-Kasas (1972).
   Control: Agrell and Lundquist (1973); Bodenstein (1971); Chen (1971); Counce (1973); Sander (1976).

POSTEMBRYONIC MORPHOGENESIS

Various Aspects: Agrell and Lundquist (1973); Chen (1971); Etkin and Gilbert (1968); Gehring and Nöthiger (1973); Hinton (1946a, 1948, 1963b); Kafatos (1972); Kafatos et al. (1967); Lawrence (1970, 1973); L'Helias (1970); Ursprung and Nöthiger (1972); Waddington (1973); Whitten (1968); Wigglesworth (1954); Wyatt (1968).

Instar Definition Controversy: Hinton (1946b, 1958, 1971, 1973, 1976); Whitten (1976); Wigglesworth (1973a).

Control: Ashburner (1970, 1972); Beerman (1972); Doane (1973); Etkin and Gilbert (1978); Gilbert and King (1973); Highnam and Hill (1977); Kroeger (1968); Menn and Beroza (1972); Novák (1975); Riddiford and Truman (1978); Schneiderman and Gilbert (1964); Sláma et al. (1974); Thomson (1976); Wigglesworth (1954, 1959, 1964, 1970); Willis (1974).

Polymorphism: Kennedy (1961); Lüscher (1976).

Aging: Clark and Rockstein (1964); Rockstein and Miquel (1973).

6

Sensory Mechanisms; Light and Sound Production

Sensory Mechanisms

In Chapter 3 we discussed the system responsible for the collection, integration, and interpretation of information from the external and internal environment (the nervous system) and the systems that enable an insect to respond and hence adjust to environmental changes (the muscular and endocrine systems). We shall now consider the structures that carry on the actual collection of environmental information: the various sense organs or *sensilla* (sing. *sensillum*).

The basic function of any sense organ is to receive some form of energy (*stimulus*) from the environment and initiate a chain of events that ultimately results in a nerve impulse (Dethier, 1963). The transformation of energy from one form to another is called *transduction*. Exactly how transduction of different types of environmental energy into nervous impulses is accomplished is not known.

A large portion of the energy "sensed" by insects is in the form of various mechanical changes, either gross changes such as the bending of a hair or the stretching of a portion of the body, or molecular movements in the form of sound waves propagated through a solid, liquid, or gaseous medium. The sensation of these changes falls under the general heading of *mechanoreception*.

Another form of energy perceived by insects is "potential energy existing in the mutual attraction and repulsion of the particles making up atoms" (Dethier, 1963). The perception of this form of energy is referred to as *chemoreception*. If the molecules happen to be water, *hygroreception* is the appropriate term. In other instances, the energy stimulating a given sense organ may be in the form of electromagnetic waves (or photons) such as light and heat. The sensation of these forms of energy will be considered under the headings *photoreception* and *thermoreception*.

Morphology of Sense Organs

The majority of sense organs are composed of two types of cells: a *receptor cell* or cells and *accessory cells*. Receptor cells are usually bipolar neurons that perform the actual detection of stimuli and generation of the nervous impulse. They are modified epithelial cells that during the development of an insect, send out a process (the axon) which eventually communicates with the central nervous system. Accessory cells surround the receptor cells and are usually involved with or actually secrete the specialized cuticular structures that make up the most obvious parts of a sense organ. In addition to the type of sense organ just described, there are multipolar receptor neurons having no contact whatsoever with the cuticle. These multipolar neurons are associated with the various muscles, the alimentary canal, and entad surface of the integument.

Sense organs that communicate with the external world can be classified morphologically on the basis of the differences in associated cuticular structures. It is thought that most external sensilla were originally derived from setae and hence are homologous structures. However, photoreceptors do not fall into this grouping and will therefore be discussed separately.

Since the various sensilla are considered to be derived from setae, the hair sensilla (*sensilla trichoidea*) should be the first type discussed. Recall from Chapter 2 that each hair is formed by two cells, the "hair-forming" *trichogen cell* and the surrounding "socket-forming" *tormogen cell*. The addition of one to several bipolar receptor cells to this picture produces the basic *trichoid sensillum* (Figs. 6-1A and 6-2). Other closely related sensilla include those with bristlelike processes (*sensilla chaetica*), scalelike processes (*sensilla squamiformia*), and peglike or conelike processes (*sensilla basiconica;* Fig. 6-2A). A further modification consists of processes such as those just described, which are sunken in shallow pits (*sensilla coeloconica;* Figs. 6-1B and 6-2) or comparatively deep pits (*sensilla ampullacea;* Figs. 6-1C and 6-2).

Two other types, *campaniform sensilla* (*sensilla campaniformia;* Figs. 6-1 and 6-2) and *placoid sensilla* (*sensilla placodea;* Figs. 6-1 and 6-2), lack hairs, pegs, cones, or bristles. Campaniform sensilla are shallow round or oval pits and in section consist of a bell-shaped cuticular cap innervated by a single receptor cell. Placoid sensilla

Fig. 6-1
Types of sensilla related to the trichoid type (diagrammatic). A. Trichoid. B. Coeloconic. C. Ampullaceous. D. Campaniform. E. Placoid.

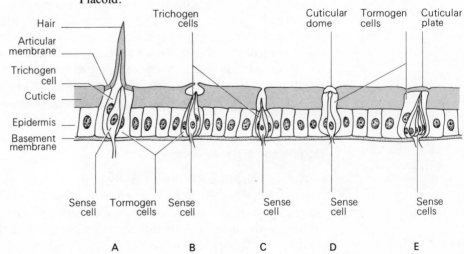

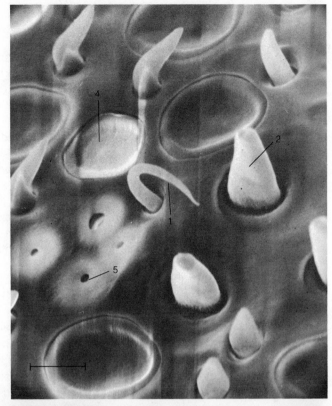

Fig. 6-2
Scanning electron micrographs of
sensilla on the antennal flagellum
of a worker honey bee. 1, trichoid;
2, basiconic; 3, campaniform; 4,
placoid; 5, coeloconic or
ampullaceous. (Scale lines = 10
μm.) [Courtesy of Alfred Dietz
and Walter J. Humphreys.]

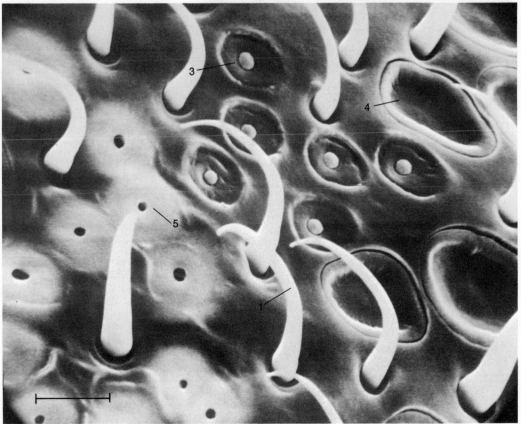

are platelike structures made up of a round or oval cuticular plate surrounded by a narrow membranous ring. In contrast to the campaniform type, a placoid sensillum is innervated by a number of receptor cells. A type of sensillum that is rather dramatically different from those already described is the *scolopophorous* or *chordotonal organ* (Fig. 6-3). This kind of sensillum usually occurs in bundles of structures called *scolopidia* (or *scolophores*). Each scolopidium consists of a bipolar neuron enveloped by a scolopale cell, and an attachment cell. Scolopidia, and hence chordotonal organs, are usually stretched between two internal integumental surfaces. Burns (1974) describes the ultrastructure of a chordotonal organ in the femora of the locust.

A given morphological type of sensillum varies both in the appearance of the associated cuticular structures and the numbers of receptor neurons. In addition, a morphological type does not necessarily imply a particular function, since a given sensillum may have different functions in the same insect or may contain two or more receptors, which receive different forms of stimuli. For example, a single hair on the labellum of the blow fly, *Phormia,* may have four chemoreceptor cells and a mechanoreceptor cell associated with it (Dethier, 1963). However, there are certain sensilla that always seem to be associated with the same general function. For example, the campaniform type has always been found to be a mechanoreceptor, which is stimulated by deformation of the cuticle. The same is true for the chordotonal sensilla, which are sound, vibration, and stretch receptors.

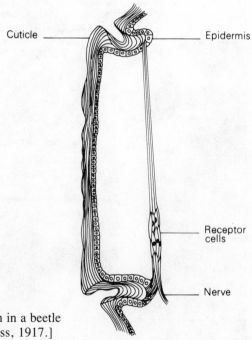

Cuticle — ——— Epidermis

——— Receptor cells

——— Nerve

Fig. 6-3
Simple chordotonal organ in a beetle larva. [Redrawn from Hess, 1917.]

Methods Used to Study Insect Sense Organs

Earlier work with insect sense organs generally consisted of a combination of behavioral and morphological approaches. Based on their gross appearance and appearance in stained sections under the light microscope, structures were identified as sensory or not. This approach did not, of course, enable investigators to be certain of the function of a given sense organ, but it did allow (sometimes accurate) speculation. The use of behavioral criteria allowed the process to be carried somewhat further since innate or learned responses to various stimuli could be taken advantage of in different ways. For instance, an insect may be naturally attracted to, repelled by, or respond in some other way to a given stimulus. Once this fact is established, a structure or structures suspected of receptor activity can be removed or blocked in a variety of ways to determine if they are, in fact, active with regard to the stimulus being tested.

When a specific receptor or specific receptors are identified, further experiments can be made to determine their sensitivity by varying the concentration or intensity of the stimulus. An excellent example of this kind of work has been carried out on the tarsal receptors of the blow fly (Dethier, 1955). The tarsal sense organs are ''taste'' or contact chemoreceptors and are sensitive to various sugars. When a solution of a ''tastable'' sugar is applied to them, the fly extends its labellum in response. Using this labellar response, the sensitivities to various concentrations of several sugars have been worked out.

These earlier approaches are still used today, but have been complemented by the much more detailed observation possible with the electron microscope, both transmission and scanning, and by the use of highly sophisticated electrophysiological techniques. The latter have enabled investigators to make critical studies of the functions of the sensory structures at the cellular level.

Mechanoreception

Organs that are sensitive to the actions of stretching, bending, compression, torque, and so on applied to the integument or some internal organ are the *mechanoreceptors*. These sense organs are responsible for the maintenance of posture, stability during locomotion, and body position with respect to gravity. In addition, many are designed so that they enable the insect to detect sound waves or vibrations in a solid substrate. In certain instances they provide information as to the state of certain internal organs (e.g., the alimentary canal).

Insects are known to possess the following mechanoreceptive senses: *tactile* (touch), *proprioceptive,* and *sound* or *vibration.* Mechanoreceptors can also be grouped as Type I, those associated with the cuticle or cuticular invaginations, and Type II (*stretch receptors*), which are associated with the viscera, but not the cuticle (McIver, 1975).

McIver (1975) and Schwartzkopff (1974) provide recent reviews of the structure, physiology, and biological functions of mechanoreceptors.

The Tactile Sense

The external receptors involved in the sense of touch are typically hair sensilla. Movement of a hair affects the associated bipolar receptor cell(s). The distal process (dendrite) of a receptor cell is in very close contact with the base of a mechanoreceptive hair and contains an array of microtubules, the *tubular body*. Deformation of the tubular body is evidently what causes a receptor cell to fire. In other words, movement of the base of a hair initiates a nervous impulse or train of impulses.

Many hair sensilla are strictly mechanoreceptive in function and have only a single receptor cell. However, others may possess a number of other kinds of receptor cells (e.g., chemical) in addition to the mechanoreceptive ones. An example of the latter was given earlier—certain hairs on the labellum of the blow fly, *Phormia*.

Mechanosensitive hairs appear to fall into two groups: *velocity sensitive* and *pressure sensitive*. In the first group neural impulses are generated only in the presence of constantly changing stimuli, for example, in an insect during flight. Not surprisingly, these hairs have been found on the anterior edges of wings of various insects. Hair sensilla that fall into the second category initiate a steady train of nervous impulses when statically deformed. This situation would occur, for example, when an insect is at rest on a solid substrate.

Hair sensilla are commonly found on the legs, mouthparts, and antennae, all of which frequently come into direct contact with the substrate or other surfaces. They are also found on the cerci and here may initiate an escape response resulting from air movement or something suddenly touching these appendages and consequently bending the associated hairs (e.g., in cockroaches and grasshoppers). Hair sensilla on the anal papillae of silkworm moths (*Bombyx mori*) provide information on the unevenness of the oviposition site and are thus important in regulating the position of eggs within egg clusters (Yamaoka et al., 1971). The hair sensilla described in this paragraph are usually velocity sensitive.

In addition to being found in locations in which they come into contact with other than the insect body itself, many hair sensilla are located between joints, between body segments, and in other areas in which there is direct contact between two body surfaces. Their function in these situations is proprioceptive, and they are generally of the pressure-sensitive type.

The Proprioceptive Sense

Proprioceptive structures are stimulated by changes in various parts of the insect body. The changes may be in length, tension, and so on. These receptors provide the insect with continuous information as to the position of the various body parts and the tensions of the various muscles. It follows, then, that these types of sensilla are of critical importance in the maintenance of "proper" orientation of the body parts with respect to one another or of the entire body with respect to gravity in both the stationary and the moving insect. You will find examples of this later in the section on locomotion. The role of proprioceptors in orientation is complemented by the actions of other receptors (e.g., tactile and photoreceptors). Mill (1976) provides a series of reviews on many aspects of invertebrate proprioceptors, including those of insects. Weis-Fogh (1971) and Wendler (1971) consider proprioceptors in insects.

A number of different types of sensilla function as proprioceptors. Among these are clusters of tactile hairs or hair plates, campaniform sensilla, stretch receptors, and chordotonal organs (including Johnston's organ).

Hair plates are very common in insects and appear as clusters of tiny trichoid sensilla. They may be looked upon as tactile receptors that respond to the insect "touching" itself since they are usually located in appressed or overlapping areas of the body. For example, in the ant hair plates are located between certain segments of the antennae, in the neck region between the head and thorax (Fig. 6-4), at the bases of the coxae and the trochanters, in the petiolar region between the first and second abdominal segments, and in the ventral gaster region between the second and third abdominal segments. Since a given position of one body part relative to another would "stimulate" particular hairs of specific plates to specific degrees, a definite complex sensory pattern would impinge on the central nervous system for every possible body position and movement. These hair plates in the ant are then important in the maintenance of "proper" posture, whether the insect is stationary or moving. In the praying mantis, hair plates in the neck region function in the process of prey capture. As the mantis turns its head, visually following potential prey, the changing pattern of impulses from the hair plates is critical in determining the accuracy of the strike with its grasping forelegs. If the operation of the hair plates is experimentally interfered with by denervation or immobilization of the head, the accuracy of the strike is seriously impaired or destroyed altogether. (Roeder, 1967). Hair plates on the vertex of the head of locusts (*Schistocerca* and *Locusta*, Orthoptera) sense air flow and are involved in the regulation of flight.

Fig. 6-4

Transverse section of the head of the ant *Formica polyctena,* showing hair plates on the prothorax (diagrammatic). [Redrawn from Markl, 1962.]

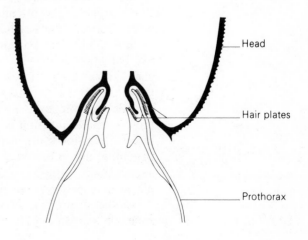

Head

Hair plates

Prothorax

Campaniform sensilla serve as compression and stretch receptors and are located only in areas on the integument that are exposed to strains of various sorts. They are concentrated particularly in areas where compression and stretching occur as a result of muscular activity, for example, in the legs, wings, halteres, ovipositor, and the bases of the mandibles. Their role in association with the legs, wings, and halteres will become clear when we discuss locomotion.

Multipolar neurons associated with muscles, the alimentary canal, and entad surface of the integument have been found to act as *stretch receptors*. They respond with a nervous impulse when the tissue in which they are embedded is subjected to a change in length. They have been found in dragonfly nymphs and members of the orders Orthoptera (grasshoppers and relatives), Hymenoptera (ants, bees, wasps, and relatives), and Lepidoptera (moths and butterflies) (Dethier, 1963).

Chordotonal organs, as explained earlier, are usually stretched between two internal integumental surfaces. They have been found in the pedicel of the antennae in all insects studied, and are commonly found in the mouthparts, wing bases, halteres, legs, and abdominal segments. They have also been found closely associated with tracheae and pulsatile structures and in the hemocoel. Early investigators ascribed an auditory function to them, and, in fact, many of them are auditory in function. However, a number are known to be proprioceptive, and as Dethier (1963) points out, very likely "all not associated with tympanic membranes or grouped to form subgenual and Johnston's organs . . . will eventually be proved to be proprioceptors." Several different proprioceptive functions have been suggested for them, for example, sensation of body orientation, passive body movements, and muscular movements. Their close associations with tracheae, pulsatile organs, and the various hemocoelic cavities suggests that they may respond to changes in intertracheal air pressure and in blood pressure.

In virtually every insect studied, a specialized group of sense organs, which are quite similar in structure to chordotonal sensilla, are found in the pedicel (second segment from the base) of each antenna (Fig. 6-5). These structures are attached to the pedicellar wall and to the membrane between the pedicel and the third antennal segment and are in a radial arrangement. They make up the *Johnston's organ*. This structure varies in complexity depending on the insectan species and reaches its greatest development in two families of flies (Culicidae, mosquitoes; Chironomidae, midges). In these two Dipteran families it completely fills the pedicel. Johnston's organ is known to function as a proprioceptor in some insects, although it has different functions in others (e.g., sound reception in mosquitoes and midges). An example of an insect in which it has a proprioceptive function is the honey bee. During the flight of this insect, Johnston's organ responds to movements of the antennal flagellum and in this way provides the bee with a measure of the stream of air passing over it. The amplitude of the wingbeat is regulated in response to this measurement (Wigglesworth, 1972).

Another example is the use of the antennal movements and hence Johnston's organ by certain aquatic Hemiptera (true bugs). Some of

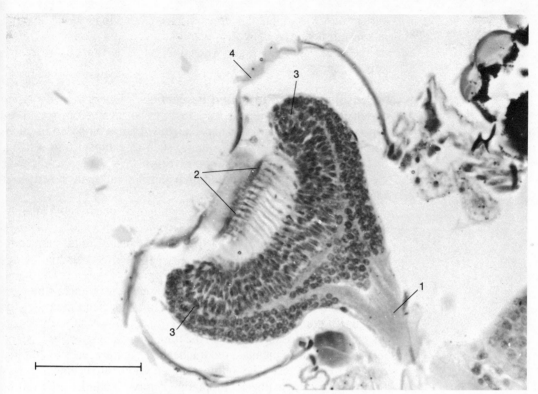

Fig. 6-5
Photomicrograph of a sagittal section of Johnston's organ in the mosquito *Aedes triseriatus*. 1, Nerve; 2, scolopophores; 3, cell bodies of neurones; 4, pedicel. (Scale line = 50 μm.)

these (*Corixa* and *Naucoris*) swim dorsal side up, while others (*Notonecta* and *Plea*) swim dorsal side down. In either case, the proper body orientation during swimming is maintained because the insect is able to sense when its dorsum is up or down. This is accomplished by the buoyant action of a small air bubble trapped between the ventral part of the head and each antenna. Any change in the position of the insect results in a change in the direction of the buoyant force of the bubble relative to the insect and hence results in a movement of the antennae, which in turn causes a change in the sensory patterns generated by each Johnston's organ.

Sound Perception

For the purposes of this discussion, we define sound as longitudinal waves of kinetic energy propagated through a continuous medium (gas, liquid, or solid), and divide perception of sound into two senses: the sense of vibration, perception of sound via the substratum, and the sense of hearing, perception of sound via air or water.

The perception of sound is important in a number of ways. Many of the stimuli that impinge on an insect from the external environment are in the form of sound. Some of these sounds are produced by other insects of the same or different species (see the section on sound production in this chapter), and other sounds come from a variety of environmental sources. Sound perception may be of value in sensing potential danger, a potential mate, prey, other members of the same species (e.g., when individual territories are maintained), and so on.

Sensilla Involved in Sound Perception. Only two basic types of sensilla have definitely been shown to be involved in sound reception, trichoid sensilla and specialized organs composed of chordotonal receptors. However, others, such as campaniform sensilla or stretch receptors, may also be involved. We shall consider first the organs composed of chordotonal sensilla and then the trichoid sensilla.

Those sound-sensitive structures composed of chordotonal sensilla are the *tympanic organs, subgenual organs,* and *Johnston's organ,* which was described earlier. Most tympanic organs are composed of the same fundamental parts. However, the degree of development of these parts varies from group to group. The basic structures (Fig. 6-6) involved are a thin integumental area (the *tympanum*) and a group of chordotonal sensilla attached directly or indirectly to the entad surfaces of the tympanum. Usually a tracheal air sac is closely associated with the tympanum and sensilla. In some insects (e.g., male cicadas) the tracheal air sac may serve to amplify certain frequencies. The number of chordotonal receptors varies from 2 in moths in the lepidopteran family Noctuidae to 1500 or more in cicadas (Dethier, 1963).

Tympanic organs have been identified in a number of different locations in a variety of insects. In the order Orthoptera (grasshoppers and relatives), they are found on the tibiae of the forelegs in the families Tettigoniidae (long-horned grasshoppers; Fig. 6-7A, B) and Gryllidae (crickets) and on either side of the first abdominal segment in members of the family Acrididae (short-horned grasshoppers). Tympanic organs occur in the metathorax of Noctuid moths and in the abdomen in geometrid and pyralid moths. In cicadas (Hemiptera, Homoptera) these structures are located in the abdomen. In addition, tympanic structures have been identified in a few members of the suborder Heteroptera (e.g., in water boatmen, family

Fig. 6-6
Section of a metathoracic tympanal organ of a noctuid moth. ✕ indicates a tracheal air sac. [Redrawn from Roeder, 1959.]

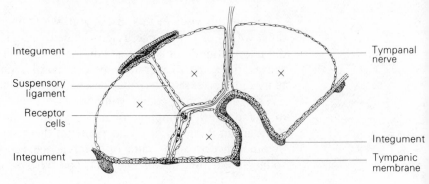

Integument

Suspensory
ligament

Receptor
cells

Integument

Tympanal
nerve

Integument

Tympanic
membrane

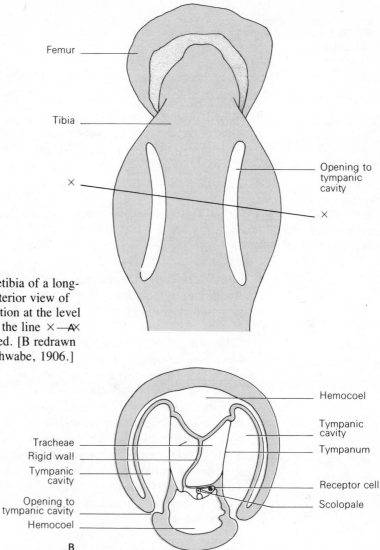

Fig. 6-7
Tympanic organ on the foretibia of a long-horned grasshopper. A. Anterior view of the tibia. B. Transverse section at the level indicated approximately by the line ×—A× in A, epidermal cells omitted. [B redrawn with modifications from Schwabe, 1906.]

Corixidae). Examples of specific functions of hearing via the tympanal organs will be discussed in Chapter 8.

Subgenual organs (Fig. 6-8) are groups of chordotonal sensilla located in the basal portion of the tibial leg segment. They are not associated with any joints. They vary considerably in degree of development from group to group, being somewhat weakly developed in the true bugs, more developed in the members of the orders Lepidoptera and Hymenoptera, and most highly developed in the beetles and the true flies.

Vibration Perception. Among the insects in which vibration sensitivity has been measured, those showing the greatest sensitivity have been found to possess subgenual organs (Schwartzkopff, 1974). These organs are thought to be specifically involved in the perception of vibration. Subgenual organs have not been found in the less sensitive species, the trichoid and small chordotonal sensilla

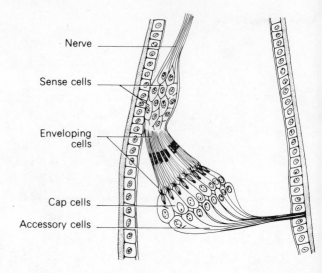

Nerve

Sense cells

Enveloping
cells

Cap cells

Accessory cells

Fig. 6-8
Subgenual organ of an ant exposed by
section of the tibia. [Redrawn with
modifications from Schön, 1911.]

in the legs being the vibration receptors. In *Locusta* (Orthoptera, Acrididae) trichoid sensilla on the sternites are sensitive to substrate vibration (Dethier, 1963).

Hearing. Schwartzkopff (1974) describes the use of aquatic surface waves by water striders (Hemiptera, Gerridae), backswimmers (Hemiptera, Notonectidae), and whirligig beetles (Coleoptera, Gyrinidae) as "preliminary stages of hearing." Water striders and backswimmers are predaceous and orient to their prey by the surface waves. These waves are sensed by receptors in the tarsi. The tarsal organ in the backswimmer *Notonecta glauca* is composed of scolopidia. Water striders maintain tarsal contact with the water's surface from above and backswimmers from below. Whirligig beetles utilize surface waves in navigation, avoiding moving and stationary objects, as they dart to and fro along the water's surface. The surface waves are sensed by Johnston's organ in the antennae.

Tympanic organs, trichoid sensilla, and Johnston's organ have been shown to perceive sound in a number of insects. The tympanal organs are, as far as known, always used for hearing. However, trichoid sensilla and Johnston's organ have been demonstrated to have other functions.

Tympanic organs in the nocturnal noctuid and related moths detect the ultrasonic emissions used by echo-locating, predatory (insectivorous) bats (Roeder, 1965, 1966, 1967). Detection of these sounds stimulates avoidance behavior of various sorts (see Chapter 8). In the orders Orthoptera and Hemiptera tympana receive sounds produced by members of the same species, and these sounds are involved with sexual behavior. In the water boatmen, *Corixa* (Hemiptera, Corixidae), males produce waterborne sounds that are perceived and responded to by the females. Evidently males also hear each other produce these sounds, since they chirp as long as the tympanic organs are intact. These are the only insects known to be sensitive to waterborne sounds (Dethier, 1963). Since tympanal organs occur bilaterally on some part of the insect body, they facilitate localization of a sound through differential stimulation.

Certain trichoid sensilla on exposed regions of the body in a number of insects have been shown to be sensitive to airborne sounds. Sensilla on the cerci of cockroaches (Blattaria) are especially sensitive, and stimulation by sound may elicit the characteristic "alarm" reaction mentioned earlier.

Johnston's organ in the antennae of some species of mosquitoes and midges has been shown to be a sound receiver. The males are able to detect the sounds produced by the rapidly beating wings of the females. Response to these sounds can be elicited by using a tuning fork that sounds at an appropriate frequency.

The green lacewing, *Chrysopa carnea*, has a swelling of the radius of each forewing that contains two scolopophorous structures. These sensilla respond to the ultrasonic chirps of echo-locating bats (Miller, 1970, 1974).

Hawkmoths (Lepidoptera, Sphingidae, Choerocampinae) detect ultrasound by means of structures associated with the mouthparts (Roeder, 1972). The second palpal segment is bulbous and is composed almost completely of an air sac. The medial region of this papal segment rests against the distal lobe of the *pilifer*, a small appendage associated with the labrum. Ultrasonic vibrations are translated via the palps to the pilifer, which contains the sensory transducer.

Chemoreception

As defined earlier, chemoreception is the process by which the "potential energy existing in the mutual attraction and repulsion of the particles making up atoms" is detected (Dethier, 1963). Thus chemoreceptive organs are responsive to direct contact with chemicals. Chemical cues from the environment are useful to insects in several ways, for example, food (host plant or animal, prey, decaying organic material, and so on) procurement, mediation of caste functions in social forms, mate location, identification of noxious stimuli that are a potential threat to survival, selection of oviposition site, habitat selection, and others.

In general terms the chemoreceptive activities of insects may be divided into three "chemical senses": distance chemoreception or *olfaction*, contact chemoreception or *gustation*, and "general" or "common" chemical sensitivity.

Distance chemoreception is mediated by chemoreceptors that are responsive to molecules or ions of a chemical in the gaseous state at comparatively low concentrations. These receptors are very sensitive and may show a high degree of specificity with regard to the kind of chemical that elicits a response.

Contact chemoreceptors are excited by direct contact with molecules or ions of a chemical in solution at a concentration usually somewhat higher than olfactory chemostimuli. Generally, these receptors are less sensitive than the distance chemoreceptors and are commonly associated with feeding activities.

The "general" chemical sense involves receptors that are comparatively insensitive except to relatively high concentrations of a stimulating chemical. These receptors are much less discriminating than either contact or distance chemoreceptors and are usually associated with an avoidance or escape response. No general chemical receptors have been positively located, and nonspecific effects on neurons may be involved (Hodgson, 1974).

Although the classification of the "chemical senses" presented above is quite useful, it is not without difficulties. First, the distinction between distance and contact chemoreception must be qualified by pointing out that in both types the molecules or ions of the stimulating chemicals always come into direct contact with the receptor cell membranes. Second, in aquatic and subterranean insects the same receptors may respond to chemicals either in the gaseous state or in aqueous solution. Third, there are instances among terrestrial insects (e.g., blow fly taste receptors) of contact receptors reacting to certain volatile substances.

A wide variety of the morphological types of sensory structures may be involved with chemoreception. Among those for which there exists evidence of olfactory activity in certain insects are the sensilla trichodea, sensilla basiconica, sensilla placodea, and sensilla coeloconica. Sensilla trichodea and sensilla basiconica have been identified as contact chemoreceptors in a number of insects. Chemical stimuli reach the nerve endings via pores in the cuticle. Olfactory sensilla typically bear many of these pores (Fig. 6-9A), while contact chemoreceptors usually have only one or two located distally (Fig. 6-9B). The number of receptor cells associated with a chemoreceptor varies from a very few—three to five neurons in *Phormia regina* (Diptera, Calliphoridae)—to fifty or more (Hodgson, 1974).

Proved or suspected contact chemoreceptors have been found in several parts of the insect body. They probably exist in the mouthparts of all insects, for example, the hypopharynx and epipharynx in caterpillars, the tips of the maxillary and labial palps and in the buccal cavity of the cockroach *Periplaneta* sp., and possibly the buccal cavity (cibarium) of mosquitoes (Day, 1954). Chemoreceptors probably also exist in the foregut; for example, sensilla in the pharynx of the house fly apparently have to do with the control of the passage of ingested food.

In honey bees and certain wasp species contact chemoreceptors are located on the distal segments of the antennae, enabling the insects to differentiate between sweetened and unsweetened water (Wigglesworth, 1972). The distal portion of the tibia and tarsi of the forelegs of many insects bear contact chemoreceptors. In many butterflies, true flies, and bees the mouthparts are extended in response to stimulation of the fore tarsi with sugar water. Contact chemoreceptors may be located in the ovipositors of parasitic Hymenoptera, which deposit their eggs directly into host insects. For example, the wasp *Venturia canescens* (Hymenoptera, Ichneumonidae) can distinguish a parasitized host from an unparasitized host with its ovipositor (Ganesalingam, 1974). Similarly, distance (olfactory) receptors have been identified in the antennae and mouthparts of a variety of insects and in the ovipositor of at least one.

Considerable work has been done with the contact chemoreceptive

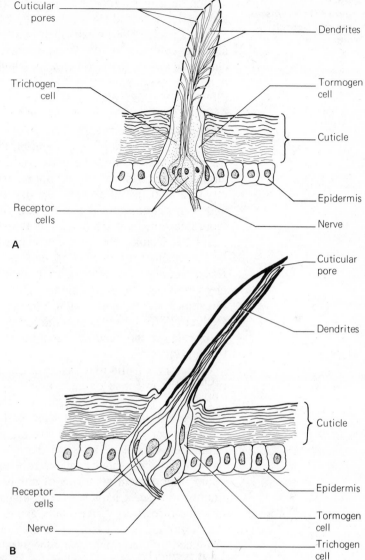

Fig. 6-9

Chemoreceptors. A. Olfactory sensillum. B. Contact chemoreceptor.
(Diagrammatic; based on several sources.)

abilities of the honey bee. This insect is capable of differentiating
the qualities sweet, bitter, acid, and salt (von Frisch, 1950). The
honey bee and man have been compared with regard to their sugar-
tasting abilities. According to Wigglesworth (1972), "out of 34
sugars and related substances tested, 30 appear sweet to man, only
9 to the honey bee; all these nine being present in the natural food
of the bee and capable of being metabolized by it." Honey bees are
a little more sensitive to bitter substances such as quinine, whereas
acetylsaccharose is exceedingly bitter to man but not to the honey
bee. This substance has been suggested for addition to cane sugar

so that the sugar might be sold more cheaply to beekeepers and not used as human food (Wigglesworth, 1972).

Von Frisch (1950) established two thresholds for honey bees relative to solutions of the sugar sucrose (cane sugar). The first he called the *threshold of acceptance,* which he defined as the minimum concentration of sugar that the bees could be induced to ingest. This threshold was established at about 40% when many of the bees' foraging plants were in bloom and approximately 5% in the fall when flowers were scarce. The second threshold von Frisch measured was the *threshold of perception,* that is, the minimum concentration of sucrose that the bees could perceive. This threshold, if accurately determined, would not be expected to vary with the scarcity or abundance of flowers. By starving the bees for several hours and then determining the minimum concentration of sucrose they would accept, he established this threshold as somewhere between 1 and 2%. Contact chemoreception has also been extensively studied in a number of true flies (*Phormia* sp., *Calliphora* sp., *Musca* sp., *Drosophila* sp., and others). In those instances, both behavioral and electrophysiological techniques have been utilized to establish thresholds and to gain insight into the basic mechanisms involved. Dethier (1976) provides information on contact chemoreceptors and their role in the feeding behavior of *Phormia regina* (Diptera, Calliphoridae).

Olfactory sensilla may occur in tremendous numbers. For example, in the antennae of male polyphemus moths (*Antheraea polyphemus,* Lepidoptera, Saturniidae) there are more than 60,000 sensilla with about 150,000 sense cells. Most of these sensilla are receptors for the female sex pheromone (approximately 60–70%; Schneider, 1969).

Olfactory cues play a major role in the lives of many insects, being important in all forms of behavior involving chemical communication (see Chapter 8). The distance chemoreceptive abilities of some insects are fantastically acute. For example, the male silkworm moth, *Bombyx mori,* reacts to the sex pheromone (*bombycol*) produced by the female at a concentration as low as 100 molecules of attractant per cubic centimeter of air (Wilson, 1970). A single molecule of female sex pheromone is sufficient to trigger an impulse in the male receptor cell (Schneider, 1974).

As with contact chemoreception, the distance chemoreceptive abilities of the honey bee have also been widely investigated. One would, of course, expect olfactory abilities to be great in these insects since their life depends to a large extent on flowering plants, and it would be of obvious advantage for a bee to be able to visit repeatedly a particular kind of flowering plant that was currently in bloom. Evidently, the olfactory abilities of honey bees and man are similar in terms of threshold concentration of various scents; however, honey bees seem to have a great ability to discriminate among many different scents (Wigglesworth, 1972).

For further information on chemoreception, see Hodgson (1958, 1965, and 1974), Lewis (1970), Schoonhoven (1977) and Slifer (1970). Schneider (1969) and Seabrook (1977) deal specifically with olfaction.

Thermoreception

Based on definite behavioral responses, it is well established that many insects are sensitive to changes in temperature. For example, honey bees trained to visit warm places are able to detect temperature differences as small as 2°C (Wigglesworth, 1972). In some insects, heat sensitivity seems to be somewhat generalized over the entire body. In others, specific locations have been identified. Temperature receptors have been found on the antennae, maxillary palps, and tarsi of many insects. For example, in the bloodsucking bug *Rhodnius prolixus* (Hemiptera, Reduviidae) the antennae are extremely sensitive to small differences in air temperature. The sensilla presumed to be involved are the thick-walled sensilla present in very large numbers on the antennal segments. In *Rhodnius* and other bloodsucking insects (mosquitoes, lice, bed bugs, etc.), perception of warmth is important in host location. A few insects may have cold receptors.

All insects will move away from high temperatures, but this is probably a generalized sensitivity with no particular sensilla being involved. In addition, there is evidence that spontaneous activity within the central nervous system is influenced by temperature (Chapman, 1971). Some insects are evidently able to perceive the radiant heat of the sun or other light source; for example, stink bugs will turn their dorsal sides toward a light source when the ambient temperature is low. By thus exposing the largest surface to the light, they are able to receive the maximum possible radiant heat.

Behavior associated with changes in temperature is discussed further in Chapter 8.

Hygroreception

As with thermoreception, the ability of insects to perceive moisture in the air is well known through observation of specific behavioral responses. Springtails, like other small soil-dwelling insects, are very sensitive to moisture, both in the air and the substratum. They are attracted to regions of high humidity. Other insects, such as earwigs and mealworm beetles, avoid very moist areas. Some insects, the honey bee and others, can perceive water from a distance. The sensilla that are sensitive to moisture have been identified in only a very few insects, and these have been found on the antennae and maxillary palps. In the human louse, *Pediculus humanus,* antennal "tuft organs" composed of several small hairs have been shown to be specific hygroreceptors (Wigglesworth, 1941). Basiconic sensilla on the antennae and maxillary palps of the mosquito *Aedes aegypti* respond electrophysiologically to water vapor (Kellogg, 1970).

Photoreception

Photoreception may be defined as the ability to perceive energy (light) in the visible or near visible (near ultraviolet) range of the electromagnetic spectrum. In order for an organism to perceive light, there must be a pigment capable of absorbing light of a given wavelength and a means of producing a nervous impulse as a result of this absorption.

Many different kinds of environmental information are available to an insect in the form of light stimuli. For instance, an insect may perceive, to a greater or lesser extent, form, pattern, movement, distance, color, relative brightness, the polarization plane of light, light versus dark, and the length of a light period.

Generally speaking, three types of photoreceptive structures have been found in insects. These are the *compound eyes* and *stemmata* (or *lateral ocelli*); and *dorsal ocelli*. Each of these kinds of photoreceptive structures will be considered in turn. Refer to Chapter 2 for information regarding the location and numbers of compound eyes and ocelli. In addition, the larvae of certain higher dipterans (true flies) have specialized photoreceptive organs that do not readily fit into any of these three groups of photoreceptors. These consist of photosensitive cells (Fig. 6-10), which are located in small cavities in the anterior end of a larva (the larvae or "maggots" of higher Diptera do not have well-defined heads). The negative phototactic response (i.e., the tendency to move away from a source of light) is presumably accomplished by the larva orienting its direction of movement so that the light-sensitive cells receive minimal stimulation. This orientation would obviously be with the anterior end of the larva away from the light source, with most of the body interposed between the light source and the sensitive cells.

In addition to discrete organs associated with light perception, many insects apparently possess a light sensitivity over the general body surface. This is evidenced by the fact that certain insects will give definite responses to light even when the operation of the photoreceptive structures listed above is disrupted in some way. For example, cockroaches continue to demonstrate a preference for dark situations even after being totally blinded. Similarly, decapitated mealworm larvae (*Tenebrio* sp.) continue to avoid light.

Photoreceptors may be reduced or absent in cave-dwelling (*cavernicolous*), burrowing, and other species that live in dark situations.

The following review various aspects of insect vision: Bernhard (1966), Carlson and Chi (1979), Goldsmith and Bernard (1974), Horridge (1975, 1977), Mazokhin-Porschnyakov (1969), Ruck

Fig. 6-10
Photoreception in the house fly larva; cephalopharyngeal skeleton with pocket that contains photoreceptive cells, indicated by hatching.

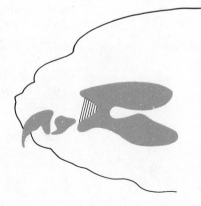

(1964), and Wehner (1971). Hoeglund et al. (1973) discuss the biochemistry and physiology of insect photopigments.

Compound Eyes

The compound eyes (Fig. 6-11A) are the major photoreceptive organs of adult insects. When present, there are two, one located on either side of the head. Each is composed of a number of individual sensory units or *ommatidia* (Fig. 6-11B). Externally, these ommatidia are marked by hexagonal cuticular facets. The facets, and hence the ommatidia, may vary in number from a very few to several thousand, for example, from 12 to 17,000 in some Lepidoptera and from 10 to more than 28,000 in some Odonata (Richards and Davies, 1977).

Structure of the Compound Eye An individual ommatidium (Fig. 6-11C, D) is divisible into two parts: the *dioptric apparatus,* which acts as the "lens," and the *receptor apparatus,* in which the events leading to the initiation of a nervous impulse occur.

The dioptric apparatus is composed of the *cornea,* the *crystalline cone,* and *corneal pigment cells.* The cornea is a cuticular structure and is continuous with the cuticle of the integument. It has the shape

Fig. 6-11
Compound eye structure (diagrammatic). A. Head with compound eye. B. Four ommatidia removed and magnified. C. Apposition ommatidium. D. Superposition ommatidium.

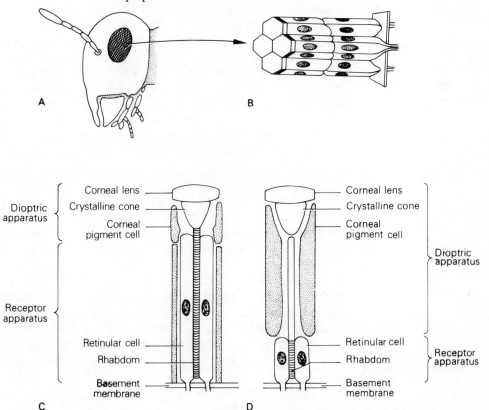

of a planoconvex lens, the convex portion forming the outer surface. In many insects a hexagonal array of very small conical projections (approximately 0.2 nanometer from tip to base and center to center) may be found on the outer surface of the cornea. These "corneal nipples" are thought to act as an antireflection coating, which reduces the reflection for the air–cornea interface (Miller, Bernard, and Allen, 1968). It is suggested that these "nipples" may also serve in insects active during light periods as camouflage by cutting out mirrorlike reflections from the cornea, which might attract predators. In insects active at night, they may help in some way to increase the sensitivity of the eyes. In addition to corneal nipples, some insects, such as the house fly, *Musca domestica,* and the honey bee, *Apis mellifera,* have interommatidial hairs (i.e., hairs between the facets). In flying insects, these hairs are thought to function as aerodynamic organs. For example, removal of these hairs causes loss of flight control in honey bees. They may also provide sensory input that elicits cleaning behavior.

The crystalline cone lies immediately beneath the cornea and is composed of a translucent material. The darkly pigmented corneal pigment cells are usually located on the periphery of the crystalline cone, except in very primitive insects such as thysanurans (silverfish and relatives), in which they lie beneath the cornea. They are considered to be the cells that originally secreted the cornea.

The receptor apparatus (Fig. 6-11C, D) is composed of six or seven *retinular* (nerve) *cells* arranged in one (usually) or two layers. If arranged in two layers, one is proximal to the other. Like the crystalline cone, the group of retinular cells is usually surrounded by rather darkly pigmented cells. Usually each retinular cell gives rise to an axon that passes through a basement membrane and enters the brain and contributes to the formation of a centrally located retinal rod, or *rhabdom*. The contribution of each retinular cell to the rhabdom is called a *rhabdomere*. The rhabdomeres are considered to be the receptive surfaces of the retinular cells. Use of the electron microscope has revealed that the rhabdomeres are made up of tiny, closely packed fingerlike projections (microvilli) from the retinal cells. These microvilli project at right angles to the long axes of the retinal cells. It is generally thought that the microvilli contain the light-absorbing pigment(s), the *rhodopsins* and *metarhodopsins,* that are directly involved with photoreception.

There are a variety of different ways the dioptric apparatus and the receptor apparatus are associated with one another. Generally, however, based on the association of these components, ommatidia fall into one of two rather broad categories. If the retinal cells lie immediately beneath the crystalline cone (Fig. 6-11C), an ommatidium is of the "apposition" type, which is characteristic of diurnal insects (those active during the daylight hours). If there is a clear space between the retinal cells and the crystalline cone (Fig. 6-11D), an ommatidium is referred to as "superposition," and this type is characteristic of nocturnal or crepuscular insects (those active during dark or dusk periods, respectively).

In addition to the dioptric apparatus and receptor apparatus, groups of tracheal branches lie in the vicinity of the basement membrane. These form a surface from which light that has traversed the rhabdoms from distal to proximal is reflected back along the rhabdoms, giving these "receptors" a double exposure to the light and hence probably helping to increase the light sensitivity. These tracheal branches are sometimes referred to collectively as the *tapetum* since their function seems analogous to this structure in the vertebrate eye. The eyes of many insects, particularly certain nocturnal Lepidoptera, when illuminated will appear to glow as a result of reflection from the tapetum.

Image Formation. the *mosaic* theory of insect vision initially proposed by Muller in 1826 and elaborated by Exner in 1891 is still generally accepted today. However, in light of more recent work, it has undergone considerable modification. According to the mosaic theory, each ommatidium "sees" only a small portion of the insect's surroundings. The combination of the images sensed by individual ommatidia supposedly together forms a composite or mosaic view of the external environment. This situation is somewhat analogous to looking at the surroundings through a handful of soda straws. Only a small part of the total view is seen through any one straw, but the combination of these "small parts" gives a mosaic image of the surroundings (Wigglesworth, 1964a).

There are two basic types of compound eyes, *apposition* (or *photopic*; Goldsmith and Bernard, 1974) and *superposition* (or *scotopic*; Goldsmith and Bernard, 1974). Apposition eyes (Fig. 6-12A) are composed of ommatidia of the apposition type and are thus, as mentioned earlier, characteristic of diurnal insects. Similarly, superposition eyes (Fig. 6-12B, C) are made up of superposition ommatidia and are characteristic of nocturnal or crespuscular insects. In the apposition type of eye there is little or no movement of the pigment in the pigment cells surrounding the crystalline cone in response to changes from light to dark or vice versa. Thus the pigment remains rather uniformly distributed. However, in superposition eyes, there is considerable pigment movement in response to a light–dark change. In a lighted situation (Fig. 6-12B), the pigment in the pigment cells tend to migrate proximally, producing the light-adapted condition. On the other hand, in a dark situation (Fig. 6-12C), the pigment migrates distally (dark-adapted condition).

According to the mosaic theory, in the apposition eye and the light-adapted superposition eye, the rhabdom receives only light rays entering parallel (or nearly so) to the long axis of an individual ommatidium. Oblique rays are absorbed by the pigment, which thus optically isolates adjacent ommatidia from one another. The image formed in this situation is an *apposition image* since the light reaching a single rhabdom enters only via the dioptric apparatus of the same ommatidium. In the dark-adapted superposition eye, distal movement of the pigment has the effect of removing the optical isolation between adjacent ommatidia. In this instance, the light rays reaching the rhabdom of a given ommatidium enter via several ommatidia. The image thus formed is a *superposition image*. One would not expect a superposition image to be as sharply defined as an apposition one because of the increased amount of light reaching

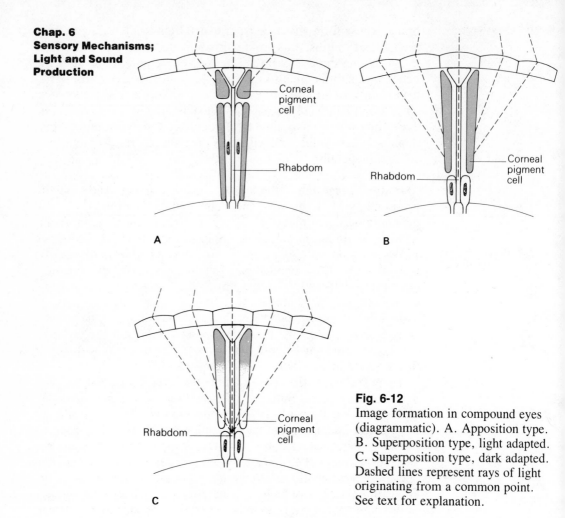

A

B

C

Fig. 6-12
Image formation in compound eyes
(diagrammatic). A. Apposition type.
B. Superposition type, light adapted.
C. Superposition type, dark adapted.
Dashed lines represent rays of light
originating from a common point.
See text for explanation.

each rhabdom. Also the formation of a sharply defined superposition image would require that all the light rays reaching a single rhabdom impinge at exactly the same point. It follows, then, that eyes in which pigment optically isolates individual ommatidia would probably have a greater resolving ability than those in which pigment does not form an optical boundary. Exner theorized that the dark-adapted condition in a superposition eye increases the light sensitivity since it enables more light to reach each rhabdom. The adaptation of the eye to light or dark situations is probably also partly due to a change in the retinal cells or in the central nervous system.

In recent decades details of the mosaic theory have been modified. Ommatidial visual fields have been found to overlap in all species studied. Few, if any, insects with superposition eyes form superposition images. In addition, there is evidence that the resolving ability of eyes is, in fact, not decreased by the distal migration of pigment. However, Exner's idea that distal pigment migration (dark adaptation) in superposition eyes increases sensitivity still is thought to be true.

The crystalline tracts found in the eyes of many insects may act as light guides in that once light has entered them it does not escape but is carried directly to the rhabdoms. Besides acting as a light guide, a crystalline tract surrounded by pigment cells is thought to act as a longitudinal pupil (Miller, Bernard, and Allen, 1968). When pigment envelops the tract (light-adapted), it absorbs some of the light; however, when the pigment migrates distally and no longer envelops the tract (dark-adapted), much more light is transmitted to the rhabdom. Other recent additions include the possibility that in at least certain instances the retinal cells are the individual functional units instead of the whole ommatidium. Also, recently, images that are formed proximal to the retinal image have been discovered. They are essentially of the superposition type, and their significance is unclear.

Perception of Form and Pattern.   There is strong behavioral evidence that insects are at least capable of simple perception of pattern and form. In particular, much work has been done with honey bees. For example, honey bees were trained to associate a particular pattern of colored paper with the presence of sugar water in a cardboard box. Subsequently, they were presented with several boxes marked with new pieces of colored paper cut in the pattern to which they had been trained and others with a rather similar pattern that did not contain sugar water. It was judged that the bees could discriminate between these two patterns since only a small number visited the boxes marked with the second pattern (von Frisch, 1950). Other work, again with training to a particular pattern, has demonstrated that bees can discriminate between the shapes shown in the upper row of Figure 6-13 and those in the lower row, but cannot distinguish among the shapes within each row. This suggested that the bees perceive form on the basis of the "brokenness" of pattern (von Frisch, 1950). It is supposed that broken or interrupted patterns produce a flickering visual impression as the bee passes during flight. It is not surprising, therefore, that bees tend to visit flowers that are being shaken by the wind more readily than those that are not.

It is well known that bees, ants, and wasps are able to locate their nests on the basis of various landmarks. In addition, flying bees have been shown to be able to distinguish right from left, before from behind, and above from below (Wigglesworth, 1972). However, the form and pattern perception abilities of insects are considered to be less than those of humans.

Intensity and Contrast Perception.   Optomotor responses have been used to study intensity perception. An optomotor response is

Fig. 6-13
Figures used to study form perception in honey bees. [Redrawn from Hertz, 1929.]

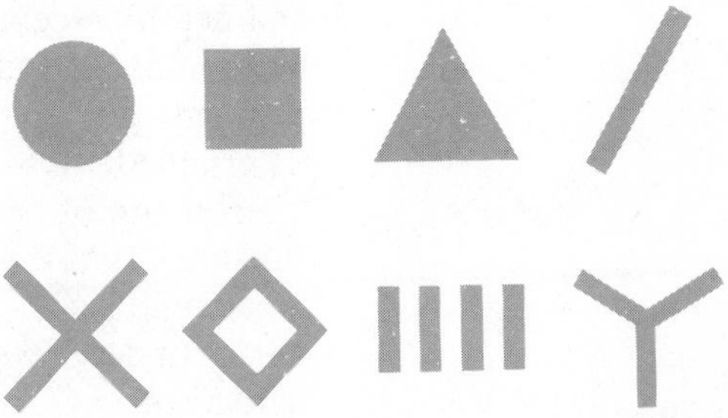

behavior stimulated by a moving visual pattern. When an insect flies, the apparent movement of the surroundings tells it the direction and rate of movement. Certain orientation movements, maintenance of flight, and changes in velocity of flight and landing may be in response to a change in the rate or direction of the moving visual pattern. In the laboratory, a moving visual pattern can be produced by surrounding a stationary insect with a cylinder on which are painted vertical stripes. Rotation of the cylinder gives the insect the sensation that it is moving.

In the study of perception, optomotor responses to patterns of rotating stripes have been used to determine the ability of insects to discriminate between different levels of intensity of adjacent stripes. Measurements indicate considerable variation among different insects. Discrimination of different intensities apparently depends on whether those being compared are relatively high or low. For example, in the house fly, at low intensities the more intense light must be 100 times brighter than the less intense, while at high intensities the magnitude difference may decrease to 2.5 times (Wigglesworth, 1972).

The brightness of the background may influence intensity perception. For example, the hummingbird hawk moth will favor a dark disc on a white background over an equally dark disc on a gray background. In other words, the moth "prefers" the situation in which there is greater contrast. It is interesting that under natural circumstances these moths enter dark crevices to spend the night.

Movement Perception. As was the case for form and pattern perception, there is behavioral evidence for the perception of movement. In fact, certain responses are elicited only by movement. For example, dragonfly nymphs will not attack prey with their labial jaws unless it is moving. The optomotor response elicited by the moving pattern of stripes is based on the perception of movement by an insect.

Distance Perception. It is not difficult to think of several instances in which it would be essential for an insect to possess the ability to judge the distance of an object from itself. An excellent example would be in prey capture. To catch prey, particularly on the wing, distance perception must be especially acute. One needs only to watch a dragonfly capture its prey in flight to be convinced. Binocular vision must be involved since the ability to judge distance accurately is lost when one compound eye is blocked in some fashion. However, unlike those of humans, the eyes of insects are fixed, and depth perception depends not on convergence of the eyes on a fixation plane, but upon the equal, simultaneous stimulation of corresponding retinal points. The distance of an object will then be determined by the location of the object relative to the points of intersection of projections of axes of the corresponding ommatidia (Dethier, 1953). A schematic representation of how depth perception is accomplished by a dragonfly is shown in Figure 6-14. For this means of depth perception, the insect must, of course, face the object perceived.

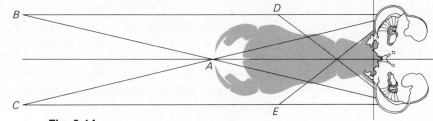

Fig. 6-14
Depth perception in a dragonfly nymph. The lines represent the visual axes of selected ommatidia. The distance and position of objects are determined by the points of intersection of the visual axes. The extended labium is shown in grey. Potential prey at point *A* is within reach of the extended labium, but not at points *B, C, D,* and *E*. [Redrawn from Baldus, 1926.]

Color Vision. The range of the electromagnetic spectrum perceived by insects is from about 253 nanometers (near ultraviolet) to approximately 700 nanometers (infrared; Dethier, 1963). Although there is a large amount of variation in the sensitivities of different insects to different wavelengths, insects in general are particularly sensitive to the ultraviolet and blue-green regions of the spectrum. Some insects, such as the nocturnal stick insect, *Dixippus* sp., are apparently colorblind. In insects that do possess color vision, part of an eye may be color sensitive and another part colorblind [e.g., in the water boatman, *Notonecta* (Hemiptera, Heteroptera)]. Evidence exists that there is more than one type of color receptor in the cockroach eye (Dethier, 1963). The dorsal area of the eye is more sensitive to ultraviolet than to blue-green light; the reverse is true in the ventral area. The spectral sensitivity of an insect may vary with its physiological state. For example, cabbage butterflies (*Pieris brassicae*) seem to prefer blue or yellow flowers; gravid females ready to oviposit seem to prefer green and blue-green.

Both electrophysiological and behavioral techniques have been used to establish the color-perceiving abilities of insects. Different wavelengths have been tested to see if they elicit a particular electroretinogram (ERG) pattern or a specific behavioral response. Both the various optomotor responses and training experiments have been utilized. For example, von Frisch (1950) used training techniques to demonstrate the color-perceiving capacities of the honey bee. He first placed a dish of honey on a blue card where bees could get to it. After several hours the bees were "trained" to associate the color blue with the presence of honey. The bees were then presented with a fresh (unscented) blue and a red card where only the blue one had been previously. The bees visited the blue card and ignored the red one. Although this experiment shows that bees can distinguish between blue and red, it does not prove that they actually perceive color, since red and blue differ in relative brightness. To determine whether the bees were distinguishing brightness or color, several cards of different gray shades between white and black and a blue card were placed where only the blue card had previously been. In addition, to discount the possible role played by scent the cards were covered with a glass plate. In this situation the bees still visited only the blue card; therefore, they are able to perceive color.

Similar results were obtained when the bees were trained to orange, yellow, green-violet, and purple. However, the color red was confused with the black and dark gray cards; therefore, the bees are red-blind. Further experiments in which bees were presented with several different-colored cards showed that they were unable to distinguish certain colors from others. As a result of these and other experiments, it has been demonstrated that bees can accurately distinguish only 4 colors as opposed to the 60 or so that can be distinguished by the human. The portion of the spectrum in which we can recognize several distinct colors between orange and green appears yellow to bees, and so on (Fig. 6-15). These characteristics of color vision in the honey bee have very interesting ramifications when we consider the relationship between bees and flowers.

We have mentioned that the spectral range in which insect eyes are generally most sensitive is in the ultraviolet and blue-green regions. We also established that bees are red-blind. However, some insects are highly sensitive to wavelengths in the red region. For example, certain butterflies are capable of recognizing red flowers or red models of flowers, and the firefly, *Photinus* sp., is able to perceive flashes of light up to 690 nanometers, which is well into the red region of the spectrum.

Since insects can perceive light into the ultraviolet zone of the spectrum, it is of significance that many flowers display patterns based on differential reflectance and absorbance of ultraviolet light. Thus flowers that appear to be a solid color to us may appear patterned to a given insect. The same is true for insect body coloration; that is, the different patterns displayed by insects are due to differential reflectance and absorbance of ultraviolet light.

Polarized Light Perception. That insects could recognize the direction of polarization of light was first discovered in honey bees and ants. These insects were found to be able to detect the polarization pattern of the sky, which varies with the position of the sun and enables them to determine direction. This directional ability is, of course, important in finding the hive or nest after a foraging or hunting trip. According to von Frisch, it is also of importance in the orientation of the bees communication dances used to inform other

Fig. 6-15
Comparison of the spectra perceived by humans and honey bees.
[Redrawn from von Frisch, 1950.]

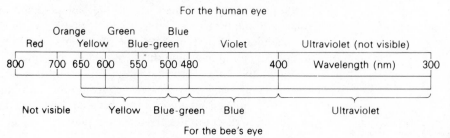

For the human eye

| Red | Orange | Green | Blue | | |
| | Yellow | Blue-green | Violet | Ultraviolet (not visible) | |

800   700 650 600   550   500 480   400   Wavelength (nm)   300

Not visible   Yellow   Blue-green   Blue   Ultraviolet

For the bee's eye

members of the hive as to the location of food. The ability to detect the polarization plane of light has been found in all insects examined in this regard.

In the desert ant (*Cataglyphis bicolor*) and the honey bee, the receptors involved in the perception of polarized light are sensitive to ultraviolet light. This is significant because the ultraviolet wavelengths are much less affected by atmospheric conditions (e.g., clouds) than the longer wavelengths visible to humans. Thus sensitivity to the polarization pattern of ultraviolet light gives a particularly stable navigational point of reference. For a fascinating account of polarized light navigation, see Wehner (1976).

Stemmata

Structurally, *stemmata* (or *lateral ocelli*) are variable. Some types are similar in structure to an individual ommatidium of a compound eye. For example, in larval butterflies and moths (Fig. 6-16A), each eye consists of a cornea, a crystalline body, and a number of retinal cells forming a rhabdom.

Stemmata function in the manner of eyes. Typically they are the only eyes found in holometabolous larvae. In various insects they have been shown to be involved with color, form, and distance perception. Like compound eyes, they receive nerves from the optic lobes of the brain.

Although a detailed pattern of the external surroundings is not likely to be perceived at any one time by an insect possessing only a few stemmata, the movement of the head back and forth, "scanning," may allow much greater detail perception than would otherwise be possible. For this type of activity to be effective, the insect must be able to convert spatial patterns into temporal patterns. In other words, the external scene is viewed as a sequence of events in time.

Dorsal Ocelli

Dorsal ocelli vary somewhat in structure. Typically, they are made up of a *corneal lens;* a layer of *corneagen cells,* which secretes the lens; and the *retina,* which is composed of up to 1000 photosensitive cells depending on the species (Fig. 6-16B). In some species, pigment cells are associated with the dorsal ocelli. Dorsal ocelli have been shown to be light sensitive, but are apparently not important in image perception. They may be "stimulatory organs" that increase the sensitivity of the compound eyes to light, for when they are blocked, reactions of the insect to light are diminished somewhat. Alternatively, they may be involved along with the compound eyes in orientation behavior relative to light. However, according to Carlson and Chi (1979), "The ocellus remains an enigma as to its specific role in insect behavior."

There is a correlation between the presence of dorsal ocelli and flight, since they are not found in apterous insects. On the other hand, there are strong fliers that lack dorsal ocelli, for example, horse flies (Diptera, Tabanidae) and relatives and sphinx moths

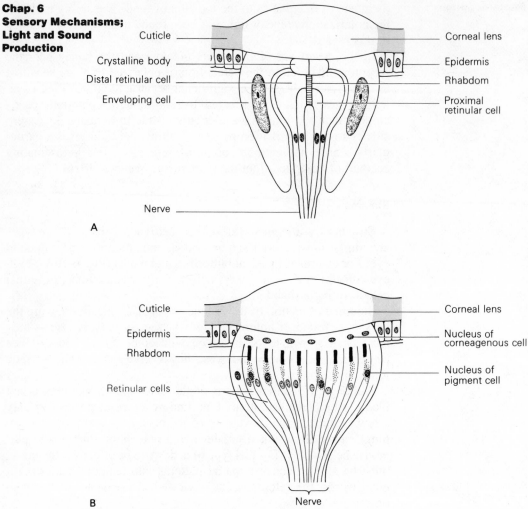

Cuticle — Corneal lens

Crystalline body — Epidermis

Distal retinular cell — Rhabdom

Enveloping cell — Proximal retinular cell

Nerve

A

Cuticle — Corneal lens

Epidermis — Nucleus of corneagenous cell

Rhabdom —

— Nucleus of pigment cell

Retinular cells —

B

Nerve

Fig. 6-16
A. Section of a stemma (diagrammatic). B. Section of a dorsal ocellus (diagrammatic).

(Lepidoptera, Sphingidae). However, recent histological and electrophysiological studies provide evidence for the presence of an internal ocellus in sphinx moths (Eaton, 1971).

Goodman (1970) reviews the literature on the structure and function of dorsal ocelli.

Light Production

The production of light by living organisms, *bioluminescence,* has fascinated scientific investigators and curious lay observers for centuries. However, it has been only during the last hundred years that a fundamental understanding has begun to develop. The phenomenon of bioluminescence has been described in several groups of plants, microbes, and animals. For example, it has been found in the clam *Pholas dactylus,* the crustacean *Cypridina hilgendorfii,* several marine annelids, a number of insects, marine dinoflagellates, and several fungi and bacteria. Luminous bacteria commonly exist symbiotically with animals. For example, they are responsible for the luminescent body regions of several marine fish.

Insects with specific light-producing mechanisms are found among the Collembola (springtails), Homoptera (cicadas, leafhoppers, and relatives), Coleoptera (beetles), and Diptera (true flies). Bioluminescence in other insects has been found thus far to be due to the presence of bioluminescent bacteria.

When certain rare species of springtails (e.g., *Achorutes muscorum*) are stimulated, luminescence occurs over the whole body.

There is only one well-established case of bioluminescence among the Homoptera (McElroy et al., 1974). This is in the species *Fulgora lanternaria*. Luminescence of the head in this species has apparently been observed only when males and females are together and evidently has something to do with mating behavior.

Bioluminescence has been described in more beetles than any other group of insects. Several families are known to be involved, among them Lampyridae (fireflies), Elateridae (click beetles), Drilidae, and Phengodidae. Members of the family Lampyridae have been the most extensively studied. The light-producing organ is in the abdomen and may occur in both sexes, or only in females, and in the larval stage. Luminescent larvae and some females in this family are often called "glowworms." Not all species of lampyrids are luminescent, but in those that are, the light flashes are usually involved in mating activities, attracting members of the opposite sex. The flashing patterns are species specific. A well-known American species of fireflies is *Photinus pyralis*, the mating behavior of which has been extensively studied. In the family Elateridae, those in the genus *Pyrophorus* are especially well known. Some of these beetles have luminous green spots on the prothorax and orange ones on the abdomen. The orange spots are visible only during flight.

Pyrophorus is particularly significant historically because these were the first insects in which the biochemical nature of the light-producing reaction was studied. In the late nineteenth century, DuBois ground up their light-producing organs in cold water. The homogenate glowed for a short period of time, and then the glow gradually faded away. Addition of an extract obtained by boiling light-producing organs in water briefly restored the glow. DuBois later carried out similar experiments with a luminescent clam and named this reaction the *luciferin-luciferase reaction*. More recently, the active principle in the hot water extract has been found to be the heat-stable substance adenosine triphosphate (ATP).

A rather bizarre example of the luminous beetles is in the family Drilidae among larvae in the genus *Phrixothrix*. These larvae are commonly known as "railroad worms" because of the eleven pairs of green luminous spots on the thorax and abdomen and the pair of red luminous spots on the head. The movement of these larvae when stimulated to luminesce gives the impression of a string of railroad cars at night.

Truly luminescent dipterans are found among larvae of the families Platyuridae and Bolitophilidae. The members of both these families are called *fungus gnats*. The larvae of the New Zealand glowworm, *Bolitophila luminosus,* occur in well-shaded humid areas, for example, the environmental situation found in certain caves. The light-producing organ has been found to be modified portions of the Malpighian tubules, which are located near the end of the abdomen. Adult female flies deposit eggs in a mucous glue on the ceiling and larvae spin silken sheaths from which they hang. The network of sticky threads traps small flying insects, which are eaten by the larvae. The prey are presumably attracted by the bioluminescent light. Luminescent larvae of the species *Platyura fultoni* have been found in moist environments in the Appalachian Mountains.

The morphology of the light-producing, photogenic organs is extremely complex and variable and will not be considered here. Similarly, the biochemical aspects are also very elaborate and far from being completely understood. For further information on bioluminescence, see the review by McElroy et al. (1974) and the article by McElroy and Seliger (1962). The role of bioluminescence in behavior is discussed in Chapter 8.

Sound Production

The ability to produce sound is widely distributed among insect groups and has evolved independently many times. These sounds are commonly correlated with well-developed organs of hearing and often play an important role in various types of behavior. Sound as a mode of insect communication is discussed in Chapter 8.

A useful classification of sound-producing mechanisms has been presented by Haskell (1974).

1. Sounds produced as a by-product of some usual activity of the insect.
2. Sounds produced by impact of part of the body against the substrate.
3. Sounds produced by special mechanisms.

In the first category no specifically adapted structures are involved in sound production. Sounds of this type include by-products of flight produced by wingbeats, vibration of the thoracic sclerites, the wings striking one another, and similar mechanisms. This category also includes sounds produced as a result of movement during copulatory behavior, cleaning, feeding, and so on (Haskell, 1961). In many instances the sounds have no specific function. On the other hand, some do have behavioral significance. A good example is found among the mosquitoes, where the sound produced by the wingbeat of the flying female elicits a mating response in males. Another possible example is the "piping" of queen honey bees, which may result from the vibration of the thoracic sclerites. This sound is produced when a colony possesses a number of virgin queens and has been suggested to be a sound of challenge (Butler, 1963).

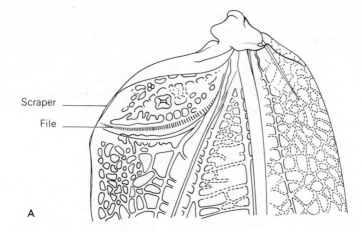

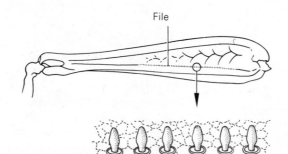

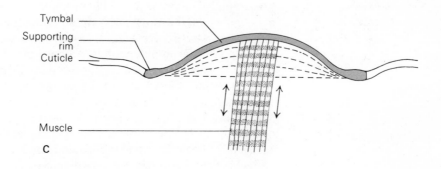

Fig. 6-17

Examples of sound-producing mechanisms. A. File and scraper on underside and posterior edge, respectively, male katydid, *Neduba carinata*. The file of one wing and scraper of the other are rapidly rubbed together, producing a chirping sound; B. Hind femur of a short-horned grasshopper, *Stenobothrus* sp., with a portion of the file enlarged. In sound production the file is rapidly rubbed against a thickening (the scraper) of the basal portion of the forewing. C. Section of a tymbal of a cicada. The dashed lines and arrows indicate the pattern of movement of the tymbal during sound production. [A redrawn from Essig, 1942; B redrawn from Comstock, 1940; C redrawn with modifications from Haskell, 1961.]

A number of insects are known to produce sound by tapping the substrate with some part of the body. The "death watch" beetles (*Anobium* and *Xestobium,* Anobiidae) tap the walls of their galleries in wood (unfortunately sometimes furniture) with their heads, producing a characteristic sound. Other insects in this category include the subterranean termite (*Reticulitermes flavipes*), book lice (Psocoptera), and several members of the Orthoptera.

Among the special sound-producing mechanisms are frictional mechanisms, vibrating membrane mechanisms, and mechanisms involving air movement.

Frictional mechanisms are found among several groups of insects, and although these mechanisms are structurally diverse, they consist of similar parts. The production of sound by means of a frictional mechanism is commonly called *stridulation*. Frictional mechanisms (Fig. 6-17A, B) are located in areas where two surfaces (two wings, a leg and a wing, etc.) may be rubbed together. One of these surfaces, the "file," bears a row of regularly spaced ridges, and the other bears the "scraper," a ridge or knoblike projection. When the file and scraper are rapidly rubbed together, a sound is produced, the quality of which is based on the rate of rubbing, the spacing of the ridges on the file, and the resonance characteristics of the surrounding cuticle. Frictional mechanisms have been described in several life stages of several orders and involve many different combinations of body parts.

Sound production in the mole cricket, *Gryllotalpa nivae* (Orthoptera, Gryllotalpidae), is of particular interest. This insect constructs a burrow such that it acts as a resonating chamber (Bennet-Clark, 1970).

Vibrating-membrane or tymbal mechanisms have been found only in members of the orders Hemiptera and Lepidoptera. Although tymbal mechanisms have been described in several homopterans, the best-known examples are found among the cicadas (Fig. 6-17C). In these insects the tymbal organs are paired structures on the dorsolateral surface of the base of the abdomen. The tymbal organs of Lepidoptera have been found in certain arctiid moths (tiger and footman moths) and others. In the moths these organs are paired and are located on either side of the metathorax. In addition, tymbal mechanisms have been identified in the hemipteran family Pentatomidae.

Mechanisms involving the movement of air are uncommon and poorly understood. One well-established example is the death's head hawk moth. *Acherontia atropos,* a European species, which produces a sound by the forcible inhalation and exhalation of air through the proboscis by means of the pharyngeal muscles. Passage of air over the epipharynx apparently produces the sound. Air released forcibly through the spiracles may result in sound production in certain Diptera, Hymenoptera, and Blattaria. The cockroach *Gromphadorhina portentosa* of Madagascar is capable of producing an audible hissing sound via the spiracles when disturbed.

Selected References

GENERAL

Chapman (1971); Richards and Davies (1977); Roeder (1967); Snodgrass (1935); Wigglesworth (1972).

SENSORY MECHANISMS

General: Dethier (1953, 1963, 1976).

Mechanoreception: McIver (1975); Mill (1976); Schwartzkopff (1974); Weis-Fogh (1971); Wendler (1971).

Chemoreception: Dethier (1976); Hodgson (1958, 1965, 1974); Lewis (1970); Schneider (1969, 1974); Schoonhoven (1977); Seabrook (1977) Slifer (1970).

Photoreception: Bernhard (1966); Carlson and Chi (1979); Goldsmith and Bernard (1974); Goodman (1970); Hoeglund et al. (1973); Horridge (1975, 1977); Mazokhin-Porschnyakov (1969); Ruck (1964); Wehner (1971, 1976).

LIGHT PRODUCTION

McElroy et al. (1974); McElroy and Seliger (1962).

SOUND PRODUCTION

Haskell (1961, 1974).

7
Locomotion

The ability to change position within the environment is of essential importance to the survival of all nonsessile organisms. Escape from predators, feeding, dispersal, mate finding, and adjustment to temperature and humidity changes all depend to some degree upon the ability of an organism to move about. Insects were originally terrestrial organisms that subsequently invaded both the aerial and aquatic environments. They are the only group of invertebrates that contains members capable of flight.

Terrestrial Locomotion

Walking and Running

Walking and running are accomplished by the six thoracic legs. Actually these limbs, unless modified for some function other than walking, serve two basic purposes; they suspend and support the insect body off the ground, and they exert the necessary forces to propel the insect. The body literally "hangs" close to the ground, resulting in a low center of gravity and thus a high degree of stability. Also, the tip of the abdomen may be in contact with the ground, providing additional stability. The use of legs as a mode of terrestrial locomotion is found in the adults of nearly all flying and nonflying insects and many nymphal and larval forms as well. In many insects walking legs are the only source of locomotion or the only source used to any great extent. For example, apterygote insects such as silverfish and proturans are solely walkers and runners; cockroaches, although they possess functional wings, seldom fly. In several beetles the elytra (forewings) are fused and hence these insects are grounded for life. Larval and nymphal forms are entirely dependent upon legs for locomotion.

Although several insects are rather sluggish in their walking and, in fact, may use their legs more as organs for clinging to a surface than for actual locomotion, many are rapid, agile runners. For example, cockroaches are well known for their running ability. The fastest speed measured for the cockroach *Periplaneta americana* is 2.9 miles per hour or 130 cm/sec (McConnell and Richards, 1955). In the absolute sense this is a much slower speed than many vertebrates are able to attain. However, in relation to body size, it is remarkably fast. Insects also have the ability to accelerate rapidly over short distances and to change direction suddenly. The speed of walking and running is proportional, within certain limits, to temperature.

Functional Morphology of the Insect Leg

In the discussion of the insect skeleton the leg has already been described as being composed of a series of segments that articulate with one another. The nature of these intersegmental articulations is of paramount importance in the capabilities for movement of the entire leg. Two basic types of articulations are found: *monocondylic* and *dicondylic*. A condyle is a prominence of exoskeleton upon which an adjoining segment articulates. A monocondylic joint (Fig. 7-1A) is one that contains a single condyle. This type of joint allows considerable movement and has been said to be the nearest single-articulation equivalent to the ball-and-socket joint found in vertebrate animals. A dicondylic joint (Fig. 7-1B, C) consists of two condyles and restricts movements to a single plane.

The insect leg is composed of four main regions (Hughes and Mill, 1974).

1. The coxa, which articulates with the thorax by a dicondylic joint or sometimes a single pleural articulation.
2. The trochanter and femur, which are fused, the trochanter articulating with the coxa by means of a dicondylic joint.
3. The tibia, which forms a dicondylic joint with the femur and a monocondylic joint with the tarsus.
4. The tarsus, which is composed of tarsomeres joined by monocondylic joints.

The dicondylic joints between the coxa and thorax and between the femur and tibia are so oriented that they enable movements in different planes and therefore allow the entire leg to move in all direc-

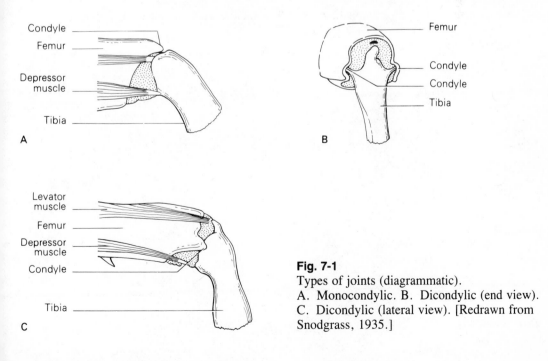

Condyle

Femur

Depressor muscle

Tibia

A

Femur

Condyle

Condyle

Tibia

B

Levator muscle

Femur

Depressor muscle

Condyle

Tibia

C

Fig. 7-1
Types of joints (diagrammatic).
A. Monocondylic. B. Dicondylic (end view).
C. Dicondylic (lateral view). [Redrawn from Snodgrass, 1935.]

tions about the articulation with the body. This system of two dicondylic joints provides a range of movement equivalent to that of a ball-and-socket joint.

It was mentioned in Chapter 3 that leg movements are accomplished both by intrinsic muscles, which originate within the leg itself, and extrinsic muscles, which are attached between the leg and the thoracic wall. The movement of the leg segments relative to one another is possible because the muscles cross one or more joints between their origins and insertions.

Patterns of Leg Movement During Walking and Running

The sequence of movements or gait of the legs during walking and running has been a topic of considerable interest. Attempts to detect details of gait by direct observation are extremely difficult, if not impossible, because of the general rapidity of leg movements. Even if one were able to discern the gait of a slow-moving insect, the results would not necessarily apply to the same insect moving more rapidly or to other species. The use of cinematography has afforded an excellent solution to this problem. Insects may be filmed at whatever speed or under whatever conditions the experimenter desires; then the film may be projected at reduced speed, and enlarged prints of different frames may be produced.

The order or sequence of leg movement during walking seems to be fairly constant at a given speed for a given insect. However, variations occur at different speeds and between different species. The classical description of forward walking of insects is that of alternating tripods of support (Fig. 7-2A). The first and third legs on one side and the middle leg on the other advance while the other three legs remain stationary and provide a tripod of support. The

Fig. 7-2
Walking patterns A. Classical alternating triangles of support.
B. Typical pattern.

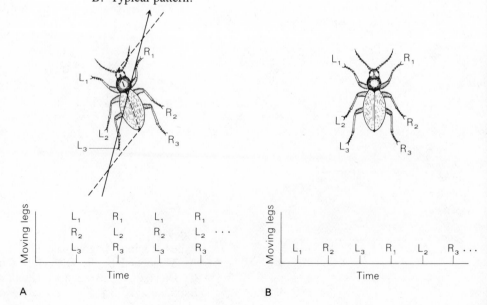

A   B

cycle then repeats itself with the previously stationary three legs
becoming the moving ones. An insect progressing in this fashion
follows a zigzag course. Many deviations from this classical description are known; thus it is useful only in a very general way.
Probably the most common pattern is a modification of the alternating triangles of support where the three legs of each triangle do
not move simultaneously but in sequence (Fig. 7-2B). In addition,
accidental or experimental amputation of one or more legs immediately results in a new coordinated pattern of walking suitable for
the remaining legs. Also, not every insect uses all six legs in walking;
for example, the praying mantis, when walking slowly, uses only
the meso- and metathoracic legs. The reviews of Hughes and Mill
(1974) and Wilson (1966) consider the topic of gait in detail.

Coordination of Legs During Walking and Running

Insect walking and running may be resolved into two basic components, the movements of each leg and the coordination of movements of all the legs. It is pretty well established that each thoracic
segment contains afferent and efferent pathways necessary for the
individual movement of the two legs associated with it. However,
the coordination of all the thoracic legs is intersegmental, as evidenced by the fact that section of the longitudinal connectives disrupts walking. This coordination lies within the thoracic ganglia
since decapitation (removal of brain and subesophageal ganglion)
does not disrupt the ability of an insect to walk. Currently, it is
thought that locomotor rhythms are centrally generated and that these
central patterns are modified, according to circumstance, via sensory
feedback (see Chapter 8). Thus, changes in walking patterns with
experimental amputation of different legs can be explained by the
difference in the patterns of nervous signals from the remaining legs.
Although, as previously mentioned, decapitation does not affect the
coordination of walking, it may result in increased locomotor activity. As a result, the brain is thought to exert a degree of inhibitory
influence on locomotor activity.

Aids to Walking and Running

For the propulsive force applied by the legs to move an insect
there must be a certain amount of friction between the leg and the
substrate. Although tarsal claws generally suffice on rough or dirty
surfaces, which afford ample points for grasping, there are situations
in which tarsal claws fail. For example, the very smooth surface of
an inclined piece of glass or a window pane. Yet many insects, such
as the house fly, are able to walk with ease on such surfaces. This
ability is due to the presence of a variety of adhesive structures: the
pulvilli and tarsal pads described in Chapter 2 or pads at the base of
the tibiae. These adhesive structures are usually covered with dense
mats of tiny hairs with expanded tips. The expanded tips are covered
with a secretion from glands located at the bases of the hairs and are
the parts of the adhesive pads that come into direct contact with the
substrate. Apparently, molecular forces among the expanded tips,

the glandular fluid, and the substrate account for the adhesion of the pad. Since it is the tiny hairs that are ultimately responsible for the "clinging" ability, they are commonly referred to as *tenent hairs*. Gillet and Wigglesworth (1932) advance the following explanation of the functioning of tenent hairs in *Rhodnius prolixus* (Hemiptera-Heteroptera; Reduviidae).

The tenent hairs that comprise the climbing organ (Fig. 7-3) on the distal end of the tibia are wedge-shaped at their tips (Fig. 7-3C), and only the hindmost part comes into contact with the substrate. When the insect moves in an anterior direction (toward *X*, Fig. 7-3C), or an external force (e.g., gravity) pulls the insect in an anterior direction, the secretion between the hairs and the substrate acts as a lubricant, reducing the friction. However, if the insect is moving or being forced to move in the opposite direction (posterior or toward *Y*, Fig. 7-3C), the surfaces of the hair tips come into very close contact with the substrate and "seizure" occurs, the surfaces being held together by the adhesive forces of the molecules of the glandular secretion. Thus *R. prolixus* can walk up a pane of carefully cleaned glass at an incline of 80° but slips at an angle of 22° when walking down the same pane. The holding power of adhesive pads is well illustrated by the fact that one bug in the same family as *R. prolixus* can support a tension of greater than 50 grams (Wigglesworth, 1972).

Jumping

Under this heading we include all the means, other than wings, by which insects are able to propel themselves through the air. Many insects are capable of jumping, but only in certain groups has it become a pronounced specialization. The suggestion has been made that perhaps wings first appeared on a jumping insect that used the jumping ability for a propulsive force and early wings merely as gliding surfaces.

Most jumping insects use modified hindlegs in jumping (see Fig. 3-15B), although other specialized mechanisms occur in some species. Those that use their legs in jumping typically have enlarged, muscular femora on the hindlegs. Examples are the members of several orthopteran families (e.g., grasshoppers and katydids), flea beetles (Coleoptera, Chrysomelidae), and fleas (Siphonaptera). Relative to their size these insects are capable of rather astounding leaps. For example, the trajectory of a jumping flea may reach a height of 8 inches and cover a distance of 13 inches; if proportionally carried out by a human being, this would result in a jump of 800 feet high (Wigglesworth, 1965)! However, as explained in Chapter 3, strength of contraction is a function of cross-sectional area of muscle (linear dimension squared) and increase in mass a function of volume (linear dimension cubed). So a flea the size of a man would have relatively much more mass per unit cross-sectional area of muscle than a normal-sized flea and would thus probably still be able to jump only 8 inches high. Grasshoppers have been reported to leap up to 30 cm high over a distance of 70 cm.

Springtails (Collembola), click beetles (Elateridae), and cheese

Fig. 7-3
Climbing organ of *Rhodnius prolixus* (Hemiptera; Reduviidae). A and B. Proximal portion of tibia and tarsus showing position of climbing organ. C. Tenent hairs which comprise the climbing organ. (See text for explanation.) [Redrawn from Gillett and Wigglesworth, 1932.]

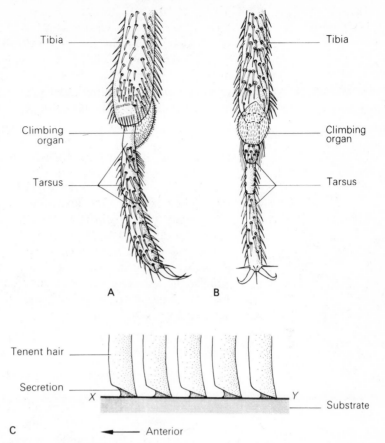

maggots (Diptera, Piophilidae) have unique methods of jumping.

Springtails (see Fig. 2-42B) are minute wingless insects that possess a fork-shaped structure, the *furcula,* on the posterior part of the abdomen and a structure called the *tenaculum* on the anteroventral portion of the abdomen. The furcula is bent anteriorly and its tip engaged by the tenaculum. Prior to a jump, tension is built up in the muscles of the furcula. When released by the tenaculum, the furcula "snaps" posteriorly against the substrate, propelling the insect into the air.

Click beetles utilize a similar principle, an elongated spine on the posterior part of the prosternum being inserted into a receptacle on the mesosternum. The jump produced by the sudden release of this prosternal spine is elicited only when the beetle is on its back and hence serves as a righting mechanism.

Cheese maggots use their entire body in jumping. The last abdominal segment is held by the mouth hooks, muscular tension is built up within the body, and the last abdominal segment is suddenly released against the substrate to propel the insect.

It has been assumed for a long time that direct contraction of leg muscles accounts for the force released when an insect jumps. However, recently, the role of cuticular elasticity has come to be appreciated (Bennet-Clark, 1976). Cuticular elasticity is associated with regions of the cuticle that contain high concentrations of rubbery protein, *resilin*. The Oriental rat flea, *Xenopsylla cheopis*, has been studied extensively in this regard (Rothschild et al., 1973; Bennet-Clark, 1976). In this flea, resilin is concentrated in the pleural arch of the metathorax. It has been known for a long time that the large metathoracic legs are the most important in jumping. The pleural arch is considered to be homologous with the wing hinge ligament of winged insects. The metathoracic leg is so constructed that it can be "cocked." When this leg is in the cocked position, the femur is raised up and engaged by a catch and the trochanter is in contact with the substrate. A catch mechanism also engages plates of the mesothorax to the metathorax. Also when the metathoracic leg is in the cocked position, the resilin in the pleural arch is compressed (i.e., elastic energy stored in it). When the catches are released, the elastic energy in the resilin is released and translated to the trochanter via a tendon. The trochanter thrusts against the substrate, accelerating the flea from the substrate. Subsequently, the thrust of the tibia against the substrate provides additional acceleration. Jumping in locusts (Orthoptera, Acrididae) and click beetles is also thought to involve energy stored in resilin.

Crawling

Crawling (or creeping) is generally associated with the locomotion of larval insects, particularly those that propel themselves by means other than the six thoracic legs alone. However, many larval and most nymphal insects walk like adults. Some of these hexapodous immature forms, for example, several beetle larvae, are aided in their locomotion by eversible *pygopodia*, which arise from the terminal abdominal segment. A pygopodium is everted by means of hemolymph pressure and aids the thoracic legs in the progression of the insect.

The immature forms of butterflies and moths (caterpillars) and certain wasps (Hymenoptera, Symphyta) bear, in addition to the three pairs of thoracic legs, accessory legs, or *prolegs*, on certain of the abdominal segments. The movement of the prolegs depends on the integrated activity of the body musculature of the abdomen. Since these insects are typically soft-bodied, hemolymph pressure is important in maintaining a "hydrostatic skeleton." The pressure of the hemolymph is maintained by the turgor muscles of the insect body. Puncturing one of these insects results in an immediate shriveling, owing to the contraction of the turgor muscles. This indicates that these muscles respond to a reduction in hemolymph pressure by contraction and gives some insight as to how they function in an intact larva. In addition to the turgor muscles, there are locomotor muscles that fail to contract when a larva is punctured and are involved with the actual movement of the prolegs. These muscles are arranged in longitudinal, dorsoventral, and transverse patterns.

Walking patterns in larvae with prolegs have been best described in caterpillars. As in walking solely with thoracic legs, the rhythms involving prolegs are probably generated in the central nervous system. Usually single peristaltic waves of contraction pass anteriorly from the posterior end of the insect. As each wave passes, the two legs of each segment always move simultaneously. Each wave involves three main phases. Hughes and Mill (1974) describe these three phases as follows (Fig. 7-4).

> In the first, the dorsal longitudinal and dorsal intrasegmental muscles contract, together with the large transverse muscles. This results in the segment becoming shortened dorsally and consequently its posterior end becomes inclined forward so that the segment behind is lifted from the ground. In the next phase the segment contracts dorsoventrally and its feet are released simultaneously from the substratum. After the legs have been moved forward, the ventral longitudinal muscles contract so as to bring the segment down toward the ground and the feet become fixed. The wave of peristalsis passing over the body is therefore not limited to a single segment but involves a simultaneous contraction of muscles in at least three segments. Contraction starts at the last segment with the release of the terminal appendages. They are lifted and placed on the ground at varying distances forward.

This description is simplified, since other sets of muscles (e.g., the transverse muscles) are no doubt involved. In addition, contractions of the turgor muscles must be in coordination with the contractions of the locomotor muscles. An interesting variation of this type of crawling is shown by the larvae of geometrid moths. These larvae are often referred to as *inch worms* or loopers. In these larvae much of the trunk is out of contact with the substrate as they move along (Fig. 7-5).

Terrestrial larvae having no legs whatsoever, for example, fly maggots, depend entirely upon peristaltic waves of contraction for their progression. Many possess spines or other structures that increase the friction between their bodies and the substrate. In most

Fig. 7-4
Caterpillar walking. Arrow indicates direction of peristaltic waves and direction of progression. See text for explanation. [Redrawn from Barth, 1937.]

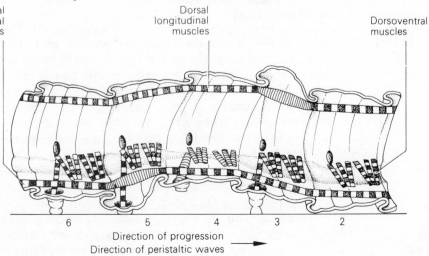

Ventral longitudinal muscles

Dorsal longitudinal muscles

Dorsoventral muscles

6   5   4   3   2

Direction of progression

Direction of peristaltic waves

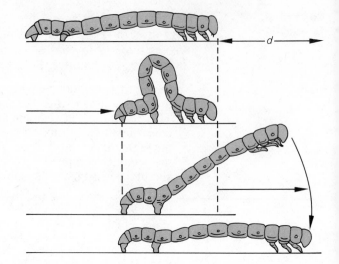

Fig. 7-5
Locomotion of ''inchworm,''
geometrid moth caterpillar. $d =$
distance advanced in one cycle.
[Redrawn from Snodgrass, 1961.]

apodous larvae, waves of peristaltic contraction, like those in larvae with prolegs, pass from posterior to anterior, that is, in the direction of progression. However, in certain ones adapted to burrowing through the soil, for example, burrowing crane fly larvae and bibionid larvae (Diptera), the waves of contraction proceed from anterior to posterior, or in the direction opposite that of progression. In this method, the insect's posterior serves as a stable point while the anterior portion of its body is pushed through the soil (Fig. 7-6).

Aquatic Locomotion

We shall examine the locomotion of aquatic insects from the standpoint of adaptations, first, those that enable insects to propel themselves on the surface of the water and, second, those by which insects ''swim'' beneath the water's surface. Nachtigall (1974a) reviews the literature on aquatic locomotion.

Surface Locomotion

To appreciate fully the various adaptations associated with insect locomotion on the water's surface, we must first consider some of the characteristics of that surface and its interaction with other surfaces. At temperatures and pressures typical of the environments in which insects are found, the molecules of water are strongly attracted to one another, a phenomenon known as *cohesion*. In the absence of outside forces upon the molecules, a droplet of water tends to contract so that it has the smallest possible surface area. Theoretically this would result in a sphere. The molecules of the gases that compose air have little attraction for water molecules, and little attraction for one another. Hence any surface of water in contact with air tends to contract. This force of contraction is called *surface tension,* and its presence makes the water's surface behave somewhat like a thin, elastic membrane. Some surfaces or surface coatings attract water molecules, and when this attractive force is equal to or greater than that between water molecules themselves, the water spreads out on the surface, which is said to be wettable or *hydrophilic* (water loving). Other surfaces have the opposite effect, exerting little or no attractive force on water molecules. Such surfaces are water repellent or *hydrophobic* (water fearing).

The magnitude of the forces associated with surface tension seems rather small when we think of large animals, which have a comparatively small ratio of surface to volume and hence comparatively little surface area per unit body mass. However, surface-tension forces are quite significant in small organisms, such as insects, that have relatively large ratios of surface to volume and thus a comparatively large amount of surface area per unit body mass. Since, in general, the insect cuticle is hydrophobic, many small insects are able to be supported against the force of gravity by the forces of surface tension. However, if the cuticle is made hydrophilic by the application of a "wetting agent," such as a detergent, the insect will immediately sink or become trapped by the water. Many surface-dwelling forms secrete a waxy material that makes their tarsi hydrophobic and allows these insects to walk on the water's surface. The surface-dwelling bugs (several families of Hemiptera) are able either to walk on the water as other insects would walk on the ground (e.g., the water measurer, *Hydrometra,* Hydrometridae) or to use the middle legs as synchronous oars, rowing from place to place (e.g., water striders, Gerridae and others). The *collophores* and tarsal claws of certain springtails (e.g., *Podura aquatica*) are hydrophilic while the rest of the integument is hydrophobic. This enables these insects to be anchored to the water and yet easily move from place to place by means of their furculae. Beetles in the genus *Stenus* (Staphylinidae) secrete a substance from anal glands that lowers the surface tension of the water behind the insect. As a result they are drawn forward by the effects of the contractive forces of the water in contact with

Fig. 7-6
Locomotion in a crane fly larva.
[Redrawn with modifications
from Kevan, 1963, after
Gilyarov.]

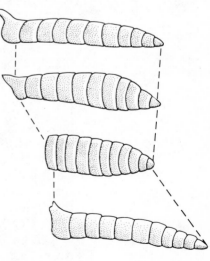

→ Direction of progression

← Direction of peristaltic waves

their hydrophobic bodies. These beetles are able to attain speeds of 45–70 cm/sec. Water striders in the genus *Velia* (Veliidae) are similarly propelled by a surface-tension-reducing substance discharged through their posteriorly directed mouthparts. This occurs as an escape response (Blum, 1978).

Subsurface Locomotion

Many insects live beneath the surface of the water. These insects can be conveniently divided into four groups.

1. Those utilizing appendages in swimming.
2. Those moving by various undulations of the body.
3. Those that are able to produce jets of water through the anus.
4. Those that walk over the substrate like terrestrial insects.

Many of these insects have air stores of various sorts, which increase buoyancy and aid in the maintenance of body position in the water (*hydrostatic balance*). In addition, in many, particularly the rapid-swimming, hard-bodied forms, the body is flattened and tapered, a shape that minimizes friction during movement through the water.

Insects that utilize appendages in swimming include representatives from the orders Coleoptera (Dytiscidae, predaceous diving beetles; Gyrinidae, whirligig beetles; Haliplidae, crawling water beetles; Hydrophilidae, water scavenger beetles; and others), Hemiptera (Notonectidae, backswimmers; Corixidae, water boatmen; Belostomatidae, giant water bugs; and others), Trichoptera (caddisflies), Neuroptera (dobsonflies, lacewings, and relatives), and even Lepidoptera. Other than insects that merely crawl about on the bottom of a body of water or on submerged vegetation, the vast majority of the insects in this category have legs that are flattened like oars, covered with "swimming hairs," or commonly both. For example, in whirligig beetles in the genus *Gyrinus* the rowing legs are so flattened that the broad side has five times the area of a comparable round leg.

The swimming legs of adult dytiscid beetles bear large numbers of hairs. These hairs (Fig. 7-7) are movable; when the insect swims, they open out, increasing the surface area applied against the water during the thrust stroke, and they fold down on the legs during the recovery stroke. Rowing or swimming legs operate in various combinations. Whirligig beetles swim with the oarlike meso- and metathoracic legs. Their forelegs are long and slender and are used in prey capture. These beetles are the ones commonly seen darting to and fro, the hydrophobic dorsum of their bodies breaking the water's surface. Beetles in the family Dytiscidae may use all three pairs of legs in locomotion. However, in the large members of this family the forelegs are used in prey capture and the mid- and hindlegs for propulsion. Hydrophilid and haliplid beetles swim with all three pairs of legs, but in contrast to the beetles just mentioned, the thrust and recovery strokes of legs on the same thoracic segment are in opposite phase.

Among the aquatic Hemiptera members of the families Nepidae

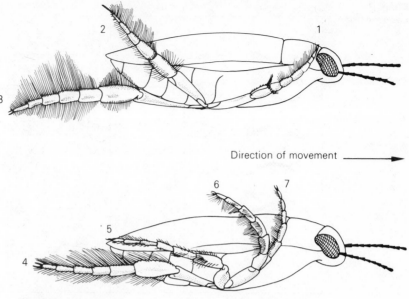

Direction of movement ⟶

Fig. 7-7
Swimming movements of a predaceous diving beetle *Acilius sulcatus*.
Numbers indicate the temporal sequence of leg positions during the
swimming stroke. Top, thrust stroke; bottom, recovery stroke. Note the
positions of the hairs at various points in the stroke cycle. [Redrawn
from Nachtigall, 1960.]

(water scorpions), Naucoridae (creeping water bugs), and Belostomatidae (giant water bugs) swim with the last two pairs of legs. The
strokes of the two legs of each segment are out of phase by 180°.
The forelegs of these insects are generally used in grasping prey,
and swimming legs are modified in ways similar to those of the
aquatic beetles. Backswimmers have a rather different manner of
swimming. They swim beneath the water with the ventral side up
and the head directed downward. Swimming is accomplished with
meso- and metathoracic legs, which move in the same phase. In
water boatmen the third pair of legs provide propulsion, the second
pair act as grasping and steering organs, and the first pair to scoop
up food-containing materials.

In addition to legs, certain insects, such as the adult females of
Hydrocampa nymphaeata (Lepidoptera, Pyralidae) and a few hymenopterous insects (*Caraphractus Cinctus,* Mymaridae; and *Lim-
nodite,* Proctotrupidae), are propelled through the water by their
wings.

The legless aquatic larvae and pupae of several Diptera are propelled through the water by twitches and undulations of their bodies.
Figure 7-8 shows some of the variations in this type of locomotion.
Certain of these insects are buoyed by air stored in "air bladders"
(e.g., *Chaoborus* larvae) or gas in external spaces resulting from the
shape of the body (e.g., the ventral air space formed by developing
legs, wings, and mouthparts in mosquito pupae).

The nymphs of most dragonflies and the nymphs of the mayfly
genus *Cloeon* draw water into, and expel it from, the rectum as part
of the ventilatory process. If alarmed, these insects can expel water

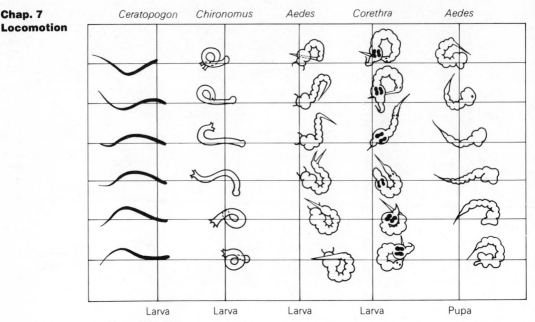

Ceratopogon	Chironomus	Aedes	Corethra	Aedes
Larva	Larva	Larva	Larva	Pupa

Fig. 7-8
Swimming movements of legless aquatic larval and pupal diptera.
[Redrawn from Nachtigall, 1963.]

from the rectum with considerable force. The expulsion of water propels the insect several centimeters, and speeds of 50 cm/sec have been recorded. When the insects propel themselves in this fashion, the thoracic legs are held close to the body, which has the effect of reducing the friction between the water and the insect. In addition, the forcibly expelled water may disturb bottom material (especially silt) and throw a protective "smoke screen."

Bottom-dwelling insects, dragonfly and damselfly larvae, larval caddisflies, and so on, walk over the substrate with thoracic legs much as the terrestrial forms do.

Aerial Locomotion

As mentioned earlier, insects, alone among the invertebrates, possess the ability to fly. This ability is perhaps one of the most important reasons for their tremendous success relative to the rest of the animal kingdom. Flight has enabled insects to take advantage of environmental situations that are virtually untouchable by their non-flying rivals.

Functional Morphology of Wings

The skeletal structure, thoracic musculature, and insect wings have already been discussed in rather general terms in Chapters 2 and 3. As explained in Chapter 3, a generalized wing-bearing segment contains both indirect and direct wing muscles (Fig. 7-9). There are two sets of indirect wing muscles, neither of which attach directly to the wing bases. The longitudinal indirect muscles run between dorsal apodemes (phragmata), and contraction causes arching of the tergum with the consequent depression of the wings. The dorsoventral (tergosternal) indirect muscles run between the notum and sternum. The structural relationship between the notum and pleura (see Fig. 3-16) is such that contraction causes elevation of the wings.

Direct wing muscles attach to various regions on the pleuron and coxa and to the axillary sclerites (Fig. 7-9) and basalar and subalar sclerites (Fig. 7-9). Contractions of these muscles are invovled with wing movement during flight and with *wing flexion*. The muscles associated with the basalar and subalar extend and depress the wings. Contraction of muscles attached to the third axillary sclerite (Fig. 7-10) results in wing flexion (the opposite of extension); that is, the wing is folded posteriorly over the abdomen. Folds in the wing (Fig. 7-10) facilitate this process. With various modifications of the generalized scheme presented here, winged members of all pterygote orders, except the Odonata (dragonflies and damselflies) and Ephemeroptera (mayflies), are able to flex the wings over the abdomen. The origin of wings and evolutionary significance of wing flexion will be discussed in Chapter 10.

There has been a tendency during insectan evolution toward the reduction of the wings to a single unit. The pressures in this direction have likely been due to the inefficiency of a hind pair of wings operating in the turbulence produced by the movements of the forewings. The wings of certain groups of insects, such as the Orthoptera, Megaloptera (dobsonflies, fishflies and alderflies), Isoptera (termites), and others, operate separately, and these insects are rather poor fliers as a result. One evolutionary solution to this problem is found in dragonflies and damselflies. Although their pairs of wings

Fig. 7-9
Generalized wing-bearing segment. A. Lateral view. B. Cross-sectional view. [Redrawn from Snodgrass, 1935.]

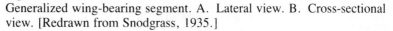
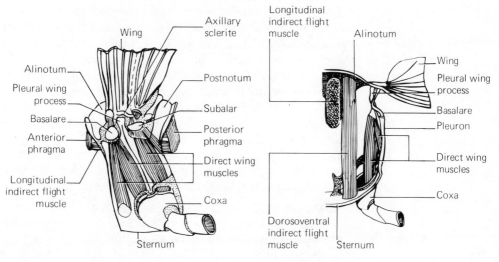

A

B

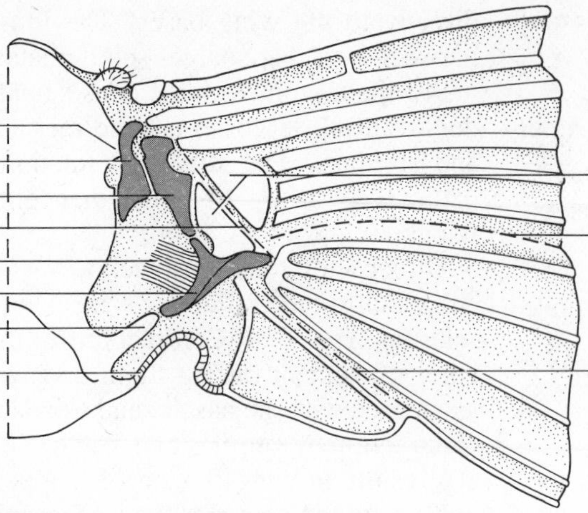

Anterior notal process
Axillary sclerite 1
Axillary sclerite 2
Fold line
Flexor muscle
Axillary sclerite 3
Posterior notal process
Axillary cord

Median plates
Vannal fold
Jugal fold

Fig. 7-10
Generalized wing base. [Redrawn from Snodgrass, 1935.]

operate separately, their flight is quite efficient because the sequence of wingbeat is reversed, the hindwings moving before the forewings during each cycle.

Along the lines of developing a single unit, there have been two basic solutions. One solution has been the development of a variety of mechanisms that enable the fore- and hindwings to be coupled to one another (see Fig. 2-40A–C). Correlated with the presence of wing-coupling mechanisms is a reduction of the size of the hindwings. Coupling mechanisms are found among the Hymenoptera, Trichoptera, Hemiptera, many Lepidoptera, and others. The other evolutionary solution has been the loss or extreme modification of one pair of wings so that they no longer function as propulsive organs of flight. For example, in the true flies, the hindwings have become the halteres (balancing structures). The forewings of beetles have similarly lost their function as organs of flight.

Flight and Its Control

Both "direct" and "indirect" muscles are associated with wing movement. Probably the muscles attached directly to the wings were the original sources of the major propulsive force of wing motion. Direct muscles are still dominant in cockroaches (Blattaria) and very important in members of the Orthoptera as well as Odonata and many Coleoptera. In other orders the main propulsive forces are furnished by the indirect muscles of each wing-bearing thoracic segment. In these insects the indirect muscles are quite large relative to the direct ones, although the direct muscles are always important in wing motion. There is a direct correlation between the weight of flight muscle as a percentage of body weight and the strength of flying. For example, according to Wigglesworth (1972),

> In the relatively weak flying orthopteran *Oedipoda* the flight muscles comprise only 8% of the total body weight, but in strong fliers they make up far greater proportions: *Musca* [fly] 11%, *Apis* [honey bee] 13%, *Macroglossa* [sphinx moth] 14%, *Aeshna* [dragonfly] 24%.

Wing-movement patterns in flying insects have been analyzed by means of a variety of techniques. Earlier methods included attaching light-reflecting devices to the wing tips, which would make the trajectory of each wingtip visible, and observation of wing positions in freshly killed insects. More recently, high-speed cinematography has proved very useful. Another important technique has been the use of the strobe light, a device that can produce regularly intermittent light flashes over a wide range of rates. At appropriate rates of flashing the entire wing is illuminated regularly at various points in its cycle, which makes the wing appear to be moving very slowly. This makes possible direct observation of the changes in wing tilt and so on as they occur. When the flash frequency is a whole-number multiple of the wingbeat frequency, the wing is caught in the same position during each cycle and appears to stand still. This is one of the methods for determining wingbeat frequency. Recording and analysis of the pitch of the sound produced during wingbeat is another important method for determining frequency of wingbeat. Highly sophisticated mechanical and electronic recordings of neuromuscular activities during flight have also been used extensively.

In modern insects, the wings are responsible for both lift and propulsion and usually are of little importance in gliding. Naturally, for bilateral appendages to produce these effects, the movements must be complex. These movements consist of "elevation and depression, promotion and remotion (fore and aft movement), pronation and supination (twisting) and changes of shape by folding and buckling" (Pringle, 1957).

A further complexity is that the velocity of wing motion varies in different parts of the cycle. In many insects movements like wing twisting are entirely under the control of direct wing muscles. However, in some of the better fliers, such as members of the orders Diptera and Hymenoptera, the basal wing articulations are designed mechanically such that the appropriate twisting of the wing occurs at the proper time during a wingbeat cycle. This is not to say that direct muscles do not exert some degree of influence, even in these insects. The propulsive action of the wings is quite efficient. According to Wigglesworth (1972), "the flying insect produces a polarized flow of air from front to rear during approximately 85% of the cycle." In typical forward flight, a wing traces out a figure "8" relative to the body at its base (Fig. 7-11). Relative to a point past which an insect is flying, the wing traces out a pattern resembling a sinusoidal curve. Many insects can hover by changing the inclinations of the figure 8 relative to the body (Fig. 7-11). Many of the especially good fliers (Diptera, Hymenoptera, and Lepidoptera) are able to fly backward (Fig. 7-11), sideways, or rotate around the head or tip of their abdomens. Rotation and flight sideways are apparently produced by unequal activity of the wings on either side. Besides being able to vary the patterns of wingbeat, insects may vary the power of wingstroke by varying the number of actively contracting muscle fibers.

The number of wingbeat cycles per second varies tremendously from one group of insects to another (Table 7-1). Not only does wingbeat frequency vary in different insectan groups, but it also

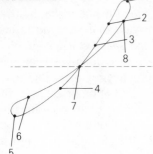

A   B   C

Fig. 7-11
Wingbeat patterns traced by the tip of a wing during flight. Dashed lines
indicate longitudinal axis of body. A. Forward flight. Numbered lines
indicate position and inclination of the wing at different stages of the
wingbeat cycle. B. Hovering. C. Backward flight. [A redrawn with
modifications from Mangan, 1934; B, C redrawn with modifications
from Stellwaag, 1916.]

varies with age, sex, season, humidity, load, air resistance, air density, air composition, wing inertia (moment of inertia), muscle tension, and fatigue. However, variation in frequency of wingbeat is
apparently not generally used by insects to control flight. Assuming
a single series of muscle contractions for each wingbeat cycle, many
frequency measurements seemed in the past to be fantastically high.
Electrophysiological recordings from flight muscle and associated
nerves showed that in certain insects the number of nervous impulses
impinging on a wing muscle were much fewer than the number of
muscle contractions observed. Thus the high wingbeat frequency
observed could not be explained on the basis of a one-to-one relationship between nervous impulse and muscle contraction. In other
cases there was a one-to-one relationship between nerve impulses
and muscle contraction. These differences remained a puzzle until
the elastic nature of the insect thorax and the action of resonating
flight muscles were discovered.

In addition to the actions of direct and indirect muscles, it has
been found that in many insects the elasticity of the thoracic skeleton
plays a significant role in the movement of the wings. As the wings
move in one direction, some of the energy of muscle contraction
may be stored as elastic energy in the thoracic skeleton and antagonistic muscles and used in the movement of the wings in the opposite direction. This can be seen in the locust *Schistocerca*. When
the major flight muscles are relaxed, the wings are completely depressed. This is the stable position for them in relation to the elasticity of the thoracic skeleton. Any movement away from this stable
position is opposed by the elastic forces.

In several insects, such as certain Diptera and Coleoptera, the
wings have two stable positions, completely elevated and completely
depressed. Movement of a wing away from one of these stable positions is opposed by the elasticity of the thorax to a point at which

Table 7-1
Representative Wingbeat Frequencies[a]

Insect	Wingbeats per Second[b]
Odonata	
Libellula	20
Aeshna	22, 28
Coleoptera	
Melolontha	46
Coccinella	75–91
Rhagonycha	69–87
Lepidoptera	
Pieris	9, 12
Colias	8
Saturnia	8
Macroglossa	72, 85
Acidalia	32
Papilio	5–9
Diptera	
Tipulids	48, 44–73
Aedes, male	587
Culex	278–307
Tabanus	96
Musca	190, 180–97, 330
Muscina	115–220
Forcipomyia	988–1047
Hymenoptera	
Apis	190, 108–23, 250
Bombus	130,240
Vespa	110

[a]Data from Wigglesworth (1965) based on various sources.
[b]Two or more sets of numbers indicate data obtained by different investigators.

the direction of the elastic force reverses. At this point the wings are suddenly driven by the elastic energy into the opposite stable position. This has been called the "click mechanism." In Chapter 3 it was pointed out that insects possess both resonating and nonresonating muscles and that, under appropriate mechanical conditions, the resonating type of muscle is capable of undergoing several successive contractions when stimulated with a single nervous impulse. The insect thorax with sufficient elasticity to produce a click mechanism is an example of such an appropriate mechanical situation. The resonating muscles involved are the mutually antagonistic dorsal longitudinal and dorsoventral flight muscles. As one set of these muscles is stimulated via nerve impulse to contract, a point is reached where the "click" occurs (i.e., where the elastic forces reverse). When this happens, the tension on the contracting muscles is suddenly released and the antagonistic pair is suddenly stretched. This sudden stretch stimulates the antagonistic pair to contract and the cycle then repeats itself. Thus several oscillations of the wing mechanism are produced before another nervous impulse initiates the series all over again. These oscillatory contractions have been identified in many of the higher Diptera and Coleoptera, and others.

Bearing in mind the influence of unavoidable experimental inter-ference and consequent measurement inaccuracies, see Table 7-2 for examples of speed of flight in different insect species.

The sensory cues pertinent to flight control undoubtedly come from a number of different sources, both from sensilla in the wings themselves and from sensilla located in other parts of the body. However, no sensory endings have ever been found in insect flight muscles. Wing sensilla fall into three main categories.

1. Bristles and hairs.
2. Campaniform sensilla.
3. Chordotonal sensilla.

The bristles and hairs may be widely distributed over the wing's surface and respond to tactile stimulation. Thus they may respond to the movement of air over the wings during flight. Campaniform sensilla are arranged in definite patterns in the wings. Apparently different ones respond to different torques acting on the wings during flight, depending on their specific orientation. Since chordotonal sensilla are sensitive to changes in length, those in the wings prob-ably respond to displacements of the cuticle at the wing base re-sulting from various torques. Like the campaniform sensilla, they are arranged in definite patterns. In certain instances they may ac-tually respond to changes in air pressure or even have an auditory function.

Sensilla that collect information useful in flight control have been described not only in the wings, but also in the head, neck, halteres, legs, and other locations. Sensilla in the head are commonly in the form of hairs on the frontal region and Johnston's organ in each antenna. Both kinds of sensilla are sensitive to airstreams across the head. The compound eyes must not be overlooked as sense organs important to flight control. Dragonflies have hair plates on the neck that respond to changes in head position.

Probably any stimulus that in some way indicates danger would

Table 7-2
Average Speed of Sustained Flight in Various Insects[a]

Insect	Speed, km/h
Mayflies, small field grasshoppers	1.8
Bumble bees, rose chafers	3.0
Malaria mosquitoes	3.2
Stag beetle, damselfly, *Ammophila* (a fossorial wasp)	5.4
House fly	6.4
Cockchafer, cabbage white butterfly, garden wasp	9.0
Blow fly	11.0
Desert locust	16.0
Hummingbird hawk moth	18.0
Honeybee, horse fly	22.4
Aeschna (a big dragonfly), hornet	25.2
Anax (one of the biggest of European dragonflies)	30.0
Deer bot fly	40.0

[a]Data from Nachtigall (1974b) based on various sources.

cause an insect to take flight as a means of escape. However, a number of specific stimuli that induce flight have been identified. The most widespread and important of these is loss of contact with the substrate. This may be easily demonstrated by suspending a cockroach or other insect from a string by means of a drop of paraffin. As long as the legs of the insect are in contact with a surface, there will be no flight movements, but if the insect is suddenly lifted into the air and surface contact lost, the wings will immediately commence to beat. This is the "tarsal reflex" since sensilla in the tarsal segments are involved. Other flight-initiating stimuli, such as a change in the relative positions of the pro- and mesothorax of the cockroach *Periplaneta,* have been described. It is interesting to note that some insects, for example, moths in the family Sphingidae (sphinx moths), are unable to fly until the wing muscles reach a temperature of 30°C. If the ambient temperature is 30°C or above, they have no trouble taking flight, but if it is below, these moths must rapidly vibrate the wings until the necessary temperature is reached.

During flight, constant corrections must be made for any forces that tend to disrupt the stability of the insect. Instabilities can be resolved into tendencies of the insect to move in one or a combination of three planes: movement about the longitudinal-horizontal axis, or *roll;* movement about a transverse-horizontal axis, or *pitch;* and movement about a vertical axis, or *yaw.* An unstabilizing force or combination of such forces undoubtedly affects different sensilla to different degrees and in different ways and thus results in some change in the pattern of stimuli that impinge on the insect during stable flight. In response, the insect modifies its flight movements in such a manner that the impinging stimuli again are in the pattern that indicates stability. These stabilizing responses can probably be explained by simple reflexes, but flight muscle and wingbeat rhythms are apparently generated in the central nervous system (see Chapter 8).

The halteres of true flies are a specific example of stabilizing organs (see Fig. 2-39E). These "modified hindwings" move in synchrony with the forewings. Groups of campaniform and chordotonal sensilla are arranged in specific patterns at the base of each haltere. Since the halteres are constantly vibrating during flight, they act like a gyroscope, which generates forces when there is any tendency to change direction. You will probably recall from basic physics the simple demonstration of gyroscopic action in which a rapidly rotating bicycle wheel held by two hands exerted resistive forces when you attempted to turn it away from its original plane of rotation. The forces generated by the vibrating halteres when subjected to turning forces are "sensed" by the campaniform and chordotonal sensilla in a manner similar to the way the resistive forces were "sensed" by the sensilla in your hands and arms. The halteres apparently monitor changes in all three of the planes described above.

Other sensilla that respond to unstabilizing forces during flight are the hairs on the neck of the dragonfly. During stable flight a characteristic distribution of pressure from the head is exerted on these sensilla and causes a "normal" pattern of signals. However, if a force tends to turn the insect in some direction, the pressure on these hairs becomes changed as a result of the inertia of the head, causing the insect to tend to resist the directional change. This change in pattern of stimulation of the hairs results in compensatory changes in the wing movements and hence correction for the unstabilizing forces. In addition to the neck hairs, the flight stability of the dragonfly is maintained by a dorsal light response, by which the insect moves to keep the source of illumination dorsal to itself, and an optomotor reaction to the general visual pattern. Johnston's organ in each antenna and the sensitive hair plates on the head may also detect changes in air flow and thus collect information pertinent to guidance as well as flight maintenance.

For further information on insect flight see Nachtigall (1974b), Pringle (1957, 1968, 1974, 1975), Rainey (1976), and Wilson (1968, 1971). Dalton (1975) provides superb stop-action photographs of insects in flight.

Selected References

GENERAL
Chapman (1971); Richards and Davies (1977); Roeder (1963); Snodgrass (1935); Wigglesworth (1972).

TERRESTRIAL LOCOMOTION
Bennet-Clark (1976); Hughes and Mill (1974); Rothschild et al. (1973); Wilson (1966).

AQUATIC LOCOMOTION
Nachtigall (1974a).

AERIAL LOCOMOTION
Dalton (1975); Nachtigall (1974b); Pringle (1957, 1968, 1974, 1975); Rainey (1976); Wilson (1968, 1971).

Behavior

In the preceding chapters organ systems have been discussed more or less individually. In this chapter, insects are considered as whole organisms, and the results of the very complex, integrated actions of the various systems in response to changes in the external and internal environments are examined.

Insect behavior has been extensively investigated under natural and laboratory conditions, both approaches being valid, appropriate, and useful. The necessary first step is to describe an insect's behavior as accurately and completely as possible (*ethogram*). This should include consideration of the time of appearance of specific behavioral variations associated with physiological state (nutritional, etc.).

Following description, there are three primary objectives.

1. Determination of the control of behavior from the nervous, endocrine, and genetic points of view.
2. Elucidation of the function of the various behaviors in the insect's life, that is, the adaptiveness of behavior.
3. Determination of the probable phylogenetic origin of behavior, the rationale being that behavioral traits, like morphological and physiological traits, are genetically based and represent adaptations that have arisen through natural selection.

One of the pitfalls encountered during the study of insect behavior has been the temptation to ascribe human purposiveness or goal seeking to many behavior patterns. This is commonly referred to as *anthropomorphism* and should be avoided. Anthropomorphic usage tends to obscure the fact that adaptive behavior is the result of natural selection and anticipation of a goal is unnecessary.

The approach of this chapter is to consider the basic kinds of behavior characteristic of insects, the control of behavior (nervous, endocrine, and genetic), and the biological functions of behavior.

A number of textbooks are available on animal behavior (see Selected References). In addition, recent texts (Atkins, 1980; Matthews and Matthews, 1978) and the review by Markl (1974) deal specifically with insects. Carthy (1965) considers the behavior of arthropods, including insects. Richard (1973) traces the historical development of studies of insect behavior.

Kinds of Behavior

In very broad terms, insects exhibit two kinds of behavior, innate and learned. *Innate behavior* to a large extent consists of a more or less fixed response or series of responses to a given stimulus or pattern of stimuli. The "more or less" in the preceding sentence should be emphasized, since innate behavior is usually somewhat flexible and may be modified by experience (learning). Innate behavior is generally considered to be based upon the inherited properties of the nervous system. On the other hand, *learned behavior* is not inherited but is acquired through interaction with the environment during the life of the individual. Obviously, although specific patterns of learned behavior are not inherited, the potential for learning is. In many instances it is difficult to determine whether an observed behavior pattern is inherited or learned. There is no evidence of the ability to reason among insects, the observed behavior patterns being explainable on the basis of innate patterns and learning.

Innate Behavior

Some innate behavior patterns seem to be comparatively simple, for example, the *reflexes*. Reflexes may involve only a part of the body, as in proboscis extension, or the whole body, as in the righting reflex, which occurs when an insect is placed on its dorsum. Reflexes can be grouped into two classes, *phasic* and *tonic*. Phasic reflexes are comparatively rapid and short-lived and are involved in rapid movements such as proboscis extension. Tonic reflexes are slow and long-lived and are involved with the maintenance of posture, body turgor (see the discussion of crawling in Chapter 7), muscle tone, and equilibrium. Reflexes vary in complexity, the simplest being mediated by a single afferent impulse from a receptor to an interneuron and then along an efferent neuron to an effector (see Fig. 3-1), the *reflex arc*. Probably all reflexes are more complex than this, involving many more neural connections. Individual segmental ganglia may show considerable reflex autonomy. For example, in the silkworm moth (*Bombyx mori*) the oviposition reflex, which results from the ovipositor making contact with a surface, resides entirely in the caudal (last abdominal) ganglion.

More complex innate behavior includes the various orientation patterns. Orientation may be defined as "the capacity and activity of controlling location and attitude in space and time with the help of external and internal references, i.e. stimuli" (Jander, 1963). Birukow (1966), Fraenkel and Gunn (1961), Jander (1975), Markl (1974), and Matthews and Matthews (1978) are all useful in their consideration of this topic. The terminology used in describing the kinds of orientation is confusing. Frankel and Gunn (1961) offer a classification scheme for the various kinds of orientation. They first recognize two broad kinds of orientation, primary and secondary. *Primary orientation* is the assumption and maintenance of the basic body position in space—that is, the normal stance—either in a stationary or in a moving insect. For example, reflexes that keep a flying insect on an even course are responsible for the primary orientation. *Secondary orientation* is superimposed upon primary orientation and has to do with the positioning of the insect in response to various stimuli. These stimuli may be external (e.g., light, humidity, temperature) or internal (e.g., the presence of a particular hormone). Secondary orientation may be of value to an insect by leading it toward potential prey or a potential host, or away from potential danger or an unsuitable environment, or toward a favorable one.

In addition to the primary and secondary divisions, Fraenkel and Gunn (1961) also offer a classification based on the mechanisms of orientation, dividing them into *kineses, taxes,* and *transverse orientations.*

Kineses are undirected locomotor reactions to stimuli, and hence there is no particular orientation of the long axis of the body relative to a stimulus source. For example, in tsetse flies, *Glossina* spp., there is an increase in activity in an arid atmosphere relative to activity in a humid one. This behavior would result in the insect "finding" and remaining in a humid atmosphere. The body louse, *Pediculus,* makes fewer and fewer directional changes as temperature, humidity and odor increase as the louse moves closer to a potential host.

Taxes are different from kineses in that they are directed responses relative to a stimulus source. The long axis of the body takes on a definite orientation relative to the stimulus. The insect may move toward (positive) or away from (negative) the stimulus source. Theoretically, when an insect is maintaining a course relative to a stimulus, angular deviations from this course result in the initiation of a "turning tendency," that is, the tendency for the insect to correct for angular deviation by turning back to its original course. The magnitude of this turning tendency increases with increasing angular deviation. Hence, if the magnitude of this turning tendency is plotted as the ordinate and angular deviation as the abcissa, a sine curve is generated (Fig. 8-1).

Taxes may be classified according to the stimuli involved (e.g., *phototaxis,* light; geotaxis, gravity; *anemotaxis,* air currents; *rheotaxis,* water currents; and *thigmotaxis,* contact).

A good example of a taxis is *klinotactic orientation.* An insect orienting klinotactically swings all or part of its body back and forth across a stimulus field and moves toward or away from the region of maximum stimulation, depending upon whether it is attracted or repelled by the particular kind of stimulus. For example, when fly maggots have completed feeding and will soon pupate, they move the head region back and forth until they are heading directly away

Fig. 8-1
Relationship between magnitude of turning tendency and angular deviation. [Redrawn with modifications from Jander, 1963.]

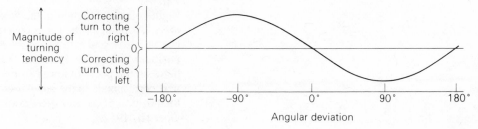

from a source of light. You will recall from Chapter 6 that this is the position in which the light-sensitive cells are the least affected by the light stimulus. These fly maggots are then photonegative.

Transverse orientations are those in which the body is oriented at a fixed angle relative to the direction of the stimulus. Locomotion may or may not be involved. Frankel and Gunn (1961) discuss two types of transverse orientations found in insects: the *light-compass reaction* and the *dorsal light reaction*. The light-compass reaction consists of an orientation such that locomotion occurs at a fixed angle relative to light rays. This orientation was first demonstrated in ants and subsequently in caterpillars, bees, and certain beetles and bugs. The light-compass reaction serves, for example, to aid an ant on a return trip to its nest. One simple but informative experiment (Fig. 8-2) demonstrating that ants use the sun in this manner involved placing an ant on the way back to its nest in a dark box for a period of time during which the sun's position in the sky would change significantly. Subsequently, the ant was released and its course was then the same as before relative to the sun, but incorrect in terms of its nest because the sun's position had changed. Similar orientations to gravity have also been described.

In the dorsal (or ventral) light reaction, both the long and transverse axes of the body are kept perpendicular to a directed source of light (Frankel and Gunn, 1961). Although symmetrical receptors (the eyes) are involved, this type of reaction is different from phototaxis because orientation is transverse to the light rays and not parallel to them. Examples of both dorsal and ventral light responses are found in certain aquatic members of the order Hemiptera. When the air bubbles trapped between the head and the antennae are removed, backswimmers, *Notonecta* and *Plea,* depend entirely upon a ventral light reaction for the maintenance of their normal swimming position while the converse is true for the water boatman, *Corixa,* and the creeping water bug, *Naucoris* (Rabe, 1953). Some

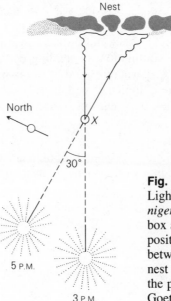

Fig. 8-2
Light-compass reaction in the ant *Lasius niger.* At point *X,* the ant is covered with a box at 3 P.M. and released at 5 P.M. (sun positions indicated). Note that the angle between the paths followed from and to the nest corresponds to the angular change in the position of the sun. [Redrawn from Goetsch, 1957.]

caterpillars are shaded such that, provided they are oriented properly with respect to light, they are effectively camouflaged. Dorsal and ventral light responses come into play in these insects, ensuring "proper" orientation. The dorsal (ventral) light response is a good example of a means for maintaining primary orientation.

Jander (1975) discusses the ecological aspects of orientation behavior and provides a classification of these behaviors developed on ecological bases.

The most complex innate behavior patterns are the *fixed action patterns*. These patterns are characterized by being unlearned, species specific, and adaptive. They differ from reflexes and orientation mechanisms in that they frequently require an internal readiness before they can occur and the eliciting stimulus (*releaser*) is commonly not required to act during the entire course of the pattern. The releasing stimulus may be internal in the form of hormone secretion or an excitatory signal from a higher neural center. The stimulus may also be external in the form of a physical factor—light, temperature, mechanical contact, and so on. Or the behavior, appearance, odor, and so on, of another animal may be involved (more on this in the section on communication).

A variety of factors may affect the development of a state of internal readiness or motivation necessary for a given behavior pattern to occur. These may be either internal or external in origin (Markl, 1974; Markl and Lindauer, 1965).

Internal motivational factors are hormonal or neural in nature. For example the relative concentrations of the juvenile hormone and the molting hormone (ecdysone) determine the spinning behavior of the caterpillars of certain moths (*Galleria* or wax moths and others). A hormone secreted by the ovaries is necessary for the female grasshoppers (Acrididae) in the subfamily Truxalinae to sing (Markl and Lindauer, 1965, citing the work of Haskell, 1960).

Internal receptors are commonly involved when a behavior pattern appears spontaneously without any discernible external stimulus. An excellent example of this has been found in the blow fly, *Phormia*, in which receptors associated with the crop carry stimuli that inhibit food ingestion as long as the crop is full. No inhibitory stimuli arise from these receptors when the foregut is empty and feeding can occur. In other instances of spontaneous behavior, regular cycles of behavioral activity of internal origin (endogenous rhythms) determine when a particular pattern may be released. For example, crickets are active only during a certain period each day and this daily rhythm continues for several days when the insects are kept in constant darkness in the laboratory. Periodicity in behavior is considered later in this chapter. The last internal motivational factor listed above was the stage of development.

External factors—light, temperature, substances produced by or activities of other insects of the same species, and others—have definite effects on motivation in addition to serving as releasers. Light and temperature commonly determine whether an insect is active or inactive. Recall that when we discussed the dorsal ocelli as stimulatory organs, we gave examples in which insects failed to demonstrate certain light responses when the ocelli were blocked.

Every insect has a temperature below which and one above which it becomes inactive. Within the range of these temperatures, specific temperatures may be necessary for particular activities. For example, the readiness of certain butterflies (e.g., *Pieris*) to copulate is determined by temperature.

Many instances are known where a substance secreted by one insect influences the behavior of another of the same species. The term *pheromone* is commonly used to denote such a substance. Pheromones may serve as releasers or as motivational factors or both. A good example of a pheromone affecting behavior is found in the honey bee. The queens produce a substance known as the queen substance, in the mandibular glands which is passed from worker to worker orally and which inhibits ovary development as well as the behavior involved with queen-cell construction. Queen cells, in which new queens develop, are constructed when a hive has lost its queen and hence the effects of the substance of her mandibular glands. Pheromones will be discussed in the section on communication.

When an insect is in a state of internal readiness (motivated), it may begin to behave in a manner that increases the probability of exposure to a releasing stimulus. This has appropriately been called *appetitive behavior*. Probably the most common appetitive behavor is increased locomotion. Appetitive behavior does not reduce the motivation and is likely to continue until the insect comes into contact with the appropriate releasing stimulus. At this time the specific behavior pattern involved with the motivation is manifested. This occurrence has been called the *consummatory act* and reduces or lowers the motivation. Male cicadas that have been sexually aroused move about randomly (appetitive behavior), thus increasing the probability of encountering a female. This random movement will continue until contact is made with a female. At this time copulatory behavior (consummatory act) is released and sexual arousal reduced. There are probably a number of behavioral steps between the initial motivation and the consummatory act. It is generally felt that behavior is organized into a hierarchical sequence of patterns ultimately based on a hierarchical organization of nervous centers. From this point of view a given fixed action pattern (e.g., sexual behavior, food getting, etc.) then consists of a definite sequence of patterns, each with a specific releaser, the final pattern being the consummatory act that reduces the motivation. A sequence of behavior patterns may be aborted at any time if a given releaser is missing along the way. Markl and Lindauer (1965) cite an excellent example from the work of Baerends, who studied the sequence of patterns involved in brood care in the digger wasp, *Ammophila adriaansei*.

In a definite sequence, three groups of activities follow one another: (1) digging of a nest, hunting for a caterpillar, and oviposition; (2) bringing more caterpillars and temporary closing of the nest; (3) fetching still more caterpillars that are required and the final closing of the nest. The process is complicated by the fact that several nests are simultaneously cared for. At a second nest, work can only proceed in the spaces between the three phases of supplying the first. Therefore, a characteristic pattern of activities appears in the supplying of the nests. The single parts of the process are also organized hierarchically. In the process of bringing in the first caterpillar this sequence is always as follows: the depositing of the caterpillar in the front of the entrance of the nest; digging up of the entrance; turning around at the open nest; and pulling in of the caterpillar. The sight of the nest situation releases the dropping of the caterpillar and scraping, the half-open nest site releases digging, the open nest releases the turning around, and the sight of the caterpillar releases the pulling in.

Learned Behavior

Although there are a number of possible definitions of learning, we shall use that of Thorpe (1963): learning is "that process which manifests itself by adaptive changes in individual behavior as a result of experience." Thus learning involves the accumulation and storage (memory) of environmental information and the subsequent effects of this stored information on an animal's behavior. Extreme care has been necessary in calling a given pattern of behavior learned, since an innate pattern that appears at some point during the development of an animal may give the distinct impression of having been learned.

There is no generally agreed upon classification of the kinds of learning. However, Markl (1974) outlines the following designations which have been described among insects and which are useful for our purposes: *habitation, associative learning* (including *classical conditioning, instrumental conditioning,* and *shock-avoidance conditioning*), and *latent learning*.

Habituation has occurred when a stimulus that initially evoked an escape or avoidance response no longer elicits such a response. This type of learning is common in insects and has been demonstrated relative to substrate vibrations, noise, visual stimulation, chemical repellents, etc. The ability of insects to learn in this fashion avoids unnecessary, energy-consuming behavioral responses.

Classical conditioning was originally discovered by the Russian scientist Pavlov. In this type of learning, two stimuli are involved. One of the stimuli (the unconditioned stimulus, UCS) elicits a response (the unconditioned response, UCR). The other, the conditioned stimulus (CS), prior to conditioning, does not elicit the UCR. However, when these two stimuli are repeatedly presented to an animal in very close succession (CS briefly preceding the UCS), the animal gradually begins to respond to the CS in the absence of the UCS.

Classical conditioning has purportedly been shown to occur in honey bees. In these insects, the proboscis extension reflex can supposedly be conditioned to respond to the essence, coumarin, in the absence of sugar. In this case, proboscis extension is the UCR, the sugar is the UCS, and the odor of coumarin the CS. However, Alloway (1972) explains that the experiments carried out to demonstrate the conditioning of the proboscis extension reflex were insufficiently controlled. Thus he concludes that classical conditioning has yet to be found among insects.

In instrumental (trial and error) learning, a response or series of responses is induced or inhibited by the presentation of a stimulus or pattern of stimuli [reinforcer(s)]. Maze learning is a good example of instrumental learning. A maze is placed between an insect and its nest, food, or some other strong positive reinforcer; or a negative reinforcer, such as an electric shock, is applied for an incorrect response. If an insect is capable of learning the maze, the number of errors (wrong turns) will decrease with the number of trials. For example, ants are capable of learning a rather complex maze placed between themselves and their nest (positive reinforcer). Another example of instrumental learning is the "training" of honey bees to associate a particular color with a sugar (reward) source (see Chapter 6). Honey bees can also learn to associate odors with a food reward.

There are many instances where an insect has been conditioned in an inhibitory way. For example, a mantis can learn not to strike an object it associates with an electrical shock.

Horridge (1962a, b) showed that a single leg of a decapitated cockroach suspended above a container of electrified saline solution can be trained to stay above a level where it would receive a shock ("shock-avoidance learning"). The experiments were designed so that a test animal received shocks according to the level of the test leg, while a control animal received the same number and intensity of shocks, but in a random fashion, without regard to leg position. Later investigators (Eisenstein and Cohen, 1966) demonstrated that even a single isolated ganglion in the ventral nerve cord could mediate the shock-avoidance learning of a single leg. Eisenstein (1972) provides a review of the literature on this fascinating topic.

Latent learning occurs without apparent reward or punishment. Social insects learn characteristics of the immediate vicinity of their nest such that they are able to find their way back from foraging flights. The predatory wasp, *Ammophila*, mentioned above, can find its nest despite forced detours and displacements (Thorpe, 1950).

Olfactory conditioning may be a form of habituation or latent learning. An example of this phenomenon is a situation in which exposure to an ordinarily repellent odor during a certain period of development results in the loss of repellency of this odor to adults. This is the case in *Drosophila melanogaster*, the adults of which are normally repelled by the odor of peppermint oil. However, if the larvae have been reared in the presence of peppermint, they are strongly attracted to it as adults.

Although the ability to learn is widespread in the Animal Kingdom, the mechanism of memory is not known.

Alloway (1972, 1973) and Markl (1974) review learning in insects. Wells (1973), Erber (1975), and Menzel and Erber (1978) deal specifically with learning in the honey bee.

Periodicity in Behavior

Insects, along with all other organisms, have evolved in an environment characterized by regularly occurring, cyclic changes. Many of the activities of insects (locomotion, feeding, mating, oviposition, eclosion, etc.) also occur at regular intervals. Further, the timing of these recurring activities is adaptive; that is, insects, under natural circumstances, behave in ways that maximize chances for survival. Thus herbivorous (plant-eating) insects hatch from eggs in synchrony with the availability of host plants, insects become dormant and cold-hardy or migrate at a time that favors survival of an upcoming cold or dry season, and so on. The term *periodicity* is applied to such recurrences of behavior. The patterns of recurrence of particular activities may be every 24 hours, monthly, annually, in coordination with the lunar cycle, and so on. Periodicity in behavior involves both endogenous and exogenous components. Endogenous components arise from within the insect and are ultimately related to its heredity. These include, in particular, rhythms, but the stage in a given physiological-behavioral cycle (e.g., gonotrophic cycle), developmental stage, and other endogenous factors may also play a role. Exogenous components arise from the external environment.

Rhythms are processes that are controlled by an "innate time-measuring sense" or "biological clock." They characteristically continue even when external conditions (temperature, light, etc.) are kept constant. Most rhythms that have been identified have a period of approximately 24 hours and are referred to as *circadian* ("about a day") *rhythms*. Several examples of circadian rhythms have been identified in insects. For example, cockroaches are nocturnally active; *Drosophila* emerge from the pupal stage at dawn; *Tettigonia* and other orthopterans sing at a particular time of day or night.

A number of noncircadian rhythms are known, for example, the lunar emergence of certain chironomid flies and mayflies and the annual emergence of a species of dermestid beetles (Corbet, 1966). A well-known example of noncircadian rhythmicity is found in the species of periodical cicadas (*Magicicada* spp.), the various broods of which emerge en masse at 13- or 17-year intervals. These rhythms are sometimes called *gated rhythms* and are evident as population, as opposed to individual, phenomena. Thus the members of a population are coordinated and reach and pass through a behavioral "gate" at the same time.

Some investigators have maintained that rhythms are under the control of undetected exogenous factors rather than endogenous biological clocks. However, evidence favors an endogenous source, particularly the fact that the phase of a rhythm does not change even if the animal is transported to a place on the earth far from the locality of entrainment ("clock setting"). In addition, two insects can be maintained on different light regimes in the same room, all other factors being identical. Since rhythmic activities are found throughout the spectrum of living organisms (plant and animal), the general feeling is that the ultimate clock mechanism lies within cells, although its exact nature still remains a mystery.

Although rhythms are not controlled by exogenous factors, some exogenous factors are known to influence rhythms. Particularly significant are the external time cues *(Zeitgebers)*, which set the phase of the rhythmic patterns of the individuals in a population. In a population under artificially constant environmental conditions (e.g., constant illumination) the rhythmic activities of the individuals may be out of phase with one another, but with the introduction of an appropriate time cue, they are set in phase and the activities of the individuals become synchronous. For example, mosquitoes (e.g., *Aedes aegypti*) held under constant illumination oviposit arrhythmically, but when exposed to a dark period, even a very brief one, oviposit synchronously at 24-hour intervals thereafter. Under laboratory conditions, the phase of a rhythmic activity can be shifted about at the investigator's will. Among the exogenous factors that may act as time cues are length of the light or dark phase of a photoperiod ("a cycle consisting of a period of illumination followed by a period of relative darkness"—Beck, 1968), light intensity, temperature, time at which food is available, and others.

Exogenous factors may influence rhythms in ways other than as time cues. For example, in the dragonfly *Anax imperator*, as described by Corbet (1966):

> Emergence (probably rhythmic) usually occurs abruptly soon after sunset; if, however, the air temperature falls below 10°C., after larvae have left the water but before ecdysis has begun, some of them will return to the water and emerge the next morning after temperatures have risen again. For these individuals the diel periodicity of emergence has been changed by a short-term response to unfavorable exogenous factors, but the phase-setting of their rhythms has remained unaltered (as evidenced by their attempting to emerge at the normal time).

Many external environmental factors display a 24-hour periodicity, which may result in corresponding periodicity in insects without the involvement of biological clocks. Among these are light intensity, moisture, temperature, wind velocity, and various possible biotic factors. An example of how one of these factors may cause periodic behavior is in the swarming of many species of mosquitoes, which is determined by light intensity. These insects continually possess the "drive" to swarm, but do so only under appropriate conditions of light intensity and can be induced to swarm for hours if the light intensity is held artificially at the appropriate level. Under natural conditions, the appropriate light intensity may be periodic (e.g., that which occurs during crepuscular periods). In this situation the mosquitoes would swarm "periodically."

For more information on biological clocks and insect periodicities, see Brady (1974), Corbet (1966), Danilevskii (1965), and Saunders (1974, 1976a, b). Bünning (1967) discusses biological clocks in general.

The Control of Behavior

An insect is continually bombarded with multitudinous stimuli in space and time. Many stimuli provide information pertinent to the insect's life; many do not. In order to survive, an insect must respond to the "right" stimuli and do the "right" thing at the "right" time: escape, locate a food source, feed, locate a mate, copulate, oviposit, migrate, become dormant, and so on. That is, an insect's behavior must be controlled.

How the environment is perceived and what behavior is exhibited is a function of the nervous and endocrine systems. The functional characteristics of the nervous and endocrine systems are in turn determined by hereditary mechanisms. In this section, the involvement of the nervous and endocrine systems in behavior and the role of heredity are discussed.

Nervous Control

Insects are especially good subjects for studying nervous function and behavior since their behavior tends to be stereotyped, but with some learning, and their nervous systems contain a relatively small number of neurons. For detailed information on the nervous control of insect behavior see Barton Browne (1974), Howse (1975), Hoyle (1970), Huber (1974), Miller (1974b), and Roeder (1967, 1970, 1974).

Aspects of nervous control of several motor patterns (especially flight, walking, and sound production) have been studied, but very little is known about the neural control of complex sequences of behavior. Techniques that have been used include observation of behavior before and after decapitation, severance of nerves, ablation or destruction of all or parts of ganglia; electrophysiological recording from various nerves and ganglia *in vivo* and in isolated preparations; and use of various stains to elucidate structural connections. A rather ingenious approach has been addressed to the establishment of the location of the sites within the nervous system directly associated with various abnormal (mutant) behavior patterns. This approach involves the use of genetic mosaics and the establishment of "embryonic fate maps" (Hotta and Benzer, 1972; Benzer, 1973).

The operation of the nervous system is dependent upon the properties of individual neurons and the ways these neurons are associated with one another both spatially and temporally. Behavior is the result of interactions between sensory input, reflexes, and centrally generated neural patterns. More on reflexes and central patterns will follow. See Chapter 3 for a discussion of the properties of neurons and the organization of the nervous system and Chapter 6 for information on sensory mechanisms.

Behavioral studies provide strong evidence for the filtering of sensory input *(stimulus filtering);* that is, from the many, diverse stimuli impinging on an insect, only certain ones or only certain aspects of a given stimulus are involved in eliciting a particular behavior.

Reflexes depend on sensory input for their occurrence. At one time, insect behavior was thought to be explainable solely on the basis of reflexes where one reflexive response provided the stimulus for the next reflex in line. For example, the repetitive motor patterns associated with flight and sound production were viewed as being controlled by cyclic reflexes, which would continue until inhibited by some other reflex. Although reflexes are undoubtedly important in behavior, more recent research has revealed evidence for the generation of patterns within the central nervous system independent of sensory input.

It is known from behavioral and neurophysiological studies that patterns of behavior may occur spontaneously without any external sensory triggering. Thus the nervous system must be capable of spontaneously generating organized patterns of behavior from within (*endogenously*). A decapitated mantid will continuously turn and perform copulatory movements. The central nervous system in a locust with all sensory input from the head and wings removed still produces motor output nearly identical to the motor output recorded during normal flight (Wilson, 1972). These activities are thought to arise from the spontaneous, preprogrammed activity of *neural pattern generators* in the central nervous system.

If copulatory behavior arises endogenously in the mantis, why doesn't the mantis behave this way all the time? How does the locust control flight if environmental conditions (wind, obstacles, etc.) vary? Why don't insects display all of their behavior patterns simultaneously? Obviously endogenous activity must be regulated. Endogenous activity is thought to be regulated through excitatory and inhibitory stimuli from "higher centers" and by modulating input from peripheral sensilla, and probably directly or indirectly by hormones. In particular, the brain and subesophageal ganglion have both been found to exert excitatory and inhibitory influences on endogenously controlled motor patterns. These patterns seem to reside largely in the particular segments involved (e.g., the thoracic ganglia in flight). Thus the mantid brain (which was removed with the head in the decapitated individual mentioned above) sends out impulses that inhibit the endogenous activity of the ventral chain ganglia relative to copulation. In the locust, there are wind-sensitive sensilla on the head, which help sustain flight, and also tarsal receptors, which stimulate flight when contact with the substrate is lost. In addition, there are stretch receptors, which aid in the regulation of flight, on the wing hinges and other parts of the wings and pterothorax. Thus, although locust flight pattern is centrally generated, the basic pattern is modifiable by peripheral, sensory input.

Something like neural pattern generators is probably involved in sequences of stereotyped behaviors more complex than simple motor patterns. For example, courtship in the grasshopper *Gomphocerus rufus* occurs in a very definite sequence of simple behavior patterns and this sequence is not interrupted by surgical action designed to block potentially controlling sensory feedback. Similarly, in *Drosophila,* an excited male will display all the intricacies of courtship in the presence of an anesthetized, unresponsive female. Thus complex behaviors appear to be programmed into the central nervous system as "tapes" of behavior sequences and a given insect must have a complete bank of "tapes" to be played as appropriate. According to Hoyle (1970):

> For purposes of functional description the nervous system can be considered to have three major functional divisions. In one of these the enormous mass of incoming sensory input is reduced to a small series of sets of information which can serve to drive motor output. The second 'selects' from the sets the one which shall at a given time be operative, and inhibits others. The third division contains neural generators whose purpose is to produce sequences of motor nerve impulses.

If the brain is involved in excitation and inhibition of the rest of the central nervous system, how is the brain, in turn, excited? Although little is known, there is evidence for excitatory input to the brain from the ocelli (mentioned earlier in this chapter and in Chapter 6), compound eyes, and antennae. Spontaneous, endogenous activity and endocrine factors probably also play a role in the levels of excitation of the brain.

Little is known about the neurophysiological basis of learning, but destruction of parts of the brain has shown that some long-term memory must reside there. Also, recall that isolated ganglia displayed shock-avoidance learning. Even habituation, probably the simplest form of learning, is not understood at the neural level. In any case, learning provides a way to modify or "rewrite" behavior "tapes" through experience.

Endocrine Control

Hormones play major roles in insect behavior. Through acting directly on the nervous system they can cause major overall changes in patterns of behavior. Whereas strictly neural mechanisms usually mediate the short term, sometimes split-second, behavioral responses to rapidly changing external and internal environmental demands (potential attack by a predator, an obstacle encountered during flight, etc.), endocrine mechanisms typically exert their influence over the longer term, coordinating behavior with less immediate environmental demands (changing seasons, changes in the availability of food, etc.). Thus strictly neural mechanisms are involved with escape responses, maneuvering during flight, and the like; endocrine mechanisms with development of migratory behavior or dormancy, changes in feeding behavior, development of adult behavior patterns, development of sexual receptivity, and so on. In the terminology of the section on the nervous control of behavior, hormones are commonly involved in determining which reflexes can be elicited and which behavioral "tapes" are played in a given circumstance.

Interfacing of the nervous and endocrine systems where the latter acts in a "switching" fashion on the former may explain, at least in part, how the very large array of behavior patterns displayed by insects can be packed into their comparatively small, simple nervous systems.

The determination of hormonal influences on behavior have usually been based on one or more of the following criteria.

1. Correlation of hormone secretion and a particular behavior pattern.
2. Application of hormone or implantation of active glands inducing a behavior pattern sooner than it would ordinarily occur.
3. Correlation between removal of an active gland and disappearance of a specific behavior.
4. A combination of "3" and subsequent restoration of the specific behavior by hormone application or implantation of active glands.

Obviously, the fourth criterion provides the strongest evidence for hormonal control of a given behavior.

Riddiford and Truman (1974) recognize two categories of hormonal effects on behavior: hormones can act as *modifiers* or as *releasers* of behavior.

When a hormone acts as a modifier, it alters the responsiveness of an organism. Where a given stimulus previously elicited one kind of behavior, a different behavior occurs. The hormonal effects on behavior as described earlier are as modifiers. A recent, excellent example of a hormone acting as a modifier is the influence of ecdysone on the biting behavior of female *Anopheles freeborni* (Culicidae, mosquitoes, Diptera) (Beach, 1979). This mosquito does not take blood meals once ovarian development has been initiated. Ecdysone, which is secreted by the ovaries, inhibits biting. This was concluded on the basis that biting inhibition does not occur after removal of the ovaries and is restored by either replacing ovaries or injecting ecdysone.

When a hormone acts as a releaser of behavior, a specific behavior pattern occurs within a few minutes of secretion. This situation is analogous to the action of nervous input eliciting a behavioral response, but is much slower in occurring (minutes vs. milliseconds). An example is the *phallic nerve stimulating hormone* in the cockroach *Periplaneta americana* (Milburn, Weiant, and Roeder, 1960; Milburn and Roeder, 1962). This hormone can be extracted from the corpora cardiaca and when injected into male cockroaches causes the abdominal movements characteristic of copulation.

The "eclosion hormone" (eclosion meaning pupal–adult ecdysis in this context) secreted by the brain (neurosecretion) and corpora cardiaca in the moth *Antheraea pernyii* acts both as a releaser in triggering pre-eclosion behavior and as a modifier by "turning on" adult behavior (Truman, 1973).

Many examples and discussions of endocrine mechanisms and behavior can be found in Barth and Lester (1973), Truman and Riddiford (1974), and Riddiford and Truman (1974). Additional examples may be found in other sections of this chapter.

Genetic Control of Behavior

There are many examples of the demonstration of genetic control of insect behavior. Not surprisingly, a great deal of work has been done with *Drosophila melanogaster* and other *Drosophila* species as well as a number of other kinds of insects. An extensive treatment of the genetics of behavior is far beyond the scope of this text, and the reader is referred to Ehrman and Parsons (1976), Ewing and Manning (1967), Fuller and Thompson (1960), Hirsch (1967), McClearn and DeFries (1973), and Rothenbuhler (1967) for more information on this subject.

It is hoped that one day we will understand the chain of events occurring between the action of genes in protein (enzyme) synthesis and the expression of behavior. Although there has been some progress, really very little is known. Investigators are approaching genes and behavior in a number of different ways. A few examples follow.

There are several examples of gene mutations in *Drosophila* that change a behavior pattern (Benzer, 1973). An early example was a sex-linked mutant with yellow body color instead of the normal gray (Bastock, 1956). Males of these yellow mutants were less successful in mating than the gray wild type.

In rare instances, it has been possible through breeding experiments to account for particular behavior patterns on the basis of the action of single genes. An outstanding example of this is the work of Rothenbuhler (1964). Rothenbuhler studied two inbred lines of honey bees, one that had been selected for resistance to American foulbrood (a disease of honey bee larvae) and another that was susceptible. He noted a distinct behavioral difference between the two lines; the resistant line removed foulbrood-killed larvae relatively soon after death (hygienic behavior), and the susceptible line did not (nonhygienic behavior). Genetic analysis revealed that the action of two sets of dominant/recessive genes on different chromosomes would account for the results of breeding experiments between the two lines. One set of alleles involved uncapping of the cells in which diseased larvae were developing: recessive allele, uncapping and dominant allele, no uncapping. Thus homozygous recessives displayed uncapping behavior. Homozygous dominants and heterozygotes did not. The other set of alleles involved removal of diseased larvae: recessive allele, larvae removed and dominant allele, diseased larvae not removed. Homozygous dominants and heterozygotes did not remove diseased larvae. Figure 8-3 shows the two genes assorting independently and the resulting behavioral phenotypes. The "remove only" group was identified by artificially removing the caps from cells containing diseased larvae and subsequently observing the removal of diseased larvae.

Fig. 8-3
Genetic hypothesis to explain hygienic and nonhygienic behavior in the honey bee (see text for explanation). u = gene for uncapping (recessive), and r = gene for removing diseased larvae (recessive). + = dominant alleles; that is, no uncapping or no removal of diseased larvae. [From Rothenbuhler, 1964.]

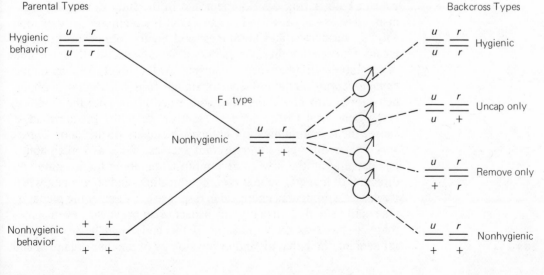

Parental Types   Backcross Types

Most behavior patterns are probably under polygenic control, and *directional selection experiments* have been useful in establishing whether a given behavior has a genetic basis (Ehrman and Parsons, 1976). In selection experiments, "individuals are selected at high or low extremes of a distribution in the hope of forming separate high or low lines in subsequent generations" (Ehrman and Parsons, 1976). If a trait does not have a genetic basis, then attempts to select for extremes would fail. A good example of a selection experiment is the work of Hirsch and Erlenmeyer-Kimling (1961). They showed that with selection, *Drosophila melanogaster* that had previously been assumed to be negatively geotactic could be separated into both positively and negatively geotactic lines. Following establishment of "extreme" strains, it has been possible in some cases to evaluate the degree of involvement of different chromosomes. For example, in the case of geotaxis in *Drosophila*, genes on each of the three major chromosomes were found to exert an influence (Hirsch and Erlenmeyer-Kimling, 1962). In *Drosophila,* several other behavioral traits have been shown to be under polygenic control, including locomotor activity, chemotaxis, duration of copulation, optomotor response, phototaxis, and preening.

Some researchers are approaching behavioral genetics from a neurophysiological point of view. For example, Bentley and Hoy (1972, 1974, 1975), Hoy and Paul (1972), and Hoy (1974) have found that the sounds produced by adult male crickets are not learned, but genetically controlled. Further, hybrids produced by crossing *Teleogryllus* and *Gryllus* species exhibit song patterns that are intermediate between those of the parents. *Teleogryllus* songs are controlled by several different genes on different chromosomes.

Communication

Communication in one form or another is part of every organism's life at all levels of complexity. Thus, for example, the nucleus communicates with the cytoplasm in directing cellular activities; cells communicate with one another as cooperators in a multicellular organism both during development and in the fully developed organism; tissues, organs, and organ systems communicate with one another via nerves and hormones; and finally, organisms communicate with one another. At this point we are specifically interested in communication between organisms. At this level, we may define *communication* as the influence of signals from one organism on the behavior and/or physiology of another organism, with the outcome being beneficial to the *sender, receiver,* or both. In its broadest sense, then, communication involves members of the same (*intraspecific*) or different (*interspecific*) species. Thus a worker honey bee dancing on the hive to give information about the location and direction of a nectar source is communicating, and so is a bug when it releases a repugnant chemical in response to a threatening predator. One could say that virtually any aspect of an organism, even something as basic as color or form of the body, can be involved in influencing the behavior and/or physiology of another organism and is hence a means of communication. Communication often involves the use of specialized structures or methods by the sender and/or the receiver. Communication systems have evolved independently many times in insects.

Although systems of communication can be grouped in various ways, such as the type of information conveyed (Otte, 1974), we will discuss this topic from the standpoint of modes of communication utilized by insects, specifically *tactile, acoustical, visual,* and *chemical*. Sensory mechanisms (discussed in Chapter 6) are channels of reception. Mechanisms of sending (signal output) include sound and light production (see Chapter 6) and chemicals produced by exocrine glands (see Chapter 3). Discussions directly pertinent to insect movement, also a form of signal output, may be found in Chapters 2, 3, and 7. Books and reviews that consider communication in general include Frings and Frings (1977), Montagner (1977), Otte (1974), Sebeok (1968, 1977), Smith (1977), and Wilson (1975). Reviews that deal with communication in particular groups include Hölldobler (1977), Ishay (1977), Lindauer (1974), Stuart (1977), and Wilson (1965) on social insects; and Ewing (1977), Otte (1977), Schneider (1975), and Silberglied (1977) on various other insect groups.

As mentioned above, communication may be between members of the same or different species. Intraspecific communication is involved in species recognition, sexual recognition, alarm, and social coordination. Interspecific communication is also involved in species recognition (e.g., signals from a plant attracting a pollinating insect) as well as aggression, defense, etc. Many examples are included in the following section on the biological functions of behavior and also in Chapter 9.

Tactile communication has been the least studied. This form of communication is limited to situations where direct contact is possible, as during courtship and copulation. Wilson (1975) describes a form of communication to explain nest building in termites. Instead of direct touch, information is passed from one individual to another by a product of work. Thus a termite responds to a portion of already constructed nest by adding more material. This response is released by the presence of a physical structure. Wilson includes this form of communication as part of a broader category, *sematectonic communication,* which he defines as "the evocation of any form of behavior of physiological change by the evidence of work performed by other animals, including the special case of the guidance of additional work."

The ability to perceive sounds is found in several groups of insects (see Chapter 6). However, members of at least five orders are known both to "hear" and to produce distinctive sounds: Orthoptera, Hemiptera (both Heteroptera and Homoptera), Lepidoptera, Coleoptera, and Diptera. Interspecific acoustical communication tends to be mediated by unpatterned sounds, such as "warning" sounds, while intraspecific communication typically involves highly patterned sounds (Matthews and Matthews, 1978). As intraspecific signals, sounds are involved in aggregation, sexual behavior, aggression, alarm, and social interactions. Acoustical signals are very effective in that they can operate over relatively long distances, can be modulated in many different ways, can operate in the dark, can provide directional information, and so on. Acoustical signals in insects are monotonal, but vary in intensity and pattern of pulses. Useful references on acoustical communication include Alexander (1967, 1975), Haskel (1974), Ishay (1977), Otte (1977), and Swartzkopff (1974).

Visual communication is common in insects. Visual signals have a great advantage in that they can provide a large amount of information. Recall the various aspects considered under photoreception in Chapter 6: perception of form and pattern, intensity and contrast, movement, distance, color, and polarized light. Consider the possibilities for variation in all of these parameters. Some insects are bioluminescent and use light production and reception as a mode of communication. However, most insects use light indirectly, that is, reflected or transmitted by something else. Visual signals are involved in reproductive behavior, aggressive displays (e.g., in territorial behavior of dragonflies), in defensive startle displays, etc. For further information on visual communication, see Hailman (1977) and Lloyd (1971, 1977).

Chemical communication is by far the most common form of communication among insects (in fact among animals in general). Chemical signals act over large distances, but they are not easily modulated and thus are used mainly as on-off signals, for example, to indicate sexual readiness in a female. They may also give directional information as in the case of a chemical trail. A given chemical signal may be a single compound or a mixture of compounds and may have been synthesized entirely by the insect or may have been obtained from the environment, especially from food.

Chemical signals can be classified on the basis of whether they mediate intra- or interspecific communication. Chemicals that function intraspecifically are called *pheromones*. For example, when ants are threatened or injured, they may release chemicals that induce dispersal from an area among foragers or aggressive behavior in the vicinity of the nest. These chemicals are acting as *alarm pheromones*. Chemicals involved in interspecific communication are commonly grouped as *allomones* or *kairomones*. An allomone is emitted by a member of one species and induces a response that is adaptively favorable to the emitter. For example, the repugnatorial substance released by a bug in response to a predator would be acting as an allomone. A kairomone is adaptively favorable to the receiving organism. For example, the mosquito *Aedes aegypti* is attracted by lactic acid (as well as other compounds) in human perspiration. In this situation, the mosquito (receiver) benefits, and hence the lactic acid is acting as a kairomone (Galun, 1977b). In many ants, bees, wasps, and termites the same chemical can function both as an allomone and a pheromone. For example, in ants formic acid can act as a defensive allomone and alarm pheromone that alerts nestmates to an intruder (Blum, 1978). Likewise a volatile or mixture of volatiles may act as both a pheromone or a kairomone. In Scolytidae (Coleoptera), for example, the aggregating pheromone may also act to attract predators and parasites of the beetles.

The following references deal with chemical communication: *General*—Beroza (1970), Ebling and Highnam (1970), Pasteels (1977), Roelofs (1975a), Shorey and McKelvey (1977), and Wilson (1965, 1970, 1975); *Pheromones*—Birch (1974), Blum and Brand (1972), Jacobson (1965, 1972, 1974), Karlson and Butenandt (1959), Law and Regnier (1971), Roelofs and Cardé (1977), Seabrook (1978), Shorey (1973, 1976, 1977a, b), Weaver (1978a), and Wilson (1963); *Allomones and Kairomones*—Blum (1978), Brown, Eisner, and Whittaker (1970), Duffey (1977), Rosenthal and Janzen (1979), Weaver (1978b), and Whittaker and Feeney (1971). See Chapter 9 for additional references.

The Biological Functions of Behavior

This section considers how insects are behaviorally adapted for the maintenance of the individual through sexual maturity and for those activities that culminate in the production of offspring. Gregarious, subsocial, and social behavior is discussed at the end of this section.

Feeding Behavior

Here we consider the sources of food, how insects are able to locate these sources, and the control of feeding.

Insects may be broadly grouped on the basis of the biological source of nutriment as follows: *detritivores* (*saprophagous*); *herbivores* (*phytophagous*, "plant-eating"); *mycetophagous*, "fungus-eating"); and *carnivores* (*zoophagous*, "animal-eating"). Insects that fall into more than one of these categories are called *omnivores*. A qualification from Dethier (1976) before proceeding:

> No two animals have precisely the same diet. Each species posseses some of the qualities of a gourmet. The common categorization as herbivore, carnivore, and omnivore depicts only the outline of the story. The details of diet reveal the enormous complexity of the feeding processes and diverse and precise specifications to which the physiological and biochemical machinery must be designed. (Dethier, 1976)

Detritivores. Detritivorous insects that feed on decaying organic materials: leaf litter, carrion, dung, and so on. Such materials were probably the basic food source for the most primitive insects, which lived on the forest floor. Many contemporary insects utilize these materials throughout all or a part of their lives. These insects are important in the progressive degradation of decaying organic material. Cockroaches, for example, subsist on a wide variety of dead plant and animal matter. They are commonly described as scavengers. Other examples of detritivores are dung beetles, carrion beetles, and many of the soil-inhabiting apterygotes (e.g., Collembola). Actually many insects that upon cursory examination appear to be detritivores are found upon closer scrutiny to be dependent upon the microorganisms associated with decaying organic material.

The fruit fly, *Drosophila melanogaster,* is an example of such an insect. The larvae cannot be maintained on a sterile culture medium. Thus, although one associates them with decaying organic matter, they are actually "microphagous" (Brues, 1946).

Herbivores. Living plants are the source of nutriment for the greatest number of insects, and the habits of many insects have evolved closely with plants (coevolution is discussed in Chapter 9). Members of the orders Orthoptera, Hemiptera-Homoptera, Lepidoptera, Thysanoptera, Isoptera, and some other smaller orders are predominately or partly herbivorous. In addition, large numbers of insects in the orders Coleoptera, Hymenoptera, and Diptera are also herbivores. Although flowering plants are by far the most common plant hosts, there are many insects that feed on fungi, algae, ferns, and so on.

Some herbivorous insects may show very little feeding preference, while others show some degree of preference, and still others are very specific as to food choice. Insects such as the migratory locust (*Schistocerca gregaria*) are *polyphagous* and will eat almost any green plant in their path. On the other hand, the milkweed bug and many others are *monophagous,* feeding only on plants of a single species or genus. Most phytophagous insects lie somewhere between these two extremes (*oligophagous*). Variation also exists with regard to what part of a plant is utilized as food. No major part of plant anatomy has been ignored by insects. There are those that feed in or on roots, stems, wood parts, buds, flowers, and fruits. However, a given insect is usually rather specific as to which part of a plant it eats. Thus there are leaf rollers, leaf miners, rootworms, root borers, stem borers, and so on. In addition to providing nutriment for the phytophagous insects, plants usually serve as a place to live, particularly in the case of larval and nymphal forms.

Many fungus-eating insects maintain fungus gardens within their nests. These are mutualistic relationships in that both the fungi and the insects derive benefit (see Chapter 9). Outstanding among these are the fungus-growing ants in the tribe Attini (Formicidae; Weber, 1972). Weber (1966) provides a brief description of their activities in the summary of a paper dealing with these insects.

Fungus-growing ants (Attini) are in reality unique fungus-culturing insects. There are several hundred species in some dozen genera, of which *Acromyrmex* and *Atta* are the conspicuous leaf-cutters. The center of their activities is the fungus garden, which is also the site of the queen and brood. The garden, in most species, is made from fresh green leaves or other vegetable material. The ants forage for this, forming distinct trails to the vegetation that is being harvested. The cut leaves or other substrate are brought into the nest and prepared for the fungus. Fresh leaves and flowers are cut into pieces a millimeter or two in diameter; the ants form them into a pulpy mass by pinching them with the mandibles and adding saliva. Anal droplets are deposited on the pieces, which are then formed into place in the garden. Planting of the fungus is accomplished by an ant's picking up tufts of the adjacent mycelium and dotting the surface of the new substrate with it. The combination of salivary and anal secretions, together with the constant care given by the ants, facilitates the growth of the ant fungus only, despite constant possibility for contamination. When the ants are removed, alien fungi and other organisms flourish.

Certain species of scale insects (Hemiptera-Homoptera, Coccidae), a wood wasp (*Sirex*), ambrosia beetles (families Scolytidae, Paltypodidae, and Lymexylomidae), and termites (*Macrotermes* and *Odontotermes*) also carry on mutualistic relationships with fungi (Batra and Batra, 1967).

Carnivores.   Carnivorous insects are those that use living animals as a source of food. They may be divided somewhat arbitrarily into *predators* and *parasites,* although there is no sharp line of demarcation between the two. Predatory insects capture and eat on the spot living animals, usually other insects. Predators are usually larger than the captured individual, the *prey,* and many prey individuals are eaten during the life of the predator. Most parasites differ from predators in several respects. They live on or within the bodies of their food sources, or *hosts*. Parasites often do not immediately kill the hosts, and if the host is killed, only one host is usually required during the life of the parasite. A parasite is smaller than its host, but an immature insect parasitizing another insect may grow and attain the same size as the host.

Predatory behavior is widespread among the insects, having arisen independently on many different occasions in widely divergent groups. The orders Odonata, Hemiptera-Heteroptera, Neuroptera, Coleoptera, Diptera, and Hymenoptera include large numbers of predatory forms, and many are found scattered among other orders, for example, praying mantids (Orthoptera).

Predatory insects usually possess speed and agility and/or great stealth. For example, adult robber flies and dragonflies are noted for their impressive demonstrations of agility, capturing prey in midair. The praying mantid and many dragonfly nymphs provide examples of stealth in prey capture. These insects remain motionless and strike at passing prey with incredible speed and accuracy. A dragonfly nymph can only extend its prehensile labium directly in front of itself and must do so completely or not at all since the force for extension is hydrostatic. Thus potential prey must be a certain distance directly in front of the nymph in order to be caught. The compound eyes serve as the ranging sensors in these insects (see Fig. 6-14). The praying mantid has a more flexible pre-capture mechanism, being able to judge the distance and position of prey relative to the longitudinal axis of its body. Thus prey can be caught in many different positions relative to the mantid. This mechanism is based on an interaction between the compound eyes, sensory spines in the neck region that monitor head position, and the muscles of the raptorial forelegs.

Certain insects, such as antlion larvae (Neuroptera, Myrmeleontidae) and the larvae of some caddisflies (Trichoptera), literally trap their prey. The antlion forms a shallow cone-shaped pit in a sandy area and buries itself just beneath the surface at the bottom of the pit. When an ant or other small insect passes near enough to the pit to cause grains of sand to roll down the sides, the antlion responds by creating a miniature landslide. This brings the prey within reach of the antlion's prehensile jaws, through which it injects enzymes and then sucks out the digested insides of the prey. The aquatic larvae of certain species of caddisflies construct silken nets in which they are able to catch small organisms.

The eggs of some carnivorous insects are placed by the parent insect in proximity to prey species or the offspring may depend upon sluggish, easily captured prey. Many examples of parent insects locating prey for their offspring are found among the Hymenoptera (e.g., the case of the female *Bembix* wasp described in the section on reproductive behavior. In the higher Diptera, the larvae have poorly developed mouthparts and move about very slowly. Thus predaceous species like many syrphid flies (Syrphidae) depend on weaker, more sluggish prey, such as aphids, scale insects, mites, other dipteran larvae, and spider and insect eggs.

Most predaceous insects eat other insects, particularly herbivorous species. This is not surprising since herbivorous insects (primary consumers) are the most abundant. Small crustaceans (especially aquatic), mites, and occasionally spiders may also fall prey to insects. Certain members of several families of beetles (Carabidae, Silphidae, Drilidae, and others) and some species of giant water bugs (Belostomatidae) feed upon snails. Protozoans compose a significant proportion of the diets of many aquatic insects. Members of the more specialized groups (e.g., solitary wasps) tend to be restrictive in prey choice, while members of the more generalized, primitive orders exhibit a wider range of prey choice.

Parasitic behavior, like predatory behavior, has arisen independently on several occasions among insects. Parasitic insects are probably best viewed on the basis of hosts, in particular as parasites of vertebrates and as parasites of other insects and related arthropods. Askew (1971) provides a comprehensive treatment of parasitic insects.

Most insects that are parasitic on vertebrates are found in the orders Anoplura (sucking lice), Mallophaga (chewing lice), Hemiptera (Heteroptera, true bugs), Diptera (true flies; Fig. 8-4), and Siphonaptera (fleas). *Ectoparasitic* insects derive all or part of their sustenance from the host and live entirely on its external surfaces. *Endoparasitic* insects invade the tissues of the host. The majority of insect ectoparasites of vertebrates, save most Mallophaga, are blood feeders (*hematophagous*). Among the members of this group, some remain on the host throughout their life cycle, some live on the host only during a particular stage of the life cycle, and others visit the host intermittently during a particular stage of the life cycle and are otherwise free-living. Parasitic insects that remain on the host (e.g., the lice) tend to be host specific. They have close evolutionary ties with the host species and are most likely to be entirely dependent on it. Fleas spend only one stage of their life cycle on the host. Flea larvae are free-living and feed on organic debris that accumulates in the host nest or burrow. Adults tend to remain on the host, taking intermittent blood meals. However, adult fleas are active and commonly move from host to host. They are less host specific than the lice.

Fig. 8-4
Blood feeding by *Phlebotomus longipes* (Diptera; Psychodidae) on a
human host. This species is a vector of *Leishmania tropica,* a protozoan
parasite of man in Ethiopia. (Length of fly approximately 3.0 mm)
[Courtesy of W. A. Foster.]

Hematophagous insects like adult female mosquitoes, "no-see-ums" (Diptera, Ceratopogonidae), and some other adult Diptera, a
few dipteran larvae, and nymphal and adult bed bugs and conenose
bugs (Hemiptera-Heteroptera; Cimicidae and Reduviidae, respectively) visit host vertebrates only to take a blood meal and exhibit
varying degrees of host specificity. Most feed on a single or very
few vertebrate species and usually cause little or no harm. This is
not meant to imply that they never harm the host, as they obviously
can when they occur in large numbers or introduce disease-causing
organisms. Insects that parasitize vertebrates are discussed by specialists in Smith (1973). The feeding behavior of mosquitoes is described by Jones (1978). Nelson et al. (1975) discuss several aspects
of vertebrate-host–ectoparasite relationships. Almost all insects that
are endoparasites of vertebrates are dipteran larvae (maggots), in
several families (Zumpt, 1965), which may invade, for example,
the alimentary canal, nasal cavities, and open sores. These larvae
attack a wide variety of vertebrates, including man. An animal infested with fly larvae is said to be suffering from *myiasis.*

Some insects act as ectoparasites of other insects, taking intermittent blood meals from host insects in the same way those mentioned above do from vertebrate hosts. In fact, many members of
the same orders and families are involved. For example, "no-see-
ums" have been observed feeding on caterpillars, along the wing
veins of lacewings, dragonflies, and so on. Insects that feed in this
fashion probably individually do little damage to their host.

Most insects that parasitize other insects do so as larvae and eventually destroy their host. These insects begin life as "typical" para-

sites in that they are much smaller than the host and live in intimate association with it. However, they eventually grow, sometimes nearly as large as the host itself, and the host is eventually killed. Thus they lie somewhere between parasites and predators. For this reason, the term *parasitoid* has been suggested to distinguish these insects from the more typical parasites and predators.

Most parasitoids live and feed as larvae within or on the host and become free-living adults. Exceptions include most female Strepsiptera (twisted-wing parasites), which spend their entire life within the host insect. The majority of parasitoids are in the order Hymenoptera in the superfamilies Ichneumonoidea, Chalcidoidea, and Proctotrupoidea, although a few occur scattered among other hymenopteran groups. A number of Dipteran families also contain parasitoid forms (e.g., Tachinidae and Sarcophagidae). These flies attack a wide variety of insects, including grasshoppers, caterpillars, true bugs, beetle grubs and adults, and hymenopterous larvae. Rhipiphorid beetles attack such insects as hymenopteran larvae and cockroaches.

Some parasitoids exhibit a high degree of host specificity, attacking a single or very few species of insects. In most instances only a single parasitoid can exist in a host. However, there are instances where several and rarely several hundred or a thousand can develop in a single host. It is of interest that a host survives for any length of time after invasion by one of these parasitoids. The parasitoids presumably feed on nonvital tissues until they are ready to pupate, and then they devour vital tissues. Some parasitoids attack other parasitoids. Thus one may find a caterpillar harboring a larva of a tachinid fly, which in turn harbors a chalcid wasp larva. This phenomenon is referred to as *hyperparasitoidism (hyperparasitism)*. Many *entomophagous* ("insect-eating") insects, especially the hymenopteran parasites, have been utilized in attempts to control various pest-insect species (see Chapter 12).

The feeding behavior of certain insects falls under the heading *social parasitism*. Brues (1946) describes the behavior of several species of wasps in the genus *Vespula*, which parasitize other members of the same genus, as follows.

> All the parasitic species have lost the worker, or infertile female, caste and consist only of males and fertile females, the latter corresponding to the "queens" in the host species. The parasitic females enter the nests built and maintained by their hosts where they lay their own eggs in the paper cells already provided. Their young are then fed and reared in cells, intermingled with those containing the brood of the host, all of the feeding being done by workers of the host species.

Several other social Hymenoptera display this kind of parasitism.

Certain beetles in the family Meloidae invade the nests of solitary bees (Askew, 1971). The beetle larvae destroy the bee's eggs and consume the food stored in the cells of the nest. Such nest invasions are also found among the Hymenoptera, for example, the cuckoo wasps (Chrysididae). This phenomenon has been called *cleptoparasitism* (Matthews and Matthews, 1978).

Many insects may be found living and feeding as more or less permanent residents in the shelters of other insects (both social and nonsocial). These forms are sometimes called *inquilines*. For example, certain galls "constructed" by one species are shared by larvae of another species without any apparent damage to the original owner.

Location of Food Sources. Dethier (1966) describes feeding behavior as "a complex and interacting sequence of responses to a variety of stimuli culminating in ingestion to repletion." He outlines the basic sequence of events involved in feeding: " locomotion bringing the insect to its food, cessation of locomotion on arrival, biting or its equivalent (probing, sucking, etc.), continued feeding, termination of feeding." In this and the following section we shall consider some of the factors involved in food location and initiation and termination of feeding.

In many herbivorous insects, the female parent oviposits on or in the vicinity of a food source for her offspring. The same is true for most parasitic insects. In these cases location of oviposition site, host location, and food location are not distinctly separate activities.

Much research has been carried out to determine the stimuli involved in food location by different insects. Visual and olfactory stimuli are probably the main ones utilized in food location by insects in general.

As an example of the use of visual cues, recall the use of color, form, and movement in the location of flowers by honey bees (see Chapter 6). Many butterflies also respond to these stimuli. As mentioned in the preceding section, dragonfly larvae and adults and praying mantids rely heavily on visual stimuli in prey capture. Visual stimuli are important in host location in mosquitoes, tsetse flies, and other hematophagous flies and likewise in insects that are parasites of other insects.

Olfactory stimuli (volatile chemicals) are involved in food location in dung-feeding insects such as dung beetles (*Geotrupes* and others) and certain flies (e.g., many in the family Calliphoridae). These insects are attracted by the volatile substances skatol and ammonia, which are found in their food. These substances are referred to as *token* or *sign stimuli* since they serve as indicators of substances with nutritive value but have none themselves. Token stimuli are very important in the attraction of herbivorous insects to their host plants. For example, the butterfly *Papilio polyxenes* is attracted to several of volatile oils that impart an odor to their characteristic food plants, members of the Umbelliferae. However, these oils have no food value for the butterflies. Also, they have no known role in plant metabolism. These oils and many other are called *secondary plant substances* (more on these in Chapter 9). Secondary plant substances may also act as repellents to some insects. Thus some act as kairomones and some as allomones. Olfactory stimuli play a role in host location by mosquitoes. Carbon dioxide, steroids, amino acids, and other volatile substances characteristic of vertebrates are involved.

In addition to visual and olfactory cues, tactile, thermal, and hygrostimuli are involved in host location in the ectoparasites of warm-blooded vertebrates. For example, human body lice, which live between the skin and clothing of their host, show a marked preference for rough-textured materials. They show an increased frequency of turning on smooth surfaces, and this frequency is inversely proportional to the stimulus strength (*orthokinesis*). Lice also demonstrate a preference for the 26.4–29.9°C range in a temperature gradient and 76% relative humidity in a humidity gradient.

Backswimmers and whirligig beetles respond to vibrations produced by the swimming or struggling of potential prey. The backswimmer locates prey by orienting itself such that it receives equal stimulation of the sensory hairs on its oarlike legs. The larvae of ladybird beetles catch their prey, aphids, by crawling along the veins of a leaf and periodically stopping and "scanning" the vicinity by turning their body to and fro about the temporarily secured abdomen. When they are successful in capturing an aphid, they show an increase in the frequency of turns, a response that keeps them in the immediate locality. This behavior increases the probability of encountering additional prey and provides a good example of *klinokinesis*.

Once potential food is located, specific stimuli are involved in the induction of feeding. These stimuli are often referred to as *phagostimulants* (a category of kairomones). They may be purely token stimuli, having no nutritional value, or they may be substances that have definite nutritive value. Many of the volatile oils (secondary plant substances) characteristic of certain species of plants are phagostimulants. For example, the cabbage aphid, *Brevicoryne brassicae*, is induced to feed even on an abnormal host when the substance sinigrin, a material extracted from Cruciferae (mustard family) and other plants, is present. Sucrose, a common sugar in the food of many insects, is an effective phagostimulant in the majority of insects. Amino acids, glucose, certain proteins, ascorbic acid, and several other chemicals have also been shown to be feeding stimulants.

The larval and/or adult stages of some parasitic, parasitoid, and predatory insect species are transported to the host or prey species' nest by adults of the host or prey species (Clausen, 1976). The insect doing the transporting is unharmed. Such biological "hitchhiking" is called *phoresy*. For example, a female of the Australian wasp, *Synoditella fulgidus* (Hymenoptera, Scelionidae), clings to the abdomen of an ovipositing host grasshopper. When host oviposition is complete, the wasp descends into the hole made by the host and deposits her eggs on the grasshopper's egg mass. These wasps may be carried by the grasshoppers when they swarm.

An example involving phoresy by larvae is found among the meloid beetles mentioned above. Depending on the species of beetle, a female may oviposit in the soil near a host nest, in flowers, or at the entrance to the host nest. The first instar, *triungulin*, beetle larva is very active. Those species hatching in the soil climb vegetation and tend to aggregate in flowers. These, along with species that hatch on flowers, respond to the vibration produced by a visiting bee and some crawl onto the bee's body. If the bee is of the "correct" species, the beetle larva, which is transported by the bee to the bee's nest, will enter a larval cell. Triungulins that hatch near the entrance to a host nest climb onto passing bees and gain entry to the nest. The larvae do not harm the adult bees, but simply gain transportation.

For more information on food location in herbivorous insects, see Dethier (1970), Kennedy (1977), Kogan (1977), Schoonhoven (1972a), and Staedler (1977). Galun (1977a, and b), Friend and Smith (1977), and Hocking (1971) deal specifically with host location in hematophagous insects. Askew (1971) contains much information on host location in parasitic insects in general. Vinson (1976) deals specifically with host selection in parasitoids.

Control of Feeding. The control of feeding has been studied extensively, but there is still a great deal to be learned (Barton Browne, 1975). Insects deprived of food typically behave differently from those that have recently obtained food; that is, they behave in a fashion that increases the probability of locating food. Such behavior usually includes increased locomotor activity and increased responsiveness (i.e., decreases in thresholds of response) to various stimuli. For example, the tendency for the desert locust, *Schistocerca gregaria*, to move upwind in the direction of the odor of grass increases with the duration of food deprivation. Similarly, the response of tsetse flies, which feed on vertebrate blood, to a slow-moving visual stimulus increases with duration of food deprivation. Responsiveness in food-deprived insects also increases to stimuli that do not emanate from food. For example, the photonegative (tending to move away from light), nocturnally feeding larvae of the cutworm, *Tryphaena pronuba,* tend to become less photonegative with starvation. Such modification of behavior may result in these larvae feeding above ground in the daylight.

Contact chemoreceptive sensilla are important once contact is made with food. These sensilla are responsible for the perception of stimuli that induce subsequent contact between the mouthparts and food. For example, stimulation of the chemoreceptors on the tarsi of the blow fly, *Phormia regina,* and many other flies and also on the tarsi of butterflies results in extension of the proboscis and contact with food. Sensilla on the mouthparts are then involved with the initiation and maintenance of feeding. The thresholds of responsiveness of sensilla on the tarsi and mouthparts decrease with food deprivation.

In many insects, the size of a meal is a function of the state of food deprivation. As food is ingested and assimilated, excitability (evidenced by increased locomotion and responsiveness) decreases, and inhibitory influences increase. These inhibitory influences may include adaptation of external sensilla (i.e., receptors become progressively less responsive to continuous stimulation), internal monitoring of nutritional state, and the action of stretch receptors associated with the alimentary canal or body wall of the abdomen. External receptors responding to feeding-deterrent molecules may also exert an inhibitory influence from the very first, even to the point of inhibiting feeding.

The control of feeding has been especially well studied in the blow fly, *Phormia regina* (Dethier, 1976). Gelperin (1971) devised a very useful scheme for depicting the regulation of uptake of sugars (potential energy) from the environment. This scheme (Fig. 8-5) summarizes nicely much of the information presented here.

Several factors other than food deprivation also influence feeding behavior: molting, temperature, dormancy (to be discussed later in this chapter), the presence of other insects, food previously ingested, reproductive state, etc.

Escape and Defense

Throughout the evolutionary history of insects there has been a significant selective advantage in favor of the development of protective mechanisms against predators and parasites (see Chapter 9). We shall examine several examples of such mechanisms from two standpoints: escape and defense.

Fig. 8-5
Regulation of sugar-feeding in the blow fly *Phormia regina*. "Slug" refers to a mass of food that moves as a unit. [Slightly modified from Gelperin, 1971.]

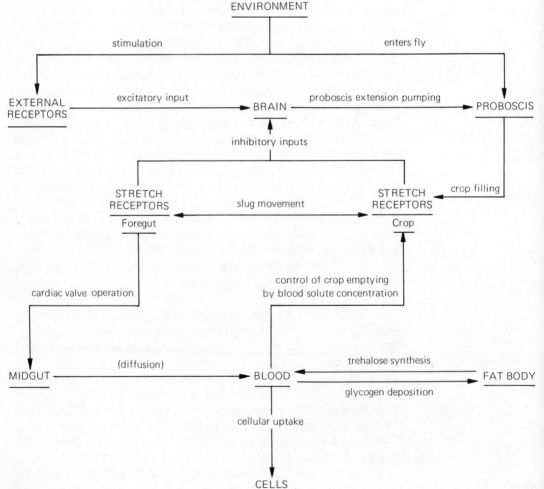

Escape. Many insects will exhibit an escape reaction when threatened. Recall the alarm response of cockroaches mentioned earlier. Even those insects that have protective means other than escape will usually attempt to flee if the stimulus is great enough. A coupling of defensive behavior with escape is shown by some. For example, the beetle *Chlaenius* (Carabidae) discharges a secretion that is repellent to ants and remains invulnerable to them for 8–13 min following a discharge. This beetle can run an estimated 100 meters during this period of time!

Particularly interesting examples of insect escape reactions are those that are elicited in certain moths (especially in the family Noctuidae) in response to ultrasonic impulses from bats (Roeder, 1965, 1966, 1967). The tympanic organs of hearing have already been described (see Chapter 6). Insectivorous bats locate nocturnally airborne insects by emitting ultrasonic impulses and locating the sources of echoes. The moths detect these impulses beyond the range of the bats' echo-locating sensitivity and evasive behavior is initiated. Patterns of evasive behavior were studied by the use of an electronic sound source that produced sound in the range of the bat "chirps." The sound was emitted from a loudspeaker mounted on an upright pole. The paths of moths arriving in the vicinity of the sound source were recorded photographically by means of a stroboscopic (regularly blinking) flash that illuminated the moths at regular, very closely spaced intervals of time. Each time an approaching moth was illuminated, the reflected light, indicating the position of the moth at that moment, was recorded on the same photographic plate. With this technique a number of different evasive tactics were detected. These included turning and flying directly away from the sound source, oblique turns upward, "power dives," "passive drops," complex "looping dives," and zigzag movements. Some individuals landed on the ground and remained motionless for a period of time. When moths were tossed into the air in the vicinity of bats, similar responses were recorded with the same technique.

Defense. Insects display great variation in their form, coloration, and patterns of coloration. Instances in which one or all of these characteristics provide protection of the insect are legion. The close resemblance of wings and other body parts to leaves, twigs, thorns, and other plant parts; insects with "eyespots" or bright colors on the wings, which are suddenly displayed in response to a threat; *mimicry* of species that vertebrate predators learn to avoid, through experience with poisonous stings, irritating sprays, disagreeable tastes, and so on; and insects that blend in so well with their normal surroundings that they become nearly invisible to predators (*crypsis*) are but a few examples.

An insect must behave in an appropriate way if it is to gain any protective benefit from its form and coloration. Thus an insect that closely resembles a leaf will tend to orient itself on a stem in an appropriate manner, one with "eyespots" or other brightly colored markings will display these characteristics when threatened by a potential predator, and one that resembles other species may also adopt similar behavior. Insects that blend with their surroundings tend to remain motionless when threatened. Protective form and coloration will be discussed further in Chapter 9.

Skeletal structures, although not always used for defense, may function in this capacity. The raptorial forelegs of a praying mantis, although functioning primarily in prey capture, are sometimes used to strike out at an attacker. The stingers of many ants, bees, and wasps also serve such a dual purpose. Many, if not all, of the sound-producing orthopterans produce sounds of warning and aggression. Any skeletal structure of sufficient strength and/or hardness can serve in a passive defense capacity. Cleptoparasitic cuckoo wasps (Hymenoptera, Chrysididae) roll up into a ball when threatened. In this position, the exposed areas of exoskeleton apparently provide protection from their hymenopteran host (Matthews and Matthews, 1978).

A large number of insects produce nonskeletal barriers of various types. Nests, cocoons, cases, and so on no doubt function in defense. The cast skins and fecal material of the larvae of the beetles in the genus *Cassida* (Chrysomelidae) remain on a "fork" of posterior abdominal appendages that are highly maneuverable. This waste material dries and forms a sturdy "fecal shield," which is used by the larva as a protective device that is said to be highly effective in blocking the bites of ants (Eisner, van Tassell, and Carrel, 1967). The predaceous larvae of the green lacewing, *Chrysopa slossonae*, feed on the wooly alder aphid, *Prociphilus tesselatus*. The lacewing plucks the waxy "wool" from the bodies of the aphids and applies this material to its own body, coming to look like its prey. This "disguise" protects it from the ants that "herd" the aphids. Artificial removal of the waxy "wool" results in attack by ants (Eisner et al., 1978).

The primary defense mechanism of many insects is chemical. A large variety of noxious substances are secreted by the repugnatorial glands associated with the integument (see Chapter 3). These substances represent "the means by which predators and other potential enemies are 'told' to desist or withdraw" (Eisner and Meinwald, 1966). The glands producing repugnatorial substances are variable in number, location, and morphology and have no doubt arisen independently many times during the course of insect evolution. Repugnatorial material is discharged from storage reservoirs associated with the glands by a variety of mechanisms. Basically these mechanisms fall into three categories (Eisner and Meinwald, 1966).

1. The secretion oozes onto the integumental surface.
2. The gland is evaginated and the secretion allowed to volatilize.
3. The secretion is forcibly discharged.

Caterpillars of the species *Papilio machaon* (and probably all swallowtail butterfly larvae; Papilionidae, Lepidoptera) possess glands behind the head, *osmeteria,* which they evert when threatened, that produce a secretion effective against ants (Eisner and Meinwald, 1965).

The forcible discharge of material is effected in a variety of ways. Contraction of muscles associated with the glands, hydrostatic pressure of the hemolymph, and air pressure are among them. Examples of insects that are capable of forcible discharge of odoriferous material are particular species of cockroaches, stink bugs, earwigs, stick insects, caterpillars, and carabid and tenebrionid beetles. Some of these are able to spray several feet, and many have a very accurate aim. A particularly impressive discharge mechanism is found in the bombardier beetles (*Brachinus* spp.). These beetles eject a hot spray, containing quinones, from glands in the posterior part of the abdomen with a distinctly audible "explosion." This is the result of a chemical reaction in which oxygen is liberated. The beetle's raised posterior at the time of the explosion adds to the impressiveness of this display. One may wonder why the defensive secretions liberated by insects do not harm the insects themselves. Apparently the gland cells allow two or more harmless precursors to be mixed in tubes or chambers isolated from the rest of the cell. This principle operates at the multicellular level in the repugnatorial glands of the bombardier beetle (Fig. 8-6). Each gland, which is two-chambered, contains phenolic precursors of quinones and hydrogen peroxide in an inner chamber and the enzyme catalase in the outer chamber. When these substances come into contact with one another, oxygen is liberated as a result of the mixing of catalase and hydrogen peroxide, and the phenols are oxidized to quinones, producing the reaction described above (Eisner and Meinwald, 1966).

Blum (1978), Duffey (1977), Roth and Eisner (1962), and Eisner and Meinwald (1966) contain tables listing many specific chemicals and the arthropods in which they have been found. Besides functioning as defense allomones, some of these chemicals are thought to have antimicrobial effects. Some also act as alarm pheromones (Blum, 1969). For example, soldiers of certain termites discharge a spray that not only works against nest invaders but also alerts other

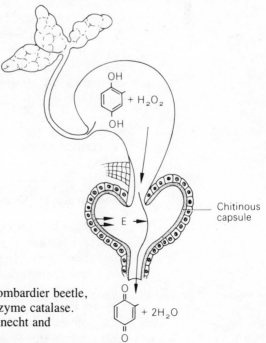

Fig. 8-6
Defensive gland of a bombardier beetle, *Brachinus* sp.; E = enzyme catalase. [Redrawn from Schildknecht and Holoubek, 1961.]

soldiers. Alarm pheromones are also found in several groups of Hymenoptera (ants, bees, wasps, etc.). Eisner (1970) and Blum (1978) are good sources for further information on the chemical defenses of arthropods.

Some insects are behavioral mimics of those that produce defensive secretions. For example, *Eleodes longicollis,* a species of tenebrionid beetle that possesses defensive glands, stands with its posterior well raised and discharges material from the glands. A second species, *Megasida obliterata,* mimics the stance of the first but lacks the defensive glands.

The products of venom glands associated with the hymenopteran stingers (modified ovipositors) and with the stinging hairs of many lepidopteran larvae (and some pupae and adults) and the toxic saliva of some Hemiptera may also be placed in the chemical defense category.

Not all defensive chemicals are directly associated with integumental glands. Some are found in the hemolymph and impart a bad taste or render an insect indigestible. For example, the Monarch butterfly, *Danaus plexippus,* possesses such substances, called *cardenolides,* in its body tissues (Reichstein et al., 1968). Cardenolides are derived from the characteristic food plants (i.e., milkweeds) of this species. The Viceroy butterfly, *Limenitis archippus,* is a well-known mimic of the monarch and is avoided by "experienced" vertebrate predators even though it does not possess any distasteful substances. Some Lygaeid bugs (e.g., the milkweed bug, *Oncopeltus fasciatus*) also obtain cardenolides from their food sources (again milkweeds). Rothschild (1972) provides many more examples of insects that have been found to utilize secondary plant substances (see "Adaptations Associated with Interspecific Interactions" in Chapter 9) as defensive chemicals.

Many species of beetles in the families Coccinellidae, Chrysomelidae, Meloidae, Lampyridae, and Lycidae discharge blood (*reflex bleeding*) in response to a threat. The blood may serve as a physical deterrent by clotting and entangling an attacker, or like the blood of many "blister" beetles (Meloidae), it may contain noxious substances. Many grasshoppers and other herbivorous insects may regurgitate or defecate when disturbed.

Many insects that produce repugnatorial substances, taste bad, or sting are brightly colored. This *aposematic* or *warning coloration* "advertises" their "undersirability" as food (more on this in Chapter 9).

Behavior and the Fluctuating Environment

Most habitats are characterized by both short- and long-term (seasonal) fluctuations in their suitability for living organisms. Although few habitats are constant, many are "predictable"; that is, they vary, but with a degree of consistency. Others vary with little consistency. This variation in suitability is related directly and indirectly to abiotic factors, especially temperature and humidity. Thus for a given insect, a given habitat at one time may provide adequately for its life requirements (food, shelter, favorable temperature, favorable moisture, etc.) and at another time be entirely unsuitable. In this section, the role of behavioral adaptations associated with variation in environmental suitability is discussed.

Dispersal and Migration. Dispersal has been defined in a number of ways. For our purposes it can be looked upon as the result of several mechanisms, probably the most important of which is migration. Other probable mechanisms of dispersal include those associated with the search for food or a mate, phoresy, and responses to gradients of environmental factors (e.g., temperature, moisture, and CO_2). Insects may be carried passively by the wind or by water currents. As Johnson (1966) points out, "It is extremely difficult to measure how much dispersion occurs accidentally, incidentally, or adaptively."

Migration may be viewed as an active mass movement adapted to displace populations. Kennedy (1975) used the term "adaptive traveling." Migration is characterized behaviorally "as an accentuation of locomotor function with a depression of vegetative function" (Johnson, 1966). In other words, migrating insects are persistent in their movement and usually do not respond to "vegetative" stimuli (stimuli associated with food, potential mates, etc.). Kennedy (1975) adds to increased locomotor function and decreased vegetative function the following characteristic of migratory behavior: the straightening out of "the flier's track over the ground so that it travels, in the sense of traversing new ground instead of frequently changing its direction in a series of 'trivial,' station-keeping flights." In addition, the female sex is always present in a migration and may or may not be accompanied by males. Migrant females are generally sexually immature, but there are exceptions. Typically, many individuals migrate simultaneously, and migrant species tend to produce large numbers of offspring.

Migration is accomplished mainly by flight, and the direction of displacement for many is influenced by the wind. Even among migrants in which wind plays a major role, there is evidence that behavior plays a role, that is, behavior associated with straightening out the track over the ground (Kennedy, 1975). Migration may occur by means other than flight. For example, army ants *(Eciton hamatum)* migrate on foot *(pedestrian migration)* (Kennedy, 1975; Schneirla, 1971).

Insect migration has been studied by a variety of techniques, including nets, airborne suction traps; aerial balloons; aircraft; photography; and marking with dyes, paint, radioisotopes, or genetic markers. Since 1966, when radar began to be used to migration of insects, that technique has contributed a great deal to our understanding of insect migration (Schaefer, 1976). Incredibly, in addition to being able to obtain data on sizes of migrating populations and their flight performance, "radar entomologists" have also been able to determine species and sex (up to a distance of 1.5 km) by echo "signature" analysis.

Three types of migration may be recognized on the basis of adult life span (Johnson, 1966).

1. Short-lived adults that emigrate and die within a season.
2. Short-lived adults that emigrate and return.
3. Long-lived adults that hibernate or aestivate.

Members of the first group usually leave the breeding site, oviposit elsewhere, and die. Examples include locusts, termites, aphids, thrips, and many butterflies. The distance covered is variable; in locusts and butterflies it may be several hundred miles, in termites only a few feet or yards. The flights of many of these are windborne. Relatively short-lived adults, which emigrate and return, depart from the breeding site to feeding sites, where the eggs mature. The females then fly back to the vicinity of the original breeding site and oviposit. This emigration and return may be repeated in a given season by the same individual. For example, such behavior is characteristic of many dragonfly species. Insects in the third category fly to hibernation or aestivation sites and return to the original breeding site the following season. The monarch butterfly, *Danaus plexippus* (Urquhart, 1976; Brower, 1977), many noctuid moths, and several beetles fall into this category.

The result of migration is the transference of a group of insects from an old to a new site. This is of particular importance when the "old" site was only temporarily suitable for survival of the insects. Temporary sites include pools of water that tend to dry out, seasonal food plants, and so on. Schneider (1962) points out, "In the course of evolution, a low level of dispersive movement has been associated with the colonisation of permanent habitats and a high level has been closely correlated with the adoption of a temporary one." He further explains that this idea also applies to species in which the adults and larvae feed in quite different habitats (e.g., mosquitoes) and to those that hibernate or aestivate in habitats different from the breeding sites. Thus migration has come to be viewed as an adaptive strategy of insects that utilize temporary habitats. It was formerly thought that only a few species of insects migrated. Now, it appears that some migratory behavior is probably characteristic of most species (Johnson, 1976). The details of migratory behavior vary greatly among species.

A couple of examples will illustrate how migratory behavior is adaptive. A particularly well-known migrant is the desert locust, *Schistocerca gregaria* (Johnson, 1976). These insects breed in an area from North Africa to East Africa to Pakistan in the spring. As these areas dry out in the summer, young locusts begin to migrate in swarms (some estimated to contain from 10^5 to 10^{10} individuals). Their migration paths lead to a zone where the generally westerly winds of Africa collide with the moisture-laden monsoon winds from the Indian Ocean (i.e., the intertropical convergence zone). In this zone there is much rainfall, and lush vegetation results. Later in the season, when the convergence zone moves northward, new swarms migrate northward. Clearly, the migratory behavior of the desert locust is adjusted to ecological changes in a way that favors survival.

The monarch butterfly, *Danaus plexippus*, is a well-known migrant in the United States (Urquhart, 1976; Brower, 1977). Monarchs breed far into the north during the summer, feeding and ovipositing on several species of milkweed plants. In the fall, they migrate southward. In the west, they fly to southern California, and in the midwest and east, they fly into central Mexico and Florida. The evolution of migratory behavior in these insects has enabled them to expand their feeding and breeding range in the summer and avoid the cold of winter.

Two hypotheses have been advanced to explain the cause of migration (Johnson, 1966). The first one suggests that insects respond to the onset of adverse conditions by flying away. The second explains migration on the basis of endocrine changes correlated with particular environmental effects, such as crowding, food deficiency, and short days. For example, crowding in aphids results in the production of winged (alate) forms instead of the nonmigrant wingless (apterous) forms, and this seems to be associated in some way with activity of the corpus allatum. Available evidence pertinent to the cause of migration favors the second hypothesis. It has been suggested that migratory behavior might depend on a particular balance between ecdysone and juvenile hormone in the blood.

For further information on various aspects of insect migration, consult Dingle (1972, 1978a), Johnson (1963, 1969, 1974, 1976), Kennedy (1975), Kring (1972), Rainey (1976), and Schneider (1962).

Dormancy. Insect development (including embryogenesis and larval and ovarian development) and general activity are subject to two extreme kinds of suppression *(dormancy)* relative to changes in the abiotic environment. At one extreme, an insect responds to adverse environmental conditions by a slowdown in metabolism and development. When conditions are no longer adverse, metabolism and development resume immediately. This type of dormancy is commonly referred to as *quiescence*. At the other extreme, an insect enters into a state of metabolic and developmental arrest in response to certain environmental conditions. These conditions may or may not be adverse, but serve as indicators of the imminence of the onset of adverse conditions. Development does not necessarily resume immediately with the return of favorable conditions. This type of dormancy is called *diapause*. Cold-hardiness, increased resistance to desiccation, and so on (see Chapter 9) may be associated with quiescence and diapause.

Diapause is looked upon as a physiological timing mechanism that provides for

1. The induction of a comparatively resistant state during periods of adverse environmental conditions, particularly low temperatures.
2. The resumption of development when adverse conditions have disappeared and food is available (e.g., synchrony with host plants).
3. Synchrony of adult emergence in species with a short adult life.

Diapause may occur in any of the life stages in different insect species. In the adult stage it is characterized by the failure of the gonads to enlarge.

In many insects, diapause is *obligatory* (i.e., insects enter this state during every generation in spite of variation in environmental conditions). Other species display *facultative diapause* and can go on for several generations before entering diapause. Species or strains that exhibit obligatory diapause have only one generation per year (i.e., they are *univoltine*). On the other hand, those with facultative diapause may complete two or more generations per year (i.e., they are *bivoltine, trivoltine, quadrivoltine,* etc.). As in the milkweed bug, *Oncopeltus fasciatus,* diapause may occur in association with migration (Dingle, 1978b).

The stage at which an insect enters diapause is a species characteristic, and even closely related species may enter diapause at different stages. Usually diapause occurs only during one life stage, but there are exceptions where diapause is entered twice. In a given species, there may be strain differences with regard to diapause. One strain may show an obligatory diapause, another facultative diapause, and still another not enter diapause at all. For example, univoltine, bivoltine, and quadrivoltine races of the silkworm *Bombyx mori* (Lepidoptera, Bombycidae) are known in addition to races that undergo no diapause whatsoever. It is not uncommon for geographical races to show differences in the number of generations per year, depending on locality. An example of such racial differences has been found in the European cornborer, *Ostrinia nubilalis* (Lepidoptera, Pyralidae). Several years after the introduction of this species into the eastern United States it was discovered that the race in the Great Lakes states was univoltine with an obligatory diapause, whereas the New England race was bivoltine.

As mentioned above, the environmental conditions responsible for induction of diapause are not necessarily adverse, but may signal the imminent onset of adverse conditions. The most important environmental factors that induce diapause appear to be those associated with the daily alternation of light and dark and the relative durations of light and dark periods. Compared to other environmental factors, these factors are the most accurate and invariable cues relative to seasonal changes in temperate zones. The mosquito *Aedes triseriatus* completes several generations during the summer, but in the fall the eggs of the last generation enter diapause in response to shortening day length despite the fact that temperature may actually be higher than in the spring when diapause is terminated and development resumes. In a few insect species the length of the dark period is the determining factor in the initiation of diapause.

Temperature is another important factor in the induction of diapause in certain species. Generally, low temperatures favor the initiation of diapause, although this is not always the case (e.g., high temperatures favor diapause in *Bombyx mori*). Temperature usually acts in association with photoperiod in producing diapause. Other environmental factors that may be involved include unfavorable nutritional conditions, low moisture content of food, desiccation, crowding, maternal diet, maternal age at oviposition, and maternal exposure to low temperatures during oogenesis. Diapause in parasitic insects may be induced in synchrony with diapause in the host. In some parasitic species development is closely attuned to the hormonal changes in the host, and in some host species the presence of a parasite disrupts a diapause that would otherwise have occurred.

Various factors may be involved in the termination of diapause. For example, several species (e.g., *Bombyx mori*) must be exposed to low temperatures for certain intervals of time before diapause can be terminated. When diapause occurs during a hot dry season, exposure to high temperatures for a period of time may be necessary before diapause can be terminated. Wounds and various kinds of shock, such as pricking with a needle or electrical stimulation, may cause the termination of diapause in individuals of certain species.

Diapause is probably controlled by the endocrine system. Williams (1946–1953) showed that the activity of the prothoracic glands is necessary for the termination of postembryonic diapause in the silk moth *Platysamia cecropia*. In some other moths (*Bombyx mori*, Fukuda and Takeuchi, 1967; *Phalaenoides glycinae*, Andrewartha et al., 1974) a secretion of neurosecretory cells in the subesophageal ganglion of the female parent induces diapause in her eggs.

Müller (1970), Mansingh (1971), and Thiele (1973) consider various classifications of types of dormancy. Dingle (1978a), Stoffolano (1974), Tauber and Tauber (1976), and Lees (1955) review various aspects of diapause.

Behavior, Temperature, and Humidity. It has been known for a long time that insects placed in a temperature gradient will demonstrate a "preferred" temperature by locating themselves at a particular point along the gradient and, further, that this preferred temperature is roughly correlated with habitat preference. Otherwise insects have generally been viewed as *poikilothermic* ("cold-blooded"). Recently, however, evidence has been accumulating for the existence of *thermoregulation*—maintenance of body temperature different from ambient temperature—in insects (May, 1979). Behavior, along with physiological mechanisms, plays an important role in thermoregulation. Mechanisms of thermoregulation involve heat exchange with the environment (via conduction, convection, radiation, and evaporation) or variation in the production of metabolic heat. *Ectotherms* rely on heat from the external environment, in particular directly from the sun. *Endotherms* generate most of the required heat metabolically. Some insects may be both ecto- and endotherms.

Common mechanisms of ectothermic regulation include control of solar input posturally, selection of "appropriate" microhabitats (see Chapter 9), and shifting of activity rhythms (May, 1979). Many insects, for example, several grasshoppers and locusts, butterflies, and cicadas, bask in the sun, assuming postures that facilitate heating. Conversely, ground-dwelling insects may "stilt," raising their bodies away from hot ground. Insects commonly move about and come to rest in a microhabitat with appropriate temperatures. For example, many insects tend to remain preferentially in shaded areas. Many desert insects burrow in response to high temperatures. In several diurnal species, which display midday activity peaks under cool conditions, the activity peaks may be shifted to early morning or evening in hot weather. Color may not be especially important in thermoregulation in most insects, but a few are known to undergo color changes in association with changes in temperature.

A good example of endothermy is found in the sphinx moth *Manduca sexta* (Lepidoptera, Sphingidae), which has been found to regulate body temperature just prior to and during flight (Heinrich and Bartholomew, 1972; Heinrich, 1974). These insects, which bear a striking resemblance to hummingbirds, cannot fly unless their bodies are at a temperature between 35 and 38°C, but they have been observed flying when the ambient temperature was as low as 10°C. When ambient temperature is less than 35°C, these moths heat the thorax by assuming a characteristic stance and vigorously vibrating the wings through a small arc. This "shivering" is accomplished by the simultaneous stimulation of both the tergosternal and longitudinal flight muscles. The duration of this warmup period varies indirectly with ambient temperature. These moths are also able to prevent "overheating" (i.e., reaching temperatures above 46°C). This is accomplished by increased heartbeat rate, which increases hemolymph circulation from the thorax to the abdomen. Loss of excess heat occurs across the abdominal wall, which is not insulated against heat loss as is the thorax.

Raising body temperature by fluttering the wings without flight has been observed in several insects in addition to *Manduca sexta*. There is also evidence that some "singing" insects warm up before singing.

A few insects are known to cool their bodies by enhancing evaporation. For example, honey bees may release fluid from their mouthparts and spread it on the venter (May, 1979).

Thermoregulation in many insects is probably under central nervous control, but little is known about this topic.

Insects may cooperate in thermoregulation. Many insects are gregarious and tend to form clusters under conditions of low temperature; overwintering honey bees form such clusters within their hives. The combined effects of the metabolic heat of the component individuals may actually accelerate the rate of development, as in the case of certain caterpillars (Bursell, 1974a). The nests of social insects provide good examples of temperature control. For instance, the nests of some species of termites consist of cells having a structure that permits them to act as an air blanket, the effect being analogous to that of storm windows. During a hot day, the outer surface of a nest may be too hot to be touched by a human hand, whereas deep within the nest the temperature may be somewhere around 30°C. On a cold winter day, the internal temperature of the same nest may be 9 or 10°C warmer than the air surrounding it (Skaife, 1961). Lüscher (1961b) describes the evolution of such "air conditioning" in termite nests. The nests of some wasps (e.g., hornets) are constructed with an outer "paper" envelope covering the layers that contain the rearing cells, resulting in an air-blanket effect. Other social wasps (e.g., *Polistes*), do not construct such an envelope. If a nest gets too hot, a number of its inhabitants will fan their wings vigorously at the nest entrance and may even bring in drops of water, which has the effect of increasing the cooling from fanning (Richards, 1953).

It is well established that insects will respond to a moisture gradient by moving into an "optimal" zone (Bursell, 1974b). This behavior obviously promotes survival. Other factors, of course, may influence exactly how an insect behaves in a moisture gradient. If the flour beetle *Tribolium,* has been starved, it will show a preference for moist air, but if it has been given access to food and water, it prefers dry air. Most insects offset the influence of dryness and other factors promoting water loss by drinking water or taking in food.

Reproductive Behavior

Reproductive behavior involves first the location of a mate, followed by courtship and mating, oviposition, and sometimes brood care. Territoriality may arise out of competition for mates or food. These aspects of reproductive behavior are considered in this section. Examples have been drawn primarily from the reviews of Carthy (1965), Markl (1974), and Matthews and Matthews (1978). The literature cited in the section on communication contains much information on reproductive behavior. Reproductive behavior is typically complex and highly variable. The evolutionary role of this complexity and variability is generally assumed to be reproductive isolation.

Mate Location. A wide variety of mechanisms function in bringing together the sexes. Initially, over relatively long distances, these mechanisms usually involve the visual, olfactory, and auditory modes of communication, singly or in combination.

The visual stimuli associated with mate location and the responses they elicit vary in complexity. The simplest pattern, observed among water striders (Gerridae, Hemiptera), involves a male insect approaching any moving object of appropriate size that happens to enter its visual field. Other insects require more specific stimuli to induce approach. For example, male damselflies in the family Lestidae (Odonata) approach any insect that has transparent wings and flies in a manner similar to that of damselflies. Damselflies in the genus *Calopteryx* (Calopterygidae) require even more specific stimuli, the males of different species being able to recognize members of their own species by the amount of light allowed to pass through the wings.

More complex visual stimuli involved in mate location include the use of luminescent organs in signaling for a mate. This is best exemplified by the fireflies in the beetle family Lampyridae (see Lloyd, 1966, 1971, 1977). In some species both the males and females produce light and have a very complicated signaling system; in other species only the males produce the light signals. In either case the light-signaling systems are species-specific. In fact, several species have been discovered through observation of differences in signaling systems. The mating behavior of *Photinus pyralis,* a well-known American species, is described in the following quote from McElroy and Seliger (1962).

At dusk the male and female emerge separately from the grass.
The male flies about two feet above the ground and emits a

single short flash at regular intervals. The female climbs some slight eminence, such as a blade of grass, and waits. Ordinarily she does not fly at all, and she never flashes spontaneously. If a male flashes within three or four yards of her, she will usually wait a decorous interval, then flash a short response. At this the male turns in her direction and glows again. The female responds once more with a flash, and the exchange of signals is repeated—usually not more than five or 10 times—until the male reaches the female, waiting in the grass, and the two mate.

In some tropical species of fireflies, several males congregate in a single tree and flash synchronously (Buck and Buck, 1968, 1976). McElroy et al. (1974) describe this phenomenon as follows.

In Burma and Siam and other eastern countries, all the fireflies on one tree may flash simultaneously, while on another tree some distance away this same synchronous flashing would be apparent, but out of step with those of the first tree. Observers have been particularly impressed by the display, which is one of the interesting sights of the Far East.

In some cases visual markers in the environment may be involved in mate location. For example, among the true flies males may swarm or sit near a marker and wait until a female approaches the marker (Fletcher, 1977).

The use of olfactory cues in mate location is widespread among the insects. Sex pheromones are produced by specialized glands in males, females, or both sexes of a given species. Sex pheromones are very potent substances; "Detected by the insect in fantastically minute amounts, these attractants are undoubtedly among the most potent physiologically active substances known today" (Jacobson, 1965). The sex pheromone *bombykol,* produced by the female silk moth, *Bombyx mori,* can elicit a response from a male in a concentration of 1000 molecules per cubic centimeter. Electrophysiological studies of antennal receptors have revealed that a single molecule of bombykol can initiate a single nerve impulse (Schneider, 1974). Most sex pheromones are apparently species-specific; however, there are several known examples of nonspecificity

The production by female insects of pheromones that attract males has been observed and described for several species in a number of orders. The combined lists of Jacobson (1965), and Jacobson (1974) are composed of 12 species of cockroaches (Orthoptera); 4 species of true bugs (Hemiptera, Heteroptera); 4 species of homopterans (Hemiptera, Homoptera) 184 species of moths and butterflies (Lepidoptera); 34 species of beetles (Coleoptera); 31 species of bees, wasps, sawflies, and ants (Hymenoptera); 8 species of true flies (Diptera); and 3 termite species (Isoptera) in which the females are known to produce sex pheromones. Some examples follow. In the Gypsy moth, *Lymantria dispar* (Lymantriidae), virgin females produce a pheromone that is capable of luring males over a distance of 100 meters. The containers in which virgin females have been held apparently absorb some of the odorous substance since they remain attractive to males for 2 to 3 days following the removal of females. Virgin females of various species of silk moths (Saturniidae) "call" males by protruding posterior segments of the abdomen and exposing pheromone-secreting glands to the atmosphere. This "calling" posture only occurs at certain times of day or in response to certain stimuli and is controlled by the release of a "calling" hormone from neurosecretory cells in the corpora cardiaca (Riddiford and Truman, 1974). Female cockroaches produce a sex pheromone that stimulates male alertness, antennal movement, searching locomotion, and vigorous wing flutter (Jacobson, 1965). Filter paper that has been in contact with virgin females has the same effect.

As mentioned earlier, males in many species produce sex pheromones. The combined lists of Jacobson (1965), and Jacobson (1974) are composed of 5 species of cockroaches, 4 species of true bugs, 61 species of moths and butterflies, 11 species of beetles, 10 hymenopteran species, 12 species of true flies, 4 species of scorpionflies (Mecoptera), and 1 neuropteran species in which males produce sex pheromones. The pheromones produced by the males of some of these species serve both as attractants and excitants or excitants alone. Two examples of male production of sex pheromones follow. Males of the greater wax moth, *Galleria mellonella,* secrete from glands on their wings an odorous substance that is very attractive to females. The odor is dispersed by the male vibrating its wings and dancing around. Bumble bee (*Bombus terrestris*) males produce an attractant in their mandibular glands that lures females. This substance has been extracted from bumble bee mandibular glands with pentane and identified as farnesol, a substance present in the flower oils of many plants. These flower oils may be the bee's source of farnesol (Jacobson, 1965).

Three species of beetles in the genus *Dendroctonus* (Scolytidae) provide an interesting variation in the function of sex pheromones. In these species a substance that is attractive to both sexes is produced by sexually mature, unmated females feeding on fresh Douglas fir phloem. Since both sexes are attracted by this odorous substance, it serves to bring them together, which eventually leads to courtship and copulation. *Trypodendron lineatum,* an ambrosia beetle, produces a substance that has a similar effect. Such substances have been called *assembling scents.*

Shorey (1977a, b) considers sex pheromones as well as pheromones in general. Fletcher (1977), Borden (1977), and Bartell (1977) include discussions of the roles of pheromones in the sexual behavior of Diptera, Coleoptera, and Lepidoptera, respectively.

Acoustic signals serve to bring the sexes together for a large number of insects—many Orthoptera, Hemiptera-Homoptera, and Diptera; a few Lepidoptera, Hemiptera-Heteroptera, and Coleoptera. Calling songs may be quite elaborate and are usually species specific. In general, a response to one of these songs depends in part on the state of internal readiness for courtship and copulation. In the majority of cases, specialized structures are involved in the production of sounds (see Chapter 6), but several insects produce sounds as a direct result of wing movements in flight. Examples of sounds produced in this fashion that serve to bring the sexes together have been found in members of the Diptera, Hymenoptera, and Orthoptera. For example, the flight sound of a mature female mosquito is very attractive to a male and elicits copulatory behavior. The males react similarly when the frequency of the flight sound is produced by a tuning fork. In some insect groups the acoustical counterpart of the "assembling scent" has been found. For example, cicadas in the genus *Magicicada* sing an aggregating song in chorus that is responsible for assembling both males and females (Alexander and Moore, 1962).

In many cases, aggregation of males and females may occur in association with biological activities other than sex, for example, at emergence, oviposition or feeding sites. Such aggregation can subsequently lead to sexual activities.

Once an insect has been induced to approach a possible mate, other stimuli are usually required to release further pursuit. These are generally olfactory, but they may also be a characteristic behavior on the part of the pursued. An insect that recognizes a member of its own species may or may not be able to distinguish the sex of the individual it pursues. Males of *Pyrrhocoris* and *Gerris* (Hemiptera-Heteroptera) and *Leptinotarsa* (Coleoptera, Chrysomelidae) recognize members of their own species by specific odors but will attempt to copulate with either sex. On the other hand, some fruit flies are able to recognize the sex of an individual by its odor.

Courtship and Mating. Following mate location, variable and often elaborate behavior is released which ultimately leads to copulation. Courtship displays function in species and sex identification, the meeting of solitary individuals, and in the stimulation and maneuvering involved in copulation.

During courtship, escape and attack responses are usually inhibited, at least to the extent that copulation may occur. However, they may not be inhibited completely. For example, as mentioned earlier, the female praying mantid very often devours the head and part of the thorax of a male that is attempting to copulate. This is not as tragic as it seems since removal of the head, in particular the subesophageal ganglion, increases the vigor of the male's copulatory movements. Signals from the intact subesophageal ganglion are thought to inhibit various endogenous nervous patterns in the rest of the body. Thus decapitation removes inhibition of copulatory movements (Roeder, 1967) (see the section in this chapter on control of behavior). In addition to releasing vigorous copulatory movements, the female mantid also obtains a highly nutritious meal.

Many of the cues that serve in mate location act further as releasers of courtship and copulation. For example, the flight sound of the female mosquito not only attracts males but also releases copulatory behavior. In many insects the sex pheromone produced by one of the sexes also serves as an excitant (aphrodisiac), stimulating courtship or attempts at copulation. For instance, the sex pheromone produced by female American cockroaches (*Periplaneta americana*) not only attracts males but induces a wing-raising display and attempts to copulate. In the absence of a female, males may attempt to copulate with one another if the female odor is present. This female sex pheromone is also capable of eliciting courtship behavior in males of other *Periplaneta* species and males of *Blatta orientalis* (oriental cockroach). In some insects, the sexually excited male or the female may produce a substance that acts as an excitant for the opposite sex. Such is the case in *Lethocerus indicus,* a giant water bug, and a family of beetles (Jacobson, 1965).

> During sexual excitement the male is easily recognized by its odor, for its abdominal glands secrete a liquid with an odor reminiscent of cinnamon. . . . This substance, produced in two white tubules 4 cm long and 2–3 mm thick, occurs to the extent of approximately 0.02 ml per male and is used in southeast Asia as a spice for greasy foods. The female does not secrete the substance, which is believed to act as an aphrodisiac to make her more receptive to the male.
>
> . . .
>
> Males of Malachiidae, a family of tiny tropical beetles, entice females first with a tarty nectar and then expose them to an aphrodisiac. The males possess tufts of fine hair growing out of their shells (in some species on the wing covers, in others on the head). These hairs are saturated with a glandular secretion that the females cannot resist. During the mating season, the male searches for a female; when he finds one he offers his tuft of hair, which the female then accepts and nibbles upon. In so doing, her antennae come in contact with microscopic pores in his shell, through which the aphrodisiac substance is excreted, thus putting her in a state of wild excitement.

Insects show a seemingly endless variety of sexual patterns. Some are simple, the male and female simply coming together and copulating with little or no courtship maneuvers. On the other hand, many have elaborate and sometimes bizarre patterns of courtship. Two examples will give you an idea of how elaborate some of these courtships are. The silverfish, *Lepisma saccharina,* guides the female to an externally deposited spermatophore by spinning a series of threads that restrict her movements to those which bring her closer and closer to making contact with the spermatophore (Fig. 8-7). Jacobson (1965) cites the work of Bornemissza (1964) in the following description of courtship and copulation of two species of scorpionflies (*Harpobittacus,* Mecoptera).

> Males of both species hunt for the soft-bodied insects on which they feed; females have never been observed hunting, capturing, or killing prey in the field. When the male holds its prey and begins to feed, two reddish-brown vesicles are everted on the abdomen between tergites 6–7 and 7–8 and begin to expand and contract, discharging a musty scent perceptible to humans. This scent attracts females to the vicinity of the male, moving upwind in his direction. As soon as the female is within reach, the male retracts his vesicles and brings the prey to his mouthparts. The female attempts to get hold of the prey but is prevented by the male, whose abdomen seeks out the tip of the female's abdomen and copulation takes place. Once in copula, the male voluntarily passes the prey with his hind legs over to the female.

Rivalry and Territoriality.   Rivalry between males may arise over the courtship of the same female and may result in direct physical aggression or displays. Reactions to members of the same spe-

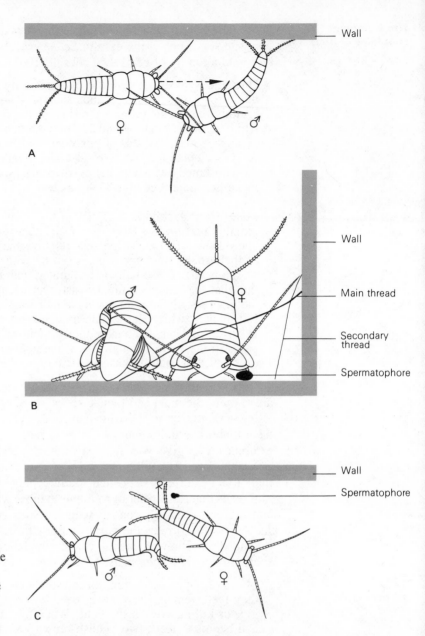

Fig. 8-7
Courtship in the silverfish, *Lepisma saccharina*.
A. Approach.
B. Male affixes threads to wall and floor and deposits spermatophore (only the main thread and one secondary thread are shown, although many secondary and irregular threads are produced). C. Female is guided to the spermatophore by the threads. [Redrawn from Sturm, 1956.]

cies are different from those involved with escape or defense, which are stimulated by threats of danger. When a male grasshopper approaches another male that is serenading a female with a courtship song, they face one another and sing a characteristic "rivalry" song (Haskell, 1974). Eventually one of the males leaves, and the other continues the courtship. Male silverfish, *Lepisma* (Lepismatidae, Thysanura), will fight over a female.

Territorial behavior is not common among insects, but there are some very definite examples. Males of certain odonates in the genera *Calopteryx* (damselflies) and *Pachydiplax* (dragonflies) defend territories against other males. A male entering another's territory is met with an attack. Females are recognized by sight since there is distinct sexual dimorphism in these particular species. Instead of aggression, these females are met with attempts to copulate. In some other genera of dragonflies (e.g., *Aeshna* and *Libellula*), when the sex of an approaching member of the same species is not recognized, males will attempt to copulate with males. The tendency for males to avoid such encounters results in their spreading out into individual territories. The suggestion has been made that such sexual encounters between males preceded the evolution of territorial behavior (Manning, 1966). Male crickets are territorial; when one male enters another's territory, he is greeted by the "rival" song of the other. This is followed either by the exit of the intruder or holder of the territory or by fighting. A rank order may become established among males whose territories are in close proximity to one another. A female entering a male's territory is also greeted with the "rival" song, and will either leave or remain quietly. Price (1975) provides a list of examples of territoriality in several insect orders.

Competition in mating often occurs even after insemination (Matthews and Matthews, 1978). Since sperm are commonly stored in a spermatheca, it would seem likely that more than one male could successfully inseminate a given female. A number of mechanisms that help prevent additional inseminations have evolved. For example, male dung flies "protect" females from the copulation attempts of other males (Fig. 8-8). Other examples include the tandem flight of dragonflies (Odonata); prolonged copulation characteristic of the love bug, *Plecia nearctica* (Diptera, Bibionidae); and mating plugs formed in the genital tract of the female from secretions of the male accessory glands.

Males of some species—certain *Drosophila* (Diptera, Drosophilidae), dragonflies, and others—congregate and display at a particular place. These aggregations of males, called *leks,* are presumably more effective in attracting females than isolated males.

Oviposition. The survival, growth, and development of immature insects depends to a great extent on oviposition in an appropriate environment. This is especially important to insects with specific diets (e.g., a particular plant, or in the case of many parasitic insects, a particular host). For example, female mosquitoes that are ready to oviposit must do so in a place where the eggs are in water or will eventually be. Although, some insects merely drop their eggs wherever they may happen to be (e.g., some mayflies), more often they are specific as to their choice of oviposition site, locating it by means of a variety of stimuli, depending upon the kind of insect. For example, the beetle *Hylotrupes* (Cerambycidae) is attracted to the terpene odor of the wood in which it deposits its eggs. The parasitic wasp *Nasonia* locates the puparial cases of host blow flies (Diptera, Calliphoridae) by the odor of the decaying flesh in which blow fly larvae and pupae are commonly found. Insects ready to oviposit may respond to stimuli hat previously elicited no response whatsoever. For example, *Pieris* (Pieridae) butterflies show a definite preference for objects with a green color when they are ready to oviposit, but when they are searching for food, they demonstrate no such preference. When an insect is highly selective in the choice of a site for oviposition, this does not mean that it has foreknowledge

A

Fig. 8-8
Copulation and oviposition in the yellow dung fly, *Scatophaga
stercoraria* (Diptera; Anthomyiidae). A. Copulation. Female on
substrate; copulating male above; competitive male partly out of picture.
B. Oviposition with the male protecting the female from other males.
This behavior prevents further mating activity from interfering with
oviposition. [Courtesy of W. A. Foster.]

B

of the needs of its offspring. Oviposition in response to specific
stimuli that are associated with an environmental situation in which
the young can survive has, during the course of evolution, no doubt
been selected for.

Brood Care. Once a parent insect has fulfilled its responsibility
for placing the egg (or larva in some instances) in an appropriate

environmental situation, it may simply leave. However, some continue an association with the eggs and immature stages. This association is most highly developed in the social insects (ants, bees, wasps, and termites) in which the brood form a core around which all activity is centered, and these brood are "reared" from egg to adult by workers. Social insects differ from social vertebrates in that a vertebrate colony is composed of a number of mating pairs and offspring, whereas an insect colony is usually the product of a single female or single male–female pair.

Many nonsocial insects also do more for their offspring than merely deposit the egg. Earwig females (Dermaptera) lay their eggs in burrows in the ground and guard them until they hatch. Female beetles in the genus *Omaspides* protect their brood from the ravages of invading ants. Some nonsocial insects go to the extent of preparing elaborate nests and stocking them with food for their young. For example, a female solitary wasp in the genus *Bembix* digs a nest in the soil with her legs and mandibles. She then captures an adult fly (or sometimes another insect) and brings it to the nest. Evans (1957) describes the prey capture.

> The capture and stinging of the prey occur with great rapidity. The female wasp proceeds slowly through the air or hovers over a source of flies, pouncing upon the flies either in flight or at rest; then she descends to the earth or to some other solid object, where she quickly bends her abdomen downward and forward, inserting the sting on the ventral side of the thorax or in the neck region of the fly. . .

Following capture, the wasp returns to the nest and deposits an egg on the fly. Thereafter, a number of flies are brought to the nest as the larva grows. Eventually, the larva spins a cocoon, pupates, and emerges from the ground as an adult. Depending on the species and the time of year, it may remain over the winter in the cocoon.

Insects in Groups

Insects display many gradations between solitary behavior and complex, organized social behavior. However, comparatively few insect species are truly social *(eusocial)*, a few thousand perhaps, and these are found in only two orders, Isoptera (termites) and Hymenoptera (ants, bees, wasps, and relatives).

Matthews and Matthews (1978) provide a useful classification of intraspecific insect associations other than sexual interludes. They divide interactions into aggregations, simple groups, primitive societies, and advanced societies *(eusocial* insects).

Group behavior is thought to provide a number of different benefits; including protection as a result of such things as collective displays and more efficient detection of potential predators, increased efficiency in detection and utilization of food, and moderation of adverse physical environmental factors. Insects that produce defensive secretions and/or display warning coloration no doubt derive increased protection by pooling their defensive capabilities. Wooly apple aphids, *Eriosoma lanigerum* (Homoptera, Eriosomatidae), secrete waxy material, which gives their bodies a whitish, bushy appearance. Distrubance of aggregations of these insects causes them to rhythmically move their bodies at the same time. This collective display gives the impression of a much larger organism. Honey bees are very efficient food users by virtue of their ability to communicate the locations of food sources to one another. During the winter, honey bees form into tight clusters, which provide protection from freezing temperatures.

Many insects form into loose, temporary *aggregations* under certain circumstances. These aggregations may result from common attraction to a particular habitat. Attracting stimuli include a particular temperature, light condition, food, and particular chemicals. Aggregations may also result from mutual attraction independent of external conditions. Examples of aggregations are hibernating groups of ladybird beetles (Coccinellidae) commonly found in human habitations during the winter; cockroaches, which are often found in feeding aggregations; and nighttime clusters of usually conspecific (members of the same species) bees or wasps.

Simple groups are characterized by coordinated movements. Examples are migrating groups of butterflies and locusts.

Groupings in the *primitive society* category range from simple parental care of offspring beyond oviposition (*brood care*) to interactions that border on true social behavior. Reciprocal communication is typical, and cooperation in nest construction and defense is common. Several examples of brood care and nesting behavior are given in the preceding section. Additional examples of communal nesting include webspinners (Embioptera), which construct and inhabit networks of silken tunnels, and tent caterpillars (Lepidoptera, Lasiocampidae), which cooperate in the construction of silken shelters in which they pass the night. During the daytime, tent caterpillars forage along the branches of the host tree. They communicate by depositing chemical trails in association with feeding success. Other individuals follow these trails preferentially.

Three traits are generally agreed upon as characterizing eusocial insect behavior (Wilson, 1975).

1. Conspecific individuals cooperate in brood care.
2. There is a division of labor based on reproduction; that is, one or more fecund individuals reproduce while essentially sterile individuals serve as a labor force.
3. At least two generations serve in the labor force; that is, offspring assist parents during part of their life.

Behavior that includes only one or two or these characteristics is viewed as *presocial*. Many of the examples of primitive societies given earlier can be viewed as presocial.

A requisite for the development of social behavior has been the evolution of systems of communication between the members of a society. In most cases communication in social insects involves chemical signals, although other sensory cues may also be important. Communication makes possible complex levels of behavior involving groups of individuals, such as construction of complex nests. However, these complex patterns of behavior result from the integration of relatively simple individual behavior patterns, which are evident in the behavioral repertoires of solitary and presocial insects. Wilson (1975) outlines nine categories of responses to communicatory signals found in social insects.

1. Alarm; 2. Simple attraction (multiple attraction = "assembly"); 3. Recruitment, as to a new food source or nest sites; 4. Grooming, including assistance at molting; 5. Trophallaxis (the exchange of oral and anal liquid); 6. Exchange of solid food particles; 7. Group effect: either increasing a given activity (facilitation) or inhibiting it; 8. Recognition, of both nestmates and members of particular castes; 9. Caste determination, either by inhibition or by stimulation.

Communication in social insects is discussed in von Frisch (1967, 1971) and Wilson (1971, 1975).

The members of insect societies are divided into forms specialized for different functions within the colony. Differences among the various specialized forms or *castes* vary from solely behavioral (e.g., many bees and wasps) to both behavioral and morphological. Morphological differences, in turn, range from slight to extreme. Further, the particular role played by a given individual may vary with age, *temporal polyethism*. For example, an individual might be involved with brood care and nest maintenance during the first part of its life and with foraging during the later part (e.g., honey bee workers).

In the broadest sense, a social insect colony may be divided into reproductives and nonreproductives. Reproductives may include one or more fecund females (*queen* or *queens*) and males. One or more *worker* castes are the numerically most abundant nonreproductives. In addition, forms specialized for colony defense, *soldiers*, may be present.

In the termites, there are typically the colony-founding primary reproductives, the *king* and *queen, workers* (immatures in the lower termites and fully differentiated forms in the higher termites), and *soldiers*. The soldiers show a wide array of variation in the structure of the head, especially the mandibles, depending on species. In addition, termites usually have the potential for producing *supplementary reproductives* should the king and queen be removed.

Whereas the termite worker castes are composed of both males and females, those of the social Hymenoptera are all females. Unlike the termites, male hymenopterans contribute little to the colony save insemination of the queen. In the ants, reproductives (in particular, queens), workers, and soldiers are morphologically well defined. In a few species workers are divided into morphological subcastes, but in most they are divided into temporal subcastes (i.e., temporal polyethism). Castes are only behaviorally differentiated in most bees and wasps, but in a few species, for example, the honey bee, *Apis mellifera,* there are clear morphological differences between queens and workers.

Details on the biology of termites and social Hymenoptera may be found in Chapter 11. Caste differentiation is discussed in Chapter 5. Table 8-1 compares some of the aspects of the social biology of termites and social Hymenoptera. Further information on social insects may be found in Andrewes (1971), Evans and Eberhard (1970),

Table 8-1

Comparison of Aspects of Social Biology: Termites vs. Eusocial Hymenoptera[a]

	Differences	
Similarities	**Termites**	**Eusocial Hymenoptera**
1. The castes are similar in number and kind, especially between termites and ants.	1. Caste determination in the lower termites is based primarily on pheromones; in some of the higher termites it involves sex, but the other factors remain unidentified	1. Caste determination is based primarily on nutrition, although pheromones play a role in some cases.
2. Trophallaxis occurs and is an important mechanism in social regulation.	2. The worker castes consist of both females and males.	2. The worker castes consist of females only.
3. Chemical trails are used in recruitment as in the ants, and the behavior of trail laying and following is closely similar.	3. Larvae and nymphs contribute to colony labor, at least in later instars.	3. The immature stages (larvae and pupae) are helpless and almost never contribute to colony labor.
4. Inhibitory caste pheromones exist, similar in action to those found in honeybees and ants.	4. There are no dominance hierarchies among individuals in the same colonies.	4. Dominance hierarchies are commonplace, but not universal.
5. Grooming between individuals occurs frequently and functions at least partially in the transmission of pheromones.	5. Social parasitism between species is almost wholly absent.	5. Social parasitism between species is common and widespread.
6. Nest odor and territoriality are of general occurrence.	6. Exchange of liquid anal food occurs universally in the lower termites, and trophic eggs are unknown.	6. Anal trophallaxis is rare, but trophic eggs are exchanged in many species of bees and ants.
7. Nest structure is of comparable complexity and, in a few members of the Termitidae (e.g., *Apicotermes, Macrotermes*), of considerably greater complexity. Regulation of temperature and humidity within the nest operates at about the same level of precision.	7. The primary reproductive male (the ''king'') stays with the queen after the nuptial flight, helps her construct the first nest, and fertilizes her intermittently as the colony develops; fertilization does not occur during the nuptial flight.	7. The male fertilizes the queen during the nuptial flight and dies soon afterward without helping the queen in nest construction.
8. Cannibalism is widespread in both groups (but not universal, at least not in the Hymenoptera).		

[a]From Wilson (1971).

Free (1977), Goetsch (1957), Herman (1979), Krishna and Weesner (1969, 1970), Michener (1974), Oster and Wilson (1978), Richards (1953), Schneirla (1971), Skaife (1961), Spradberry (1973), Sudd (1967), von Frisch (1950, 1967), Weber (1972), Wheeler (1910), Wilson (1971).

Origins of Social Behavior. Eusociality is considered to have evolved once in the termites and at least eleven times within the Hymenoptera. Although there are many differences between the various groups, there are also many similarities as a result of convergent evolution (Table 8-1).

The termites (especially the "lower" termites) have many traits in common with the cryptocercid cockroaches. Outstanding among these similarities is the fact that the lower termites and cryptocercid cockroaches are the only wood-eating insects that depend on cellulase-producing symbiotic intestinal protozoans. These symbionts are passed from one generation to the next by means of anal *trophallaxis* (trophallaxis is the exchange of alimentary liquids between individuals). This behavior involves interaction between members of overlapping generations and hence rudimentary social behavior. Termites may be viewed as "social cockroaches," bound together originally by the contact required to pass along the symbiotic protozoans and subsequently forming into complex societies.

Two possible evolutionary pathways to eusocial behavior among the Hymenoptera have been envisioned (Wilson, 1975). The *parasocial* sequence (which may have been followed by most social bees) began with *communal* behavior involving cooperation in nest construction, but with separate brood rearing. Subsequently, cooperation in brood care was added (i.e., *quasisocial* behavior). This was followed by the development of a nonreproductive worker caste (*semisocial* behavior). With the development of cooperation of two or more overlapping generations the path to *eusocial* behavior was complete.

The *subsocial* sequence of evolution of eusocialism has been envisioned for ants, social wasps, and a few species of social bees (and also applies to termites as described above). In this pathway, the female first remained with brood for a time following oviposition, but departed before eclosion (*primitively subsocial*). Subsequently, the female remained with a newly hatched brood, and an overlap in generations occurred (*intermediate subsocial I*). In the next stage, the first generation offspring aid in the rearing of the next generation of brood (*intermediate subsocial II*). Differentiation into reproductive and nonreproductive castes then led to complete sociality.

The "prime mover" in the evolution of social behavior among Hymenoptera is considered to be *haplodiploidy* (Hamilton, 1964; Trivers and Hare, 1976; Wilson, 1975). The term *haplodiploidy* denotes that female Hymenoptera arise from fertilized eggs and are diploid, whereas males arise parthenogenetically and are haploid (see pages 158–159). As a result of haplodiploidy, female Hymenoptera are more closely related to their sisters than to their daughters. That is, on the average, sisters share 75% of their genes with each other but only 50% with their daughters. This fact has been advanced to account for the tendency for Hymenopteran societies to be characterized by *altruistic behavior* between sisters. An example of altruistic behavior is a honey bee "sacrificing" its life to protect the hive. When a honey bee stings, the barbed sting apparatus remains embedded in the victim and is ripped away from the bee's body; the bee subsequently dies. The notion of altruistic behavior is linked to *inclusive fitness,* a central concept in the rapidly developing area, *sociobiology* ("the systematic study of the biological basis of all social behavior"—Wilson, 1975). *Fitness,* in Darwin's original view, is a measure of the reproductive success of an individual, that is, an individual's relative contribution of genes to the next generation. Inclusive fitness is more encompassing than Darwinian fitness in that the contribution of an individual to the reproductive success of close relatives (*kin*) is considered in addition to an individual's own reproductive success. In this view, then, it is possible for an individual to be fit in the evolutionary sense without reproducing directly. Thus, the honey bee worker is not really making such a "sacrifice" after all since the act of apparent altruism actually enhances the reproductive success of her younger sisters, with whom she shares most of her genes and from among whom future queens will arise.

The study of insect societies is playing a major role in the development of sociobiology. For more information on this exciting topic see Barash (1977), Dawkins (1976), Hamilton (1964), Ruse (1979), Trivers and Hare (1976), and Wilson (1975).

Selected Reverences

GENERAL ETHOLOGY

Alcock (1975); Brown (1975); Dethier and Stellar (1970); Eibl-Eibesfeldt (1970); Hinde (1969); Klopfer and Hailman (1967); Marler and Hamilton (1966).

GENERAL INSECT ETHOLOGY

Carthy (1965); Markl (1974); Matthews and Matthews (1978); Richard (1973).

INNATE BEHAVIOR

Birukow (1966); Fraenkel and Gunn (1961); Jander (1975).

LEARNED BEHAVIOR

Alloway (1973); Eisenstein (1972); Erber (1975); Menzel and Erber (1978); Wells (1973).

PERIODICITY

Brady (1974); Bünning (1967); Corbet (1966); Danilevskii (1965); Saunders (1974, 1976a, b).

CONTROL OF BEHAVIOR

Nervous: Barton Browne (1974); Benzer (1973); Hotta and Benzer (1972); Howse (1975); Hoyle (1970); Huber (1974); Miller (1974b); Roeder (1967, 1970, 1974); Wilson (1972).
Endocrine: Barth and Lester (1973); Truman (1973); Truman and Riddiford (1970, 1974).
Genetic: Ehrman and Parsons (1976); Ewing and Manning (1967); Fuller and Thompson (1960); Hirsch (1967); McClearn and De Fries (1973); Rothenbuhler (1967).

COMMUNICATION

General: Frings and Frings (1977); Montagner (1977); Otte (1974); Sebeok (1968, 1977); Smith (1977); Wilson (1975).
Specific insect groups: Ewing (1977); Hölldobler (1977); Ishay (1977); Lindauer (1974); Otte (1977); Schneider (1975); Siberglied (1977); Stuart (1977); Wilson (1965).

The Biological Functions of Behavior

Acoustical: Alexander (1967, 1975); Haskel (1974); Ishay (1977); Otte (1977); Swartzkopff (1974).

Visual: Hailman (1977); Lloyd (1971, 1977).

Chemical:

General: Beroza (1970); Ebling and Highnam (1970); Pasteels (1977); Roelofs (1975a); Shorey and McKelvey (1977); Wilson (1965, 1970, 1975).

Pheromones: Birch (1974); Blum and Brand (1972); Jacobson (1965, 1972, 1974); Karlson and Butenandt (1959); Law and Regnier (1971); Roelofs and Cardé (1977); Seabrook (1978); Shorey (1973, 1976, 1977 a, b); Weaver (1978a); Wilson (1963).

Allomones and Kairomones: Blum (1978); Brown, Eisner, and Whittaker (1970); Duffey (1977); Rosenthal and Janzen (1979); Weaver (1978b); Whittaker and Feeney (1971).

FEEDING BEHAVIOR

Various aspects: Askew (1971); Brues (1946); Dethier (1966); Jones (1978); Nelson et al. (1975); Smith (1973); Zumpt (1965).

Food Location:

Herbivores: Dethier (1966, 1970); Kennedy (1977); Kogan (1977); Schoonhoven (1972a); Staedler (1977).

Blood-feeders: Galun (1977a, b); Friend and Smith (1977); Hocking (1971).

Parasitoids: Askew (1971); Vinson (1976).

Control of feeding: Barton Browne (1975); Gelperin (1971).

ESCAPE AND DEFENSE

Blum (1978); Duffey (1977); Eisner (1970); Eisner and Meinwald (1966); Reichstein et al. (1968); Roeder (1965, 1966, 1967); Rothschild (1972).

BEHAVIOR AND THE FLUCTUATING ENVIRONMENT

Dispersal and migration: Dingle (1972, 1978a); Johnson (1963, 1969, 1974, 1976); Kennedy (1975); Kring (1972); Rainey (1976); Schneider (1962).

Dormancy: Dingle (1978a); Lees (1955); Mansingh (1971); Müller (1970); Tauber and Tauber (1976); Thiele (1973).

Behavior, temperature, and humidity: Bursell (1974a, b); Heinrich (1974); May (1979).

REPRODUCTIVE BEHAVIOR

Manning (1966) (also see references cited under ''Communication'').

INSECTS IN GROUPS

Andrewes (1971); Evans and Eberhard (1970); Free (1977); Goetsch (1957); Hermann (1979); Krishna and Weesner (1969, 1970); Michener (1974); Oster and Wilson (1978); Richards (1953); Schneirla (1971); Skaife (1961); Spradberry (1973); Sudd (1967); von Frisch (1950, 1967); Weber (1972); Wheeler (1910); Wilson (1971).

SOCIOBIOLOGY

Barash (1977); Dawkins (1976); Hamilton (1964); Ruse (1979); Trivers and Hare (1976); Wilson (1975).

Insects and Their Environment

The study of organisms in relation to their environment is the modern science of ecology, which has developed in recent years from the earlier, essentially descriptive endeavor, natural history. The inclusion of this chapter under the general heading "Structure and Function" is consistent with Odum's (1971) comment that "it is more in keeping with the modern emphasis to define ecology as the structure and function of nature."

A number of general ecology texts are available (see Selected References). Price (1975) deals specifically with insect ecology. Andrewartha and Birch (1973) trace the historical development of insect ecology.

A basic functional unit in ecology is the *ecosystem,* which is composed of a *community* of living (biotic), interacting organisms (plants, animals, and microorganisms) and the nonliving (abiotic) components of their environment. Ecosystems vary in size and complexity, ranging anywhere from a woodland pond to the entire *biosphere* (that portion of our planet in which living forms exist). The biotic part of an ecosystem is made up of producers, consumers, and decomposers. *Producers* are green plants, which synthesize food (carbohydrates and other compounds) from carbon dioxide and water by utilizing energy derived from solar radiation (i.e., *photosynthesis*). *Consumers* are those organisms, the vast majority of which are animals, that obtain energy and molecular building blocks by ingesting organic matter from plants (*herbivores* and *detritivores;* see "Feeding Behavior" in Chapter 8) or from other consumers (*carnivores* and *detritivores;* Chapter 8). *Decomposers,* mainly bacteria and fungi, break complex organic molecules down to simple inorganic nutrients usable by producers, facilitating the continuous cycling of inorganic nutrients throughout an ecosystem.

The Life-System Concept

Clark et al. (1967) introduce a particularly useful concept for our purpose, the concept of the life system. A life system is "that part of an ecosystem which determines the existence, abundance, and evolution of a particular population." In other words, the life system of an insect is that part of the environment (*effective environment*) that directly influences the fate of a given population plus the population itself. Obviously not all parts of an ecosystem necessarily directly influence a given animal population, although the situation becomes rather hazy when one considers indirect influences. The essential components of a life system are presented in Figure 9-1. In view of the life system concept, an ecosystem becomes a series of interlocking life systems.

The following sections deal with various aspects of the major components involved in the functioning of a life system, i.e., population and effective environment.

Populations

Members of a given insect species are typically separated into more or less discrete groups or *populations*. Or a population may be a group of individuals of the same species delineated, for purposes of study, by a biologist. Like individuals, populations have characteristics that can be measured and described: for example, *genetic composition* (the individuals of most natural populations vary phenotypically and genotypically), *sex ratio* the proportion of females to males), *age composition* (most populations are composed of adults and immatures of varying ages), arrangement in space or *dispersion* (most natural populations tend to be composed of clumps of indi-

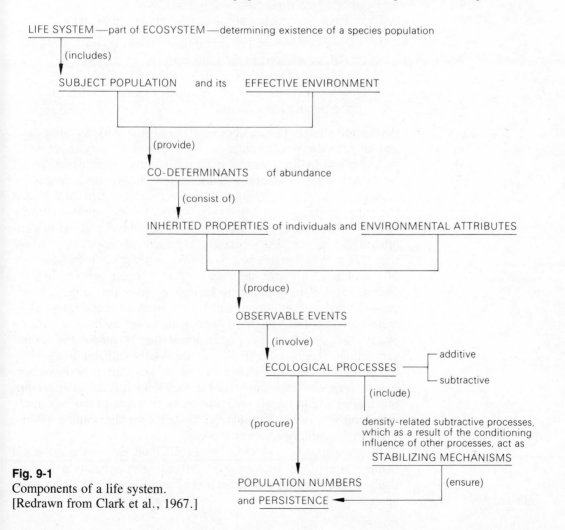

Fig. 9-1
Components of a life system.
[Redrawn from Clark et al., 1967.]

viduals due to irregularities in the distribution of food and shelter and the fact that individuals often attract one another), *population size* (expressed as total number of individuals), *population density* (the number of individuals per unit area or volume), *biomass* (the total weight of a population), and *dynamics* changes in numbers and/ or density over time).

It has been long recognized that organisms have the capacity under ideal conditions to increase by a geometric progression (*exponentially*). Since this capacity for exponential increase (*biotic potential*) is never fully realized under natural circumstances, it is clear that there are factors in the environment (predators, limited resources, etc.) that act to prevent such increases. The collective action of these factors has been called *environmental resistance*. In the broadest terms, one may view the dynamics of a given population as representing the outcome of the interaction between biotic potential and environmental resistance.

It is the goal of population studies to understand the composition and behavior of populations and ultimately to be able to make predictions. Such information is essential to the development of a detailed understanding of evolution and further of paramount practical importance when we consider pest organisms.

Three basic approaches are used to study populations.

1. Analysis of population performance under controlled, but artificial, conditions in the laboratory.
2. Evaluation of populations in the field.
3. Development of theoretical, mathematical models that describe population dynamics

Population studies in the laboratory are of necessity too simplified and only remotely representative of what must actually occur under the complexities characteristic of field conditions. On the other hand, the complexities of natural conditions make it difficult to assess the action of particular environmental factors. Further, such studies must rely on *sampling*—studying a comparatively small number of individuals from a population with the hope that the characteristics exhibited by a sample accurately represent those of the whole population. Whether this hope is realized depends upon the adequacy of sampling. Sampling methods vary with insect species, habitat, and the kind of information to be derived from the sample. Many sophisticated sampling techniques have been developed, and elaborate statistical analyses have been made easier by the use of computers. It is necessary that population studies be made over several generations. Theoretical studies are especially valuable in developing descriptive and predictive models of population performance; however, as with laboratory studies they tend to represent oversimplifications and must rely on assumptions that may or may not apply in natural situations. Fortunately, modern computers allow evaluation of increasingly complex models.

All three approaches to studying populations have some value and definite limitations. Increased understanding of populations will no doubt result from an amalgamation of ideas generated by all three approaches.

For information on population biology, see Clarke et al. (1967), Price (1975), Solomon (1969), Southwood (1966, 1975), Varley et al. (1973), Wilson and Bossert (1971), and Wilson (1975).

Relationship Between Environmental Components and Populations

Population Growth.   This section establishes a simple, but useful model, which helps in thinking about population growth. Assume we are evaluating the repopulation of an area from which all members of a given species have been previously removed. Since populations under ideal conditions have the capacity for exponential increase, the first step is to consider an equation that describes exponential increase. Such increase in a population after time t is described by the equation $Nt = N_0 e^{rt}$, where Nt = population size after time t; N_0 = initial population size; e = base of natural logarithms (a constant; approximately 2.7183); and r = the instantaneous rate of increase per individual in the population, which is the difference between the instantaneous birth rate b (*natality*) and death rate d (*mortality*). Thus $r = b - d$. If b is greater than d, then the population is growing; if b is equal to d, the population is stable ($r = 0$); if b is less than d, the population is decreasing in size. The graph depicted in Figure 9-2A shows the shape of a curve that would be generated by the exponential population growth equation if r remained at a constant positive value. The population size increases in an increasing rate.

Under natural conditions, r does not remain constant, being subject to strong environmental influence. For example, r may well decrease with increasing population density. Competition for food and shelter, rates of predation, and invasion by parasites and parasitoids are examples of environmental factors that may increase as population density increases. Looking again at the equation for exponential growth, but with r decreasing at a constant rate as population size increases, one finds that a curve like that shown in Figure 9-2B would be generated. In this case, rate of population growth decreases gradually and approaches a point where it levels off. The population size at which the leveling off occurs is called the *carrying capacity* (K) of the environment. This curve is commonly called the *logistic model* of population growth.

Although the logistic model may come close to describing population growth in some cases, it is severely limited because the

Fig. 9-2
Population growth curves. A. Exponential population growth. B. Logistic population growth.

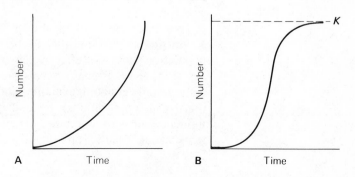

assumptions upon which it is based are rarely true in reality. Among these assumptions are equality of reproductive potential among all individuals of the population, an even age distribution with a constant proportion of individuals breeding all the time, reproduction uninfluenced by climate or other factors, and no changes in carrying capacity (Price, 1975).

In addition to birth and death rates, emigration and immigration also influence population growth. Birth rate, death rate, and emigration-immigration rates are themselves influenced by a variety of factors. For example, weather, in particular temperature, influences all these rates. In addition, it can be argued that any factor that exerts an influence on one of these rates influences the others as well. For example, an environmental change, say in temperature, might cause higher mortality among insects in one stage than in another or in insects of a certain age. If this age-specific mortality happened to occur in ovipositing females, one could say that both the birth and death rates were being affected.

Among the major factors that influence birth rate are average fecundity of the females, average fertility of the females, and the sex ratio. Average fecundity represents the average number of offspring that would be produced by each female under ideal environmental conditions; average fertility represents the average number of offspring actually produced, the difference between fecundity and fertility being related to such environmental factors as quality and quantity of available food, weather components, and population density. The sex ratio is the fraction of the total population that is female. In most insects the sex ratio is 0.5, but in species where parthenogenesis occurs there may be drastic deviations from this 50–50 balance of sexes. As with fecundity and fertility, the sex ratio may be influenced by several environmental factors.

The death rate is affected by such factors as adverse weather conditions, temperature extremes in particular; predators, parasites, and pathogens; accidents; low vitality; food shortage; and lack of adequate shelter (Clark et al., 1967).

The carrying capacity (K) of the environment also varies with many factors (e.g., availability of food, extent of predation).

More Regarding the Life-System Concept. Several different models have been proposed to explain the numerical behavior of insect populations. As previously mentioned, Clark et al. (1967) provide a useful integrating concept, i.e., the "life-system" (see Fig. 9-1). Our objective here is to elaborate somewhat on this concept.

Two kinds of "ecological events" occur as a result of the interaction of the "codeterminants of abundance" (i.e., the "subject population" and its "effective environment"): *primary* and *secondary* (Fig. 9-1). Primary events are those directly involved in the demographic equation explained above (i.e., births, deaths, emigrations, and immigrations). Secondary events are those that exert an influence on the magnitude, extent, frequency, or duration of primary events (quality and quantity of available food, weather components, etc.). Primary and secondary ecological events may be additive or subtractive. Processes that act in a positive way on a population are additive, for example, immigration and weather conditions that reduce populations of natural enemies. The planting of a monoculture, such as corn, is an additive process relative to species, such as the European corn borer, that thrive on this crop, as well as to insects that are parasitoids or predators of the corn-eating species. Subtractive processes (e.g., emigration and adverse weather conditions) have a negative effect on a population, causing the death of individuals and/or a decrease in the number of progeny produced. The planting of corn would constitute a subtractive process for insects that were originally present, if they could not survive in a corn agroecosystem.

Additive and subtractive ecological events may be either density-independent or density-related (i.e., the effects of the action of a given ecological process on a population may or may not be related to the density of that population). Subtractive density-related processes are of special interest since they may act, in some instances, as regulators of the level of abundance; that is, as a given population increases, a point is reached where further growth is inhibited by some factor.

Long-Term Numerical Changes.   Over the long term, as populations interact with their effective environments, there are fluctuations in population size. Some populations fluctuate irregularly, apparently in response to changes in environmental components such as food supply and weather. Other populations tend to fluctuate regularly about a mean level of abundance. As mentioned above, this may be a reflection of the action of some subtractive density-related process that is serving to regulate the level of abundance. Insects with a reproductive period restricted to a particular time or times of year tend to display regular, seasonal population peaks associated with reproduction.

It has become more and more apparent that, in a given life system, certain "key" ecological processes with age-specific effects determine major population trends, while other processes exert somewhat lesser effects. Elaborate "multifactor" studies have been carried out for a number of insect species. The objective of these studies has been to derive "life tables" that contain information as to population densities at different times during the life cycle and age-specific information regarding the "key" ecological processes that account for mortality. Such a life table is presented in Table 9-1. The development of a truly representative life table requires the sampling of a population over many generations. Another approach that has been used where preliminary studies have indicated a few key influences involved in a life system in the *key-factor method*. In this type of study, population density is measured at only one point in each generation. This approach requires much less sampling and enables investigators to concentrate on critical periods of time in the life cycle when the "key" process or processes are active without spending much time on other periods.

Table 9-1
Life Table for Second Generation of the Diamondback Moth on
Early Cabbage, Merivale, Ont., 1961[a]

	Numbers per 100 Plants	Mortality		
Age Interval		Causative Factors	Per 100 Plants	Per Cent
Eggs	1580	Infertility	25	1.6
Larvae				
Period 1	1555	Rainfall	1199	77.1
Period 2	356	Rainfall	36	10.1
		Parasitism by *M. plutellae*	52	14.6
Period 3	268	Parasitism by *H. insularis*	69	25.7
Pupae	199	Parasitism by *D. plutellae*	92	46.2
Moths	107	Sex (49.5% ♀♀)	1	1.0
Females × 2	106	Photoperiod	78	73.6
"Normal" females × 2	28	Adult mortality	20	71.4
Generation totals			1572	99.5

Trend index,[b] 0.55. Stable number rate,[c] 99.1%

[a] From Harcourt and Leroux (1967).

[b] Trend index $= \dfrac{\text{eggs laid in new generation}}{\text{eggs laid in old generation}}$

[c] Stable number rate = the mortality rate necessary for the population to remain the same size from generation to generation. Since the percent mortality for the generation represented in this table was 99.5%, it is apparent that the population was declining.

Environmental Components

Various components of the environment that have been found to be significant in their influence on insect or other animal or plant populations are discussed in this section. Any combination, or all of these components, may at some time act to varying degrees upon a given population. When one of them approaches or exceeds the species-specific limit of tolerance, whether expressed in terms of survival, development, fecundity, and so on, it becomes a limiting factor. That is, it becomes the environmental component that is directly responsible for limiting the extent of survival, degree or rate of development, fecundity, and so on. In the field it is often very difficult to pinpoint a given environmental component as a limiting factor since, obviously, several components are likely to be interacting to produce the "effective environment" of any population. For the same reason, it is difficult to relate studies of environmental effects under the carefully controlled conditions of a laboratory to what actually occurs in the field.

Another problem further complicates the picture. Environmental components, in particular temperature, moisture, and light, are not uniform throughout an ecosystem. For example, the temperature of one part of a plant may be quite different from that of another part. Thus, at least in many instances, to obtain a truly accurate picture, one must measure the temperature in specific parts of an ecosystem (e.g., bottom side of a leaf). Realization of this problem has led to the development of the concept of "microenvironment," which implies the recognition of distinct horizontal and vertical (spatial) and temporal differences in environmental components within an ecosystem. Figure 9-3 portrays such differences in terms of relative humidity and water vapor pressure at different levels above the ground (vertical) and at different times during the 24-hour cycle (temporal).

Environmental components may be divided into four basic groups as follows (Andrewartha and Birch, 1954; Rolston and McCoy, 1966): (1) weather; (2) food; (3) a place in which to live; and (4) other organisms.

Weather

Weather results from the combined action and influence of all the physical factors of the environment at any given time. It varies continually throughout days, weeks, months, and years and exerts

Fig. 9-3
Changes in relative humidity (A) and water vapor pressure (B) at different levels above the ground and at different times of day. [Redrawn from Chauvin, 1967.]

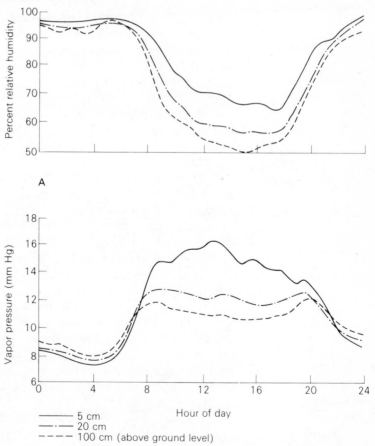

an influence on insect abundance, longevity, rate of development, and so on, from one year or season to the next. Climate, on the other hand, is the average course or condition of the weather in a locality over a period of several years. Weather changes rapidly, often violently, while climate tends to remain pretty much the same or change very slowly over a period of many years. The main elements of weather are temperature, moisture, and light, although several other physical environmental factors are known or thought to exert a degree of influence on insects. Various materials and methods used in measuring and interpreting physical aspects of the environment are described in Platt and Griffiths (1964). Insect behavior in relation to changes in the environment is discussed in Chapter 8.

Temperature. Insects are basically poikilothermic—that is, their body temperature tends to be the same as ambient temperature. However, this does not necessarily mean that an insect's body temperature is always the same as that of the environment.

For every insect species there is a fairly well-defined range of temperature within which it is able to survive. Exposure to temperatures above the high or below the low extremes of this range results in death. The range of tolerable temperatures varies from species to species, within a species, and with the physiological state of an individual. Thus there are times or stages during the life cycle when an individual may be able to survive exposure to high or low temperatures that at other times would kill it. For example, many insects are able to survive much lower temperatures in the fall and winter than in the spring or summer. Tropical species are generally less tolerant of cold than those in temperate zones. Terrestrial insects usually have a somewhat wider range of temperature tolerance than do aquatic insects, and not surprisingly the range of temperature variation in terrestrial habitats is usually substantially greater than that in aquatic habitats.

The range of survival relative to temperature for most insects probably lies somewhere between 0 and 50°C, although it is likely that no one species can thrive throughout this entire range. There are, however, exceptional species that are able to survive at temperatures well beyond these extremes. For instance, some dipteran larvae apparently thrive at temperatures of 55°C or higher, while certain species of beetles go through their entire life cycle in ice grottos at temperatures slightly below and slightly above 0°C (Andrewartha and Birch, 1954). The firebrat, *Thermobia domestica,* can live indefinitely at temperatures of 42°C and higher (Andrewartha and Birch, 1954). It typically inhabits such places as ovens, hot-water pipes, and similar "hot" environments. If insects of a given species are exposed to a temperature gradient, they will move until they reach the "preferred temperature," at which point they will tend to congregate. Near the upper and lower tolerable limits of temperature, insects become dormant. In the range between these limits, they are active.

The actual cause or causes of death at the limits of the temperature range are not clear. At the lower limit the submicroscopic structures of cells may be disrupted by the formation of ice crystals or the metabolic balance may be thrown off. At the upper limit protein denaturation, metabolic imbalance, disruption of ordered molecules (as, for example, in the wax layer of the epicuticle), and desiccation are likely to be involved.

The phenomenon of insect cold-hardiness has been reviewed by Salt (1961), Asahina (1966), and briefly by Baust and Morrissey (1977). Downes (1965) discusses adaptations of insects in the Arctic. Many insects that become dormant in temperate regions are able to survive low temperatures for considerable periods of time. Most of these insects are dormant in a stage that is more cold-hardy than the preceding one. Some are capable of long exposure to low temperatures and display a certain amount of cold acclimation, but succumb to freezing of the body fluids. For example, *Aedes aegypti* larvae reared at 30°C are killed by exposure to −0.5°C for 17 hours, but survive such an exposure if preconditioned for 24 hours at 18 or 20°C (Bursell, 1974).

Other insects are able to avoid freezing because they can withstand supercooling (i.e., being cooled below the point of freezing, but without freezing actually taking place). Individual species vary as to the temperature to which they can supercool. Cryoprotective compounds such as glycerol, sorbitol, and erythritol have been found in the tissues of overwintering insects (Baust and Morrissey, 1977). Salt (1959) showed that glycerol depresses the hemolymph freezing point of *Bracon cephi* larvae as much as 17.5°C and that this compound plays a key role in cold-hardiness. Glycerol begins to appear in the tissues in correlation with the advent of dormancy (see "Behavior and the Fluctuating Environment" in Chapter 8) and usually disappears rapidly when the insects come out of this state (Salt, 1961). A number of insects (e.g., many caterpillars) are able to survive the formation of ice within their bodies. Members of this group of "freezing-tolerant" insects have been found to possess higher concentrations of cryoprotective compounds in their hemolymph. Apparently these compounds act to prevent tissue destruction by influencing how and where ice crystals form as well as depressing the hemolymph freezing point.

The lethal effects of high temperatures are difficult to study because of the influence of moisture changes. The interaction of moisture and temperature is considered later. It has, however, been established that lethal temperatures may vary depending upon the temperature to which an insect or, more properly, a sample of a population of insects has been exposed previously. Such high temperature acclimation has been demonstrated in *Calliphora and Phormia* (Diptera, Calliphoridae), and others (Bursell, 1974). Acclimation to both high and low temperatures may or may not occur in the same species. Like cold acclimation, high-temperature acclimation is likely to be of value to insects under natural circumstances, since seasonal and daily high temperatures are usually preceded by a gradual transition from somewhat lower temperatures. Acclimation at both ends of an insect's tolerable range of temperatures may be looked upon as promoting the insect's survival against the effects of extreme daily and seasonal temperature fluctuations.

Temperature also affects the duration of life. For example, a clear relationship exists between the duration of life in male and female *Drosophila subobscura* and temperature (Bursell, 1974).

Temperature may also affect survival by influencing the rate of utilization of food reserves when food is present in limited quantities (Bursell, 1974). This is particularly evident in blood-sucking insects such as the tsetse flies, *Glossina* spp. (Diptera, Muscidae), which depend on stored food reserves between meals. Increases in temperature shorten the survival period between meals. Obviously, if a fly uses up its reserves from one meal before it is able to obtain another, it will perish.

Temperature exerts a strong influence on the reproduction and rate of development of insects. As with lethal limits of temperature, insects also have definite tolerable ranges of temperature in terms of reproduction and development beyond which neither will occur. For example, the temperature range in which the eggs of the beetle *Ptinus* (Coleoptera, Ptinidae) will develop is between 5 and 28°C (Bursell, 1974). *Pediculus* (Anoplura, Pediculidae) fails to lay eggs below 25°C (Wigglesworth, 1972). For a given species the range in which development will occur is probably somewhat broader than that in which reproduction will be successful. Within the tolerable ranges, egg development, oviposition rate, and the rate of larval and pupal development usually increase with increasing temperature. For example, the duration of pupal life of the mealworm beetle, *Tenebrio mollitor* (Coleoptera, Tenebrionidae), is decreased by 180 hours (from 320 to 140) as the temperature is increased by 12°C (from 21 to 33°C) (Wigglesworth, 1972).

The specific ranges of tolerance and influence of temperature on rates of reproduction and development vary among members of the same species. In addition, the span of ranges varies from species to species. Within each range of tolerance is an optimum zone in which the rates of reproduction and development are maximal. For example, the oviposition rate of *Toxoptera graminum* (Hemiptera-Homoptera, Aphididae) increases with increasing temperature to a maximum of approximately 25°C and then falls off. Most laboratory studies of the effects of temperature on reproduction and development have been carried out at constant temperature. The results of such studies do not necessarily reflect what would occur under uncontrolled field conditions. Periodically fluctuating temperature, which is characteristic of field conditions, tends to induce higher rates of development than would occur at constant temperature. For example, "grasshopper eggs kept at a variable temperature showed an average acceleration of 38.6% and nymphs an acceleration of 12%, over development at comparable constant temperature" (Odum and Odum, 1959).

Figure 9-4 illustrates the idea of tolerable temperature ranges relative to survival, mobility, development, and reproduction.

The distribution, horizontal and vertical, of an insect species is often greatly affected by temperature. In temperate zones the northern extreme of a given insect's distribution is commonly determined by low-temperature extremes. When northern limits are determined in this manner, there is usually a zone somewhat below the extreme limits in which the overwintering stage is killed but which is repop-

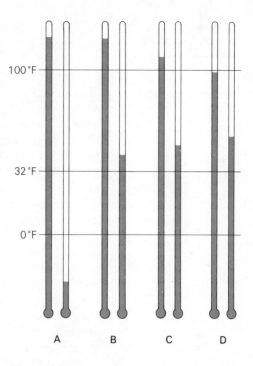

Fig. 9-4
Tolerable temperature ranges for various aspects of
a hypothetical insect's life. The left-hand
thermometer of each pair registers the maximum
tolerable temperature; the right-hand one, the
minimum tolerable temperature. A. Survival. B.
Mobility. C. Development. D. Reproduction.
[Redrawn from Rolston and McCoy, 1966.]

ulated during the warm season. Proceeding southward from such an
area, a greater and greater percentage of the overwintering individuals survive. For example, the corn earworm, *Heliothis zea* (Lepidoptera, Noctuidae), in eastern North America must become
completely reestablished during the warm season each year in Canada and to a progressively lesser extent in the United States proceeding southward (Andrewartha and Birch, 1954). The ability to
become cold-hardy also plays an important role in determining the
northernmost limits of distribution. The number of generations per
year often varies between the northern and southern regions of an
insect's range. For example, the corn earworm completes a generation approximately every 36 days as long as weather permits (Rolston and McCoy, 1966). Thus in its northern range it may only
complete two or three generations per year, whereas in the southern
portion it may complete several more, breeding the year round. One
would expect a gradation in the actual number of generations between the northern and southern extremes of the range.

Temperature also affects, to a greater or lesser extent, the rate of
dispersal of an insect species.

To this point we have been discussing the various ways temperature influences insects. We now want to consider the actual temperature of an insect and the factors that may influence it. We have
established the idea that the temperature, or any of several other
parameters, may vary substantially even within a small area (e.g.,
different parts of a plant). Likewise, the temperature of an insect in
one part of a habitat may not be anywhere near the same as one in
another and may not be the same as the ambient temperature. On
the other hand, the temperature of an insect under controlled laboratory conditions usually reflects the ambient temperature.

The temperature of an insect depends on the sources of heat gain and loss operating under a given set of circumstances. The major sources of heat gain are solar radiation and metabolic heat. Solar radiation may cause the temperature of an exposed insect to be significantly different from the ambient temperature (Fig. 9-5). The effect of solar radiation on the temperature of an insect is influenced by such factors as size, larger insects being more affected than smaller ones; color, darker colors absorbing more radiation than lighter ones; shape, the more surface directly exposed to radiation, the greater the absorption; and orientation with respect to the sun, some insects orienting such that a large or small amount of body surface is exposed.

Metabolic heat results from the breakdown of complex organic molecules. Part of this energy is stored in the high-energy bonds of ATP; the remainder is released as heat. In the absence of solar radiation, this is the sole source of heat and can be quite significant in heat balance, particularly during flight or in clusters of gregarious forms. Sources of heat loss from an insect include evaporation, convection, conduction, and long-wave radiation. Evaporation of water from an insect has a cooling effect since heat is required to propel a molecule of water from the body surface. Evaporation is the major cause of heat loss in the absence of solar radiation. You will recall from Chapter 2 that the rate of evaporation of water from an insect is dependent partly on the size of the insect. Since smaller insects have a larger ratio of surface area to volume, they have a greater tendency to lose water through evaporation than larger ones. The other causes of heat loss—convection, conduction, and long-wave radiation—are significant in the absence of solar radiation when the temperature differential between an insect and its surroundings is usually greater than in the presence of solar radiation. Air movement may accentuate the heat loss by contributing to the maintenance of a steep gradient between an insect and its surroundings. Dense coverings of hairs and scales, on the other hand, may serve as an insulating layer and retard heat loss.

Fig. 9-5

Effect of radiation on the body temperature of selected insects. A. Thermometer painted black. B. Air temperature. C. *Bombus* (bumble bee). D. *Xylocopa* (carpenter bee). E. *Apis* (honey bee). F. *Anisoplia* (beetle) G. *Asilus* (robber fly). H. Damselfly. I. Butterfly. J. Butterfly. K. Damselfly. Light grey, sunlight; dark grey, shade. [Based on data from Masek Fialla, 1941.]

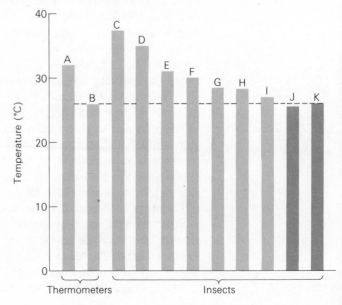

As discussed in Chapter 8, many insects may exert a certain amount of control of heat gain and loss (see the section "Behavior, Temperature, and Humidity".

Moisture. The water content of insects varies from less than 50% to more than 90% of total body weight (Wigglesworth, 1972). Variation occurs both between different species and between different life stages of the same species. Soft-bodied insects such as caterpillars tend to have comparatively large amounts of water in their tissues, whereas many insects with hard bodies (i.e., relatively thick cuticles) tend to have somewhat lesser amounts. Active stages commonly have a higher water content than dormant stages. In most instances it is critical that water content be maintained within certain limits, which are influenced by several other environmental factors (e.g., temperature, pressure, air movement, available surface water). If the limits of tolerance under a given set of circumstances are exceeded, an insect either perishes or many of its activities are seriously impaired.

Environmental moisture factors of major significance are precipitation, humidity, condensation, and available surface water. Rainfall is the most common and widespread form of fluid precipitation. Snow is the most common form of solid precipitation; hail and sleet are less common forms. The annual and seasonal amounts of precipitation are primarily determined by the movements of large masses of air and by topographical characteristics. Thus there are very "wet" and very "dry" regions and innumerable gradations between. Humidity refers to the amount of water vapor in the air and depends on temperature and atmospheric pressure.

Condensation (dew, fog, and white frost) occurs when the atmosphere becomes saturated with water vapor. Saturation is the result of the relative humidity approaching 100% or the temperature dropping below the dew point (the point at which the relative humidity becomes 100%).

Available surface water is related to all the other moisture factors and to the nature of the substrate (soil, leaf surface, bark, and so on). It is well known that soils vary in their water-holding capacity and the rapidity with which water runs off or soaks in.

All these moisture factors influence the water balance of terrestrial insects. What, then, are the environmental factors that influence the water balance of aquatic insects? The "wetness" or "dryness" of an aquatic environment is a function of the osmotic pressure. Thus insects living in fresh water must cope with a comparatively "wet" environment; those in brackish and salt water (e.g., many of the salt-marsh forms, such as the mosquito *Aedes sollicitans*) are living in a very "dry" environment (see "The Excretory System in Chapter 4).

Compounding the water-balance problem in insects is the fact that under other than carefully controlled laboratory conditions the moisture factors continually change. Obviously, an insect living in a given locale must be able to survive the extremes of these changes. As with temperature and the other weather parameters, one must think in terms of microenvironments. For example, the amount of rainfall measured at the edge of a forest hardly reflects the actual amount of water reaching an insect living in a tree hole or on the underside of a leaf. All the moisture factors vary both temporally and spatially. For example, relative humidity varies with location, time of day or year, topography, vegetation, and so on, and commonly tends to be comparatively high during the night and lower during the day. It may also be different at different heights above the ground (Fig. 9-3).

As with temperature, there is an optimal moisture range in which a given species thrives. Mortality may occur under conditions of excessively low environmental moisture content for the active stages of most insects, and under conditions of excessively high moisture content for many. Death under very dry conditions is generally due to excessive water loss. Under conditions of excessive moisture the causes of death are more variable and may be direct or indirect. For example, drowning may be the result of excessive moisture, as is the case with overwintering pupae of the moth *Heliothis zea* during wet years in the southeastern United States (Andrewartha and Birch, 1954).

Survival is indirectly affected by very wet conditions that favor the spread of viral, fungal, and bacterial diseases and by any negative effects the excessive moisture might have on the food of a given insect species. For example (N.A.S., 1969),

> The fall webworm, *Hyphantria cunea,* the armyworm, *Pseudaletia unipuncta,* and the gypsy moth, *Porthetria dispar,* succumb most readily to viruses when the weather is warm and the humidity is high. Whether high humidity affects the host or the development of the pathogen is not clear. However, many viruses do not develop rapidly unless temperatures of 21 to 29.4°C and relative humidities of 50 to 60% are experienced.

If excessive moisture does not kill an insect, it may seriously affect the length of its life. For example, newly emerged adult migratory locusts, *Locusta migratoria,* live longer the lower the humidity (Andrewartha and Birch, 1954). Since environmental moisture content may determine survival, it is commonly a major factor along with temperature in determining the geographic distribution of an insect species. For example, according to Rolston and McCoy (1966),

> The High Plains grasshopper reproduces and develops in an area of about 50,000 square miles, and from this center the adults may spread over an area perhaps twice as large. The region of endemic infestation lies in the short-grass belt of the High Plains where the average winter temperature generally falls between 28° and 38°F and the average annual precipitation ranges from about 15 to 18 inches. That the High Plains grasshopper flourishes in only a part of the short-grass belt indicates that climatic conditions, rather than host availability, limit the range of the insect.

Extremes of environmental moisture content directly influence many of the activities of insects, including feeding, reproduction, and development. Spruce budworm larvae stop feeding when the air becomes saturated with water (Graham and Knight, 1965), and the tsetse fly, *Glossina tachinoides,* does not feed on its vertebrate hosts when the relative humidity is above 88% (Andrewartha and Birch, 1954). Newly emerged adult migratory locusts do not produce eggs below about 40% relative humidity (Andrewartha and Birch, 1954). Generally low humidities adversely affect the rates of oviposition, which increase as humidity increases. The rate of development may be decreased by extremes of moisture content or development may be halted altogether. Under very moist conditions silkworm larvae fail to pupate (Andrewartha and Birch, 1954). According to Bursell (1974) the incubation time for eggs of the spider beetle *Ptinus,* under a constant temperature of 20°C, is 15 days at 30% relative humidity and 10 days at 90%. He further points out that generally higher humidities are more favorable for embryonic stages than low humidities. Two sets of hypothetical curves summarize the points made in this and the preceding paragraphs (Fig. 9-6).

The effects of environmental moisture content are often strongly modified by other weather factors. Temperature and moisture, in particular, interact to a large extent in their effects. Thus temperature exerts a relatively great effect on insects at the extremes of moisture conditions and vice versa. For example, the boll weevil is more

Fig. 9-6
Influence of humidity on various aspects of a hypothetical insect's life. A. Insect not harmed by high humidity. B. Insect adversely affected by high humidity. Zone 1, lethal dryness; Zone 2, favorable moistness; Zone 3, lethal wetness. [Redrawn from Andrewartha and Birch, 1954.]

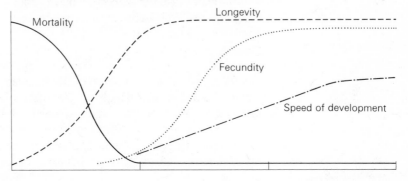

A

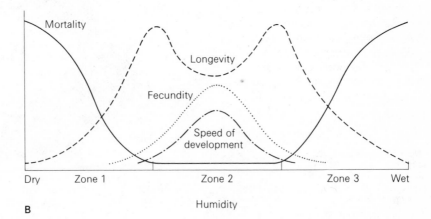

B

tolerant of higher temperatures at comparatively low humidities than at high humidities (Graham and Knight, 1965). Another good example of the interactions of temperature and moisture is their combined effect on the rate of development of insects (Fig. 9-7). As mentioned earlier, air movement, atmospheric pressure, availability of surface water, and other factors may also influence the effects of moisture conditions on insects.

Insects are adapted in a number of ways and to varying degrees to cope with changes in environmental moisture conditions. Dormant stages are often well suited for exposure to drought. The eggs of many species of mosquitoes (*Aedes, Psorophora,* and others) can withstand prolonged drying in air (Clements, 1963). Many insects become dormant in response to different environmental conditions and in this state are able to withstand long periods of drought. For example, the potato beetle, *Leptinotarsa decemlineata,* becomes quiescent in response to dryness and can survive in this resting, desiccated state for months. Such a response ensures that when eggs are laid, the larvae will be exposed to environmental conditions more favorable for survival (Andrewartha and Birch, 1954). Hinton (1951, 1960) describes a chironomid (Diptera, Chironomidae) larva, *Polypedilum vanderplanki,* that is able to survive almost complete dehydration for several years. These insects show no visible signs of metabolic activity and are described as being in a state of *cryptobiosis.*

Most insects offset the influence of dryness and other factors promoting water loss by drinking water or taking it in with food. Time is important in the survival of active insects under adverse moisture conditions. Many can survive extensive desiccation for extended periods. Some insects that live under extremely dry conditions, such as mealworm beetle larvae (*Tenebrio molitor,* Tenebrionidae, Coleoptera), are able to utilize metabolic water (i.e., water that becomes available as a result of the metabolic breakdown of food materials). Other insects, such as some Thysanura (silverfish and relatives), are able to absorb moisture directly from the atmosphere (Wigglesworth, 1972). See Chapter 2 for a discussion of the role of the cuticle in water balance, Chapter 4 for examples of several ad-

Fig. 9-7
Interrelationship of temperature and humidity as they affect the rate of development of a hypothetical insect.
A. Region of most rapid development.
B. Region of favorable development. C. Region of retarded development. D. Region of no development. [Redrawn from Graham and Knight, 1965]

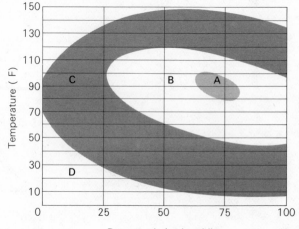

aptations that help insects conserve water, and Chapter 8 for behavior associated with variations in humidity. Cloudsley-Thompson (1975) discusses adaptations associated with arid environments.

Light. Light is of direct important to insects more as an environmental point of reference than as a survival factor since its parameters (photoperiod, illuminance, wavelength, etc.) are more or less constant and it is seldom, if ever, directly lethal under natural conditions. In the nonequatorial region of the earth there is a regular change in photoperiod due to the tilt of the earth's axis of rotation 23½° from vertical to the imaginary plane that passes through the sun and the earth's orbit. Regular change in photoperiod serves as an annual clock for insects and is used by many to maintain synchrony with the seasons and their host plants. Photoperiod is one of the major stimuli that induces diapause (see "Behavior and the Fluctuating Environment" in Chapter 8). The daily cycle of dark and light with crepuscular (dawn and dusk) periods in between also serves as a clock by which the feeding, mating, etc., are regulated (see "Periodicity in Behavior" in Chapter 8). As with temperature and moisture, the reactions of insects to photoperiod and other light parameters vary both among different species and among different life stages of the same species.

Different wavelengths of reflected light are commonly utilized by plant-feeding insects in host location. The position of the sun and degree of polarization of light in different parts of the sky are important to many insects in orientation and navigation. More information on the response of insects to various light parameters may be found in Chapters 6 and 8. The effect of light on both aquatic and terrestrial plants may indirectly influence the activities of insects. For example, the amount of light reaching submerged aquatic vegetation will affect oxygen-generating photosynthesis and in turn affect the oxygen concentration in the water, which then influences aquatic insects.

Other Factors. Other environmental factors that under some circumstances influence insects include currents in air and water, gases dissolved in water, air composition, electricity, and ionizing radiation.

Air and water currents are determined to a large extent by physiographic conditions. Air movement is modified by trees and other vegetation and by anything else that may block or redirect it. Currents in water are influenced by such factors as water volume, slope of stream bed, and temperature differences between the surface and various depths. Wind is a very effective agent in the distribution of insects—aphids, leafhoppers, and others being blown for hundreds or even thousands of miles (see "Dispersal and Migration" in Chapter 8).

Air movement may be directly responsible for the death of insects in two ways. First, severe wind and heavy rain together may cause mortality. Second, movement of air above a surface where evaporation is occurring (e.g., insect cuticle) increases the gradient of water vapor concentration and hence tends to increase the rate of evaporation. Other factors being constant, the rate of evaporation is proportional to air movement. On the other hand, air movement may be beneficial if humidity is high.

Water currents often determine which species of insects will live in a given area. For example, the various genera of mayflies (Ephemeroptera) may be classified into still- or rapid-water forms (Needham, Traver, and Hsu, 1935). The legs and bodies of these insects are appropriately adapted (e.g., legs capable of clinging and streamlined bodies associated with fast-moving water). Black fly larvae fasten themselves to stones or other stationary material in the water. Several caddisflies (Trichoptera) attach their cases to submerged objects. Many aquatic insects (e.g., mosquito larvae) are unable to survive in moving water. Another important aspect of currents in the aquatic environment involves the circulation of dissolved gases, salts, and nutrients.

The insect fauna in a given location are often determined to a large extent by the amount of dissolved oxygen. For instance, caddisfly and mayfly larvae may be found under conditions of relatively high oxygen concentration, midge and black fly larvae (Diptera; Chironomidae and Simuliidae, respectively) at somewhat lower concentrations, and certain mosquito (Diptera, Culicidae) and other fly larvae at very low concentrations (Usinger, 1956). Dissolved oxygen is determined by water movement, wind, water splashing, photosynthesis in aquatic plants, and so on. Other dissolved gases that may be of importance are carbon dioxide (a highly soluble waste product of cellular respiration) and nitrogen.

The gaseous composition of air is remarkably constant and is probably not a significant limiting factor for terrestrial insects and other animals. However, the gas content in cavities in flowers may be quite different from the surrounding air, and a large number of insects live in such cavities (Chauvin, 1967). Further, the composition of air in the middle of a thick canopy of vegetation is not necessarily the same as in open air, nor is it necessarily the same throughout the day since temperature and the photosynthetic activities of plants vary during the course of a day. For example, measurement of the carbon dioxide concentration above fields of wheat and clover at different times of day revealed that carbon dioxide decreased toward the middle of a day and reached a maximum between midnight and 6 A.M. (Chauvin, 1967).

Electrical factors have seldom been taken into account, although they may have a direct effect on insects (Folsom and Wardle, 1934). Under natural conditions, the ionization of air and atmospheric potential may vary and may affect certain activities of insects. For example, the ionization of air has been found to modify the flight activity of a blow fly, *Calliphora* sp. (Chauvin, 1967). An increase in the number of ions causes a temporary increase in activity; for example, flight in *Drosophila* is abruptly reduced for a short period of time by sudden exposures to a potential gradient of 10–62 volts/cm (Chauvin, 1967). Under natural conditions the atmospheric potential falls rapidly between the ground and a few meters above the ground (Chauvin, 1967).

Electromagnetic radiation bombards the earth continuously. Radiation in the visible and near visible region plays many important roles in life processes; for example, it affects photosynthesis and is involved in vision. Radiation of shorter, more energetic wavelengths also influences living organisms. This highly energetic radiation arises from cosmic, solar, and terrestrial sources. Most radiation from extraterrestrial sources is absorbed or reflected by the atmosphere and deflected by the magnetic fields, but a significant amount gets through none the less. Cosmic radiation probably varies considerably from time to time as a result function of changes in the earth's magnetic fields, the occurrence of nearby supernovae (exploding stars) and so on. Such changes in the intensity of cosmic radiation may have played (and may continue to play) a major role in biological evolution by inducing drastic increases in mutation rates at times during the earth's history. Most terrestrial sources of highly energetic radiation occur in trace amounts throughout the earth's crust. However, since the beginning of this century, man has learned how to release immense quantities of radiation by "splitting the atom." The inherent dangers of this activity are well known.

Entomological interest in high-energy radiation stems from

1. The usefulness of radiation as a basic research tool, for example, radioisotopes are used as tracers in the study of biochemistry, and radiation is used to induce physiological, morphological, and genetic changes with the hope of further understanding biological processes.
2. The usefulness of insects as model biological systems in evaluating the harmful effects of radiation.
3. The applied value of high-energy radiation in inducing sterilization or deleterious genetic changes in insect pests (see "Genetic Control" in Chapter 12). For information and literature sources regarding the influence of high-energy radiation on insects, see Grosch (1974).

Food

Insect feeding habits are considered in Chapter 8. This section deals with the ways insects may be influenced by variations in quantity and quality of food resources. As a group, insects are essentially omnivorous. As individual species, however, they exhibit much variation, some being extremely selective and perhaps relying on only one or a few kinds of food (including host plants and animals and prey) and others taking advantage of many kinds of food. Whatever the habits of a given species may be, abundance and quality of food may play an important role in survival, longevity, distribution, reproduction, speed of development, and so on.

Quantity of Food. Animal populations that consume all or most of their food resources are few compared with those that do not (Andrewartha and Birch, 1954). Thus "absolute" shortage of food is probably not an important limiting factor for most populations.

However, such shortages may occur in patches throughout the distribution of a given population. In these circumstances there is an "effective" shortage of food.

There are many possible causes of "absolute" and "effective" shortages of food. Among them are large numbers of individuals per unit quantity of food (intraspecific competition), more than one species consuming the same food material(s) (interspecific competition), species that influence the food of other species without consuming it, and other environmental factors. The first instance might occur, for example, among insects during the actively growing larval stages. When there is a relatively small number of individuals per unit quantity of food, there is enough for all to grow to adulthood. However, as numbers increase, a point is reached where there is insufficient food for all individuals present to grow to maturity; only those that develop most rapidly reach adulthood, the slower developers do not. Obviously, two or more species consuming the same food will reduce the total amount available for any one of them.

Species that do not actually consume the food supply of another may exert a positive or negative influence on that supply. For example, many of the sanitary practices of humans, such as incineration of garbage and treatment of waste materials, destroy the food supply of many insect species, especially cockroaches and flies. On the other hand, failure to carry out these procedures increases the same food supply, enhancing the survival of insects that thrive in such materials. Microbes that are pathogens of an insect that is the prey of another may reduce the population size of the prey species sufficiently to influence the predator species. For example, according to Andrewartha and Birch (1954),

> an epizootic of the fungus *Entomophthora* so reduced the numbers of the caterpillar *Plutella* that the predators (*Angitis* spp. and others) suffered severely from a shortage of food where it had been abundant before the outbreak of disease.

A similar relationship may exist between a plant species, the insects that feed on it, and microorganisms that are pathogenic to members of the plant species. In fact, any environmental factor, biotic or abiotic, that has the effect of reducing populations of particular plants or animals will probably cause the reduction not only of animal populations that utilize these plants or animals for food but also of the parasites and predators of these animals. Monophagous insects are more likely to be affected by such reductions in food supply that polyphagous species.

Examples of causes of "effective" shortages of food include accidental separation from the food source, certain behavioral traits, and effects of feeding activities on animal hosts. It is not difficult to imagine an ectoparasite starving to death when separated from its host. Tsetse flies (Diptera, Muscidae) remain close to tall bushes and will feed on vertebrates that are in the vicinity of these bushes, but will not feed on the same or other vertebrates in open grassland (Clark et al., 1967). Where a single or a few blood-feeding insects might successfully feed to satiation on a host, larger numbers may elicit an avoidance or destructive response on the part of the host, resulting in very few, if any, successfully feeding. Thus, although a host may have ample blood for several hundred tsetse flies, mosquitoes, and so on, it may tolerate the feeding activities of only a small fraction of that number.

An absolute shortage of food could result from the food plant or animal being completely destroyed within a localized region, but it is unlikely that total destruction of a food plant or animal could occur in this way. In response to a food shortage, individuals may survive but be much smaller than they would have been with more food. However, as numbers of individuals increase, a point is eventually reached where fewer and fewer individuals are able to survive, and finally none survive to adulthood. Another frequently encountered result of food shortage is cannibalistic behavior. One well-known species in which cannibalism is known to occur is the confused flour beetle, *Tribolium confusum* (Coleoptera, Tenebrionidae) (Clark et al., 1967). Cannibalism may also occur when there is overcrowding, accidental injury to an individual, or during the vulnerable period immediately following ecdysis but before the cuticle has become sclerotized (i.e., the teneral state). Several adaptations that promote survival under conditions of food shortage have evolved, particularly in response to regularly occurring shortages. Probably one of the most important is migration (see Chapter 8). Insects that respond to food shortages by migrating are much more likely to discover a fresh food source than those that do not. Many adult insects, fruit flies and house flies, for example, lay their eggs in situations where the food supply is in excess relative to the numbers of individuals feeding on it. The behavior of the adults in egg deposition essentially guarantees abundant food for their offspring. Polyphagy on the part of a parasite, predator, or herbivorous insect also helps ensure against the likelihood of being exposed to dangerous food shortages.

Quality of Food   Nutritional requirements of insects in general are discussed in Chapter 4. Egg production and larval development are especially susceptible to variations in the nutritive quality of food. Survival, longevity, and size may also be strongly influenced by the quality of available food. Many insect species store sufficient nutrients during the larval stage to accomplish adult activities (copulation, egg production, and oviposition). The adult lives of these species are of comparatively short duration, and they commonly do not feed at all. For example, mayflies (Ephemeroptera), which live only long enough to copulate and deposit their eggs, rely entirely on larval reserves. In other species, larvae may store nutrients sufficient for egg production, but the adults must ingest water and carbohydrates to survive.

Many mosquitoes and other true flies (Diptera) fall into a third category in which adults usually need to ingest a complete diet (especially protein) in order to sustain life and produce eggs. For mosquitoes and other blood-feeding forms, blood meals afford the necessary nutrients. The queens of the social Hymenoptera and termites also fall into this category. These insects lay eggs for almost all of their adult lives and require food constantly. Commonly, this food is provided by a worker caste, which in turn may obtain food by foraging.

An example of the effect of quality of food on the rate of development is given by Folsom and Wardle (1934). Both bananas and horse manure are suitable foods for house fly maggots. However, at approximately 21°C, the maggots complete their development on horse manure one to nearly two weeks sooner than on bananas. The quality of ingested food in some instances influences the outcome of development; for example, the differences between workers and queen honey bees depend entirely on the diet that each receives during larval development (Richards, 1953). Both receive the same food for the first two days of larval life—royal jelly, a substance secreted by glands that open into the mouths of workers. After the third day, larvae destined to become workers receive a diet containing increasing amounts of honey, whereas the queen larvae continue to be fed royal jelly. Another example of food quality affecting development is found in aphids (Hemiptera-Homoptera, Aphididae). Changes in the quality of food may induce the appearance of winged forms and migratory behavior (see "Polymorphism" in Chapter 5, "Dispersal and Migration" in Chapter 8). House (1969) discusses the effects of different proportions of nutrients on insects.

A Place in Which to Live

The place where an organism lives is its *habitat*. This "place" may be an entire field or forest, the shoreline of a stream, the undersides of leaves of a particular kind of plant, among others. A given species is typically found living in essentially the same kind of surroundings throughout its distribution, although there may be variation within the limits of tolerance of the species. The degree of suitability of a given habitat will vary with the extent to which it satisfies the needs within these limits of tolerance.

Each species of the community of organisms living in a given habitat requires certain resources (e.g., food, shelter, breeding sites, favorable temperature, favorable humidity, space, and time) in sufficient amounts and of sufficient quality to survive and reproduce. That is, every species occupies a particular physical location at a given time and does a given thing (eats, copulates, rests, etc.) at that place and time. A compilation of all needed resources, both in qualitative and quantitative terms, describes the *ecological niche* of a species. A community is composed of individuals of different species segregated into different niches.

Insects commonly occupy a wide variety of niches in almost every kind of terrestrial and freshwater habitat. About the only habitats not broadly colonized by insects are those within the ocean; although many insects from different orders are found in intertidal marine habitats, and some are found on the open ocean (e.g., several species of water striders, *Halobates*, Hemiptera, Gerridae; Cheng, 1976). However, there are no known marine insects that spend their entire life cycle submerged. Janzen (1977) provides an interesting discussion of the possible reasons why insects in particular have been able to occupy so many different niches (see Janzen's quote in the section on insect phylogeny, Chapter 10).

Other Organisms

Other organisms include individuals of the same or different species as part of each other's environment. The topic thus divides into intra- and interspecific interactions.

Intraspecific Interactions. The ways in which members of the same species interact are related in part to population density. There seem to be advantages associated both with relatively low and relatively high densities. Examples of circumstances in which high population density may be advantageous include mate finding and survival of potential predation (Andrewartha and Birch, 1954). In a situation where the probability of finding a mate is decreased owing to low population density, the result would be fewer fertilizations and hence a lower average fertility.

Whether a given population density is advantageous or disadvantageous is relative and depends upon other environmental influences. This is particularly apparent with regard to such available resources as food and shelter. The maximum population of a given species an area can support depends on the abundance of these resources in that area. When resources are limited, *intraspecific competition* may become increasingly pronounced as the maximum population is approached and may result in the death or emigration of individuals. Other environmental influences may, however, serve to keep a population well below a level where competition for limited resources would occur.

Interspecific Interactions. Interspecific interactions may be conveniently divided into competition, symbiosis, predator–prey interactions, herbivore–plant interactions, and indirect interactions.

Interspecific competition comes about when the needs of two or more different species for a given resource (food, shelter, etc.) coincide—when their niches overlap. In situations where two species with essentially identical life needs are brought together, one species may be expected to have a competitive advantage and eventually to eliminate the other. Stated another way: "no two species can occupy the same niche at the same time for very long." This is the phenomenon of *competitive exclusion,* the practical applications of which are considered in Chapter 12. The segregation of the different species in an ecosystem into different niches is explained on the basis of the operation of competitive exclusion. A good example of competitive exclusion is the outcome when two species of flour beetles (*Tribolium castaneum* and *T. confusum*) are placed in the same container of flour (Park, 1962). If the container is maintained under conditions of high humidity and temperature, *T. castaneum* invariably wins out. However, maintaining the beetles under conditions of low humidity and temperature results in a "victory" for *T. confusum.* In the absence of the other, either species can be maintained indefinitely under wet and warm or dry and cool conditions.

Symbiosis is used here to mean a close association between two different species. This association may be mutually advantageous *(mutualism)*, disadvantageous for one of the species *(parasitism)*, or advantageous for one without harm to the other *(commensalism)*.

There are many examples of mutualistic associations between insects; one is seen when ants actively care for and protect aphids, which in turn excrete honeydew that is ingested by the ants (Way, 1963). The relationships between fungus-growing ants and fungi (see "Feeding Behavior" in Chapter 8 and the next section of this chapter) are good examples of mutualism between insects and other kinds of organisms. Additional examples include the relationships between microbes and insects relative to digestion and nutrition (see Chapter 4). A recent review deals specifically with the intracellular symbiotes of the Hemiptera-Homoptera (Houk and Griffiths, 1980). Koch (1967) discusses endosymbionts of insects.

There are numerous examples of mutualistic relationships between plants and insects where the plants provide food for an insect that in turn pollinates the plants (see next section and "Beneficial Insects" in Chapter 12).

Among the organisms that parasitize insects are insects themselves (Fig. 9-8; see also "Feeding Behavior" in Chapter 8), microorganisms, mites, and nematodes. In some instances microbes and nematodes parasitize both insects and vertebrates. In these cases the insects may act as *vectors,* carrying parasites to their vertebrate hosts (Harwood and James, 1979) or plant hosts (Pfadt, 1978).

Well over 1000 microbes, most of them pathogens (disease-causers), have been described in insects, and new ones are being added regularly. According to N.A.S. (1969), the current list includes "90 species and varieties of bacteria, 260 species of viruses and rickettsiae, 460 species of fungi, 255 species of protozoa." These microbes may gain entrance into insects orally or via wounds in the integument, or they may be capable of actively penetrating the integument (particularly fungi). In some instances pathogens that have gained entrance may be passed *transovarially* (via the egg) to offspring, as are certain arboviruses in mosquitoes.

Interest in pathogenic microbes associated with insects has largely been focused on their potential use in biological control (see Chapter 12). During the past 30 years or so, insect microbiology has grown rapidly as an entomological specialty. For information on this topic, see Steinhaus (1949, 1963), Cantwell (1974), and the references cited under "Biological Control" in Chapter 12.

Among the bacteria that infect insects, the best known is *Bacillus thuringiensis,* a spore-former that has been used with some success in biological control. It is pathogenic for nearly 200 species of pest insects, particularly Lepidoptera. Another member of the same genus, *Bacillus larvae,* causes American foulbrood, a serious disease of honey bee larvae. In addition to members of the genus *Bacillus,* certain species of *Clostridium* and *Streptococcus* are also insect pathogens. For more information on bacterial diseases of insects, see Faust (1974). Many bacterial diseases of plants are mechanically transmitted by insects. Certain bacterial pathogens of vertebrates are also transmitted by insects, for example, those that cause plague *(Yersinia pestis)* in humans.

A

Fig. 9-8
Scanning electron micrographs. A. Bee
louse, *Braula coeca* (Diptera,
Braulidae), an example of an insect
that parasitizes another insect. B.
Dorsal view of the tarsus, showing the
comblike claws used to cling to the
characteristic branched hairs, C, of the
honey bee. The larva of this insect
lives beneath the honey cappings; the
adult lives on the body of an adult bee
and takes food directly from the mouth
of its host. An adult bee louse is about
1.5 mm long. [Courtesy of A. Dietz,
W. J. Humphreys, and J. W. Lindner.]

B

C

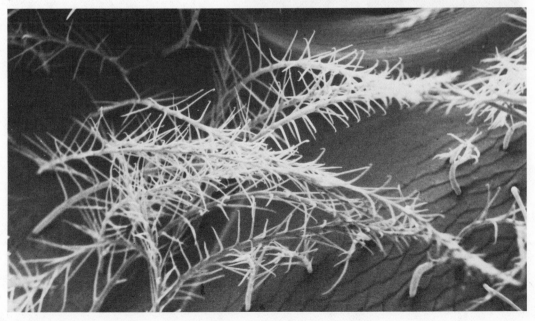

Viruses (Fig. 9-9) are obligate cellular parasites composed in part of DNA or RNA. Those associated with insects are generally not as well known or understood as the bacteria. They are classified on the basis of the part of the host cell in which they develop, the presence or absence and morphology of an inclusion body, and the kind of nucleic acid (DNA or RNA) present. Some insect viruses produce granular inclusion bodies; most produce crystalline polyhedral bodies in the nucleus or cytoplasm of an infected cell. Virus particles may be alone in an inclusion body or in packets. Several viruses that are pathogenic for insects have been identified and described. Most have been found among Lepidoptera, although several also are known from the Hymenoptera, Diptera, Coleoptera, and Neuroptera. It is interesting that a virus may act in conjunction with a bacterium to produce a disease syndrome not produced by the virus or bacterium separately. Such is the case in two silkworm diseases, gattine and flacherie, *Streptococcus bombycis* and *Bacillus bombycis*, respectively, being the bacteria involved (Rolston and McCoy, 1966). Insects, especially aphids and leafhoppers (Hemiptera-Homoptera), are important vectors of viruses pathogenic for plants (Pfadt, 1978). Insects also serve as vectors of several viruses that attack humans and other vertebrates, notably the *arboviruses* (arthropod-borne viruses), including those that cause yellow fever, den-

Fig. 9-9
Electron micrograph showing the virus of Venezuelan equine encephalitis in thoracic tissue of the mosquito *Aedes aegypti*. The virus is approximately 450 Å in diameter and is composed of RNA surrounded by a protein coat. The arrow indicates a single virus. [Courtesy of R. F. Ashley.]

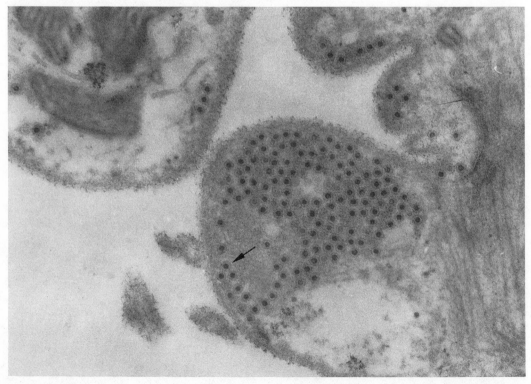

gue, and encephalitis (Horsfall and Tamm, 1965; Theiler and Downs, 1973). David (1975), Gibbs (1973), Smith (1967, 1977) and Vaughn (1974) provide extensive information on insects and viruses.

Microbes somewhat similar to bacteria, but lacking cell walls, *mycoplasmas,* have recently been found to cause the "yellows diseases" of plants, such as aster yellows (Whitcomb, 1973; Whitcomb and Davis, 1970). Leafhoppers (Hemiptera-Homoptera, Cicadellidae) are involved in the transmission of these organisms. For a review of mycoplasmas and microbes relative to plant diseases see Carter (1973) and Nienhaus and Sikora (1979).

Some rickettsiae are insect pathogens. In size they lie somewhere between viruses and bacteria, but, unlike viruses, they are capable of independent metabolism in vitro. Those rickettsiae that are pathogenic for insects are very slow in killing their hosts. They have been grown in mammalian-tissue culture and have killed white mice upon injection. Owing to slow kill of insects and potential danger for mammals, they are unlikely ever to be used in insect control (N.A.S., 1969). *Rickettsia prowazekii* (causes epidemic typhus) and *R. prowazekii mooseri* (causes murine and endemic typhus) which are pathogenic for humans, are vectored by body lice and fleas, respectively.

There are more species of fungi that are known to be pathogenic to insects than of any other group of microbes. These *entomogenous* fungi have been described from several groups of fungi. Species in the genus *Beauveria* have been used in attempts at biological control. However, the success of fungus infections in insects (i.e., whether they become established and kill an insect) depends to a great extent on weather conditions, and hence these microbes would likely be undependable as agents of biological control. Although insects are apparently not vectors of fungi that cause human or other vertebrate diseases, they do serve as mechanical vectors of fungi pathogenic to plants. An example is the fungus that causes Dutch elm disease, which is vectored mainly by the European elm bark beetle, *Scolytus multistriatus.* Madelin (1966) discusses fungi that are parasites of insects.

A large number of protozoa, especially *Microsporidians* (Fig. 9-10), are known to kill insects (see Brooks, 1974), but, like rickettsiae and fungi, they act very slowly. No doubt many of those slow-acting microbes weaken their hosts and make them more susceptible to the effects of other environmental components. Insect pathogens have been described in the protozoan phyla Sarcodina (amebas), Mastigophora (flagellates), and Sporozoa (sporozoans). The microsporidian *Nosema bombycis* (Sporozoa) causes the "pebrine" disease of silkworms. The elucidation of this relationship by Louis Pasteur in the nineteenth century stands as a classic in microbiological research. Other well-known microsporidian insect pathogens include *Nosema apis* and *N. pyraustai,* which attack the honey bee and European corn borer, respectively. A mosquito, *Culex erraticus,* parasitized by the microsporidian *Thelahania minuta,* is shown in Figure 9-10. Protozoans are common partners in mutualistic relationships with insects; we have already discussed those found in the gut of termites,

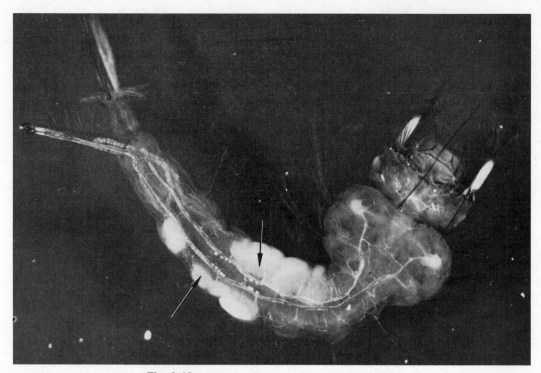

Fig. 9-10
Microsporidian *Thelahania minuta* (indicated by arrows) infecting the fat body of the mosquito *Culex erraticus*. [Courtesy of H. C. Chapman and J. J. Petersen.]

which provide cellulase and enable their hosts to digest cellulose (Cleveland et al., 1934). Several protozoa pathogenic for humans and other vertebrates are vectored by insects. Examples include the malarias (*Plasmodium* spp.) vectored by *Anopheles* spp. mosquitoes (different forms of malaria occur in birds, monkeys, and humans), American trypanosomiasis (Chagas' disease) caused by *Trypanosoma cruzi* and vectored by blood-sucking bugs (Hemiptera, Reduviidae), and the two forms of African sleeping sickness caused by *T. rhodesiense* and *T. gambiae* and vectored by tsetse flies (Diptera, Muscidae, *Glossina* spp.). The serious disease of cattle, *nagana*, which has retarded the development of major parts of Africa, is caused by the protozoan *T. brucei*, also vectored by tsetse flies.

Mites (Arachnida, Acarina) are very small arthropods, many species of which parasitize insects (Fig. 9-11). For example, the larvae of members of the family Erythraeidae are parasites of insects and other arthropods and have been found attached to wing veins and other locations on the insect body (Baker et al., 1956). *Acarapis woodi* (Scutacaridae) causes the acarine disease of bees. This species either lives in the tracheae or on the external parts of the bee's body and weakens the host to the point where it can no longer fly and eventually succumbs (Baker et al., 1956; Gochnauer, 1963).

About 100 species of nematodes (phylum Aschelminthes, roundworms; Fig. 9-12) have been described as being associated with approximately 1500 insect species (N.A.S., 1969). In most in-

stances, these "entomophilic" nematodes damage and eventually
kill their host. Nematodes show some promise in the biological control of insects. A number of species that attack humans, for example,
filarial worms, (*Wuchereria bancrofti* and others), and other vertebrates, for example, dog heartworm, spend part of their life cycles
in insects (mosquitoes for the examples here), which serve as their
vectors. See Poinar (1972, 1975) and Welch (1965) for reviews in
the area of "entomophilic nematology."

Many relationships between two or more insect species or between
insects and other organisms are commensal. For example, probably
many of the microbes associated with insects neither harm nor particularly help their host but gain food and a place to live. Phoretic
relationships (see also Chapter 8) are good examples of commensalism. In a phoretic relationship an individual of one species at-

Fig. 9-11
Scanning electron micrograph of a mite (Arachnida; Acari) on the head
of a termite (highly magnified). The mite is indicated by the arrow and is
shown at a still higher magnification in the inset. [Courtesy of Walter J.
Humphreys.]

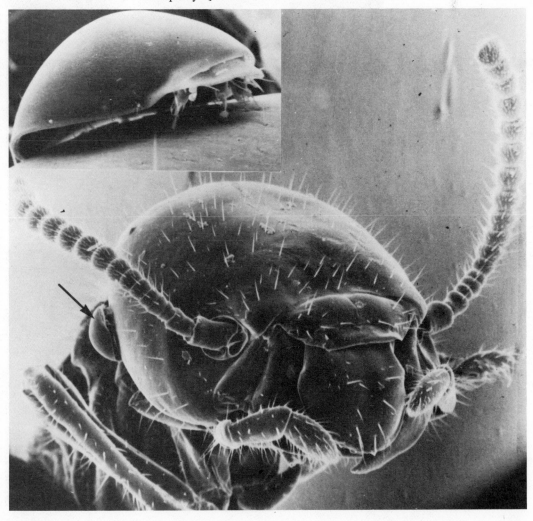

A

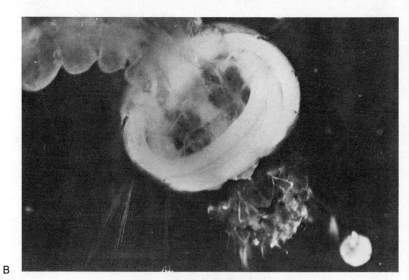

B

Fig. 9-12
Nematode parasite,
Reesimermis nielseni,
in the thorax of the
mosquito *Culex pipiens
quinquifasciatus.* A.
Several infected
larvae. B. Single
infected larva. C.
Postparasitic stage of
the nematode escaping.
[Courtesy of H. C.
Chapman and J. J.
Petersen.]

C

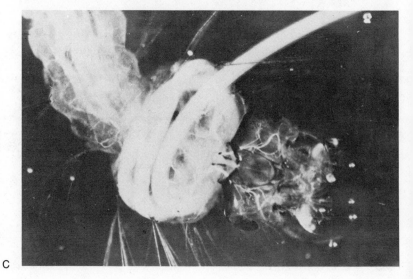

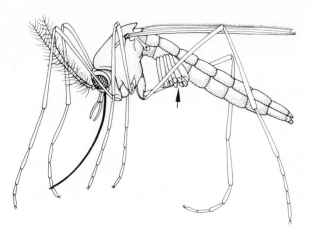

Fig. 9-13
Mosquito, *Psorophora* sp., carrying the
eggs (indicated by arrow) of the torsalo
fly, *Dermatobia hominis*. [Redrawn from
USDA photo.]

taches to an individual of another and gains a mode of transportation.
Insects may be transporters or riders, or both. Some chewing lice
(Mallophaga) attach themselves by their mouthparts to louse flies
(Diptera, Hippoboscidae), some of which are also parasites of birds,
and in this way move from one host to another. The torsalo fly,
Dermatobia hominis (Fig. 9-13), deposits eggs on other flies, such
as mosquitoes, black flies, and house flies. When these insects come
into contact with a potential vertebrate host (including man), larvae
emerge from the eggs and penetrate the skin (Harwood and James,
1979). Clausen (1976) discusses phoresy among entomophagous
(''insect-eating'') insects.

Certain insects, for example gall midges (Diptera, Cecidomyiidae) and gall wasps (Hymenoptera, Cynipidae), induce plants to
form characteristic growths in which the larvae of these insects feed,
grow, and develop (Felt, 1940). These growths are called *galls* (Fig.
9-14) and are made up entirely of plant materials. Depending upon
the effect of galls on a plant, such a relationship could be viewed as
parasitic or commensalistic. Certain fungi, nematodes, mites, and
other organisms also cause galls to develop, but insects are the most
common gall inducers and the most elaborate and specific galls are
those induced to form by insects.

Predators of insects include insects themselves; other Arthropoda,
such as mites, spiders, scorpions, and pseudoscorpions; and vertebrates, including birds, reptiles, amphibians, fish, and mammals.
Insects as predators are discussed in Chapter 8, and comments regarding vertebrate predators are included in Chapter 12 (''Biological
Control''). Sweetman (1936) discusses predatory invertebrates in
some detail. In addition to animals, a few plant species, such as the
Venus flytrap, pitcher plant, and the aquatic *Utricularia vulgaris,*
trap, digest, and assimilate insects (Frost, 1959). Heslop-Harrison
(1978) and Swartz (1974) discuss carnivorous plants.

As explained in Chapter 8, a majority of insects feed on plants
(i.e., are herbivores). In most cases the association goes much further than simple herbivory, the plants providing shelter, oviposition
sites, and so on. The plant/herbivorous insect association is discussed further in the next section.

Insects may be influenced by other organisms indirectly. No doubt

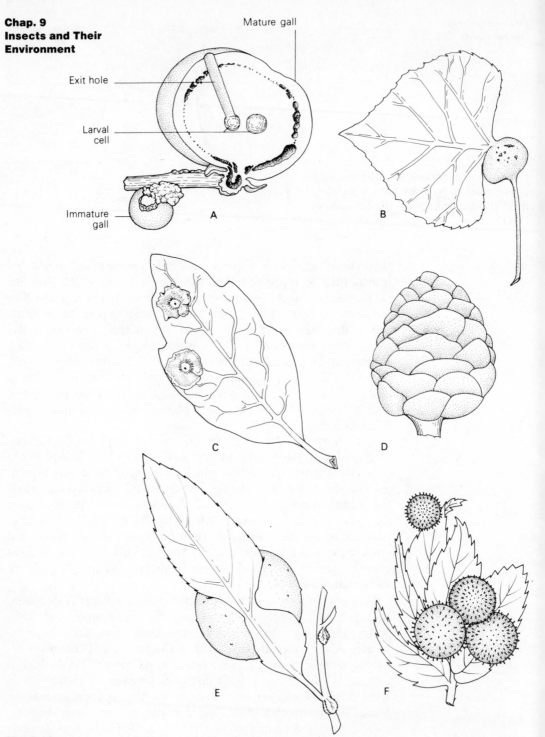

Exit hole

Larval cell

Immature gall

Mature gall

Fig. 9-14
Insect galls. A. California gallfly, *Andricus californicus* (Hymenoptera; Cynipidae). B. Transverse poplar gall aphis, *Pemphigus populitransversus* (Homoptera) on stem of poplar leaves. C. *Andricus pattersonae* (Hymenoptera; Cynipidae) on leaf of blue oak. D. Pine-cone willow gall caused by *Rhabdophaga strobiloides* (Diptera; Cecidomyiidae). E. Willow apple gall caused by the sawfly *Pontania pomum* (Hymenoptera; Tenthredinidae). F. Spiny rose gall caused by *Rhodites puslulatoides* (Hymenoptera; Cynipidae). [A–C redrawn from Essig, 1958; D–F redrawn from Boreal Labs key card.]

the greatest effects are those produced by humans purposely against pest populations and inadvertently against the vast majority of insects that are "innocent" (see Chapter 12). The effects of modern agriculture, urban sprawl, and superhighways on various habitats are obvious, and the far-reaching and unfortunate effects of highly residual pesticides have become well known. The effects of air and water pollution on the environment are beginning to get the kind of attention they deserve. Insects themselves are sometimes polluters: "Predatory anthocorid bugs enter open galls and feed on a few of the aphids therein. The aphids which are not attacked die from the fumigant effect of chemicals produced by the stink glands of the bugs" (Clark et al., 1967, citing the work of Dunn, 1960).

Adaptations Associated with Interspecific Interactions.   The "goal" of every organism is to obtain enough food (matter and energy) to maintain itself long enough to reproduce, passing its genes to the next generation. As an organism proceeds from conception to reproductive maturity, it must constantly obtain food and at the same time avoid becoming food for some other organism. Organisms that are unable to do this successfully, for whatever reason, are maladapted and are, in the long term selected against. Although a number of environmental factors are no doubt involved in the selection process, examination of specific adaptations often leads to the identification of specific selecting agents. Such is the case when we examine adaptations associated with many interspecific interactions. Organisms that are closely associated ecologically often act as agents of selection reciprocally; that is, they act as selecting agents on one another. There is good evidence that many organisms have done so for long periods of geologic time. These organisms are said to have *coevolved* or to be *coadapted*.

Examination of several examples where coevolution has played a role reveals at least two types of coevolutionary interactions. Two (or more) different species may have coevolved to the extent that they benefit one another in the quest for reproductive success; their relationship thus is mutualistic. In this case the partners have become (or are becoming) more and more successful in their interaction leading to sustained or improved reproductive success for both. Alternatively, the participants may have engaged and continue to engage in a kind of genetic warfare in which one member "evolves" a mechanism of offense providing increased access to the other as food, only to have the other "respond," in the evolutionary sense, with a defense, and so on through time. Such is the case in interactions between parasites (and parasitoids) and hosts, predators and prey, and herbivores and plants. In this section, examples of several coevolutionary adaptations are examined. For a comprehensive introduction to coevolution, see Gilbert and Raven (1975).

An excellent example of coevolution that has resulted in a mutualistic relationship is that which exists between the bull's-horn acacia, *Acacia cornigera,* and the ant *Pseudomyrmex ferruginea.* Janzen (1967) describes this relationship as follows:

> The bull's-horn acacia . . . is a representative swollen-thorn acacia with well-developed foliar nectaries, enlarged stipular

thorns, and small nutritive organs (Beltian bodies) borne at the tip of each leaf segment. The colony of *P. ferruginea* living in the enlarged stipules obtains sugars from the foliar nectaries, and oils and proteins by eating the Beltian bodies.

. . .

. . . The workers patrol and clean the surfaces of the acacia, and bite and sting animals of all sizes that contact the plant. The workers maul any other species of plant that contact the acacia and in many cases, any that grow under the acacia. The colony attains a very large size and up to 25 per cent of the workers may be active on the surface of the acacia both day and night. The larger the colony becomes, the smaller is the damage sustained by the plant from defoliators. The colony enhances its own probability of survival by protecting the acacia on which it is completely dependent for food and domatia.

Pollinating insects and the plants they pollinate also provide excellent examples of adaptations associated with long coevolutionary relationships.

Sexual reproduction in flowering plants is accomplished by the transfer of pollen from the anther of a male flower to the stigma of a female flower. When a pollen grain contacts a stigma, the male germ cell in the pollen grain unites with the female germ cell or egg, and a fertile seed develops. When the seed is exposed to the proper conditions in or on an appropriate substrate, it will give rise to a new plant. The transfer of pollen from a male to a female flower is accomplished primarily by the wind or by the activities of insects that associate with plants. Examples of wind-pollinated plants include cereal plants such as wheat and corn and many species of trees. The flowers of these plants are generally small with weakly developed petals, do not produce nectar, and produce dry pollen grains, which are easily picked up by the wind.

The relationships between plants and their insect pollinators are mutualistic in that the plants provide sugar-rich nectar and/or pollen for insects, which in turn serve as accessory plant "sex organs" in transmitting gametes from the male to the female plant (or flower). Recently, nectar of certain plants has been found to contain nutritive substances, such as certain amino acids and lipids, in addition to sugar (Baker and Baker, 1975). Insect-pollinated plants have evolved traits that make them attractive to effective insect-pollinators. In many cases the result has been a one-to-one relationship between a single plant species and a single insect species. An example of such a relationship is that between the yucca plant and yucca moths (*Tegeticula* spp.) (Fig. 9-15; Powell and Mackie, 1966). An adult female moth uses its specialized mouthparts to gather the pollenia of the yucca plant into a ball. This ball is then scraped across the plant's stigma, accomplishing cross-pollination. The dependence of the yucca plant on the pollinating activity of the moth is demonstrated by the failure of the plant to develop pods in the absence of the moth. The yucca plant provides a home for the moth larvae, which remain within the plant until ready to emerge as adults.

Distinct correlations can be made between the anatomical and physiological characteristics of the flowers of a given species and

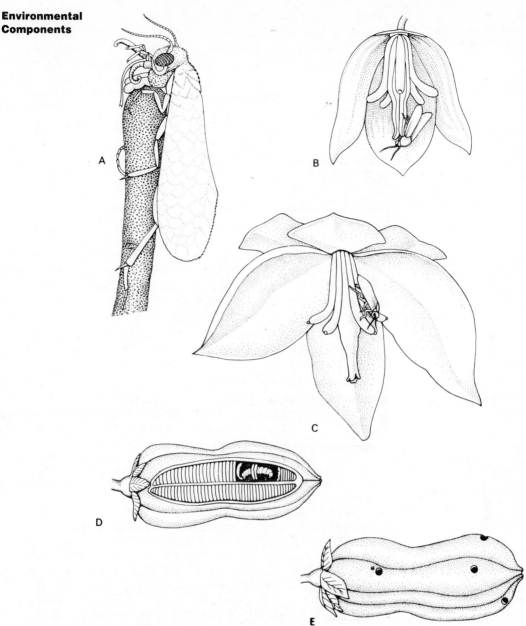

Fig. 9-15
Yucca moth and yucca plant. A. Female moth gathering pollen. B.
Female scraping the pollen ball across the plant's stigma. C. Female
ovipositing (several petals removed). D. Larva within mature pod
(portion of pod removed). E. Pod with exit holes and constrictions
caused by larvae. [Redrawn from Schneirla, 1953, after Riley, 1892.]

the anatomy, physiology, and behavior of their insect pollinators.
Among the characteristics of flowers attractive to insects are

1. The production of particular scents.
2. Color, size, and shape of petals.
3. Patterns of stripes or spots on petals.
4. Whether petals are separated or not.
5. Shape of flower.

Scent production is usually very important in attracting a potential pollinator and often in determining behavior following landing. Flowers pollinated by true flies, which breed in dung, may actually mimic the odor of dung, so "fly" flowers commonly have a disagreeable odor for humans. On the other hand, "butterfly" and "moth" flowers usually have a sweet smell. Flower color may also be correlated with the type of pollinating insect. For example, honey bees and their relatives are attracted by many blue, purple, and yellow flowers, but cannot "see" red flowers. However, butterflies are attracted to red flowers. As explained in Chapter 6, many insects (e.g. honey bees) can see ultraviolet light. In fact, many bee-pollinated flowers that appear to be unpatterned solid colors to us turn out to have characteristic stripes or spots when photographed with ultraviolet-sensitive film. These stripes, spots, and so on are visible to honey bees and may serve as visual guides into a flower nectary in addition to simply being attractive to these insects (Fig. 9-16). The shape and depth of nectaries are often such that only particular species are able to extract nectar from them. These specially constructed nectaries also serve to inhibit "undesirable" insects. The amount of nectar secreted by a nectary is such that pollinators get enough to make the effort worthwhile energetically, but not so much that they are satiated by visiting one or even a few flowers (Heinrich, 1973).

Examples of the extremes of plant adaptation that have occurred in association with pollinators include imprisonment of pollinating insects and pseudocopulation. Yeo (1973) describes the following fascinating case of imprisonment as a mechanism that culminates in pollination.

In *Arum nigrum* Schott the spathe opens overnight and the spadix emits a bad smell. In the morning dung-frequenting beetles and flies are attracted by the scent, and alight on the spadix appendage and the inner surface of the spathe. The cells of the epidermis of these organs each bear a downwardly directed papilla. These papillae are of such size that the claws of an insect can grip them, but as the grip tightens the claw is simply forced off the tip of the papilla. The entire surface is covered with an oily secretion which is believed to nullify the effects of the pulvilli. The result is that insects walk a short distance and then suddenly fall. The heavier ones take wing as they fall but the smaller ones are too slow, and drop through the ring of bristle-like sterile florets into the "prison." At this time the

Fig. 9-16
Marsh marigolds as they appear to humans (A) and to insects sensitive to ultraviolet light (B). [Drawings based on photographs in Eisner et al., 1969.]

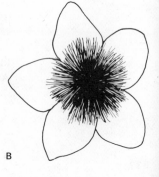

A   B

stigmas can be pollinated by pollen carried by the incoming insects. The following night the stigmas cease to be receptive; pollen is then shed into the chamber and dusts the insects. By the morning the sterile florets, previously smooth and oily, have become wrinkled so that insects can walk on them. The papillae on the spathe and spadix also shrink, and the insects can escape.

Certain orchids (*Ophrys* and others; Fig. 9-17) produce stimuli that attract their specific pollinators (particularly Diptera or Hymenoptera) by releasing copulatory behavior (Kullenberg, 1961). These orchids produce scents that mimic the effects of sex pheromones as well as stimulating the insects visually and tactilely.

The most common insect pollinators are various members of the orders Coleoptera, Lepidoptera, Diptera, and Hymenoptera. Among the common adaptations of pollinating insects are elongated mouthparts, that facilitate obtaining nectar from flowers with deep nectaries, plumose (featherlike) hairs on the body to which pollen clings, and various specialized pollen-collecting and/or transporting structures such as the corbiculae (pollen baskets; see Fig. 2-37A) on the hind tibiae of honey bees and their relatives and pollen brushes on the hindlegs of most bees.

For additional information on the fascinating topic of insect pollination and pollination ecology, see Baker and Hurd (1968), Faegri and van der Pijl (1978), Free (1970), Heinrich (1973), Heinrich and Raven (1972), Percival (1965), Price (1975), Proctor and Yeo (1972), Wiebes (1979), and Yeo (1973).

Another good example of the evolutionary "fine tuning" that can occur between mutualists is seen in the association between attine ants and their fungus gardens (see "Feeding Behavior" in Chapter 8). As ants prepare a fungus garden, they add their own fecal material, which benefits the fungus (Martin, 1970). The fecal material contains proteolytic enzymes and various nitrogenous materials that play vital roles (i.e., protein digestion and enhancement of growth) in the metabolism of the fungus. In turn, the fungus contributes cellulase, which benefits the ants by catalyzing the digestion of cellulose. Such close metabolic interdependency is characteristic of many mutualistic associations.

Parasites and hosts apparently pass through successive cycles of "parasite attack–new host defense," the parasite tending to become increasingly benign and the host more tolerant, perhaps eventually leading to a commensalistic or even mutualistic outcome. For example, parasites that invade the insect hemocoel are commonly encapsulated (see "Circulatory System" in Chapter 4). The presence of such a capsule around a parasite may severely limit the parasite's activities by inhibiting mobility or interfering with feeding and/or gaseous exchange. On the other hand, some parasites secrete materials that block capsule formation (Jones, 1977). Duffey (1977) contains information on chemicals in insects that have an antibiotic effect. Whitcomb et al. (1974) consider insect defenses against microorganisms and parasitoids.

A remarkable example of an extremely close adaptive adjustment between parasite and host is the relationship between the rabbit fleas *Spilopsyllus cuniculi* and *Cediopsylla simplex* and their rabbit hosts

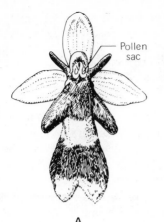
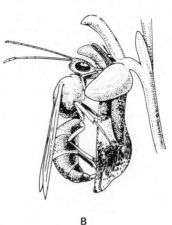

Pollen
sac

Pollen sac

A   B   C

Fig. 9-17
A. The orchid *Ophrys insectifera,* which visually resembles a female
bee. B. A bee attempting to copulate with the flower. C. As the result of
attempted copulation, pollen sacs become attached to the bee's head.
[Redrawn from Wickler, 1968.]

(Rothschild, 1965a, b; Rothschild and Ford, 1966, 1972). The reproductive behavior and physiology of these fleas is strongly influenced by the rabbit's reproductive hormones. Egg maturation only
occurs on pregnant doe rabbits and has been shown to be stimulated
by the host's corticosteroids. Estrogens and other hormones also
contribute to egg development and influence oviposition. Larvae
develop in the rabbit nests in which the young rabbits are born.
Linkage of the sex cycles of the fleas and rabbit hosts ensures that
the reproductive activity of the fleas occurs such that newborn flea
larvae end up in the host's nest.

For recent discussions of host–parasite interactions, see Nelson
et al. (1975), Galun (1977), and Vinson and Iwantsch (1980).

That predators of insects and their insect prey have exerted strong
selection pressures on one another is clearly evident in the means
predators have evolved to capture prey and the avoidance mechanisms of potential prey. Several examples of coevolutionary adaptation involving insects as predators and insects as prey have already
been given in earlier chapters (see, especially, "Escape and Defense" in Chapter 8). Here we consider examples of insect adaptations in response mainly to selection pressures by insectivorous
vertebrates, in particular protective coloration, form, and pattern;
and coevolutionary adaptations in bats and moths.

Examples of protective form, coloration, and pattern include crypsis, aposematic coloration, mimicry, and spots. As explained in
Chapter 8, such protective devices would be of diminished value
(or even detrimental) without appropriate behavior. Thus, although
we will emphasize form, coloration, and pattern in this discussion,
it must be remembered that "appropriate" behavior has also come
about as a result of selection pressure applied by predators. General
treatments of protective form, coloration, and pattern include Cott
(1957), Portmann (1959), and Wickler (1968).

Crypsis refers to various combinations of form, color, and pattern that facilitate "hiding" from potential predators. Resemblance to various inedible, even repulsive, objects like thorns or bird droppings; coloration and shading that cause an insect to blend into the background; and transparent wings that allow the background to show through all belong in the crypsis category.

A classic example of crypsis (and one that has contributed significantly to our understanding of natural selection) is *industrial melanism* (Kettlewell, 1959, 1961, 1973; Bishop and Cook, 1975). Industrial melanism refers to a phenomenon first observed in the peppered moth, *Biston betularia,* in England. These moths are nocturnally active and during the day they rest on lichen-covered tree trunks. Prior to 1845, collection records of peppered moths described them as light-colored, an appearance that provided them with camouflage when they rested on the tree trunks. In 1845 and thereafter, dark (*melanic*) forms began to appear around the industrial center of Manchester and eventually came to be more common than the light-colored forms. It was hypothesized that the change in predominant phenotype (from light to dark) was associated with the overall darkening of tree trunks due to industrial pollution. Such darkening was thought to lessen the selective advantage of light-colored moths while favoring the survival of dark forms. Birds were assumed to be the agents of selection. Kettlewell demonstrated experimentally that birds did, in fact, prey more on uncamouflaged forms; light-colored forms had a definite selective advantage in unpolluted areas. He also showed that moths tended to rest on a substrate that matched their body coloration, indicating that they could discriminate between light and dark backgrounds. With the advent of pollution controls, light moths are beginning to appear again in industrial regions.

Aposematic or *warning coloration* is sometimes called "advertising" in the sense that an insect tastes very bad (i.e., is unpalatable), stings, or does something similarly disagreeable to a predator and communicates this ability visually by colors and patterns that contrast blatantly with the background. Vertebrate predators soon learn to avoid these insects on the basis of a few unpleasant encounters. An aposematically colored insect may also be advertising falsely in that although it looks like (mimics) a harmful insect, it is really palatable and harmless. Aposematically colored insects are usually boldly patterned in shades of orange, red, or yellow contrasting with black and stand out against the greens and browns that are characteristic of the environment. Unpalatability is commonly due to the presence of chemicals in the hemolymph that have been derived from host plants. For example, the monarch butterfly (*Danaus plexippus;* Fig. 9-18A) and certain true bugs and beetles, which feed on various species of milkweed plants (Asclepiadaceae), accumulate chemicals (specifically *cardiac glycosides*) from these plants. These chemicals not only render the insects unpalatable to vertebrate predators (especially birds) but also act as emetics (i.e., they make a predator vomit) (Brower, 1969; Brower and Brower, 1964; Brower and Glazier, 1975; Reichstein et al., 1968; Rothschild, 1972; Duffey, 1980). Although all monarch butterflies are aposematically colored, an individual may or may not be unpalatable depending on the level of cardiac glycoside in the particular milkweed plant it fed

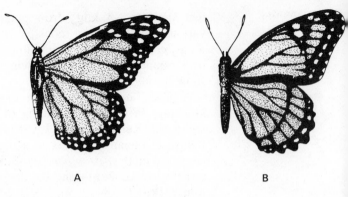

Fig. 9-18
The monarch butterfly, *Danaus plexippus* (A), serves as the model for its Batesian mimic, the viceroy butterfly, *Limenitis archippus* (B).

A   B

upon as a larva (Brower et al., 1968). In an interesting coevolutionary turnabout, certain insectivorous birds have circumvented the monarch's chemical defense by learning to reject or ingest individuals on the basis of taste (Calvert et al., 1979). In this instance aposematic coloration would tend to be disadvantageous for the insects since it would render them more visible to predators.

Mimicry, as used here, refers to the close resemblance (in form, color, pattern, or otherwise) between one insect, the *mimic,* and another insect or a plant part, the *model* (insects may mimic thorns, leaves, etc.). Various forms of mimicry have been identified in insects, among them Batesian (Bates, 1862), Müllerian (Müller, 1879), and Wasmannian (Rettenmeyer, 1970). A *Batesian mimic* is palatable (and otherwise harmless), but is protected somewhat by virtue of its similarity to an unpalatable (or otherwise harmful) model. An oft-cited example is the viceroy butterfly (*Limenitis archippus*), which is a Batesian mimic of the monarch butterfly (Fig. 9-18B). Batesian mimics are commonly less numerous than their models. This is interpreted as advantageous on the basis that large numbers of aposematically colored but palatable individuals would reduce the avoidance learning in predators. *Müllerian mimics* differ from Batesian in that both mimics and models are unpalatable (or otherwise harmful). Several species that bear close resemblance to one another may comprise a *Müllerian mimicry complex,* all members gaining advantage by virtue of the fact that the number of different colors and patterns predators are required to learn to avoid are reduced; that is, predators need learn only one pattern to result in protection for all the insects in the complex. Wasmannian mimics are insects that live in close association with ants and that have come to closely resemble their hosts. In this case the models (i.e., ants) are considered to be the selecting agents, the guests who fail to resemble their hosts sufficiently being attacked. For additional information on insect mimicry, see Rettenmeyer (1970).

Many insects (e.g., certain moths, butterflies, and caterpillars) bear spots at various locations on their body. These spots often closely resemble eyes of vertebrates and perhaps, in some instances, eyes of other insects. Many moths, which have eye spots on the hindwings, are otherwise cryptically colored when resting on their usual substrate. If the first line of defense, crypsis, fails and a predator strikes, the forewings are thrown forward, suddenly displaying the eyespots (Fig. 9-19). Such behavior has been shown experimentally to frighten and repel insectivorous birds (Blest, 1957). The eyespots resemble eyes of birds (e.g., owls) that prey on the insectivorous birds. Small spots are commonly found along the edges of the wings of many butterflies (Wickler, 1968). These have been shown to serve as pecking targets and by virtue of their location away from vital parts of the body may provide protection by diverting the peck of a predator and allowing time for escape.

A superb example of predator/prey coevolution is the association between insectivorous bats and members of certain families of night-flying moths, in particular Noctuidae, Arctiidae, and Geometridae (see "Escape and Defense" in Chapter 8). Bats locate prey by emitting pulses of sound beyond the range of human sensitivity (ultrasonic). As these sounds bounce off various objects, living and nonliving, the bats hear the echo and use this information to guide their flight and to locate flying prey insects. Details of the discovery and elucidation of this echo-locating ability are recounted in Griffin (1959). Tympanic organs that perceive the echo-locating cries of bats have evolved in several moth families (see "Sound Perception" in Chapter 6). As described in Chapter 8, the moths display various escape maneuvers in response to bat cries. The coevolutionary "battle" has proceeded a step further in certain Arctiid moths, which possess microtymbal organs on the metathorax. When a bat approaches one of these moths, trains of sound pulses are generated by the microtymbal organs. These sounds have been shown to, in effect, "jam" the bat's echo-locating system and hence interfere with its ability to catch the moth (Dunning and Roeder, 1965).

Herbivorous insects and plants have played major roles in one another's evolutionary history. Several structural and behavioral adaptations of herbivorous insects are described elsewhere (see "Feeding Behavior" in Chapter 8 and "Mouthparts" in Chapter 2). Detailed information about insect and plant coevolution may be found in de Wilde and Schoonhoven (1969), Ehrlich and Raven (1964), Jermy (1974), Sondheimer and Simeone (1970), and van Emden (1972).

Plants have responded, in the evolutionary sense, in a variety of ways to the continued onslaught of herbivorous insects. Among the adaptations that provide defense against insect herbivores are numerous structural and biological characteristics, but the major defensive adaptations involve chemicals.

Fig. 9-19
Io moth (female), *Automeris io,* with
forewings thrown forward, displaying
eyespots.

Structural characteristics that interfere with insect attack (feeding and/or oviposition) include tough cuticle, hard seed coats, velvetlike pubescence, and a variety of spines and thorns. Recently certain plants have been found to be covered, at least in part, with highly specialized, minute, sharp-pointed, hooked hairs *(hooked trichomes)* (Gilbert, 1971; Pillemer and Tingey, 1976). These hairs hook into the integument of caterpillars (*Heliconius* spp.) feeding on *Passiflora adenopoda* (Gilbert, 1971) or leafhoppers feeding on certain strains of field beans (Pillemer and Tingey, 1976), immobilizing and puncturing these insects, which eventually succumb to starvation and desiccation. More information on trichomes may be found in Levin (1973) and Rathcke and Poole (1975).

Limited growing seasons, nutrient-poor sap (relative to herbivorous insects), and mechanisms that allow rapid dispersal of seeds (which increases difficulty in food location for seed eaters) are examples of various biological means by which plants "combat" herbivores.

Many chemicals found in plants have no apparent role in plant metabolism, but are present in comparatively concentrated amounts. In several instances, these *secondary plant substances (secondary metabolic products;* see "Location of Food Sources" in Chapter 8) have been shown to be of value to the plants by influencing the behavior or physiology of herbivorous insects. That is, they sometimes act as allomones (see "Communication" in Chapter 8). General treatments of insects and secondary plant substances include Rosenthal and Janzen (1979), Schoonhoven (1972b), and Whittaker and Feeny (1971).

Secondary plant substances may have toxic, repellent, deterrent, or hormonal effects on insects. The toxic effects of certain plants have long been known, and several *botanical insecticides* still find use today (see "Chemical Control" in Chapter 12). Examples include *nicotine* (derived mainly from *Nicotiana tabacum,* the common tobacco plant), the *pyrethrins* (from various species of *Chrysanthemum*), and *rotenone* (from *Derris* spp. and many other legumes) (Fronk, 1978). Citronella and cedar oil are examples of insect repellents derived from plants. Deterrents of various sorts have been identified. These may be substances that are ingested and influence feeding or such things as resins that physically interfere with feeding.

Substances with molting-hormone activity have been found in several ferns (Polypodiaceae) and two families of gymnosperms (Williams, 1970). Many different *phytoecdysones* have been isolated and analyzed; the most common is β-*ecdysone,* the major insect-molting hormone. Many phytoecdysones are more potent than insect hormones. These substances protect plants by interfering with insect development. Juvenile-hormone activity has also been identified in certain plants, especially balsam fir and eastern hemlock (Sláma, 1969; Williams, 1970). *Juvabione,* a juvenile-hormone mimic from balsam fir, is very similar in chemical structure to juvenile hormone. Like the phytoecdysones, juvabione and other plant substances with juvenile-hormone activity cause pathological effects, such as inhibition of metamorphosis and sterilization of adults. Juvenile-hormone activity in plants was discovered by chance when the bug *Pyrrhocoris apterus* failed to complete development in cages the bottoms of which were covered with a particular kind of paper towelling (Sláma and Williams, 1966). The hormonally active paper towelling turned out to be produced from balsam fir. Recently, a secondary plant substance with antijuvenile hormone activity has been identified from the plant *Ageratum houstonianum* (Bowers et al., 1976).

Many insects have "fought back" against plant defenses. An insect (or any herbivore) that breaks through a given plant's defenses obtains a double advantage; it finds a new source of food and does not face competition from other herbivores that have not broken through the plant's defenses. One mechanism that has provided some insects with protection against secondary plant substances has been the development of microsomal mixed-function oxidases (MFOs) secreted in the midgut. These enzymes degrade a wide variety of substances; for example, they are active against pyrethrins. However, in the case of *Chrysanthemum* (producer of pyrethrins), the plant has responded with the production of a substance called *sesamin,* an MFO inhibitor (Weaver, 1978). Interestingly, MFO systems are best developed in polyphagous insect species, which are most likely to be confronted by a variety of secondary plant substances.

Some insects not only have broken through plant defenses but also have come to use secondary plant substances to their own advantage. For example, as previously mentioned, the monarch butterfly accumulates cardiac glycosides, which render this insect unpalatable to predators. For other insects secondary plant substances serve as kairomones (see "Communication" and "Feeding Behavior" in Chapter 8), that is, as attractants that identify food plants.

Selected References

GENERAL

Andrewartha (1971); Boughey (1971); Chauvin (1967); Collier et al. (1973); Kormondy (1976); Krebs (1972); Lewis and Taylor (1967); MacArthur (1972); May (1976); Odum (1971); Price (1975); Ricklefs (1973); Southwood (1979); Whittaker (1975).

POPULATION ECOLOGY

Clarke et al. (1967); Hassell (1978); Messenger (1975); Solomon (1969); Southwood (1966, 1975); Varley et al. (1973); Wilson and Bossert (1971); Wilson (1975).

INFLUENCE OF PHYSICAL ENVIRONMENT

Temperature: Bursell (1974); Wigglesworth (1972).
Insect cold-hardiness: Asahina (1966); Baust and Morrissey (1977); Downes (1965); Salt (1961).
Moisture: Cloudsley-Thompson (1975); Wigglesworth (1972).
Radiation: Grosch (1974).

FOOD

House (1969) (see also Chapters 4 and 8).

INTERSPECIFIC INTERACTIONS

Mutualism: Batra (1979); Houk and Griffiths (1980); Koch (1967); Martin (1970); Way (1963).

Parasitism:

General: Cantwell (1974); Harwood and James (1979); Steinhaus (1949, 1963).

Bacteria: Faust (1974).

Viruses: David (1975); Gibbs (1973); Horsfall and Town (1965); Smith (1967, 1977); Theiler and Downs (1973); Vaughn (1974).

Mycoplasms: Carter (1973); Nienhaus and Sikora (1979); Whitcomb (1973); Whitcomb and Davis (1970).

Fungi: Madelin (1966).

Protozoa: Brooks (1974).

Nematodes: Poinar (1972, 1975); Welch (1965).

Commensalism: Clausen (1976); Felt (1940).

Carnivorous plants: Heslop-Harrison (1978); Swartz (1974).

Coevolutionary adaptation:

General: Gilbert and Raven (1975).

Pollination: Baker and Hurd (1968); Faegri and van der Pijl (1978); Free (1970); Heinrich and Raven (1972); Heinrich (1973); Percival (1965); Proctor and Yeo (1972); Wiebes (1979); Yeo (1973) (see also Chapter 12).

Host/parasite: Duffey (1977); Galun (1977); Nelson et al. (1975); Rothschild (1965a, b); Rothschild and Ford (1966, 1972); Vinson and Iwantsch (1980); Whitcomb et al. (1974).

Predator/prey: Bishop and Cook (1975); Blest (1957); Brower (1969); Brower and Brower (1964); Brower et al. (1968); Cott (1957); Duffey (1980); Dunning and Roeder (1965); Griffin (1959); Kettlewell (1959, 1961, 1973); Portmann (1959); Reichstein et al. (1968); Rettenneyer (1970); Rothschild (1972); Wickler (1968) (see also "Escape and Defense" in Chapter 8).

Insect/plant—other than pollination: Bowers et al. (1976); de Wilde and Schoonhoven (1969); Ehrlich and Raven (1964); van Emden (1972); Jermy (1974); Rosenthal and Janzen (1979); Schoonhoven (1972b); Sondheimer and Simeone (1970); Weaver (1978); Whittaker and Feeny (1971).

PART TWO
Unity and Diversity

10

Systematics and Evolution

One need only look around to appreciate the tremendous diversity of living things, and insects are certainly no exception. Yet a close look reveals that despite the diversity, many organisms (insects) are very similar to one another. In fact, organisms can be grouped on the basis of similarities. This has, of course, been recognized for centuries. Aristotle (300 B.C.) was among the early scientists who devised a grouping (classification) of organisms. Modern classification began with the work of Carl Linné (Linneaus) and others in Europe during the eighteenth century. Tuxen (1973) and Lindroth (1973) trace the historical development of insect systematics.

This chapter first deals with the methods and principles involved in describing, naming, and classifying insects and then considers the evolution of insects beginning with the origin of the phylum Arthropoda.

Systematics

Having gained some idea of insect morphology, physiology, and behavior from the preceding chapters, you are in a position to consider the threads of unity that run through the diversity of insects. Finding these threads of unity—morphological, physiological, or otherwise—and utilizing them to produce meaningful and useful groupings of insects is the objective of *insect systematics*. Some authors define systematics simply as the science of classification, making it synonymous with another widely used term, *taxonomy*. Others expand the definition to include the study of evolution.

Regarding the place of systematics among the other specialties in the biological sciences, the following quotations are appropriate.

> It is probably safe to say that in point of years of service systematic biology is the senior branch of the whole field of the study of living things. It can hardly be otherwise, for discrimination among or between facts or things is one of the very first steps in the organization of our knowledge of these things or facts and our use of them.—(Ferris, 1928)

> Systematics is at the same time the most elementary and the most inclusive part of zoology, most elementary because animals cannot be discussed or treated in a scientific way until some systematization has been achieved, and most inclusive because systematics in its various guises and branches eventually gathers together, utilizes, summarizes, and implements

everything that is known about animals, whether morphological, physiological, psychological, or ecological.—(Simpson, 1961)

The major tasks of systematics are

1. Identification.
2. Description.
3. Concern with the proper application of the rules of nomenclature.
4. Study of speciation.

These tasks are typically carried out in sequential order for a given group of organisms. For a few groups of insects, systematics has reached the speciation level, but most groups are still in the identification, description, and classification stages of development.

Obviously, collection and preparation of specimens for study must precede identification and the other steps. Discussion of collection and preparation techniques would be inappropriate here, but see Bland (1978); Borror, DeLong, and Triplehorn (1976), Borror and White (1970), Cummins et al. (1965), Knudsen (1966), Peterson (1959), and Ross (1962). In addition to collection of one's own specimens for study, extensive collections, housed in museums throughout the world, are usually available for serious study.

It is often desirable to culture certain species. Such cultures can provide a wealth of useful biological data (e.g., biochemical, serological, physiological, developmental, genetic), that would be unavailable from preserved specimens. Information on culturing insects can be found in Peterson (1959), Needham et al. (1937), and Siverly (1962).

Identification

Purpose of Identification. The purpose of identification is to determine what kind of organism a given specimen is. The meaning of "kind" depends largely upon one's objectives. A student in a general zoology course may be happy to identify a given organism as a fly, whereas a professional entomologist may need to know the species of fly. Professional reasons for wanting to know the species of an organism (insect) fall into two categories: the nonsystematist's reasons and the systematist's reasons. The nonsystematist, for example, needs to identify a specimen in order to find pertinent literature and to have a name under which to publish his or her findings. Thus the species name of an organism serves as the "file category" under which any and all information pertinent to this species is stored.

The systematist, in addition to using the species name as a key to the literature and a category for data storage, needs to identify an organism to species or at least to the point where it is realized that the organism has not been previously described. This has been called *taxonomic discrimination* (Mayr, Linsley, and Usinger, 1953). Identification of a specimen must necessarily precede fitting this specimen into a scheme of classification.

Methods of Identification. Ultimately, all methods used to identify organisms are based on comparison. One of the most common methods of identification is the use of a key. A key is a printed information-retrieval system into which one puts information regarding a specimen-in-hand and from which one gets an identification of that specimen to whatever level the key is designed to reach. Most keys are *dichotomous* (double branching; Fig. 10-1). At the beginning of a key, one is presented with two alternatives (a *couplet*), each of which leads to another pair of alternatives. At each point the user of a key picks the alternative that best describes some aspect of the specimen being identified. Finally one reaches a terminal pair, one member of which presents the identity of the organism. The simplest keys are designed for the layman and may offer only superficial identification or deal only with insects found in a specific habitat, such as a garden. More complex keys are designed for the nonsystematist professional biologist who needs to identify an organism to a more specific level than the layman. These keys are more technically oriented and require some training in morphology before they can be used effectively. Finally, there are keys designed for the specialist, which are very technical and require detailed knowledge of the morphology of a particular group of organisms. These keys usually carry identification to the species level.

Another method of identification is the use of pictures in the form of color plates, black-and-white photos, or line drawings. These pictures may be of entire organisms or parts of organisms and are often used in conjunction with keys. "Pictorial keys" offer alternatives in the form of pictures with or without accompanying verbal alternatives. Pictures are particularly useful when an organism is highly patterned or characteristically colored.

One of the best means for identifying a specimen is to compare it with specimens that have previously been identified. This approach can be used to obtain an identification at any level. *Type specimens* are those on which an original species description is based. These specimens may be extremely useful in making species identifications. Original written descriptions of species based on type specimens are the permanent records of the attributes of a given species. Such records are particularly useful when the type specimens are available, and they constitute the only original record if type specimens are destroyed, lost, or unavailable.

Usually, combinations of the above methods are used in identification. Although the layman or nonsystematist professional biologist can—in many, if not most, instances—identify a specimen to

Fig. 10-1
Sample dichotomous key that differentiates the subclasses of insects from centipedes and millipedes.

1A	two body regions (head and trunk) 2
1B	three body regions (head, thorax, and abdomen) .... an insect, CLASS INSECTA, 3
2A	one pair of legs per body segment; poison jaws present .... a centipede, CLASS CHILOPODA
2B	two pairs of legs per most body segments; poison jaws lacking ... a millipede, CLASS DIPLOPODA
3A	wings absent .... SUBCLASS APTERYGOTA
3B	wings present or secondarily absent, i.e., winged ancestors .... SUBCLASS PTERYGOTA

family, or in some cases below the family level, the best and safest way to obtain a species identification is to consult a specialist. From experience and familiarity with pertinent literature, the specialist is in the best position to use the highly specialized keys, original descriptions, and type specimens. In view of the vast number of insect species already described, a given individual is likely to specialize in the systematics of only a very few families or even a single family.

Problems Encountered in Identification. Unfortunately, identification is not without problems. Keys may not include the organism being identified, fail to include the stage or sex of the unknown organism, or fail to take into account any of a large number of possible variations in morphological or other characters used. Pictures can be misleading, especially if an unknown organism very closely resembles the one depicted. Written descriptions can be ambiguous, particularly when such traits as color and texture are described, and such descriptions commonly require knowledge of specialized terminology to be of use. Original species descriptions are sometimes in early volumes of obscure journals that may be difficult to obtain and may be in a foreign language. A collection of accurately identified specimens is not always available. Type specimens may be very difficult or impossible to obtain and may have been lost or destroyed. The major difficulty with consulting a specialist on a given group of organisms is that such a person may be dead or perhaps have never existed. Even if a specialist does exist, there is still the problem of contacting this person.

Description

The systematist, having completed an identification, will follow one of two courses. If a species has already been described, it will probably be put aside for possible future study with other specimens of the same species. If not, it will be described as a new species and placed in whatever grouping the systematist sees as appropriate. This description will serve to identify this species for other investigators. The production of such descriptions is the job of descriptive systematics.

Subjects of Description. Ideally, specimens to be described are obtained by a carefully planned and executed procedure such that they constitute a statistical *sample* of a population (i.e., they display the same distribution of traits as do the members of the entire population). The other, and more common, alternative is that a group of specimens is obtained in a more or less erratic fashion; perhaps several individuals were taken in the same locality, or perhaps different individuals were taken from different localities at different times and under different circumstances. Such a group is termed a *series*. Although many species descriptions are based on series and samples, several are based on a single specimen or on a series of a few individuals.

Characters Described. A *character* is any possible trait an individual might possess. This could be anything from the shape of a particular sclerite (morphological character), to a particular kind of amino acid in the hemolymph (biochemical character), to a particular mechanism for excretion (physiological character), to a specific way of responding to a change in photoperiod (behavioral character). Obviously, since we have as yet no way of arriving at perfect knowledge of anything, all characteristics of a given organism are not used in a description of that organism. It follows, then, that any species description plus all information gained since the discovery of that species is only a partial picture and that a species description in this sense is never really completed. In reality the vast majority of species descriptions are based on selected morphological features, and although the modern trend is definitely in the direction of including other sorts of data (particularly physiological, biochemical, serological, and behavioral), morphological characters will likely continue to be central to most descriptions, at least of insects.

There is a wide variety of characters to choose from in a description (Mayr, Linsley, and Usinger, 1953).

1. Morphological characters
    a. General external morphology
    b. Special structures (e.g., genitalia)
    c. Internal morphology
    d. Embryology
    e. Karyology (and other cytological differences)
2. Physiological characters
    a. Metabolic factors
    b. Serological, protein, and other biochemical differences
    c. Body secretions
    d. Genic sterility factors
3. Ecological characters
    a. Habitats and hosts
    b. Food
    c. Seasonal variations
    d. Parasites
    e. Host reactions
4. Ethological characters
    a. Courtship and other ethological isolating mechanisms
    b. Other behavior patterns
5. Geographical characters
    a. General biogeographical distribution patterns
    b. Sympatric–allopatric relationship of populations

It is extremely important to know the extent to which a given character used in a description varies within a species. For characters to be of value in identification and classification they must be reasonably constant or vary predictably. Failure to recognize variation can lead to confusion. There are any number of potential kinds of variation that can occur within a population of a given species (Linsley and Usinger, 1961):

1. Extrinsic or noninherited variation
    a. Variation due to age
    b. Seasonal variation
    c. Castes in social insects

   d. Variation due to habitat
   e. Variation due to crowding
   f. Climatically induced variation
   g. Heterogonic variation
   h. Traumatic variation
  2. Intrinsic or inherited variation
   a. Primary sex differences
   b. Secondary sex differences
   c. Alternating generations
   d. Gynandromorphs
   e. Intersexes
   f. Mutations resulting in continuous variation
   g. Mutations resulting in discontinuous variation
   h. Genetic polymorphism

Classification

The many hundred thousands of species descriptions would be practically impossible to deal with if they were not organized in some fashion. Such organization is necessary because of the differences between organisms and at the same time is made possible by the similarities among them. By grouping organisms based on degrees of similarity, one can arrive at a system of classification.

Classifications serve several purposes (Warburton, 1967):

1. Classifications provide intellectual satisfaction for the taxonomists who make them.
2. They make possible the identification of species and higher groupings.
3. They ". . . provide a convenient, practical means by which zoologists may know what they are talking about and others may find out" [Warburton quoting G. G. Simpson.]
4. They provide a system of information retrieval.
5. They may reflect phylogenetic relationships [to be discussed later].
6. They may serve as summarizing and predicting devises.

The first purpose listed requires no elaboration and is probably the primary motivation for the efforts of most scientists. The ability to place a given organism at some point in a classification facilitates further identification by ruling out myriads of other organisms. For instance, if one can determine the class, order, or even family of an organism by determining where it fits in an already established classification, one is several steps closer to a complete identification.

A classification's value as a basis for communication is obvious. As mentioned earlier, an organism's name is the "file category" under which all information pertaining to that organism is stored. Within a classification one has not only the name of an organism but also the names of the successively more inclusive groups of which it is a part and hence the "file categories" of the literature pertinent to these groups.

A classification, by the nature of the way it is constructed—a descending or ascending hierarchy of successively less or more inclusive groups—is a summarizing device, since all categories are arrived at by the grouping of organisms on the basis of similarities. Thus a great deal of information is summarized merely by recognizing that a given organism is an insect. Similarly, an organism's position in a classification enables one to predict certain additional things about that organism.

Characters Used in Classifying Organisms. When developing a classification, only *homologous* characters—those that could theoretically be traced back to a common ancestor—are used as comparative bases. For example, the wings of all insects, although they vary considerably in both structure and function, are fundamentally similar and are assumed to have been derived from an ancestor common to all winged insects. Hence insect wings are considered to be homologous among the various insectan groups. In practice, homology is not always easy to establish. For example, one might note the rather close similarity between the raptorial forelegs of a praying mantid (Fig. 11-21) and those of a mantispid (Fig. 11-40A) and conclude that these two insects share a common ancestor with raptorial forelegs. However, such is not the case. Based on comparison of many other characteristics, mantids and mantispids are placed in widely divergent orders (Mantodea and Neuroptera, respectively). Thus the character "raptorial" is not homologous in these two insects and must have arisen independently in each. Similarities between nonhomologous characters are considered to be due to *convergent evolution,* that is, evolutionary responses (adaptation) to similar selection pressures. In the preceding example, then, we would view the raptorial legs as representing closely similar, but independent, adaptations associated with prey capture.

Since common ancestry is nearly impossible to prove directly, a number of pragmatic criteria for judging homology exist. For example, two characters are considered to be homologous on the basis of minutely detailed resemblance, close similarity in embryological origin, and the occurrence of many other homologous features in the same organism. Obviously, the existence of a complete fossil lineage back to a common ancestor would clinch a decision of homology. Such complete lineages unfortunately do not exist.

Kinds of Classifications. There are basically two schools of thought regarding classification, the *cladistic (phylogenetic)* school and the *phenetic* school (Scudder, 1973).

The cladistic approach assumes that the evolutionary history of organisms can be deduced (to a greater or lesser degree of accuracy) by careful comparative studies. Further, it is assumed that certain characters have undergone greater evolutionary modification than others and hence are more useful in establishing the most probable evolutionary sequences of branching lineages (such "family trees" are sometimes called *cladograms*). Thus certain features are given more *weight* than others in deducing relationships between groups of organisms. Hennig (1965, 1966) outlines the major principles of the cladistic approach. Boudreaux (1979) interprets arthropod phylogeny (insects included) using a cladistic approach.

The phenetic school (Sokal and Sneath, 1963) argues that phylogeny cannot be deduced with any certainty and that it is more appropriate to establish relationships (affinities) between organisms on the basis of as many features as possible, giving all features equal weight. A classification derived in this fashion is called a phenetic classification, and the associated "family tree" is a *phenogram* (or *dendrogram*). Interestingly, the classifications generated by the phenetic approach often resemble closely those generated by the cladistic approach. As an example of a numerical taxonomic analysis of a group of insects see Kamp (1973).

Components of Biological Classification. Four components form the basis for most systems of biological classifications (definitions from Simpson, 1961).

1. Hierarchy—"a systematic framework for zoological classification with a sequence of classes (or sets) at different levels in which each class except the lowest includes one or more subordinate classes."
2. Taxon (pl. taxa)—"a group of real organisms recognized as a formal unit at any level of a hierarchic classification."
3. Category—"a class, the members of which are all the taxa placed at a given level in a hierarchic classification."
4. Rank—a category's "absolute position relative to other categories."

Thus Insecta (taxon) is a class (category) that in a widely accepted biological hierarchy (Fig. 10-2) is ranked between superclass and subclass.

Taxonomic Categories. The *species* is the basic unit of taxonomic classification, referring directly to real populations of real animals. The *subspecies* is a subdivision of a species and also refers to real animals. On the other hand, *higher categories* (all those categories in the biological hierarchy above the species level) are essentially subjective, their nature depending upon the professional

Kingdom
Phylum
Subphylum
Superclass
Class
Subclass
Cohort
Superorder
Order
Suborder
Superfamily (*-oidea*)
Family (*-idae*)
Subfamily (*-inae*)
Tribe (*-ini*)
Genus
Subgenus
Species
Subspecies

Fig. 10-2
Hierarchy of generally accepted taxonomic categories.

opinions of the taxonomists who "recognize" them. They refer to real animals indirectly in that they ultimately include one to several species.

Despite its objective reality, a number of different concepts revolve around the term "species." Early, pre-Darwinian biologists considered each species to have arisen as a result of special creation and to be composed of nonvarying individuals. With this *type concept* of species, a single specimen was all that was necessary to describe a species, and morphological characters, sometimes supplemented with geographical data, were generally the only ones used. Note that the type species and the use of "types" in species description mentioned earlier are very different. Types in the latter sense serve a pragmatic purpose, but modern concern is with series or samples of individuals of a species, and great stress is placed on identifying the kind and degree of variation of any character used.

With modern understanding of organic evolution and heredity, a new concept of species, the *biological species*, has developed. According to this concept, "Species are groups of actually (or potentially) interbreeding natural populations which are reproductively isolated from other such groups" (Mayr, Linsley, and Usinger, 1953). However, according to Blackwelder (1967);

> Morphogeographical species are those of the ordinary taxonomist, from Linneaus to modern times. Although data other than "morphology" and geography have been increasingly used in recent years, these still remain the basic species of taxonomy.

A *subspecies* may be defined as " . . . an aggregation of local populations of a species, inhabiting a geographical subdivision of the range of the species, and differing taxonomically from other populations of the species" (Mayr, 1963). In other words, subspecies are geographical races within the same species that are sufficiently different in some regard for them to be classified in this manner. Based on the biological species definition, subspecies should be able to interbreed if given the opportunity to do so.

As mentioned earlier, the higher categories are subjective, depending on the opinions of taxonomists. This subjective nature is apparent when one realizes that a taxon that was once an order is now a family (or vice versa) or that a given taxon may be at the same time recognized as a family by some systematists and as an order by others, for example, order Blattaria (cockroaches) versus family Blattidae (cockroaches when included in the order Orthoptera). Further, the taxa of one group of animals usually do not correspond to the taxa of another group relative to the deversity in a given category. For example, the orders of birds are much less distinct than the orders of insects.

A higher category may appropriately be defined in relation to the category immediately below it in the hierarchy. For example, a genus may be defined as a group of species that have a sufficient number of features in common to warrant inclusion within a single group. In practical terms the decision regarding "a sufficient number of features" is up to the systematist making the decision. Generally, the higher the category, the greater the stability with regard to changes. In other words, genera are changed by systematists much more frequently than families or orders.

Future of Classification.   Classifications should be looked upon as organizational systems that are continually undergoing a sort of "evolutionary" development in light of new information about organisms and their relationships to one another. "The well-classified parts of the animal kingdom form no more than a hundredth part of the known fauna and probably only a thousandth part of what actually exists" (Blackwelder, 1967). Since the largest portion of the animal kingdom is held by insects, at least a major part of the job will have to be done by insect systematists.

Nomenclature

Nomenclature is involved with the naming of organisms and groups of organisms and the rules and procedures to be followed in such naming. The rules of nomenclature have been developed and gradually changed and modified over the years. The current rules are in the 1964 revision of the International Code of Zoological Nomenclature.

Scientific Names.   Every species and higher category has a Latinized scientific name. At the species level, the scientific name consists of two parts, the genus and the *specific epithet,* which comprise the *binomen.* An example would be *Aedes occidentalis,* in which *Aedes* is the genus and *occidentalis* the epithet. These two words together comprise the scientific name of a species of mosquito. The epithet is not the species name and should not be used as such. By convention, all species names are printed in italics (or underlined when written or typed without italics), the generic name is capitalized, and the epithet is all lowercase. Actually a complete species name can be more complex than a simple binomen, although the binomen is still valid to use. For example, the complete scientific name of one of the subspecies of the mosquito species just mentioned is *Aedes (Finlaya) occidentalis occidentalis* (Skuse, 1889) *(Culex).* The first and third terms are the binomen. The second term, in parentheses, is the subgenus and the fourth indicates the subspecies. Skuse is the taxonomist who first described this species, and his description was published in 1889. The parentheses indicate that Skuse placed this species in a different genus than *Aedes.* The last name, *Culex,* in parentheses, is the genus in which Skuse placed this species. Sometimes a generic name is written followed by sp. or spp., for example, *Aedes* sp., or *Aedes* spp., if the specific epithet is unknown. These abbreviations indicate a particular species or several species within a given genus, respectively.

The names of families are uninominals that end in -idae, and name changes are strictly regulated. All family names are based on a type genus; for example, the family name of tiger beetles, Cicindelidae, is based on the generic name *Cicindela.*

The names of categories above family are all uninomials and are not extensively considered by the international code. They are used in the form of plural Latin nouns (e.g., Pterygota, Insecta), and no type categories are involved. The categories used in a given classification depend entirely upon the opinion of the taxonomist preparing the classification. Like family, several other categories have standardized endings (Fig. 10-2).

Common Names. Insects that are particularly common in a given locality may be given vernacular names. For example, dragonflies have been variously referred to as "mosquito hawks," "snake feeders," and "devil's darners." Such names have little value other than in the context of local color. However, certain common names are useful when applied to agricultural or medical pests since they facilitate communication between the professional and layman. Most orders and families and many species have common names that over the years have become well established. Two particularly outstanding common names are "bugs" and "flies." Although these words are found in several ordinal and familial names (butterflies, mayflies, doodlebugs, etc.), they apply specifically to the Hemiptera-Heteroptera (true bugs) and the Diptera (true flies). When used in reference to members of these orders, "fly" and "bug" are written as separate words (e.g., house fly and conenose bug). However, if they are used for members of any other order, they are written with a modifier of some sort as a prefix (e.g., dragonfly or ladybug). When there is no common name, or when it is preferable not to use a common name, the scientific name may be used as a noun or modifier (e.g., mosquitoes may be called culicids, derived from the family name Culicidae).

Insect Evolution

Evolutionary change may be viewed as occurring in three successively longer time frames: *microevolution,* the changes that occur in populations usually in response to changes in selection pressures, such as industrial melanism (see Chapter 9) and the development of insecticide resistance in a population (see Chapter 12); *speciation,* the changes that typically occur over much longer time spans than microevolution and that result in reproductively isolated, new species; and finally *macroevolution,* the major phylogenetic patterns that develop over wide spans of geological time.

The objective of this section is to outline the major macroevolutionary events that are thought to have occurred among insects and their relatives. The discussion includes consideration of the origin of the arthropods, origin of the insects, and phylogeny within the class Insecta.

At least a rudimentary knowledge of evolutionary theory is assumed in this discussion. Readers unfamiliar with natural selection, speciation, the concept of geological time, and so on, should consult a good general biology or zoology text or any of several excellent texts available which deal specifically with organic evolution (see Selected References).

Comparison of present-day (*extant*) forms has led to the recognition of structural and developmental affinities among the Annelida, Arthropoda, and Onychophora.

Annelida.   Annelida, the segmented worms, apparently evolved from the ancient protostomes, which also gave rise to the mollusks (Meglitsch, 1967). Annelids (Fig. 10-3) are soft-bodied, elongate, cylindrical, bilaterally symmetrical animals. Their bodies are composed of repeating segments or metameres capped anteriorly by the *prostomium* or *acron* and posteriorly by the *periproct* or *telson*. The mouth is on the venter between the prostomium and first metamere. The anus opens from the periproct. Some annelids have several bilateral pairs of segmental appendages called *parapodia*, which arise as evaginations of the body wall. Segmentation of the body is evident in nearly all tissues save the alimentary canal, which is essentially a longitudinal tube running from the mouth to the anus. The circulatory system is a closed system of tubes, usually one dorsal longitudinal and one ventral longitudinal vessel, connected by lateral trunks that provide vascularization for the various body tissues. The blood commonly carries a respiratory pigment. The body cavity is a coelom divided into segmental compartments. Excretion is accomplished by paired segmental *nephridia,* tubes that communicate between the coelom and the exterior. Ventilation is cuticular or by means of gills evaginated from the body wall. The nervous system

Fig. 10-3
Representatives of four classes of segmented worms, phylum Annelida. A. Chaetogaster (Oligochaeta). B. Leech (Hirudinea). C. Earthworm (Oligochaeta). D. Dinophilus (Archiannelida). E. Lugworm (Polychaeta). [Redrawn from Stiles, Hegner and Boolootian, 1969.]

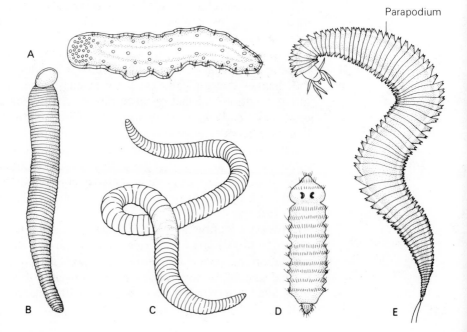

Parapodium

A

B   C   D   E

is composed of an anterior prostomial ganglion which lies dorsal to the alimentary canal. It is connected by a pair of bilateral connectives to a ventral chain of paired segmental ganglia linked by paired longitudinal connectives. Some annelids have a *trochophore* larval stage, which has played an important role in assessments of the evolutionary relationships of annelids with members of other invertebrate phyla.

Arthropoda. Arthropods, the joint-footed animals, are a highly successful diverse group of animals unrivaled by other animals in diversity of structure and function. They are found in nearly every imaginable ecological niche. A feature that has played a central role in their success is their relatively impermeable exoskeleton. Such an impermeable covering was a prerequisite for invasion of the terrestrial environment, which poses a continuous challenge to the water-holding capabilities of all animals and of arthropods in particular because of their generally small body size (see "Permeability Characteristics" in Chapter 2). Additional adaptations are discussed later in this chapter.

Arthropods (Figs. 10-4 through 10-7) have segmented bodies composed of chitinous exoskeletal plates of varying degrees of hardness, which are separated by less-hardened "membranous" areas. Paired segmental appendages are usually present on at least some body segments. Arthropods share several fundamental traits with annelids. Particularly outstanding are their segmented bodies and the ventral chain of segmental ganglia. On the other hand, the arthropods have an open circulatory system, and the body cavity is a hemocoel; coelomic sacs appear, during embryogenesis, but do not form the body cavity. Also most arthropods ventilate by means of a tracheal system and have well-developed appendicular mouthparts. For more information on the arthropods, see Boudreaux (1979), Clarke (1973), and Manton (1977).

Extant arthropods include the chelicerates, crustaceans, myriapods, and insects.

The chelicerates (Figs. 10-4 and 10-5) include the spiders, mites and ticks, scorpions, solpugids, and their relatives (class Arachnida). Horseshoe crabs (Xiphosura, e.g., *Limulus*), sea spiders (Pycnogonida), water bears (Tardigrada), and tongueworms (Linguatulida or Pentastomida) are variously classified, some authors including them with the chelicerates. Chelicerates are characterized by the presence of six pairs of appendages; the anterior pair are called *chelicerae* and are jawlike in more primitive forms. They have been modified in a variety of ways in more advanced forms. The second pair of appendages are called *pedipalps* and also vary in structure. The remaining four pairs of legs generally have a locomotor function. Antennae are lacking, and there are usually two body regions; a fused head and thorax (*cephalothorax*) and an abdomen. The appendages are borne by the cephalothorax.

The crustaceans (crayfish, lobsters, crabs, shrimp, barnacles, copepods, sowbugs, and so on; Fig. 10-6) have two pairs of antennae and a pair of mandibles. The crustaceans commonly have two body regions: a cephalothorax, which is covered by a *carapace,* and an

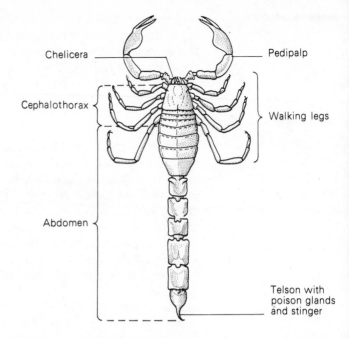

Chelicera

Pedipalp

Cephalothorax

Walking legs

Abdomen

Fig. 10-4
Scorpion, illustrating the major
chelicerate features. [After U.S.
Public Health Service.]

Telson with
poison glands
and stinger

abdomen. These arthropods are mostly freshwater or marine
animals.

Myriapods are cylindrical, have a well-defined head bearing primitive mandibulate mouthparts, and have a trunk with nine or more
bilateral pairs of legs. There are four myriapod classes: Diplopoda
(millipedes), Chilopoda (centipedes), Pauropoda (pauropods), and
Symphyla (symphylans). All are terrestrial animals.

The insects possess a head, thorax (usually with wings and legs),
and an abdomen. This is the only group of invertebrates that contains
members capable of active flight. Insects are primarily terrestrial
with many secondarily aquatic forms.

A major group of extinct arthropods, the trilobites, have played
an important role in the study of arthropod evolution. These animals
(Fig. 10-7) became extinct or very rare in the Permian period (see
Fig. 10-14) and therefore are known only from the fossil record.
They were marine animals, and their bodies were made up of three
tagmata: head, thorax, and pygidium. The name Trilobita refers to
these three tagmata. The head and first few thoracic segments were
covered by a carapace, most of the body segments had a bilateral
pair of jointed legs, and the head carried a single pair of antennae.

Onychophora. Onychophorans are a very small group of terrestrial animals, but are significant in that they show morphological
affinities with both annelids and arthropods. These animals (Fig. 10-
8) have elongate, wormlike bodies with several bilateral pairs of
lobelike legs. They are not segmented as adults, but are clearly
segmented during their embryogenesis. They have paired nephridia
(excretory organs) along the body that open near the leg bases. This
fact, together with other morphological characters, suggests an evolutionary relationship with the annelids. On the other hand, onychophorans share several characters with arthropods, including a
chitoproteinous cuticle (Hackman and Goldberg, 1975), an open

circulatory system with a hemocoel, tracheal ventilation, and appendages associated with the mouth. Onychophorans have been variously classified as annelids, arthropods, or as a separate phylum.

Origin of Arthropoda

The basic similarities among annelids, arthropods, and onychophorans strongly suggest common ancestry at some point in metazoan evolution. Some authors have, in fact, classified all three groups as members of a superphylum, Annulata (Snodgrass, 1952). Sharov (1966), in a similar vein, supports Cuvier's* concept of the phylum Articulata with annelids, onychophorans, and arthropods being rec-

*Georges Cuvier (1769–1832) was among the founders of paleontology, the study of the fossil record.

Fig. 10-5

Representatives of the subphylum Chelicerata. A. Spider, *Loxosceles* sp. (Arachnida; Araneida). B. Mite (Arachnida; Acari). C. Tick (Arachnida; Acari). D. Wind scorpion (Arachnida; Solpugida). E. Harvestman or daddy longlegs (Arachnida; Phalangida). F. Horseshoe crab, *Limulus* sp. (Xiphosura). [A–E after U.S. Public Health Service.]

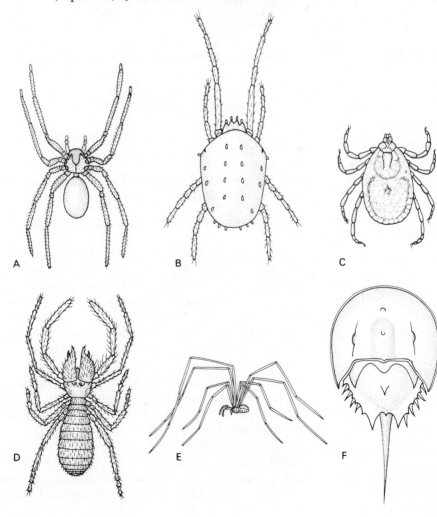

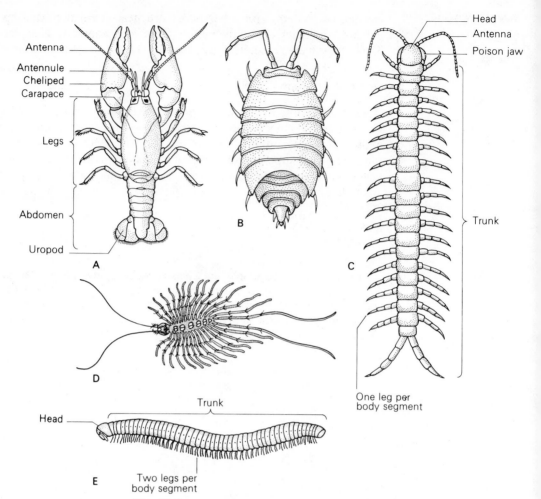

Fig. 10-6

Representative mandibulates. A. Crayfish (Crustacea). B. Sowbug (Crustacea). C. Centipede, *Scolopendra heros* (Chilopoda). D. Eastern house centipede, *Scutigera cleoptrata* (Chilopoda). E. Millipede, *Narceus americanus* (Diplopoda). [After U.S. Public Health Service.]

ognized as separate subphyla. The details of the evolutionary relationships among these three groups are purely hypothetical. Unfortunately, knowledge of the fossil record is of little help, since each group is distinctly defined in the Cambrian strata, from which the oldest "good" fossils come. This, of course, means that the supposed common ancestor arose in pre-Cambrian times (see Fig. 10-14). There is, however, general agreement that the common ancestor was an annelid or annelidlike creature.

Some have suggested that the primitive arthropod evolved directly from a primitive polychaete annelid (Meglitsch, 1967). Snodgrass (1952), later supported by Sharov (1966), conceived of a segmented wormlike animal that produced two branches, one ultimately giving rise to the polychaete annelids and the other to a form with paired, undifferentiated lobelike legs (*lobopod*). He envisioned the lobopods as subsequently branching into arthropodan and onychophoran lines. Once established, members of the phylum Arthropoda are viewed by some as having branched out into three groups (subphyla): the trilobites (Trilobita), the chelicerates (Chelicerata), and the mandibulates (Mandibulata), which include the Crustacea, myriapods, and insects (Fig. 10-9).

There is disagreement as to whether the arthropods actually constitute a single phylum (are *monophyletic*), sharing a common arthropodan ancestor, or are actually two or three groups (are *polyphyletic*) of distantly related animals in which major arthropodan traits, especially jointed appendages but also many other structures, evolved convergently (Scudder, 1973; Boudreaux, 1979). Recent cases for the latter point of view have been made by Tiegs and Manton (1958), Manton (1964, 1970, 1972, 1979), Cisne (1974), and Anderson (1973, 1979). On the basis of mandibular structure and movement and locomotor limbs, Manton recognizes three groups that became ''arthropods'' convergently (Fig. 10-10): (1) the trilobites and chelicerates; (2) the Crustacea; and (3) the Onychophora, myriapods, and hexapods (insects). Anderson (1973, 1979) corroborates this point of view embryologically. The trilobites, chelicerates, and crustaceans have primitively *biramous appendages* and *gnathobasic mandibles*. The form of the gnathobasic mandibles and biramous appendages in the trilobites and chelicerates is fundamentally different from those of the crustaceans. On these and other bases, the trilobites and chelicerates are viewed as evolving independently. The Uniramia all have similar uniramous limbs and bite with the tips of the mandibles; thus they are considered to constitute a monophyletic group. Cisne (1974) studied parts of the internal anatomy of pyritized trilobites using X rays and provides evidence in support of a polyphyletic origin of arthropods. However,

Fig. 10-7
Drawings of a trilobite, *Triarthrus becki,* which have been composed from examination of fossils. A. Dorsal view. B. Ventral view. [Redrawn from Ross, 1965 (from Schuchert after Beecher).]

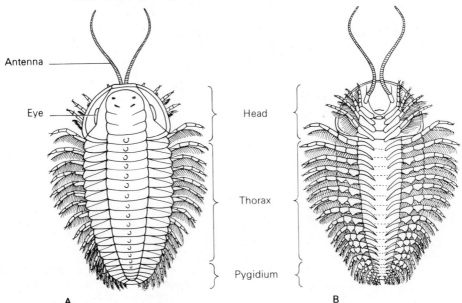

Antenna

Eye

Head

Thorax

Pygidium

A

B

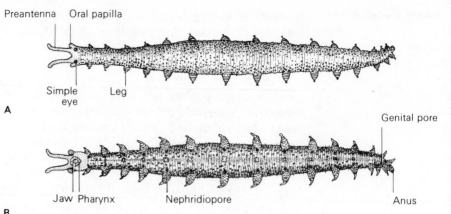

Fig. 10-8
Onychophoran, *Peripatus*. A. Dorsal view. B. Ventral view. [Redrawn from Boreal Labs, key card.]

he views the trilobites as a link between the chelicerates and crustaceans and hence recognizes two arthropod lines (Fig. 10-11).

Although there is substantial evidence for a polyphyletic origin of arthropods, acceptance of this point of view requires acceptance of the independent, convergent origin of several characters: a hemocoel; paired, jointed locomotor appendages; tracheae; Malpighian tubules; compound eyes; a chitinous cuticle, and so on. Some experts find this very difficult to accept. In a recent multiauthored volume entitled *Arthropod Phylogeny* (Gupta, 1979), a monophyletic origin of arthropods is supported by studies of the embryology of the head, olfactory mechanisms and sensilla, eye structure, visceral anatomy, intersegmental tendon systems, and sperm ultrastructure. Boudreaux (1979) also favors a monophyletic explanation of arthropod origin and provides an especially useful discussion of arthropod phylogeny. Unfortunately, the fossil record has not provided sufficient infor-

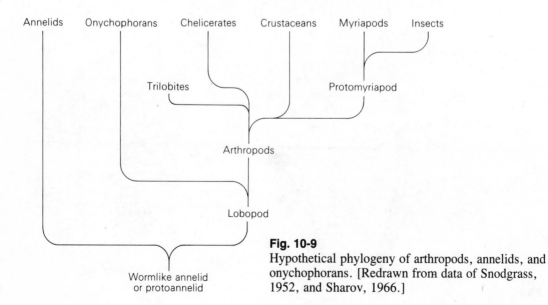

Fig. 10-9
Hypothetical phylogeny of arthropods, annelids, and onychophorans. [Redrawn from data of Snodgrass, 1952, and Sharov, 1966.]

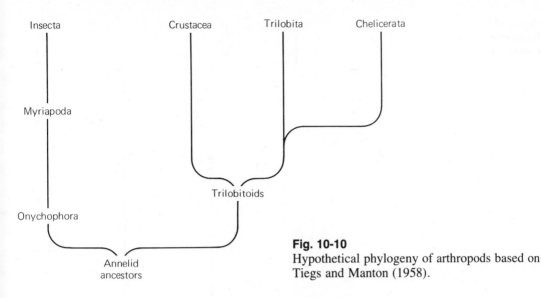

Fig. 10-10
Hypothetical phylogeny of arthropods based on
Tiegs and Manton (1958).

mation to resolve the issue of a mono- versus polyphyletic origin of
the arthropods (Bergström, 1979).

Origin of Insects

The class Insecta is generally considered to have evolved from a
myriapod or protomyriapod of some sort during the Devonian period. Based on differences in mandibles and mandibular movement,
Manton (1964) concluded that insects are not direct descendants of
myriapods and that these two groups are best looked upon as sharing
a common ancestor.

A series of diagrams originated by Snodgrass (1935) and modified
by Ross (1965) is helpful in a very simplified way in visualizing the
hypothetical origin of the insects. The first (Fig. 10-12A) represents
the segmented, legless, wormlike annelid or annelidlike stage. The
undifferentiated body is composed of a series of somites or metameres capped anteriorly by the *prostomium* (*acron*) and posteriorly
by the terminal body segment, the *periproct* (*telson*). The mouth is
located between the prostomium and the first body segment; the anus
opens in the periproct. Figure 10-12B represents the evolutionary
stage in which were developed paired, bilateral, lobelike appendages
on the somites as well as a pair of simple eyes on the prostomium
and antennae on the second somite. The appendages on the first
somite have been lost altogether. This level of organization is ony-
chophoranlike. The next stage (Fig. 10-12C) represents a proto-
myriapod/protoinsect in which arthropodization has occurred; that
is, the bilateral appendages of each body segment have themselves
become segmented. The appendages of somites 4, 5, and 6 have
been reduced and have moved into close association with the primitive head, becoming involved with the manipulation of food. The
appendages of the hindmost somite have become sense organs and
no longer function in locomotion. As mentioned earlier, there are
varying opinions as to the origin of the myriapod and insect groups,
but there is little doubt of a close relationship between these two

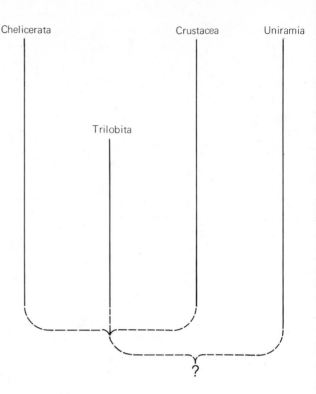

Chemicerata   Crustacea   Uniramia

Trilobita

Fig. 10-11
Hypothetical phylogeny of arthropods
based on Cisne (1974). Uniramia includes
onychophorans, myriapods, and insects.

groups. Figure 10-12D can be looked upon as representing the myriapod level of organization, bilateral appendages being retained on most body segments and the appendages of the 7th, 8th, and 9th body segments becoming the typical, primitive mandibulate mouthparts. The insectan level of organization is represented by Fig. 10-12E. The body has been differentiated into the three tagmata (head, thorax, and abdomen) characteristic of insects. The appendages of segments 7, 8, and 9 have been retained as locomotor structures, but locomotor appendages have disappeared from the remaining body segments. The appendages of abdominal segments 8 and 9 have been modified as external genitalia, and the cerci, appendages of segment 11, have been retained. The telson has been lost, and the anal opening is now within the 11th abdominal segment.

The evolution of the insect head from the prostomium plus four somites, which is implied in Figure 10-12, has been disputed. Rempel (1975) reviews the evidence regarding various theories of origin of the insect head and supports the one depicted in Figure 10-13 where the head originates from the fusion of the prostomium and six somites.

Insect Phylogeny

Carpenter (1953, 1977) recognizes four major stages in the evolution of insects. The first was the appearance of primitive wingless insects, which probably resembled contemporary Archeognatha and Thysanura. These primitive *apterygotes* are thought to have arisen during the Devonian period (Fig. 10-14). The fossil record of apterygotes is poor, probably because of their relatively soft bodies. Other insect groups with harder body parts and particularly with wings have been more amenable to fossilization processes. Despite lack of extensive fossil evidence, insects are generally thought of as originally having been wingless.

The second major step was the development of wings, which is thought to have occurred prior to the Lower Carboniferous period (Fig. 10-14). Insects were the first animals to invade the air, doing so some 50 million years before the reptiles and birds. Since predatory amphibians, reptiles, and arthropods were abundant at this time, wings no doubt provided a strong selective advantage by facilitating escape. The early winged insects had very simple wing articulations that allowed flight, but at rest the wings were held out from the body, not being flexed out of the way over the abdomen.

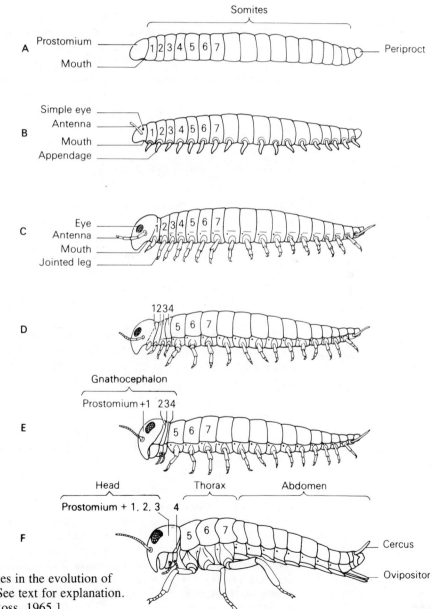

Fig. 10-12
Hypothetical stages in the evolution of the insect form. See text for explanation. [Redrawn from Ross, 1965.]

These *paleopterous* ("primitive-winged") insects became the dominant group of insects during the Carboniferous and Permian periods, but are represented today by only two orders, Odonata and Ephemeroptera.

A number of hypotheses have been advanced to explain the origin of wings in insects. These hypotheses are based mainly on studies of extant forms, the fossil record providing little definitive information since most of the earliest insect fossils known already had wings. It is generally agreed that wings and their control and propulsive mechanisms have been superimposed on segments originally adapted to walking. Wings have been added without the loss of legs. This is different from the vertebrates, in which walking appendages and the associated nerves and muscles have become the organs of flight.

A widely accepted theory of the origin of insect wings views these structures as having arisen as bilateral expansions (*paranotal processes*) of the thoracic nota (Müller, 1879; Hinton, 1963a, 1977; Flower, 1964; Hamilton, 1971). Hinton (1963) and Flower (1964) envision paranotal expansions serving first as surfaces that enabled insects to control body attitude when falling from vegetation, ensuring that they would land in an upright position and could immediately run to shelter if being pursued by a predator. Subsequently, with further lateral expansion of the paranotal processes, gliding from high vegetation or even a running jump became possible (Fig. 10-15). It is not difficult to imagine a subsequent gradual development of an articular and neuromuscular arrangement by which the tilt fore and aft of the paranotal processes could be varied, thereby exerting a degree of glide control. The next step would have been the development of basal articulations and neuromusculature that allowed the wings to exert both controlling and propulsive forces. It is possible that the original paranotal processes first functioned as other than organs of flight. For example, it has been suggested that they were originally involved in courtship displays by male insects (Alexander and Brown, 1963).

Recently, Wigglesworth (1976c) has renewed support for a theory, advanced in the late 1800s, which views insect wings as having originated from tracheal gills in a primitive aquatic form. Gills in aquatic insects are commonly articulated, contain branching

Fig. 10-13 (opposite)
Origin of the insect head from fusion of the prostomium and six somites. A. Annelidlike ancestor with prostomium containing a ganglionic mass (archicerebrum). B. Cuticle, segmental appendages, and apodemes have evolved. C. Hypothetical arthropod; ganglion of first somite (now the labral segment) has become preoral. D. More advanced arthropod; ganglion of second somite (now the antennal segment) has become preoral. E. Insect head; ganglion of the third somite (now the intercalary segment) has become preoral, and the ganglia of somites 4, 5, and 6 have formed the subesophageal ganglion. According to this theory the protocerebrum of the brain arose from the fusion of ganglionic masses associated with the prostomium (archicerebrum) and first somite; the deuto- and tritocerebrum, from ganglionic masses in the second and third somites, respectively. [Redrawn from Rempel, 1975.]

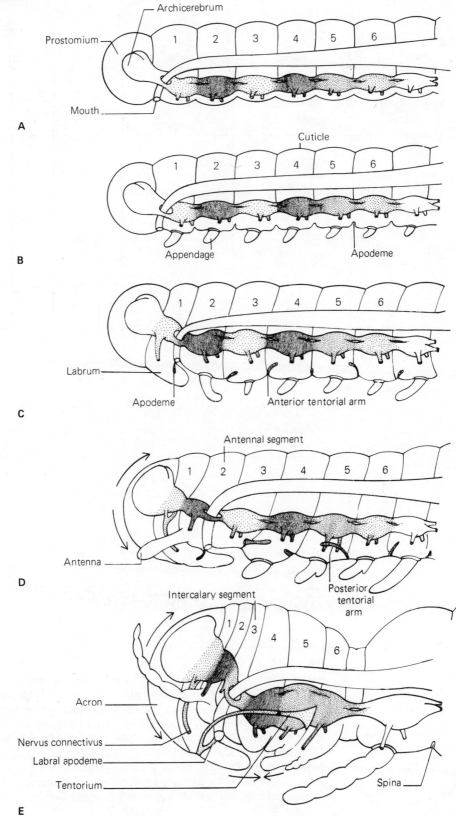

A

B

C

D

E

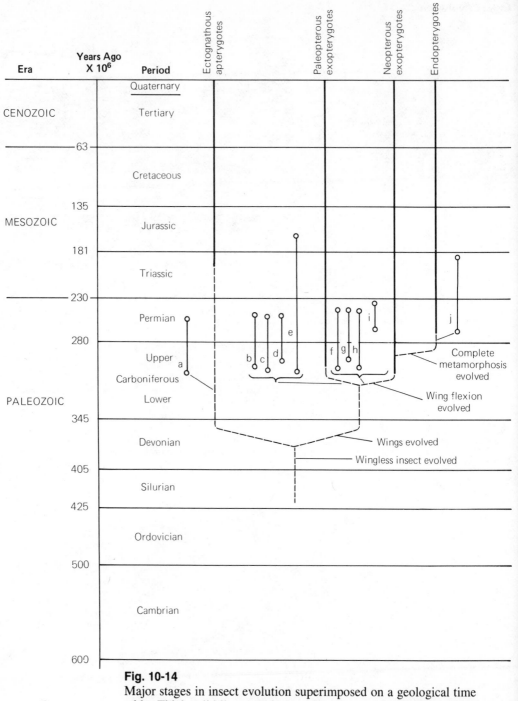

Fig. 10-14

Major stages in insect evolution superimposed on a geological time table. Thick solid lines run from earliest known fossil representative of an extant group to the present. Dashed lines represent hypothetical branches. Thin solid lines bounded by circles represent the following extinct orders: a, Monura; b, Paleodictyoptera; c, Megasecoptera; d, Diaphanopterodea; e, Protodonata; f, Protorthoptera; g, Caloneurodea; h, Miomoptera; i, Protelytroptera; j, Glosselytrodea. [Based on information from Carpenter, 1977; Sharov, 1966; Smart and Hughes, 1972. Geological eras and periods based on Villee and Dethier, 1976.]

tracheae, and are movable. For example, mayfly larvae are able to wave their gills to and fro, a movement that brings oxygenated water closer to the body (see "Ventilatory System" in Chapter 4). Gills may also aid in swimming.

The third major evolutionary advance appears to have occurred during the Lower Carboniferous (Fig. 10-14). This was the development of a wing-flexion mechanism, which enabled insects to flex the wings posteriorly over the abdomen, the *neopterous* ("new winged") condition. This added the advantage of being able to run and hide from predators and to move into niches inappropriate for forms with continuously outstretched wings. The neopterous insects subsequently radiated rapidly and became the dominant group of insects, as they are today. They comprise 90% of the contemporary orders and 97% of the total number of species (Carpenter, 1953). Only apterygotes and the Odonata and Ephemeroptera are not neopterous orders.

The final major evolutionary step was the development of complete metamorphosis (see Chapter 5), probably during the Upper Carboniferous (Fig. 10-14). The evolution of complete metamorphosis enabled insects to benefit from the favorable aspects of different habitats. For example, an insect with aquatic larval stages and terrestrial-aerial adults would benefit from the abundant and readily available nutrients of the aquatic habitat and later from the potential for dispersal (and hence enhanced genetic mixing) offered by the terrestrial-aerial habitat.

Hinton (1977) discusses several insectan features, in addition to those recognized by Carpenter (1977), the origin of which he considers as major events in the evolution of insects. Among these are the evolution of a tracheal system, a relatively impermeable cuticle (mentioned earlier), and the fat body. These major adaptations are characteristic of most terrestrial arthropods and likely preceded the evolution of insects as such. They are therefore perhaps best viewed as preadaptations, which set the stage for the major evolutionary events just outlined.

The evolution of the tracheal system enabled insects (and other terrestrial arthropods) to localize respiratory and moisture exchanges

Fig. 10-15
Hypothetical insects with bilateral extensions of the nota in the "glider" stage of insect evolution. [Redrawn from Snodgrass, 1958.]

with the environment as opposed to these exchanges occuring over the general body surface. This was followed by the development of a relatively impermeable cuticle (see "Permeability" in Chapter 2 and "Arthropoda" earlier in this chapter), a feature that enabled insects (and other arthropods) to radiate widely into terrestrial habitats. Evolution of spiracular closure mechanisms also deserves mention in this context (see "Ventilatory System" in Chapter 4). These developments made possible the evolution of flight by counteracting the drying tendency of the ambient air. Hinton views the fat body (see Chapter 4), particularly with regard to the storage of fat, as a major terrestrial adaptation that allows insects and other arthropods to withstand comparatively long periods without water. He points out that the total amount of fat in arthropods is greater in terrestrial forms than in aquatic forms.

Hinton (1977) also views the size range of insects as a particular advantage, pointing out: "Insects are large enough so that they are not imprisoned by surface tension forces; many make great use of these during locomotion. At the same time they are small enough so that they can fall to any distance without injury." Janzen (1977) sees a great ecological advantage to the typical small size of insects: "I think that there are so many species of insects because the world contains a very large amount of harvestable productivity that is arranged in a sufficiently heterogeneous manner that it can be partitioned among a large number of populations of small organisms."

During the course of their evolution, insects have clearly influenced and been influenced by other organisms. See the section "Adaptations Associated with Interspecific Interactions" in Chapter 9 for examples of close (coevolutionary) ties between insects and other organisms. For extensive information on the evolutionary relationships between insects and plants, see Smart and Hughes (1972). See Chapter 8 for comments on the evolution of behavior.

Based on Carpenter's four major evolutionary stages in insects, the class Insecta may be divided into groups as follows:

> Apterygotes—primitively wingless insects
> Pterygotes—winged insects
>> Paleopterous exopterygotes—wing-flexion mechanism
>>> lacking; simple metamorphosis
>> Neopterous exopterygotes—wing-flexion mechanism
>>> present; simple metamorphosis
>> Neopterous endopterygotes—wing-flexion mechanism
>>> present; complete metamorphosis

Figure 10-14 displays a widely held concept of the phylogenetic relationships between these various groups. This cladogram is superimposed on a geological time scale to indicate approximately when each group originated. The grouping of insects as listed above has served as the basis of organization for Chapter 11. Additional information on these categories and on the possible phylogenetic associations of the various insect orders is included in Chapter 11. Boudreaux (1979), Hamilton (1972), Hennig (1969), Jeannel (1949), Kristensen (1975), Mackerras (1970), Rohdendorf (1969), Ross (1955), Scudder (1973), and Wille (1960) all consider insect phylogeny in some depth. Handlirsch (1908) and Wilson and Doner (1937) should be consulted for early interpretations of insect phylogeny. Ross (1973) traces the historical development of concepts of insect phylogeny.

Although it has not provided definitive answers, the geological record has been of value in piecing together many aspects of insect phylogeny and in evaluating the evolutionary associations between insects and other organisms. Since fossilization occurs only under certain very uncommon circumstances, the fossil record for most organisms is fragmentary. Insect fossils consist mainly of wings, but various other body parts are also commonly represented. In addition to representation of most extant orders (save Zoraptera, Grylloblattodea, Mallophaga, and Anoplura) in the fossil record, 52 extinct orders of insects have been described by paleoentomologists (see Carpenter, 1977; Smart and Hughes, 1973). Many of these orders are based on very limited data, such as isolated wings or wing fragments. Carpenter (1977) reduces the number of recognized extinct orders significantly by applying the following criteria: ''My acceptance of an extinct order requires the knowledge of both fore- and hindwings (in the case of Pterygota) and the nature of the head, including mouthparts.'' On these bases 10 extinct orders can be recognized.

> Apterygotes
>> order Monura
> Pterygotes
>> Paleopterous exopterygotes
>>> order Palaeodictyoptera
>>> order Megasecoptera
>>> order Diaphanopterodea
>>> order Protodonata
>> Neopterous exopterygotes
>>> order Protorthoptera
>>> order Caloneurodea
>>> order Miomoptera
>>> order Protelytroptera
>> Neopterous endopterygotes
>>> order Glosselytrodea

The spans of geological history known to have been occupied by members of these extinct orders and their possible relationships to main insectan lines are depicted in Figure 10-14. Information on each extinct order listed above may be found in Chapter 11. For further information on insect fossils see Callahan (1972), Carpenter (1977), Reik (1970), and Smart and Hughes (1972). Rohdendorf (1973) outlines the history of paleoentomology.

Selected References

SYSTEMATICS

General: Blackwelder (1967); Hennig (1965, 1966); Mayr (1969); Mayr et al. (1953); Ross (1974); Simpson (1961); Sokal and Sneath (1963).

History: Lindroth (1973); Tuxen (1973).

Collecting and preparing insects: Bland (1978); Borror, DeLong, and Triplehorn (1976); Borror and White (1970); Cummins et al. (1965); Knudsen (1966); Peterson (1959); Ross (1962).

Rearing insects: Needham et al. (1937); Peterson (1959); Siverly (1962).

Insect: See below.

EVOLUTION

General: Dobzhansky et al. (1977); Futuyma (1979); Mayr (1963, 1970); Raup and Stanley (1978); *Scientific American* (1978); Volpe (1977); Windley (1977).

History: Rohdendorf (1973); Ross (1973).

Origin of Arthropoda: Anderson (1973, 1979); Boudreaux (1979); Cisne (1974); Gupta (1979); Manton (1964, 1970, 1972, 1979); Scudder (1973); Sharov (1966).

Origin of Insecta: Manton (1964); Rempel (1975); Sharov (1966); Snodgrass (1935).

Fossil history: Callahan (1972); Carpenter (1953, 1973); Reik (1970); Smart and Hughes (1972).

Origin of wings: Alexander and Brown (1963); Flower (1964); Hamilton (1972); Hinton (1963a); Wigglesworth (1973b, 1976c); Wooton (1976).

Insect phylogeny and classification: Boudreaux (1979); Wilson and Doner (1937); Hamilton (1972); Handlirsch (1908); Hennig (1969); Jeannel (1949); Kristensen (1975); Mackerras (1970); Rohdendorf (1969); Ross (1955); Scudder (1973); Wille (1960).

Survey of Class Insecta

The objective of this chapter is to provide a survey of the orders of insects. A comprehensive survey is, of course, well beyond this text; the coverage of each order here is comparatively brief.

The ordinal divisions of the class Insecta are a matter of contention, and it is not easy to decide which divisions to use. Many different ordinal groupings are recognized by different experts, and which one is "best" will no doubt remain a problem. Since there are different opinions regarding the category levels of various insect groups, especially between class and order, categorical names, such as subclass and division, have purposely been omitted. Thus the taxa between class and order described here, for example, Apterygota and Pterygota, should simply be viewed as groups within the Insecta that have certain characters in common. In most cases Boudreaux (1979) has been used as a basis for recognizing groups above order, primarily because it is the most recent analysis of insect phylogeny. However, the taxon names used by Boudreaux are not used in every case. See "Insect Phylogeny" in Chapter 10 for literature citations pertinent to various attempts at phylogenetic classification of insects.

To facilitate comparison of the sizes of the different orders, the number of species described in each order is presented in graphic form (Fig. 11-1). Figure 11-2 depicts the size standards used throughout this chapter.

Apterygota

The members of the apterygote orders of insects are all primitively wingless (i.e., none of their ancestors possessed wings). Based on the ordinal divisions recognized in this text, this group of insects includes the orders Protura, Collembola, Diplura, Archeognatha, and Thysanura.

There is disagreement as to whether these orders represent a mono- or polyphyletic group. Boudreaux (1979), Hennig (1969), Kristensen (1975), Tuxen (1970), and others view them as monophyletic. However, Manton (1970), Sharov (1966), and others support a polyphyletic interpretation, elevating the Protura, Collembola, and Diplura each to the rank of class. In any event it is generally agreed that members of Protura, Collembola, and Diplura do not show as close affinities with the rest of the class Insecta as do the Archeognatha and Thysanura. The Protura, Collembola, and Diplura all have the mouthparts somewhat pulled into the head (*entognathous*). Those who recognize these groups as separate classes view the

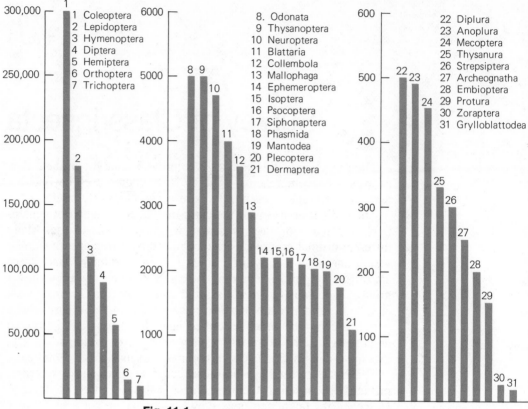

Fig. 11-1

Relative sizes of the orders of Insecta. [Data from Bland, 1978; Borror, DeLong, and Triplehorn, 1976; C.S.I.R.O., 1970.]

entognathous condition, absence or weak development of compound eyes, thin and pale cuticle, general small size, and several other traits as covergent adaptations to life within the soil. On the other hand, the presence of *ectognathous* (not pulled into the head) mouthparts, both simple and compound eyes, well-developed ovipositors, and a number of other characteristics clearly supports the hypothesis of a close relationship among the Archeognatha, Thysanura, and pterygote (winged) insects. The Thysanura, in particular, are considered to be descendents of the primitive insect from which the pterygotes evolved. The Archeognatha have a single point of articulation between the mandible and head capsule, whereas the Thysanura and pterygotes have two such points of articulation.

Figure 11-3 is a cladogram depicting one possible interpretation of apterygote phylogeny (i.e., as a monophyletic group).

Entognathous Apterygotes

Order Protura (*prot,* first; *ura,* tail); Myrientomata; proturans (Fig. 11-4A).

BODY CHARACTERISTICS minute, elongate, whitish.

MOUTHPARTS entognathous, sucking.

EYES and OCELLI lacking.

ANTENNAE lacking; "pseudoculi" may be vestigial antennae.

VL

L

Me

S

Mi

1   2   3   4   5

Centimeters

WINGS lacking.

LEGS first pair carried in elevated position suggestive of antennae; tarsi 1-segmented with a simple claw; an empodium may be present.

ABDOMEN short bilateral styli on first three segments; 12 segments; cerci lacking.

COMMENTS tracheae lacking; Malpighian tubules are small papillae.

Proturans are inhabitants of soil and litter and require moist conditions. They apparently feed on decaying organic matter. Protura are worldwide in distribution and are common, but easily overlooked.

Proturan development is ametabolous. The adults are essentially identical to the immatures except adults have fully developed gonads and a larger number of abdominal segments. During the course of development abdominal segments are added anterior to the telson, and reach a total of twelve in the adult. This type of growth is called anamorphosis.

Order Collembola (*coll*, glue; *embol*, a wedge); springtails (Fig. 11-5).

BODY CHARACTERISTICS minute, somewhat tubular (suborder Arthropleona) or globose (suborder Symphypleona).

Fig. 11-3
Hypothetical phylogeny of the apterygotes and their relationship with the pterygotes. [Based on Boudreaux, 1979.]

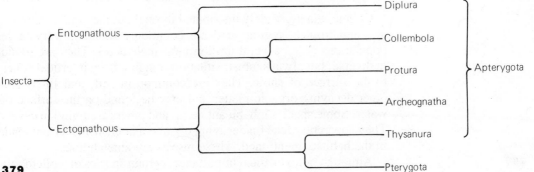

Insecta

Entognathous ─── Diplura / Collembola / Protura

Ectognathous ─── Archeognatha / Thysanura / Pterygota

Apterygota

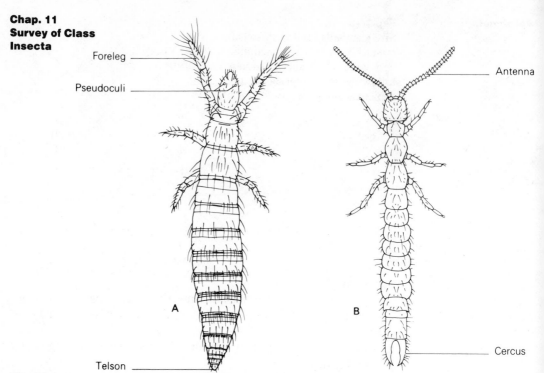

Fig. 11-4
A. Proturan, *Acerentulus barberi barberi* Ewing. B. Diplura, *Japyx diversiungus*. (Mi.) [A redrawn from Ewing, 1940; B redrawn from Essig, 1942.]

MOUTHPARTS entognathous, chewing.

EYES and OCELLI eye patches consist of 1 to several lateral ocelli comparable to those found in holometabolous larvae; dorsal ocelli lacking.

ANTENNAE short; usually 4-segmented.

WINGS lacking.

LEGS claw and empodium at distal end of tibia; tarsi fused to tibia (tibiotarsus) or 1-segmented.

ABDOMEN 6 segments; lobelike organ on venter of first segment, the *ventral tube,* which functions in water uptake and as an "adhesive organ", and possibly has an osmoregulatory function; forked structure, *furcula,* on venter of fourth segment; *tenaculum* on venter of third segment.

COMMENTS postantennal organs in many species; furcula directed anteroventrally and secured by tenaculum; "springing" accomplished by sudden release of furcula from tenaculum; tracheae present or absent; Malpighian tubules lacking. Springtails are so named because of their habit of jumping by means of the furcula.

Collembola are widely distributed throughout the Arctic, temperate, and tropical regions and are commonly found in very large populations (e.g., several million in a single acre). They are found in diverse, but always moist, situations ranging from intertidal zones to the surface of snow. They are common in soil, leaf litter, and other decaying organic matter and may be found on the surface of water. Some species inhabit ant nests, and others are found in caves. They apparently feed on decaying organic matter, which is abundant in the habitats mentioned. These insects are ametabolous.

Although of no medical importance, certain species of Collembola have been identified as pests of mushrooms, truck garden crops, forage and cereal crops, sugar cane, and in households (Maynard, 1951). The most outstanding is *Sminthurus viridis*, the "Lucerne flea," a serious introduced pest of alfalfa in Australia.

Order Diplura   (*dipl*, two; *ura*, tail); Entotrophi, Aptera, Entognatha; campodeids and japygids (Fig. 11-4B).

BODY CHARACTERISTICS minute to small; slender; whitish.

MOUTHPARTS entognathous; chewing.

EYES and OCELLI lacking.

ANTENNAE long; filiform.

WINGS lacking.

LEGS 1-segmented tarsi with 2 claws (movable spurs).

ABDOMEN 10 visible segments; cerci forcepslike or long caudal filaments.

COMMENTS tracheae present; Malpighian tubules vestigial or absent.

Diplura are found in damp situations in caves, under bark, in the soil, and in similar habitats. They undergo ametabolous development.

Ectognathous Apterygotes

Order Archeognatha   (*archeo*, ancient; *gnath*, jaw); Microcoryphia; jumping bristletails (Fig. 11-6a).

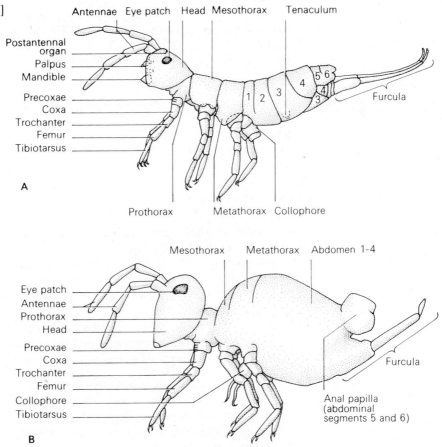

A

B

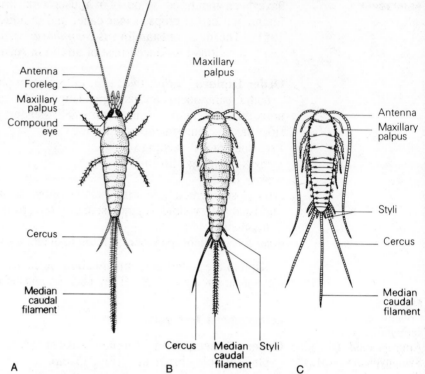

Antenna
Foreleg
Maxillary
palpus
Compound
eye

Maxillary
palpus

Antenna
Maxillary
palpus

Cercus

Styli

Cercus

Median
caudal
filament

Median
caudal
filament

Cercus   Median   Styli
caudal
filament

A

B

C

Fig. 11-6
A. Archeognatha, jumping bristletail, *Machilis* sp. B. Thysanura,
common silverfish, *Lepisma saccharina*. C. Thysanura, firebrat,
Thermobia domestica. (S to Me.) [A redrawn from Essig, 1942, after
Lubbock; B, C after U.S. Public Health Service.]

BODY CHARACTERISTICS small; elongate; many with scales on part of
body.

MOUTHPARTS ectognathous, chewing; single point of articulation
between mandibles and head capsule; elongate, 7-segmented maxillary palp.

EYES and OCELLI large contiguous compound eyes; 3 ocelli.

ANTENNAE elongate, filiform.

WINGS lacking.

LEGS coxae of meso- and methathorax often with styli; 2- or 3-
segmented tarsi with 2 claws.

ABDOMEN small bilateral styli on venter of several segments; females with elongate jointed ovipositor; elongate, multisegmented
cerci; elongate median caudal filament; eversible vesicles on many
abdominal sterna.

COMMENTS The name "Archeognatha" refers to the presence of a
single articulation between the mandible and head capsule in contrast with two such points in the Thysanura and pterygotes.

Jumping bristletails are commonly found in forest litter, under
decaying wood, rocks, and so on. They are nocturnally active and
feed on lichens, algae, and decaying vegetation. Jumping, achieved
by flexing the abdomen downward, is part of their escape behavior.

These insects undergo ametabolous development and are included
by some with the Thysanura.

Order Thysanura (*thysan*, fringe; *ura*, tail); Ectognatha, Ectotrophi; bristletails (Fig. 11-6B, C).

BODY CHARACTERISTICS small, elongate; usually covered with scales.

MOUTHPARTS ectognathous, chewing; two articulations between each mandible and head capsule.

EYES and OCELLI compound eyes usually present; 0 or 3 ocelli.

ANTENNAE elongate, filiform.

WINGS lacking.

LEGS 2- to 5-segmented tarsi; 2 or 3 claws.

ABDOMEN small bilateral styli on venter of several segments; females have elongate jointed ovipositor; cerci elongate with many segments; elongate median caudal filament present.

COMMENTS styli interpreted as vestiges of locomotor appendages inherited from myriapodlike ancestors (same for Archeognatha); Malpighian tubules present.

Most species of Thysanura live in soil, leaf litter, rotting wood, and similar damp habitats.

Bristletails undergo ametabolous development. These insects are different from the pterygotes in continuing to molt periodically after they have reached sexual maturity.

Some thysanuran species, for example, the common silverfish, *Lepisma saccharina* (Fig. 11-6B), and the firebrat, *Thermobia domestica* (Fig. 11-6C), are domesticated and share human habitations. Silverfish are found in cool, damp locations; firebrats frequent warmer locations and are generally found around steam pipes, furnaces, and similar heat-producing objects. Both species can become pests because they feed upon such things as book bindings, starched clothing, and cloth of various sorts. They have no medical significance.

Fossil Apterygotes

Distinct fossil apterygotes have been found in beds ranging from the Upper Carboniferous through the Lower Permian in Europe and North America. These fossils are various species of Dasyleptus in the extinct order Monura (see Fig. 10-14). These insects resembled modern Archeognatha in a number of ways, but lack cerci. For more information on the Monura, see Sharov (1966) and Boudreaux (1979).

Pterygota

It is generally agreed that the pterygotes comprise a monophyletic group. Although some insects classified as pterygotes are wingless (e.g., lice, bed bugs, and fleas), these insects are considered on the basis of other morphological (and developmental) grounds to have arisen from winged ancestors.

As explained in Chapter 10, the four major evolutionary steps described by Carpenter (1977) are used as a basis for classifying insects in this text. Thus the subgroups of the Pterygota are the paleopterous exopterygotes, neopterous exopterygotes, and the neopterous endopterygotes. Figure 11-7 is a cladogram depicting a

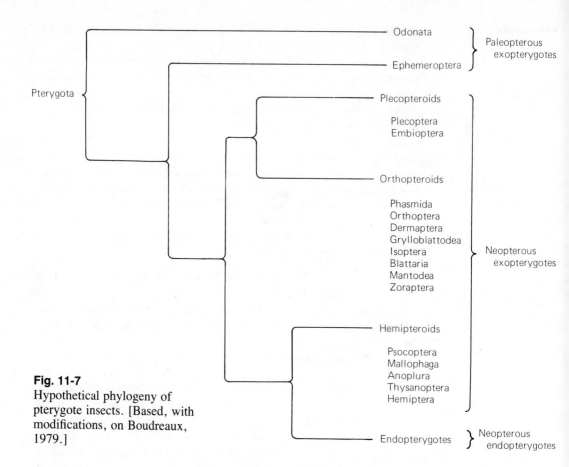

Fig. 11-7
Hypothetical phylogeny of
pterygote insects. [Based, with
modifications, on Boudreaux,
1979.]

recent interpretation of the possible phylogenetic relationships
among these groups. Note that although the pterygotes as a group
and the neopterous groups within the Pterygota are viewed as monophyletic, the paleopterous exopterygotes are not. Some view the
Ephemeroptera, on bases other than wing flexion, as being more
closely related to the neopterous insects than to the Odonata (Matsuda, 1970; Boudreaux, 1979). On the other hand, Kristensen (1973)
suggests closer ties between the Odonata and neopterous exopterygotes than between Ephemeroptera and the latter group.

Paleopterous Exopterygota

As mentioned earlier, the wings in this group are primitive in that
they cannot be flexed and laid down over the abdomen and hence
must, at rest, be held laterally or together above the thorax and
abdomen. All other pterygotes with wings have a wing-flexion mechanism or are descendants of insects that possessed such a mechanism. Although the fossil record indicates the past existence of other
paleopterous orders, Odonata (dragonflies and damselflies) and
Ephemeroptera (mayflies) are the sole survivors.
Paleopterous insects undergo simple metamorphosis (i.e., they
are hemimetabolous).

Order Odonata   (*odon*, a tooth); dragonflies and damselflies (Figs. 11-8 and 11-9).

BODY CHARACTERISTICS   medium to very large; elongate.

MOUTHPARTS   chewing; larva with prehensile labium.

EYES and OCELLI   large compound eyes; 3 dorsal ocelli.

ANTENNAE   short; bristlelike.

WINGS   2 pairs; membranous; netlike venation; characteristic pigmented cell, the *stigma*, immediately posterior to the costal vein near the apex of each wing; well developed cross vein, the nodus, near the middle of the leading edge of each wing.

LEGS   adults: basketlike arrangement, adaptation for prey capture; 3-segmented tarsi.

ABDOMEN   elongate; adult males with gonopores on ninth segment

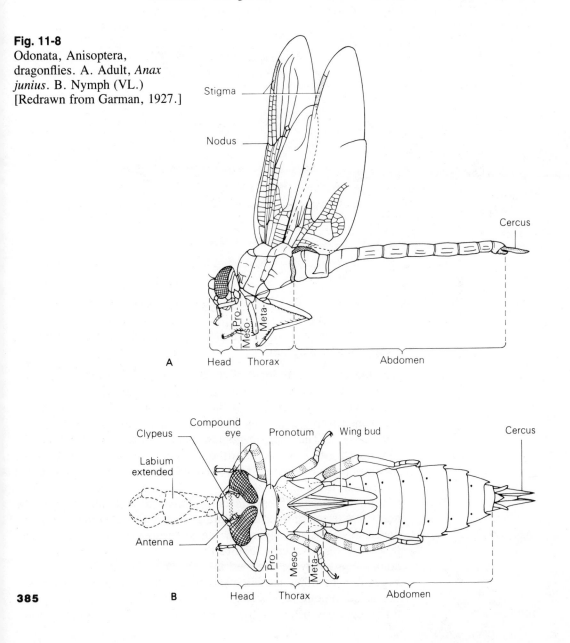

Fig. 11-8
Odonata, Anisoptera, dragonflies. A. Adult, *Anax junius*. B. Nymph (VL.)
[Redrawn from Garman, 1927.]

Stigma

Nodus

Cercus

A   Head   Thorax   Abdomen

Pro- Meso- Meta-

Clypeus   Compound eye   Pronotum   Wing bud   Cercus

Labium extended

Antenna

B   Head   Thorax   Abdomen

Pro- Meso- Meta-

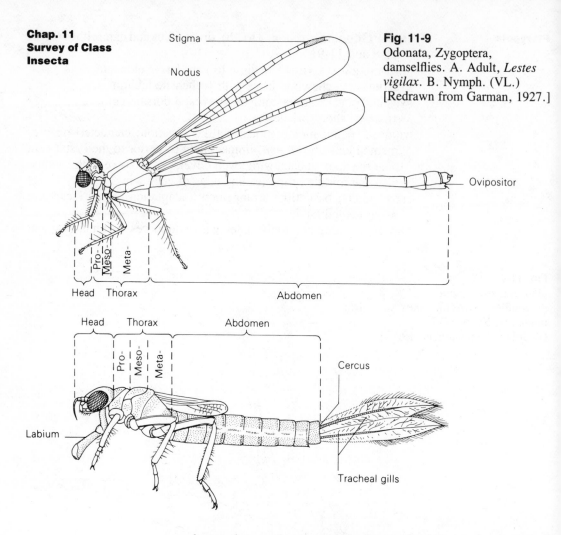

Fig. 11-9
Odonata, Zygoptera,
damselflies. A. Adult, *Lestes
vigilax*. B. Nymph. (VL.)
[Redrawn from Garman, 1927.]

and complex penis on venter of second segment; 1-segmented
cerci in males serve as claspers during copulation.

COMMENTS larvae: closed tracheal system, gill ventilators; adults:
open tracheal system.

Odonata can be divided into three suborders, Anisoptera, the dragonflies (Fig. 11-8), Zygoptera, the damselflies (Fig. 11-9), and
Anisozygoptera. Members of the first two suborders are easily distinguished from each other both in the larval and adult stages (see
Table 11-1). The Anisozygoptera share characters with both the
Anisoptera and Zygoptera. Although there are many extinct forms,
this group is represented today by only two species, one in Japan
and one in India.

Immature dragonflies and damselflies are aquatic, whereas the
adults are terrestrial-aerial but are generally found in the vicinity of
water. Nymphs usually reach the adult stage in a single season;
however, some species require two or three years to develop into
adults. Eggs may be dropped into the water or attached to aquatic
vegetation or other partially submerged objects. Rarely, eggs are
deposited in gelatinous strings or masses. In a few species, females
go beneath the surface of the water to oviposit. See "Seminal Trans-

fer'' in Chapter 5 for the remarkable method of insemination characteristic of the Odonata.

Dragonflies and damselfly larvae prey on a wide variety of aquatic organisms. Some of the larger species may capture tadpoles and small fish. Adults generally prey on other flying insects (e.g., mosquitoes and midges, moths, bees, and other dragonflies). They are able to catch, hold, and devour prey in flight.

Although these insects have been recorded as occasional pests in apiaries and sometimes attack trout fry, they are generally considered to be very beneficial. They are harmless to man.

Order Ephemeroptera (*ephemero*, for a day; *ptera*, wings); Ephemerida, Plectoptera; mayflies (Fig. 11-10).

BODY CHARACTERISTICS small to medium; fragile; soft-bodied.

MOUTHPARTS nymphs: chewing; adults: vestigial.

EYES and OCELLI compound eyes present; 3 dorsal ocelli.

ANTENNAE short; setaceous.

WINGS adults and subimagoes; 2 pairs; membranous; forewings larger than hindwings; few species with only mesothoracic wings; held vertically at rest; typically many veins.

LEGS 3- to 5-segmented tarsi.

ABDOMEN nymphs: bilateral ventilatory gills on the first 4 to 7 segments; adults: pair of long, filamentous cerci; some with additional long median caudal filament.

COMMENTS nymphs: closed tracheal system, gill ventilators; adults: open tracheal system.

The members of this order are aquatic in the nymphal stages and terrestrial-aerial as adults.

Table 11-1
Outstanding Differences Between Suborders of Odonata

Anisoptera	Zygoptera
ADULTS	
1. Fore- and hindwings unequal in size; hindwings basally broader than forewings.	1. Fore- and hindwings approximately the same size and shape.
2. Wings laterally spread at rest	2. Wings held together dorsally over the thorax and abdomen.
3. Strong, agile fliers.	3. Comparatively weak fliers.
4. Compound eyes in close proximity or meet dorsally.	4. Compound eyes widely separated.
5. Males have 3 terminal abdominal appendages.	5. Males have 4 terminal abdominal appendages.
NYMPHS	
1. Ventilatory gills in rectum, not externally visible.	1. Ventilatory gills are 3 terminal abdominal appendages, externally visible.
2. Stout, robust body.	2. Relatively slender, fragile body.
3. Able to propel themselves short distances by forcibly ejecting water from the rectum.	3. Lack ''jet-propulsion'' mechanism.

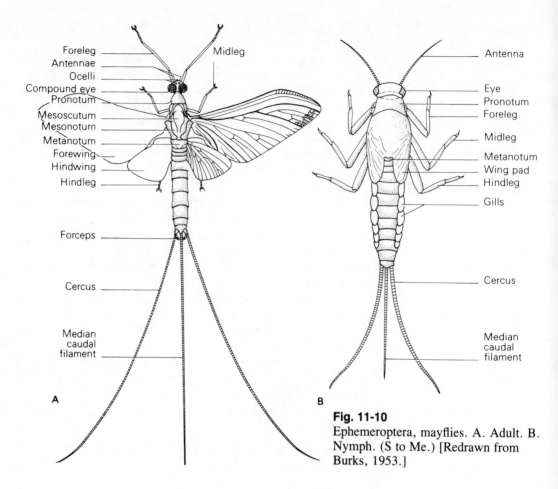

Foreleg
Antennae
Ocelli
Compound eye
Pronotum
Mesoscutum
Mesonotum
Metanotum
Forewing
Hindwing
Hindleg

Forceps

Cercus

Median
caudal
filament

A

Midleg

Antenna

Eye
Pronotum
Foreleg

Midleg

Metanotum
Wing pad
Hindleg

Gills

Cercus

Median
caudal
filament

B

Fig. 11-10
Ephemeroptera, mayflies. A. Adult. B.
Nymph. (S to Me.) [Redrawn from
Burks, 1953.]

Mayflies are unique on the class Insecta in passing through a winged, subadult (subimaginal) stage that is essentially identical to the adult stage. The subimaginal and adult stages live for a brief period of time (a few hours to a few days), and their apparent sole function is to survive long enough to copulate and deposit their eggs, which are often simply dropped into the water and subsequently sink to the bottom. Some species attach their eggs to partially submerged rocks or aquatic vegetation. The larval period is generally fairly long, a year or more, although some species may develop in a few months.

Mayflies commonly emerge from lakes, streams, and rivers in fantastically large numbers, and in areas near towns or cities the shed subimaginal cuticles and dead adults may literally accumulate in piles on streets and sidewalks. This was once a common occurrence in towns along the shores of Lake Erie until pollution took its toll.

Larval mayflies (Fig. 11-10B) are solely responsible for the intake of nutrients since the adults do not feed. Their food consists of aquatic vegetation and small aquatic organisms.

Mayfly larvae are important food for aquatic predators, particularly fish, and provide a periodic feast for insectivorous birds and dragonflies when they emerge in vast numbers. They have been likened to terrestrial herbivores in that they are close to the bottom of several food chains, converting plant material into animal material. They also provide excellent models for many of the common "tied flies" cherished by avid fly fisherman.

In areas where large populations of adult mayflies invade towns and cities, there have been reports of severe allergic responses (asthma, conjunctivitis, and so on) to airborne detritus composed of the broken casts of subimagoes and dried bodies of adults.

Fossil Paleopterous Exopterygote Orders

Carpenter (1977) recognizes four extinct paleopterous exopterygote orders: Palaeodictyoptera, Megasecoptera, Diaphanoptera, and Protodonata (see Fig. 10-14). The first three listed are apparently closely related, sharing such traits as piercing-sucking haustellate mouthparts, long cerci, and an external ovipositor. They were probably terrestrial (fossil larvae lack gills) and presumably herbivorous. The Diaphanoptera were able to flex the wings over the abdomen, but the flexion mechanism differed from that found in modern neopterous insects. Protodonata were similar to modern Odonata in appearance and predaceous feeding behavior. Among the Protodonata were the largest known insects, one genus, *Meganeura,* reaching a wing span of 750 mm.

Neopterous Exopterygota

This taxon includes three groups of orders (Fig. 11-7): the plecopteroids, the orthopteroids, and the hemipteroids.

Plecopteroid and Orthopteroid Orders. The plecopteroid orders as described here include the Plecoptera and Embioptera, and the orthopteroid orders include Phasmida, Orthoptera, Dermaptera, Grylloblattodea, Isoptera, Blattaria, Mantodea, and Zoraptera (see Fig. 11-7). There are differences of opinion about the groupings of these orders, especially with regard to the plecopteroid insects. Both Plecoptera and Embioptera have been included as orthopteroids, and the Plecoptera are sometimes listed as being separate from all other neopterous exopterygote orders. Among the orthopteroid orders, there is strong evidence and much agreement that the Isoptera, Blattaria, and Mantodea are closely related. Thus the Blattaria and Mantodea are sometimes grouped as the order Dictyoptera. Lumping all three orders into a single order has also been suggested.

The plecopteroids and orthopteroids are similar relative to a number of features including: (1) primitive, "generalized" mandibulate mouthparts; (2) complex wing venation with hindwings typically larger than forewings owing to enlarged anal areas (hindwings fold along vannal and jugal folds when wings are flexed); (3) male intromittent organ from everted ejaculatory duct; (4) cerci present; (5) typically many Malpighian tubules; and (6) ventral chain ganglia distributed segmentally.

Both plecopteroid and orthopteroid orders undergo hemimetabolous development.

Among the bases for grouping Plecoptera and Embioptera as plecopteroids are: (1) external appendages associated with male genitalia reduced or absent; (2) abdominal styli lacking; (3) accessory clasping structures in males from various sources including extension of tenth abdominal tergum and modified paraprocts; (4) forewings not leathery or sclerotized; and (5) trochantin attached to epimeron with no suture between.

Traits common to the orthopteroid orders include: (1) external genital appendages present in males; (2) abdominal styli on ninth abdominal segment of males in many; (3) vannus of hindwings forms pleats when folded; and (4) forewings leathery or somewhat sclerotized; and (5) trochantin separated from episternum by suture or membrane.

Order Plecoptera   *pleco,* twist; *ptera,* wing); stoneflies (Fig. 11-11).

BODY CHARACTERISTICS   small to medium; soft-bodied; elongate, flattened; body nearly parallel-sided.

MOUTHPARTS   chewing; commonly vestigial in adults.

EYES and OCELLI   well-developed compound eyes; usually 3 dorsal ocelli, rarely 2.

ANTENNAE   long and tapering.

WINGS   2 pairs; membranous; males of some species apterous or brachypterous (short wings); at rest, hindwings folded in "plaits" beneath forewings.

LEGS   3-segmented tarsi.

ABDOMEN   ovipositor lacking; many-segmented cerci.

COMMENTS   nymphs: closed tracheal system, gill ventilators, gills usually on venter of thorax, but sometimes on other parts of body; adults: open tracheal system.

Stonefly nymphs (Fig. 11-11B) are aquatic and are usually found around and beneath stones in fast-moving, well-aerated water. The adults (Fig. 11-11A) are terrestrial-aerial and are found in the vicinity of nymphal habitats. The nymphs are campodeiform and resemble the adults in general appearance except for the absence of wings and the presence of tracheal gills.

Stoneflies are mostly herbivorous as nymphs, feeding on various forms of small aquatic plant life; a few are carnivorous and feed on other aquatic insects. The adults with well-developed mouthparts feed on algae and lichens; those with vestigial mouthparts do not feed at all. Although most species complete the nymphal stage in a year or so, some species require two, three, or four years to develop into adults.

Comstock (1940) mentions adults of *Taeniopteryx pacifica* as fruit tree pests in the Wenatchee Valley, Washington. However, the value of stoneflies as fish food far outweighs this isolated instance of pestiferous behavior.

Order Embioptera   (*embio,* lively; *ptera,* wing); Embiidina; web-spinners (Fig. 11-12).

BODY CHARACTERISTICS   minute to small; brown to yellow; elongate.

MOUTHPARTS   chewing.

EYES and OCELLI   compound eyes present; no dorsal ocelli.

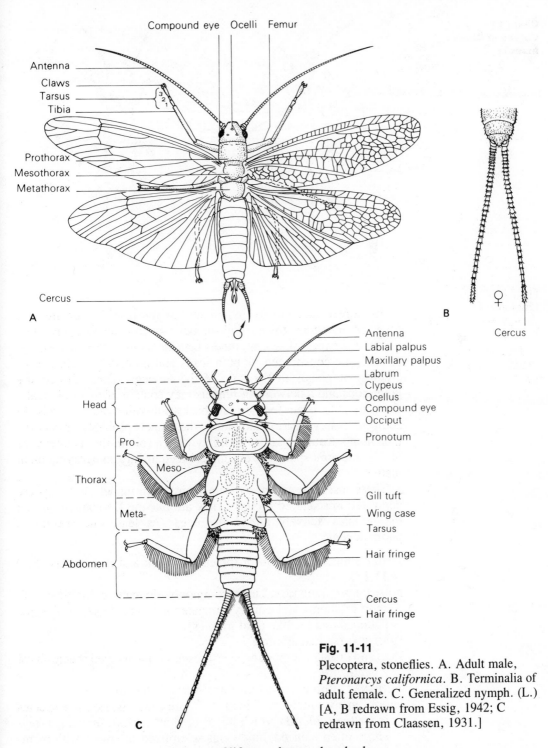

Labels in figure A (clockwise from top): Compound eye, Ocelli, Femur, Antenna, Claws, Tarsus (3, 2, 1), Tibia, Prothorax, Mesothorax, Metathorax, Cercus

Label in figure B: Cercus, ♀

Labels in figure C:
Head — Antenna, Labial palpus, Maxillary palpus, Labrum, Clypeus, Ocellus, Compound eye, Occiput
Thorax — Pro-, Meso-, Meta-, Pronotum, Gill tuft, Wing case, Tarsus, Hair fringe
Abdomen — Cercus, Hair fringe

Fig. 11-11

Plecoptera, stoneflies. A. Adult male, *Pteronarcys californica*. B. Terminalia of adult female. C. Generalized nymph. (L.) [A, B redrawn from Essig, 1942; C redrawn from Claassen, 1931.]

ANTENNAE filiform; shorter than body.
WINGS some males with 2 pairs of membranous wings of nearly equal size and shape; "smoky" in appearance; females and some males wingless.
LEGS basal tarsal segment of foreleg enlarged and contains silk-producing glands; 3-segmented tarsi.
ABDOMEN pair of cerci present and of unequal size in males.

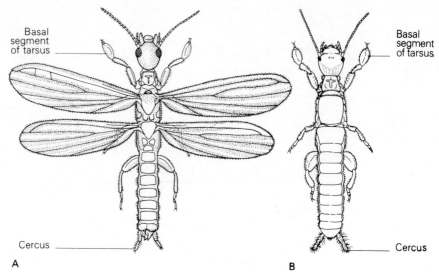

Basal segment of tarsus

Basal segment of tarsus

Cercus

Cercus

A

B

Fig. 11-12
Embioptera,
webspinners. A. Male,
Oligotoma saundersii.
B. Female. (Mi.)
[Redrawn from Essig,
1942.]

Embioptera are mostly inhabitants of the tropics, but are also found in temperate zones. Webspinners are so named because they live in a network of silken tunnels beneath stones, bark, and so on. Tunnels are constructed by both sexes and immatures from the secretions of glands in the tarsi of the forelegs. These insects are gregarious and a network of tunnels may contain many individuals. They are able to run forward and backward within these tunnels. In addition to serving as a habitation, the tunnels probably protect the inhabitants from predators and may play a role in the maintenance of a humid atmosphere. Webspinners feed mainly on decaying plant matter.

These insects reproduce within their silken tunnels and eggs are deposited in scattered groups. They are like the Dermaptera in that the females actively attend the eggs. At least one species is parthenogenic.

Order Phasmida (*phasm*, phantom); stick and leaf insects (Fig. 11-13).

BODY CHARACTERISTICS large; commonly elongate, cylindrical with short prothorax and usually elongate meso- and metathorax; some flattened and leaflike (leaf mimics).

MOUTHPARTS mandibulate

EYES and OCELLI compound eyes present; some winged species have 2 ocelli.

ANTENNAE filiform or moniliform; variable length.

WINGS many wingless (all in the U.S. are wingless, save one species in southern Florida which has very short wings); forewings (*tegmina*) often reduced; hindwings sclerotized anteriorly with membranous anal area; in leaflike forms venation mimics leaf venation.

LEGS generalized cursorial; tarsi usually 5-segmented.

ABDOMEN ovipositor small and somewhat concealed; short, unsegmented cerci.

Walking sticks are rather sluggish, solitary, herbivorous insects. Although several species occur in temperate zones, they are predominately tropical. Some species grow quite large, some tropical forms ranging up to a foot in length. *Megaphasma dentricus* is the largest insect in the United States, reaching a length of 7 inches.

Walking sticks are somewhat protected from predators by their resemblance of stems or leaves of plants. Some, for example, *Carausius morosus,* are capable of physiological color changes effected by pigment granule migration in epidermal cells. Many phasmids produce repugnatorial substances from glands in the prothorax. *Anisomorpha buprestoides,* an aposematically colored species (see ''Adaptations Associated with Interspecific Interactions'' in Chapter 9), is able to control the direction of its spray and will spray in response to an approaching potential predator such as a bird (Eisner, 1965). If a nymph happens to be caught by a leg, *autoamputation (autotomy)* may result, allowing the insect to escape. Autoamputated legs subsequently regenerate.

Walking sticks undergo hemimetabolous development. Their eggs, which may resemble host plant seeds, are generally dropped singly to the ground. Both obligate and facultative parthenogenesis is found among members of this order.

Order Orthoptera   (*ortho,* straight; *ptera,* wing); grasshoppers, crickets, and related forms (Figs. 11-14 and 11-15).

BODY CHARACTERISTICS   minute to very large; body form variable.

MOUTHPARTS   chewing; all gradations between hypognathous and opisthognathous.

EYES and OCELLI   well-developed compound eyes; 0, 2, or 3 dorsal ocelli.

ANTENNAE   variable; filiform in many (e.g., katydids and crickets); somewhat shorter in many grasshoppers and others.

WINGS   most with 2 pairs; forewings parchmentlike *tegmina;* at rest, hindwings folded pleatlike beneath forewings.

LEGS   hindlegs adapted for jumping (the hind femora are enlarged,

Fig. 11-13
Phasmida. A stick insect, *Diapheromera femorata.* (VL.) [Redrawn from Hebard, 1934.]

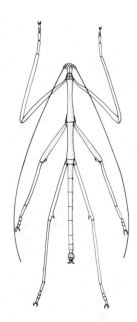

accommodating the muscles that produce the jumping force); tarsi 3- or 4-segmented.

ABDOMEN females in several families with well developed ovipositors; cerci variable: short, long, clasperlike, segmented, unsegmented.

Orthoptera includes grasshoppers, crickets, and relatives. This order can be divided into two suborders, Caelifera (grasshoppers, pygmy grasshoppers, etc.; Figs. 11-14 and 11-15 F) and Ensifera (katydids, crickets, mole crickets, etc.; Fig. 11-15 A-E). The Caelifera have comparatively short, filiform antennae (fewer than 30 segments), comparatively long cylindrical or sword-shaped ovipositors; tympanic organs (if present) are located at the bases of the tibiae of the forelegs.

Members of this order are typically terrestrial and are found in trees, bushes, and other vegetation; on the surface of, or burrowing into, the ground; and occasionally in caves. A few species are aquatic or semiaquatic. Orthoptera includes general scavengers, many voracious herbivores, and some carnivorous species.

There are generally five or more nymphal instars in an orthopteran life cycle. Eggs are deposited singly or in masses, in or on vegetation (e.g., most katydids), or in the soil (e.g., most grasshoppers).

The ability to produce and perceive sound is found among many orthopterans. Although sound production is effected in a variety of ways, it generally involves the rubbing of one part of the body against another. The females of some species are capable of pro-

Fig. 11-14
Orthoptera, a short-horned
grasshopper (Acrididae), showing
major external structures. (L.)
[Redrawn from Essig, 1942.]

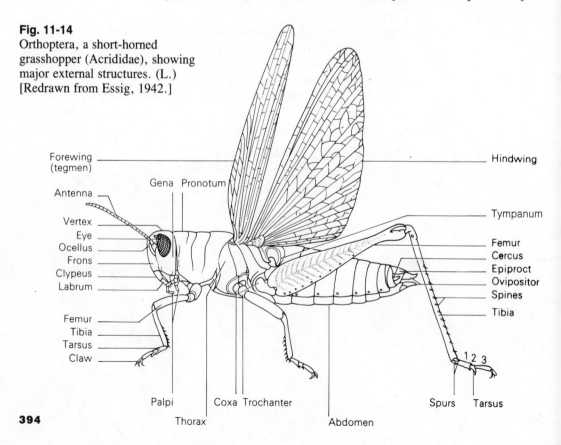

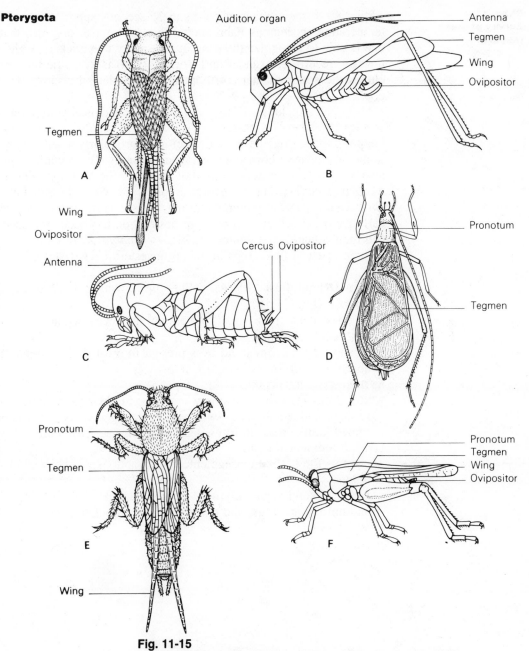

Fig. 11-15
Representative Orthoptera. A. Field cricket, *Gryllus assimilis*
(Gryllidae). B. Bush katydid, *Scudderia furcata* (Tettigoniidae). C.
Jerusalem cricket, *Stenopelmatus longispina* (Gryllacrididae). D. Tree
cricket, *Oecanthus niveus* (Gryllidae). E. Mole cricket, *Gryllotalpa
hexadactyla* (Gryllidae). F. Pygmy grasshopper, *Telmatettix hesperus*
(Tetrigidae). (A, D, and F, Me; B, C, and E, L.). [Redrawn from Essig,
1942.]

ducing sound, but it is typically the males that are the accomplished
singers. The songs are most commonly involved with bringing members of the opposite sexes together. The snowy tree cricket, *Oecanthus fultoni,* is purported to be a "living thermometer" because

adding 40 to the number of chirps in 15 seconds approximates the temperature in degrees Fahrenheit. This, of course, assumes that one is able to recognize the song of the snowy tree cricket. A phonograph record (Alexander and Borror, 1956) is available that includes the songs of many orthopterans, including the snowy tree cricket.

Orthoptera includes a large number of economically important species. Members of the family Acrididae, the short-horned grasshoppers, are serious pests of field crops, particularly small grains. Some of the long-horned grasshoppers (family Tettigoniidae) are also serious pests of field crops. For example, the Mormon cricket, *Anabrus simplex,* attacks several crops in the West. In Salt Lake City, Utah, a statue commemorating the sea gulls that saved the early Mormon settlers' crops from destruction has been erected in Mormon Square. The coulee cricket, *Peranabus scabricollis,* is a common pest of field crops in the northwestern United States.

Order Dermaptera (*derma,* skin; *ptera,* wing); Euplexoptera; earwigs (Fig. 11-16).

BODY CHARACTERISTICS small to medium; narrow; elongate.

MOUTHPARTS chewing.

EYES and OCELLI compound eyes present in most, but vestigial or absent in some; dorsal ocelli lacking.

ANTENNAE long; slender.

WINGS forewings short, leathery tegmina; hindwings semicircular in shape and membranous with radially arranged veins; hindwings folded fanlike beneath forewings; wingless species common.

LEGS 3-segmented tarsi.

ABDOMEN forcepslike, unsegmented cerci; abdomen not covered by wings.

COMMENTS cerci apparently used in prey capture, defense, folding and unfolding wings, and possibly during copulation.

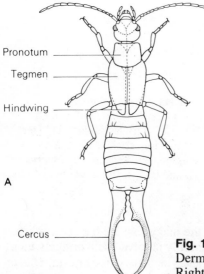

Pronotum

Tegmen

Hindwing

A

Cercus

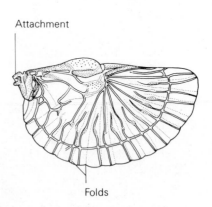

Attachment

Folds

B

Fig. 11-16
Dermaptera. A. European earwig, *Forficula auricularia,* male. B. Right hindwing unfolded. (Me.) [Redrawn from Essig, 1942.]

Most earwigs are nocturnal and prefer damp situations (e.g., soil, under bark, under stones, among vegetation, and near bodies of water). They are mostly omnivorous and feed upon small living or dead insects, decaying plant material, and tender parts of living plants. Some earwigs are parasitic (e.g., *Hermimerus* sp., which inhabits the fur of African rats).

Earwigs display parental care of offspring, which is rather uncommon among insects and may represent an early stage in the evolution of social behavior (see Chapter 8). Eggs are deposited in the soil and the females literally "roost" on them until they hatch and then care for the newly hatched young.

When disturbed, an earwig may raise the abdomen with its imposing cerci, somewhat resembling the behavior of a scorpion. An additional defense is the ability of some species to squirt (sometimes over several centimeters) a repugnatorial substance produced by glands on the dorsum of the second or third abdominal segments.

Some species (e.g., *Forficula auricularia*), when abundant, may damage flowers and tender foliage. The common name "earwig" may have come from the notion that these insects enter the ears of sleeping persons, but this has little basis in fact, and they are considered to be harmless. Alternatively, their common name may be a corruption of "earwing," in reference to the hindwing resembling a human ear (Richards and Davies, 1977).

Order Grylloblattodea (*gryll,* cricket; *blatta,* cockroach); Notoptera; rock crawlers or ice bugs (Fig. 11-17).

BODY CHARACTERISTICS somewhat elongate and slender; medium to large; light brown to gray color.

MOUTHPARTS mandibulate.

EYES and OCELLI compound eyes reduced or absent; ocelli lacking.

ANTENNAE filiform.

WINGS lacking.

LEGS generalized cursorial; 5-segmented tarsi.

ABDOMEN well-developed sword-shaped ovipositor; long cerci.

Fig. 11-17
Grylloblattodea. A rock crawler,
Grylloblatta campodeiformis. (L.)
[Redrawn from Essig, 1942.]

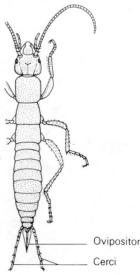

Ovipositor

Cerci

This is the smallest order of insects with only 17 species, 11 represented in North America (Bland, 1978). Rock crawlers are active at very low temperatures (around 1°C) and are found on snow, at edges of snow fields, in moss, under rocks, and so on. Some species are *cavernicolous* (found in caves). They appear to be omnivorous, feeding on mosses and dead or sluggish insects. Given the characteristic low temperature of their habitat, it is not surprising that they have low rates of metabolism and are slow in their development. They may take as long as 7 years to complete one life cycle.

Order Isoptera   (*iso,* equal; *ptera,* wings); termites (Figs. 11-18 and 11-19).

BODY CHARACTERISTICS minute to large.

MOUTHPARTS chewing.

EYES and OCELLI compound eyes present in all winged forms, present or lacking in apterous forms; 0 or 2 dorsal ocelli.

ANTENNAE short, moniliform or filiform.

WINGS when present, 2 identical pairs; membranous; shed by breakage along a basal fracture line.

LEGS 4- to 6-segmented tarsi.

Fig. 11-18
Isoptera, termites. A. Winged form, *Zootermopsis angusticollis*. B. Eggs. C. Third instar nymph. D. Last instar nymph. E. Soldier. F. Head of *Nasutitermes* sp. (A and E, Me.) [A–E redrawn from Essig, 1942.]

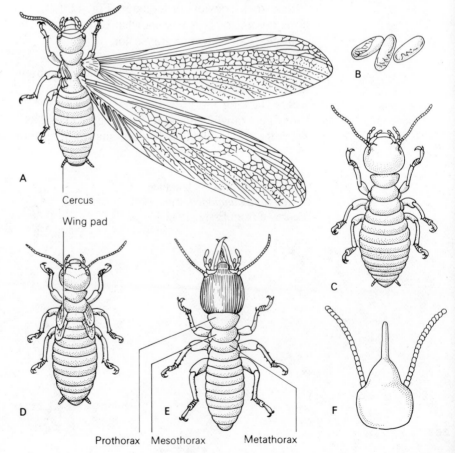

A

Cercus

Wing pad

B

C

D

E

F

Prothorax   Mesothorax   Metathorax

ABDOMEN genitalia lacking or weakly developed; pair of short cerci.

COMMENTS characteristic depression on dorsum of head called *fontanelle*.

The termites are a very primitive order of orthopteroid insects, being most closely related to the cockroaches.

They are mainly tropical and subtropical, but also occur in temperate regions. Some termites live in dry wood (e.g., that frequently used in the construction of buildings and furniture). Others are subterranean and require very humid conditions. Subterranean forms probably play an ecological role similar to earthworms in that they aerate and add nutriment to the soil. In addition to wood, termites feed on a wide variety of cellulose-containing materials, fungi, and dried animal remains. The digestion of cellulose is carried out by flagellate protozoans or bacteria, which are mutualistic inhabitants of the gut.

Termites are commonly confused with ants. This is probably the reason that termites are sometimes referred to as "white ants." However, there are several distinct differences between these two groups of insects. Some of the more obvious include the following. In ants the hindwings are smaller and have fewer veins than the forewings and at rest are held vertically over the abdomen, whereas in termites the wings are similar in size and venation and flexed horizontally over the abdomen at rest. In wingless forms, ants have relatively dark, sclerotized bodies, whereas termites are pale and soft-bodied. The abdomen of ants is basally constricted in both winged and apterous forms, but this is not the case in termites.

All termites are social insects. A colony's population is initiated and maintained by a queen that may live for as many as 50 years in some species. Two to several castes may be present, depending on the species, and all castes are composed of both males and females. Conveniently, castes can be divided into reproductives and nonreproductives. There may be two kinds of reproductives present in a colony of a given species: primary reproductives and secondary reproductives. The primary reproductives are the queen (Fig. 11-19C, H) and king (Fig. 11-19D, F) and are thought to comprise the original caste in isopteran phylogeny. They typically have dark, sclerotized (at least compared to other castes) bodies with completely developed wings and compound eyes. Secondary reproductives (Fig. 11-19G) occur in a variety of forms (e.g., ones with shorter wings, less pigmentation, and smaller compound eyes than the primary reproductives or ones with no wings and pale bodies resembling workers).

The nonreproductive castes are the workers and soldiers. The workers (Fig. 11-19E) are typically wingless, unpigmented, soft-bodied forms, without compound eyes or well-developed mandibles, that function as the labor force of the colony, maintaining and adding to the structure of the nest, tending fungus gardens, and feeding members of other castes and immatures. Soldiers (Fig. 11-18E and 11-19I), on the other hand, may or may not have compound eyes and generally have enlarged, sclerotized heads with well-developed mandibles. They function as defenders of the colony either with their mandibles or in some species by plugging up holes with their heads. In the soldier caste of *Nasutitermes* spp. (called *nasuti;* Fig. 11-

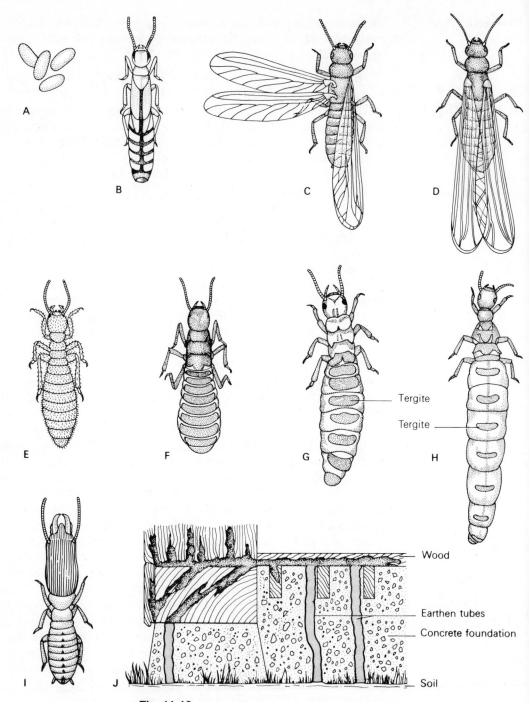

Fig. 11-19
Subterranean termite, *Reticulitermes*. A. Eggs. B. Nymph. C. Female winged form. D. Male winged form. E. worker. F. King. G. Supplementary reproductive (abdomen distended with enlarged ovaries). H. Queen (abdomen distended with enlarged ovaries). I. Soldier. J. Earthen tubes from soil across surface of concrete foundation to wooden structures. (Mi to S.) [Redrawn from Boreal Labs key card.]

18F), the duct from a gland in the head opens at the end of a long narrow snout through which a sticky secretion from the gland can be forced. This secretion is squirted as a means of defense. In the simplest social structure, characteristic of certain primitive termite species, there are reproductives and soldiers, and the immatures of both these castes function as workers (Fig. 11-18C, D).

The control of caste differentiation has long been a topic of interest and investigation and is still a long way from being understood completely. There is strong evidence that complex interactions of pheromones are involved (see "Polymorphism" in Chapter 5).

New colonies are formed when at certain times of the year, winged primary reproductives appear and swarm from the nest. Males and females pair off during this swarming, but mating usually occurs on the ground. Subsequently, their wings break off along preformed basal fracture lines, and the pair locates a site for nest construction. Another method of nest foundation is *sociotomy* (Lindauer, 1965), the process by which colonies divide by the separation of immatures and secondary reproductives from the parent colony or by division of a migrating colony into two daughter colonies. Colony size may vary from a few individuals to over 100,000.

Termite nests vary from simple cavities in soil or wood to vast subterranean complexes or elaborate edifices that project well above the ground. A nest of a species in northern Australia, *Nasutitermes triodidae,* has been reported (Richards and Davies, 1977) to reach a height of 20 feet and a diameter of 12 feet at its base! Very elaborate ventilation systems, designs that provide for maintenance of constant temperature, canopies that deflect rainwater, and other structural adaptations of nests have been described in various termite species. The means by which the behavior of individual members of a colony are coordinated to produce such complex structures has long been a source of amazement (see Chapter 8).

Termites are very significant structural pests, damaging wooden structures (e.g., furniture, building timbers, and wooden floors). Some of the most destructive termites in the United States are subterranean and in the genus *Reticulitermes.* They can construct earthen tubes on concrete and are thus able to invade a structure even though it is not in direct contact with the soil. The presence of earthen tubes (Fig. 11-19J) is one of the characteristics used to "diagnose" a termite infestation. Not all destructive termite species are subterranean; for example, colonies of *Kalotermes* spp. (dry-wood termites) and *Zootermopsis* spp. (damp-wood termites) exist entirely in wood. These termites are found along the Pacific Coast and in the southern United States.

Order Blattaria   (*blatta,* cockroach); Blattodea; cockroaches (Fig. 11-20).

BODY CHARACTERISTICS   somewhat flattened dorsoventrally; small to very large; generally dark colored, but some tropical species brightly colored.

MOUTHPARTS   mandibulate.

EYES and OCELLI   compound eyes usually well developed; 2 ocelli in some; *fenestrae* (pale regions with nerves to brain) in place of ocelli in most.

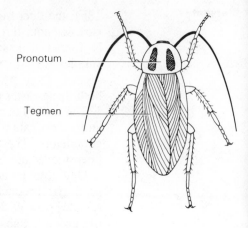

Pronotum

Tegmen

Fig. 11-20
Blattaria. A cockroach,
Blattella germanica. (Me.)
[After U.S. Public Health
Service.]

ANTENNAE   long, filiform.

WINGS   forewings tegmina; hindwings membranous, broader than
forewings and with enlarged anal area.

LEGS   generalized, cursorial; large coxae.

ABDOMEN   pair of styli on 9th sternum of males; ovipositor reduced
and concealed by 7th abdominal sternum.

COMMENTS   large shieldlike pronotum nearly covers dorsum of head.

The cockroaches are a predominately tropical group, but several
species are found in temperate zones. They are basically omnivorous
insects, mostly feeding on decaying animal matter. Certain species,
for example, *Cryptocercus* sp., feed on dead wood and are able to
digest cellulose by virtue of the presence of cellulase produced by
intestinal symbionts (protozoans in the case of *Cryptocercus,* bacteria in some others). Cockroaches are generally found in litter,
among low vegetation, or on or in the ground. Some species inhabit
caves, and a few are associated with ant colonies *(myrmecophilous).*
Cockroaches are very agile runners and usually depend on this ability, instead of flying, to escape potential predators. The cockroach
Gromphadorhina portentosa produces a hissing sound through the
spiracles when threatened. Cockroaches tend to be nocturnally
active.

Cockroaches undergo hemimetabolous development. There are
many oviparous and ovoviviparous species, and one species is viviparous. In the oviparous forms, the eggs are lined up in an egg
purse or ootheca, which is sometimes cemented to the substratum
(e.g., the American cockroach, *Periplaneta americana*).

The cockroaches are common household pests and feed on a wide
variety of household goods, but the major indictment against them
is that they are dirty, distasteful, and odoriferous creatures and are
attracted to such materials as garbage, feces, and foodstuffs consumed by humans. The most common household-invading species
(domiciliary) in the United States are (1) *Periplaneta americana,*
the American cockroach; (2) *Blattella germanica,* the German cockroach; and (3) *Blatta orientalis,* the oriental cockroach. Although
there is little conclusive evidence that points to cockroaches as dis-
seminators of pathogenic organisms, the circumstantial evidence is
strong, and it has been suggested that they may, in fact, rival house
flies in their capacity for disease transmission. Their nocturnal, se-

cretive habits have perhaps been the reason they have been somewhat overlooked in this capacity in the past. Readers interested in more information regarding cockroaches, both their biology and medical-economic significance, should consult Cornwell (1968), Guthrie and Tindall (1968), Harwood and James (1979), Roth and Willis (1957, 1960), and Ragge (1973).

Order Mantodea (*mantid,* soothsayer); praying mantids (Fig. 11-21).

BODY CHARACTERISTICS large to very large (largest over 100 mm); body usually elongate, somewhat cylindrical, with elongate prothorax.

MOUTHPARTS mandibulate; hypognathous.

EYES and OCELLI well-developed compound eyes; three ocelli or none.

ANTENNAE filiform.

WINGS usually present in males; may be reduced or absent in females; forewings are tegmina with small anal area; hindwings membranous with enlarged anal area.

LEGS raptorial forelegs at anterior end of elongated prothorax; mid- and hindlegs cursorial; 5-segmented tarsus.

ABDOMEN comparatively short, multisegmented cerci; pair of styli usually associated with 9th sternum in males.

COMMENTS head triangular with wide range of movement.

Praying mantids are solitary, exclusively carnivorous insects. They capture prey such as flies and grasshoppers with their well-developed raptorial forelegs. Some large South American species may even attack small birds and other small vertebrates. Mantids do not actively pursue prey, but wait motionless on vegetation for passing prey. They monitor the position of passing prey by turning their heads as prey passes and strike out with their forelegs when prey are in range. Few other insects are able to turn their heads so freely. Some species of mantids mimic flowers and leaves. This probably aids in prey capture and may also provide protection from predators. Mantids will run or fly and may make short jumps when disturbed. Some adopt a threat posture when disturbed, raising their sometimes brightly colored wings and striking out with their forelegs. If grasped by the meso- or metathoracic legs, they may escape by autoamputating the imprisoned leg.

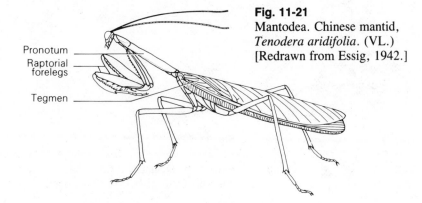

Pronotum

Raptorial forelegs

Tegmen

Fig. 11-21
Mantodea. Chinese mantid, *Tenodera aridifolia.* (VL.)
[Redrawn from Essig, 1942.]

Mantids undergo hemimetabolous development. They encase their eggs in a frothy substance that solidifies into an egg case (ootheca). Egg cases (some containing more than 200 eggs) are typically glued to twigs or bark, and their shape and size tend to be species-specific. It is common for the female to begin devouring the male during copulation (see ''Courtship and Mating'' in Chapter 8).

Order Zoraptera   (*zor*, pure; *ptera,* wings); zorapterans (Fig. 11-22).

BODY CHARACTERISTICS   minute

MOUTHPARTS   chewing.

EYES and OCELLI   compound eyes and dorsal ocelli in winged forms; compound eyes and dorsal ocelli lacking in wingless forms.

ANTENNAE   9-segmented; moniliform or filiform.

WINGS   2 pairs when present; membranous; forewings larger than

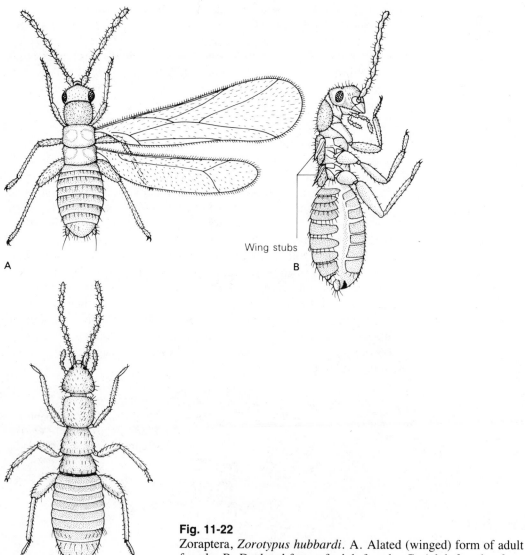

A

Wing stubs

B

C

Fig. 11-22
Zoraptera, *Zorotypus hubbardi*. A. Alated (winged) form of adult female. B. Dealated form of adult female. C. Adult female of apterous form. (Mi.) [Redrawn from Caudell, 1918.]

hindwings; shed by breakage along a basal fracture line; both sexes of a given species may have winged and wingless forms; wingless forms more common.

LEGS 2-segmented tarsi.

ABDOMEN short, 1-segmented cerci.

This is the second smallest order in the class Insecta. Only 22 species have been described in the world (Borror, DeLong, and Triplehorn, 1976). All described species are included in the family Zorotypidae.

Members of this order have been found in such places as decaying wood, humus, sawdust piles, termite nests, and under bark and stone. They tend to be gregarious but apparently lack a definitive social structure. These insects are most likely scavengers, feeding on such materials as fungal spores and the remains of other arthropods. Zorapterans occur in the southeastern United States but are uncommon.

They are significant in that they are considered by some to provide an evolutionary link between the orthopteroid orders and Psocoptera and hence the hemipteroid orders. However, Boudreaux (1979) disagrees and views them as highly specialized relatives of the cockroaches, termites, and mantids.

Hemipteroid Orders. The orders grouped as hemipteroids are Psocoptera, Mallophaga, Anoplura, Thysanoptera, and Hemiptera (Fig. 11-7). Boudreaux (1979) applies the name Acercaridae to this group in reference to the complete absence of cerci. Members of these orders share certain features, including: (1) specialized mandibulate or haustellate mouthparts (lacinia slender, often stylets; labial palps absent or with no more than 2 segments); (2) enlarged clypeus associated with musculature of cibarial pump; (3) hindwings lacking a large anal lobe; (4) cerci lacking; (5) four or fewer Malpighian tubules, and (6) a tendency toward concentration of the ventral chain ganglia.

Development is hemimetabolous in this group of insects.

Order Psocoptera (*psoco,* rub small; *ptera,* wing); Corrodentia; book lice and relatives (Fig. 11-23).

BODY CHARACTERISTICS minute; head capsule large compared to rest of body.

MOUTHPARTS chewing; labial silk glands present.

EYES and OCELLI compound eyes strongly or weakly developed; dorsal ocelli: 3 in winged forms, absent in wingless forms.

ANTENNAE long; filiform.

WINGS most with 2 pairs; membranous; reduced venation; held rooflike over body at rest; forewings larger than hindwings; vestigial in some species and absent in others.

LEGS 2- or 3-segmented tarsi.

ABDOMEN ovipositor partly and aedeagus completely concealed; cerci lacking.

Psocopterans live in a variety of terrestrial habitats: under bark (bark lice), amid vegetation, and in bird nests. Many are gregarious. Several of the wingless species occur in human habitations, partic-

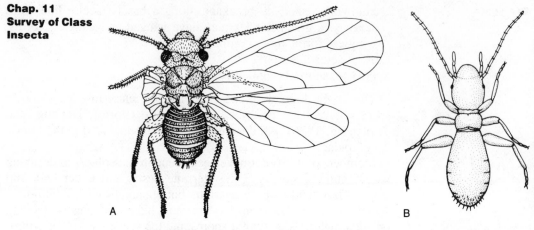

A

B

Fig. 11-23
Psocoptera. A. Winged species, *Peripsocus californicus*. B. Wingless
species, *Troctes divinatorius*. (Mi.) [Redrawn from Essig, 1942.]

ularly around books and papers, and are commonly called book lice.
The outdoor species feed on organic matter, including fungi, algae,
lichens, pollen, and fragments of decaying organic material.

Most species are oviparous, although some are viviparous. Parthenogenesis occurs in some species.

Psocopterans are considered to be somewhat intermediate between
the zorapterans (here treated as an orthopteroid group) and the remaining hemipteroids.

Psocoptera may reach large enough populations among books,
papers, stored cereal grains, insect collections, and other materials
to constitute a pest situation, but they are generally of little economic
importance.

Order Mallophaga   (*mallo,* wool; *phaga,* eat); chewing lice (Fig.
11-24).

BODY CHARACTERISTICS   minute; dorsoventrally flattened; triangular
head broader than thorax.

MOUTHPARTS   chewing.

EYES and OCELLI   reduced compound eyes; dorsal ocelli lacking.

ANTENNAE   3- to 5-segmented; usually capitate or filiform; when capitate, concealed beneath the head.

Antenna

Legs

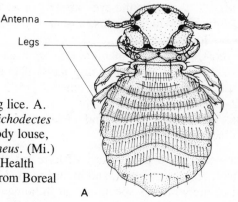
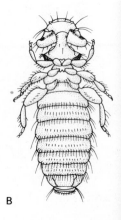

Fig. 11-24
Mallophaga, chewing lice. A.
Dog-biting louse, *Trichodectes
canis*. B. Chicken body louse,
Menacanthus stramineus. (Mi.)
[A after U.S. Public Health
Service; B redrawn from Boreal
Labs key card.]

A

B

WINGS lacking.

LEGS tarsi modified for grasping hairs in species with mammalian hosts; 1- or 2-segmented tarsi.

ABDOMEN cerci lacking.

Most Mallophaga are ectoparasites of birds, but some (members of the families Trichodectidae and Gyropidae) have mammalian hosts. They feed on various organic fragments (e.g., from feathers and skin) and epidermal secretions, and certain species, for example, *Menacanthus stramineus* (Fig. 11-24B), are known to gnaw through the skin and obtain blood. Some obtain sebum and perhaps serum by attacking hair follicles.

Some authors include the Mallophaga with the Anoplura (sucking lice) in the order Phthiraptera. Both the Anoplura and Mallophaga appear to be related to the Psocoptera.

Chewing lice pass through three nymphal instars. The eggs are attached to host feathers or hairs, and generation after generation is spent on the same host. Individuals are typically transferred from one host to another by direct contact, and they soon die if separated from an acceptable host. Some species are transferred from host to host by means of phoresy (see "Kinds of Interspecific Interactions" in Chapter 9). Different species sometimes show different preferences as to the part of the host's body attacked.

Although probably all domestic animals are attacked by one or more species, chewing lice usually do not constitute major problems. However, if an infestation is large, some damage may occur, particularly to poultry, which may suffer restlessness, decrease in egg production, and loss of feathers. Three common chewing lice that attack poultry are *Menopon pallidum,* the common chicken louse; *Menacanthus stramineus* (Fig. 11-24B), the chicken body louse; and *Menopon gallinae,* the shaft louse. Cattle, horses, sheep, goats, dogs, and cats are among the domestic animals attacked by species of chewing lice. Humans are not attacked by these lice, but may occasionally become infested by direct contact with infested animals; the lice do not survive long on a human host. Indirectly, however, the common dog louse, *Trichodectes canis* (Fig. 11-24A), does at times constitute a threat to human health because it serves as an intermediate host for the double-pored dog tapeworm, *Dipylidium caninum*. If an infected louse is inadvertently ingested by a human playing with an infected dog, the human can become infected with the tapeworm.

Order Anoplura (*anopl,* unarmed; *ura,* tail); Siphunculata; sucking lice (Fig. 11-25).

BODY CHARACTERISTICS minute; dorsoventrally flattened; head narrower than thorax.

MOUTHPARTS piercing-sucking; retracted into head when not feeding.

EYES and OCELLI compound eyes weakly developed or lacking; dorsal ocelli lacking.

ANTENNAE short; 3 to 5 segments.

WINGS lacking.

LEGS 1-segmented tarsi; tarsi usually adapted for grasping hairs.

ABDOMEN cerci lacking.

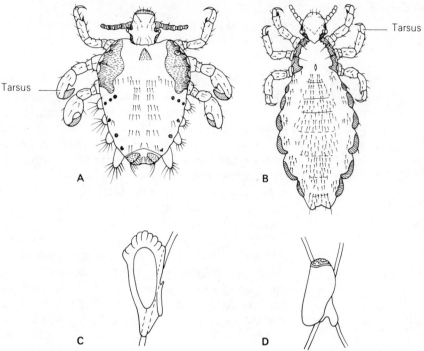

Tarsus

Tarsus

A

B

C

D

Fig. 11-25
Anoplura, sucking lice. A. Crab louse, *Phthirus pubis*. B. Human louse.
Pediculus humanus. C. Egg or nit of crab louse attached to host hair. D.
Egg of human head louse attached to host hair. (Mi.) [A and B after U.S
Public Health Service; C and D redrawn from Boreal Labs key card.]

Sucking lice are all blood-feeders. They attack a wide variety of
mammals, even marine forms (e.g., seals), and are highly host
specific.

The major pests of domestic animals in this order are in the families Haematopinidae (e.g., the hog louse, *Haematopinus suis*) and
Linognathidae (e.g., *Linognathus vituli,* the long-nosed ox louse).

Two species of sucking lice attack man, *Phthirus pubis* (Fig 11-
25A) and *Pediculus humanus* (Fig. 11-25B). Some systematists include these two species in a single family, Pediculidae; others split
them into two families, *Pediculus* in Pediculidae and *Phthirus* in
Phthiriidae.

Phthirus pubis is the human pubic louse or crab louse and infests
the pubic and armpit areas, although it may be found in other areas
of the body with coarse hair. This louse causes intense itching, and
an infestation may result from contact with an already infested individual via toilet seats, blankets, and sexual intercourse. The life
cycle ordinarily takes about 1 month to complete and the eggs, or
nits, are attached to hairs on the host (Fig. 11-25C). Pubic lice are
not know to transmit any human pathogens.

Pediculus humanus can be divided into two subspecies, *P. humanus capitis,* the head louse, and *P. humanus corporis,* the body
louse. The first infests particularly the head region, but it has been
found on other hairy areas of the body. The body louse infests regions
that come into frequent or continuous contact with clothing (e.g.,
armpits, neck, and crotch). The eggs are usually attached to clothing,

which provides an ideal vehicle for transmission to a new host. Head lice, on the other hand, attach their eggs to hairs (Fig. 11-25D), and a new infestation may result from contact with a stray hair bearing an egg. Both head and body lice are spread via direct contact. They develop quite rapidly, a complete generation occurring in about 3 weeks.

Infestation with lice is sometimes referred to as *pediculosis*. The attendant discomforts range from intense itching to anemia and pathologic changes in the skin resulting from the chronic feeding of lice and the associated blood loss. In addition to the problems caused directly by their feeding, body lice are involved in the transmission of a number of human pathogens, the most important of which are epidemic relapsing fever, caused by a spirochaete, *Borrelia recurrentis;* epidemic typhus, caused by a rickettsia, *Rickettsia prowazekii;* and murine typhus fever, caused by *Rickettsia typhi*. All three of these diseases have occurred in large epidemics and in the case of epidemic typhus have claimed thousands of lives, particularly during wartime when populations become highly concentrated and sanitary practices are lax. Several other pathogenic microbes may be transmitted by lice.

Order Thysanoptera   (*thysano,* a fringe; *ptera,* wing); Physapoda; thrips (Fig. 11-26).

BODY CHARACTERISTICS   minute to small; slender; usually darkly colored.

MOUTHPARTS   rasping-sucking; asymmetrical, the right mandible lacking or reduced.

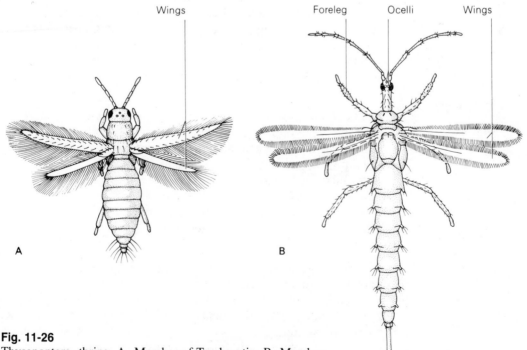

Wings   Foreleg   Ocelli   Wings

A   B

Fig. 11-26
Thysanoptera, thrips. A. Member of Terebrantia. B. Member of Tubulifera, female. (A, Mi; B, Me.) [A after U.S. Public Health Service; B redrawn from Essig, 1942.]

EYES and OCELLI small compound eyes; 3 dorsal ocelli in winged
forms; dorsal ocelli lacking in wingless forms.
ANTENNAE short; 6 to 10 segments.
WINGS 2 pairs; narrow with fringe of hairs; coupled by basal hooks.
LEGS 1- or 2-segmented tarsi are bladderlike at the tip.
ABDOMEN cerci lacking.

Thysanoptera is divided into two suborders, based mainly on the
presence or absence of an ovipositor. Females in the suborder Terebrantia (Fig. 11-26A) have well-developed ovipositors used to deposit eggs in plant tissues; wings, if present, have veins. Females
in the suborder Tubulifera (Fig. 11-26B) have either weakly developed ovipositors or none; wings, when present, are nearly or completely veinless. The posterior extremity of the abdomen is tubular,
and eggs are deposited in crevices.

Although Thysanoptera is classified as hemimetabolous, the life
cycle is strongly suggestive of holometabolism. Early instar nymphs
are active and wingless but otherwise resemble adults. However, the
last two or three instars are quiescent, sometimes in a cocoon or
earthen cell during which time wings, if they are going to be present
develop. Generation time is relatively short, and there are usually
several generations per year. Parthenogenesis is common in this
group.

Some thrips are predaceous (all in Tubulifera), feeding on aphids
and various other small insects or mites, but most are plant-feeders,
usually feeding on sap. They are widely distributed on all sorts of
vegetation, and although generally weak fliers, some may be carried
many miles by the wind.

Several species are pests, particularly on truck and floricultural
crops, although some are pests of field crops (e.g., tobacco). They
not only cause damage by feeding, but also may transmit plant disease organisms.

Thrips can cause human discomfort by attempting to pierce the
skin and may constitute a serious annoyance if present in large
numbers.

Order Hemiptera (*hemi,* one-half; *ptera,* wing); true bugs,
aphids, leafhoppers, and relatives (Figs. 11-27 through 11-32).
BODY CHARACTERISTICS minute to very large; extremely variable.
MOUTHPARTS piercing-sucking; degenerate palpi.
EYES and OCELLI variable.
ANTENNAE variable; concealed in some.
WINGS usually 2 pairs; many wingless forms.
LEGS variable; 2- or 3-segmented tarsi.
ABDOMEN cerci lacking.

This is the largest of the hemimetabolous orders. Its members
form a very diverse, heterogeneous group both in structure and habits. Hemiptera may be divided into two large suborders (Table 11-
2): Heteroptera (true bugs) and Homoptera (aphids, leafhoppers, and
relatives). Some authors place each of these suborders in separate
orders. This is apparently logical in terms of North American species, but not so when world fauna are considered (Fox and Fox,
1964). Members of the family Peloridiidae found in Australia, New

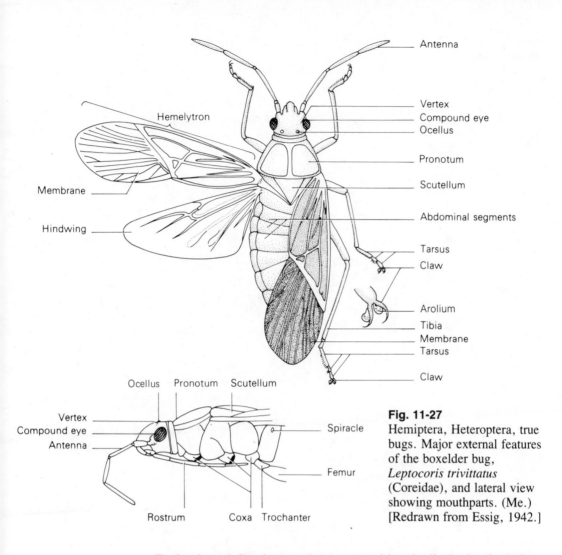

Antenna

Vertex
Compound eye
Ocellus

Hemelytron

Pronotum

Scutellum

Membrane

Abdominal segments

Hindwing

Tarsus
Claw

Arolium
Tibia
Membrane
Tarsus

Claw

Ocellus   Pronotum   Scutellum

Vertex
Compound eye
Antenna

Spiracle

Femur

Rostrum   Coxa   Trochanter

Fig. 11-27
Hemiptera, Heteroptera, true
bugs. Major external features
of the boxelder bug,
Leptocoris trivittatus
(Coreidae), and lateral view
showing mouthparts. (Me.)
[Redrawn from Essig, 1942.]

Zealand, and South America are transitional, displaying both het-
eropteran and homopteran traits (China, 1962).

Heteroptera.   These insects (Figs. 11-27, 11-28, and 11-29)
usually have compound eyes and there may be two ocelli or none.
The four-or five-segemented antennae are comparatively long and
in many are somewhat concealed. These insects usually have a pronounced pronotum with a distinct triangular scutellum on the mesothorax between the bases of the forewings. Although most are
winged, there are many wingless and brachypterous forms.

Many adult Heteroptera possess repugnatorial or scent glands located in the region of the metathoracic coxae. In some nymphs,
similar glands are located on the dorsum of the abdomen. Many
species give off a very unpleasant odor when handled. Bed bugs
(Cimicidae) have a characteristic odor, and a large infestation can
be identified by smell alone.

Many species produce sounds (Richards and Davies, 1977), and
stridulatory organs have been identified in the prosternal region
(Reduviidae and Phymatidae), abdominal sterna (some Pentatomidae), forelegs and clypeus (Corixidae), coxae (certain Nepidae), and
dorsum of the abdomen in association with the wings (some Pen-

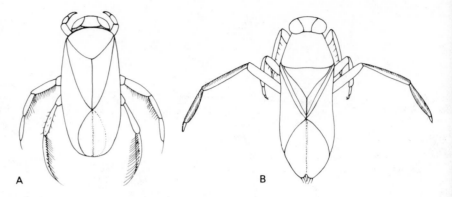

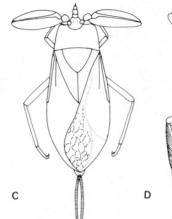

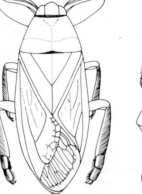

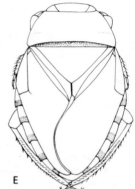

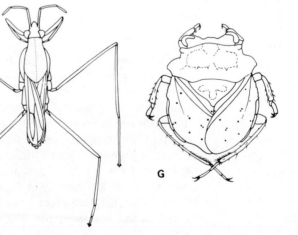

A

B

C

D

E

F

G

Fig. 11-28
Representative Hydrocorisae and Amphibicorisae A. Water boatman
(Corixidae). B. Backswimmer, *Notonecta irrorata* (Notonectidae). C.
Water scorpion, *Nepa* sp. (Nepidae). D. Giant water bug, *Lethocerus
americanus* (Belostomatidae). E. Creeping water bug, *Pelocoris
femoratus* (Naucoridae). F. Water strider, *Gerris buenoi* (Gerridae). G.
Toad bug, *Gelastocoris barberi* (Gelastocoridae). (A, S to Me; B, C, E,
and F, Me; D, L; G, S.) [B–G redrawn from Britton, 1923.]

Table 11-2
Outstanding Differences Between Suborders of Hemiptera

Heteroptera	Homoptera
1. Labial sheath arises from the anterior part of the head.	1. Labial sheath arises more posteriorly.
2. Forewings are hemelytra; i.e., basal portion leathery and apical portion membranous.	2. Forewings not hemelytra; both pairs of wings essentially uniform in texture.
3. Wings held horizontally over abdomen at rest.	3. Wings held rooflike over abdomen at rest.

tatomidae). The habitats in which heteropterans are found are highly diversified, terrestrial, aquatic, semiaquatic, and ectoparasitic forms being known. Feeding habits are likewise diverse, some being herbivorous, others carnivorous, and some hematophagous.

The Heteroptera can be divided into three groups of families, the Hydrocorisae, Amphibicorisae and Geocorisae.

The Hydrocorisae (Fig. 11-28A–E, G) have short antennae (shorter than the head) concealed in grooves on the venter of the head and are all aquatic or semiaquatic; the most common families are Corixidae, water boatmen; Notonectidae, backswimmers; Nepidae, water scorpions; Belostomatidae, giant water bugs; Naucoridae, creeping water bugs; and Gelastocoridae, toad bugs. The vast majority of these insects are predatory, and certain ones, such as backswimmers and giant water bugs, can inflict a painful bite if carelessly handled. Most are capable of flight and many appear in the vicinity of artificial lights at night. These families show many interesting variations of mechanisms involved in ventilation in an aquatic habitat (see "Ventilation in Aquatic Insects" in Chapter 4).

The Amphibicorisae are mostly aquatic, living on the surface of water or living along the shore. The antennae are not concealed and are longer than the head. This group includes the families Gerridae (water striders; Fig. 11-28F), Hydrometridae (water measurers), and others. Members of most families are predatory, but water striders feed mainly on dead insects. The seagoing *Halobates* spp. (see "A Place in Which to Live" in Chapter 9) are in this family.

The Geocorisae (Fig. 11-29) have unconcealed antennae longer than the head. There are terrestrial, aquatic, and parasitic forms. This group which contains a much larger number of families than the other two, usually is divided into several superfamilies.

Geocorisae contains many species of economic and medical significance. The chinch bug, *Blissus leucopterus* (Lygaeidae), attacks corn, sorghum, and small grains in the United States. In 1934 the chinch bug caused an estimated loss of $27.5 million to the corn crop and $28 million to wheat, barley, rye, and oats (Burkhardt, 1978). Many species in the family Miridae, plant bugs, are serious pests of leguminous forage crops, particularly alfalfa. The families Pyrrhocoridae, Coreidae, Pentatomidae (stink bugs), and others also contain important agricultural pests.

Cimicidae (bed bugs and relatives) and Reduviidae (assassin bugs

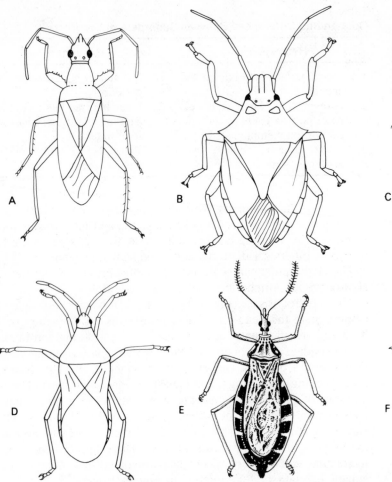

Fig. 11-29
Representative Geocorisae. A. Seed bug, *Ligyrocoris diffusus*
(Lygaeidae). B. Stink bug, *Apateticus cynicus* (Pentatomidae). C.
Tarnished plant bug, *Lygus lineolaris* (Miridae). D. Squash bug, *Anasa
tristis* (Coreidae). E. Conenose bug (Reduviidae). F. Common bedbug,
Cimex lectularius (Cimicidae). (A, C, and F, S; B, S to Me; D, Me; E,
Me to L.) [A–D redrawn from Britton, 1923; E and F after U.S. Public
Health Service.]

and relatives) are Geocorisae of medical importance. Cimicidae are
wingless ectoparasites of birds and mammals (bats, various rodents,
and others). Three species attack humans (James and Harwood,
1979): *Climex lectularius* (Fig. 11-29F), *Climex hemipterus* (tropical
America, Africa, Asia, East Indies, and certain Pacific islands), and
Leptocimex boueti. Although experimentally bed bugs can transmit
pathogens that cause a number of human diseases, there is little
evidence that they are the natural vectors of any pathogen. The
reduviids (Fig. 11-29E) are mostly predaceous, feeding on other
insects (many can inflict painful bites), but members of the subfamily
Triatominae (kissing bugs and relatives) are exclusively ectoparasites of vertebrates. Several species (e.g., *Rhodnius prolixus* and
Triatoma spp.) are vectors of Chagas disease, caused by the protozoan *Trypanosoma cruzi*. On the positive side, *Rhodnius prolixus*

has been the subject of much significant biological research, particularly that carried out by a well-known insect physiologist, Sir Vincent Wigglesworth.

Homoptera. The majority of these insects (Figs. 11-30, 11-31 and 11-32) are small with a few exceptions (e.g., some of the fulgorid planthoppers and cicadas). Compound eyes and ocelli are usually present, and antennae vary from filiform to bristlelike. The prothorax is usually inconspicuous except in the Membracidae, where it takes on a variety of bizarre shapes. Members of several groups have wax glands, which may secrete a powdery material. The females of some have distinct ovipositors. Cicadas and certain leafhoppers produce sound by means of specialized structures on the dorsolateral portion of the base of the abdomen. Aphids, whiteflies, and others produce honeydew, a liquid material excreted from the anus.

The Homoptera differ from the Heteroptera in being exclusively terrestrial herbivores.

As with all the Heteroptera, most homopterans undergo hemimetabolous development. However, in certain scale insects, whiteflies, and phylloxerans the last nymphal instar is quiescent and resembles a pupa.

Homoptera may be divided into two groups: Auchenorrhyncha and Sternorrhyncha. In the Auchenorrhyncha (Fig. 11-31) the proboscis clearly arises from the posterior of the head and the antennae are short and bristlelike; members of this group are usually very active. In the Sternorrhyncha (Fig. 11-32) the proboscis appears to arise from the sternum between the forecoxae, and the antennae are generally long and filiform; many of these lead inactive or sedentary lives, in particular certain stages of the scale insects (superfamily Coccoidea).

There are no medically important Homoptera, but many are economically important pests. These insects are destructive to agricultural crops by feeding on plant juices, causing oviposition damage, and serving as vectors of several plant disease organisms, particularly viruses. Especially significant families are Psyllidae, jumping

Fig. 11-30
Homoptera. Major external structures of a leafhopper, *Paraphlepsius irroratus.* A. Dorsal view. B. Anterior view. (S.) [Redrawn with modifications from Borror and DeLong, 1971.]

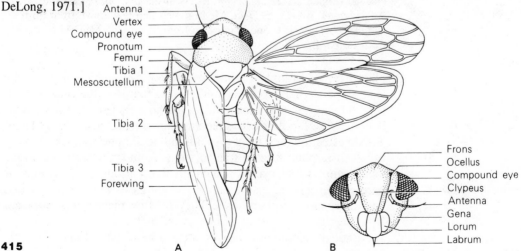

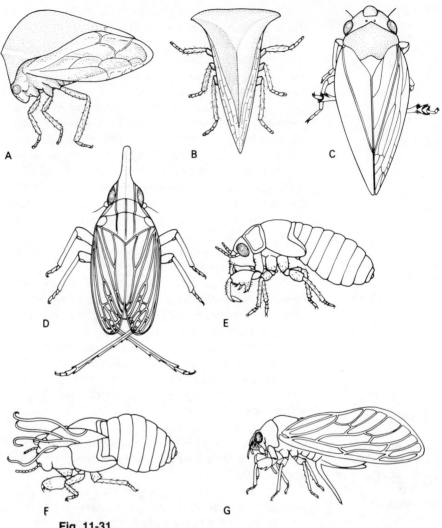

Fig. 11-31

Homoptera, Auchenorrhyncha. A. Treehopper, *Stictocephala bubalus* (Membracidae). B. Same species as A, dorsal view. C. Spittlebug (Cercopidae). D. Fulgorid planthopper (Fulgoridae). E–G. Periodical cicada, *Magicicada septendecim:* E, last instar nymph; F, cast skin; G, Adult female. (A–D, S; D–G, Me to L.) [A and B redrawn from Borror and DeLong, 1971; C and D redrawn from Britton, 1923; E–G redrawn from Boreal Labs key card.]

plant lice; Aleyrodidae, whiteflies; Aphididae, aphids or plant lice (Fig. 11-32A, B); Coccidae and related families, the scale insects (Fig. 11-32C-F); and Cicadellidae, leafhoppers (Fig. 11-30). Members of these families cause feeding damage and may also produce a toxic effect on their hosts. Aphids and leafhoppers are extensively involved in plant disease transmission. Aphids are well known for their complex life cycles, which involve various combinations of parthenogenesis, sexual generations, wingless and winged forms, and alternation of plant hosts. They feed on plant juices extracted from nearly every part of a plant, roots included.

Although many homopterans are common, the cicadas (Fig. 11-31E–G) are particularly well known. These insects range in length from about 12 mm to approximately 10 cm, and certain species are the largest homopterans. The nymphs are subterranean and feed on the roots of plants; the adults live above the ground in trees. The periodical cidadas, *Magicicada* spp., are a particularly well-known group; they spend either 13 or 17 years in the soil, depending on the species. At the end of their time in the soil, they emerge as adults, mate, and the females oviposit in slits in the twigs of trees, sometimes causing considerable damage. The emergence of a large brood may cause large patches of forest to appear brown owing to dead twigs and leaves. In a few weeks the newly emerged nymphs drop to the ground and burrow in, commencing another 13 or 17 years of feeding. There are several broods in the United States, and they

Fig. 11-32
Homoptera, Sternorrhyncha. A. Aphid, *Aphis pomi* (Aphididae), wingless viviparous female. B. Same species as A, winged viviparous female. C–F, San Jose scale, *Aspidiotus perniciosus* (Coccoidea); C, scales on a twig; D, nymph; E, adult male; F, adult female with scale removed. [Redrawn from Boreal Labs key card.]

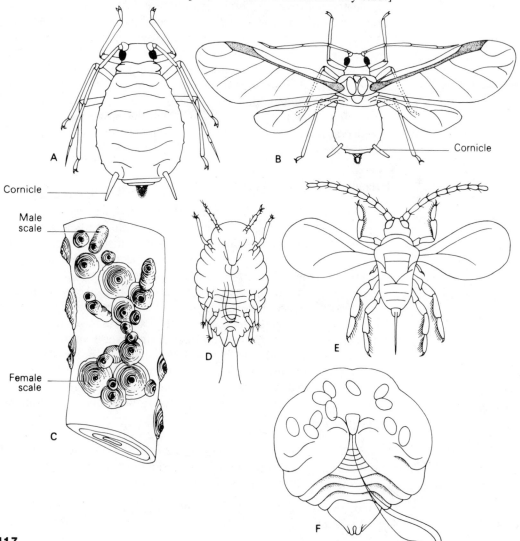

emerge in different years. Another group of cicadas (e.g., *Tibicen* spp.) are referred to as "dog-day" cicadas and emerge from their subterranean home annually in May. Cicadas are chronic noisemakers, only the males being able to produce sound.

Another commonly encountered and interesting group of homopterous insects is the spittlebugs (Cercopidae; Fig. 11-31C), so called because the nymphs surround themselves in a frothy spittlelike substance. This substance apparently protects them from potential predators and parasites and prevents desiccation (Richards and Davies, 1977). It is released through the anus. These are small insects, less than 12 mm in length, and the adults are active jumpers.

Fossil Neopterous Exopterygote Orders

Carpenter (1977) recognizes four extinct neopterous exopterygote orders: Protorthoptera, Caloneurodea, Miomoptera, and Protelytroptera (see Fig. 10-14). The Protorthoptera were the largest, most diverse extinct order and displayed several orthopteroid traits, such as leathery forewings, hindwings with enlarged anal region, well-developed cerci, and mandibulate mouthparts. The Caloneurodea were also orthopteroid in appearance, but the fore- and hindwings were identical in appearance and the cerci short. The Miomoptera were small insects with mandibulate mouthparts and short cerci. The fore- and hindwings were the same size and shape, and the venation was similar to the modern Psocoptera. The Protelytroptera had hardened forewings (elytra), beneath which the broad hindwings were folded at rest, and short cerci. The wing venation was similar to the Dermaptera.

Neopterous Endopterygota

The orders in this category are generally viewed as being monophyletic. However, there is some disagreement as to the subgroups. Here we recognize division of the neopterous endopterygotes into coleopteroid, neuropteroid, panorpoid (mecopteroid), and hymenopteroid orders (Fig. 11-33). Kristensen (1975) and others include the coleopteroid orders as neuropteroids. Within the panorpoid orders, (a) the Mecoptera, Diptera, and Siphonaptera and (b) the Trichoptera and Lepidoptera are widely seen as two well-defined groups, the members of each displaying several similar features. The hymenopteroid category includes only one order, the Hymenoptera. These insects are sufficiently different that most agree to their separation from the rest of the neopterous endopterygotes.

Members of all orders in this group undergo complete metamorphosis.

Coleopteroids. The coleopteroid orders, Coleoptera and Strepsiptera, differ from the remaining endopterygote orders in the specialization of the metathorax for flight (Boudreaux, 1979).

Order Coleoptera (*coleo,* sheath; *ptera,* wing); beetles (Figs. 11-34, 11-35, and 11-36).

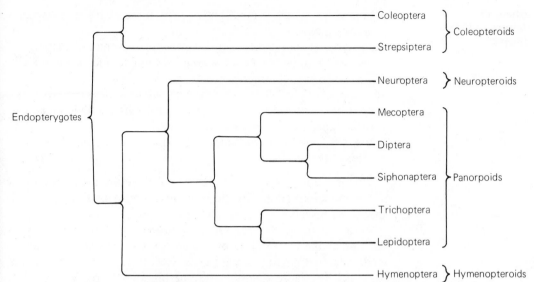

Fig. 11-33
Hypothetical phylogeny of endopterygote insects. [Based, with modifications, on Boudreaux, 1979.]

BODY CHARACTERISTICS   minute to very large; hard-bodied; larvae: campodeiform or eruciform; pupae: exarate.

MOUTHPARTS   chewing, typically prognathous; some hypognathous; larvae: chewing.

EYES and OCELLI   compound eyes present or absent; ocelli generally lacking, but 1 or 2 in certain groups; larvae: compound eyes lacking, but lateral ocelli usually present.

ANTENNAE   variable; small in larvae.

WINGS   forewings sclerotized elytra which protect the membranous

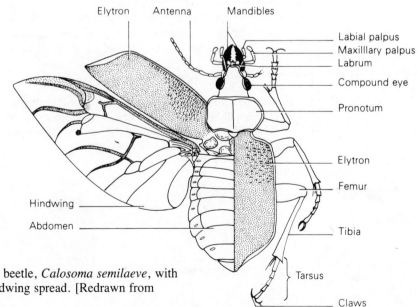

Fig. 11-34
Coleoptera, ground beetle, *Calosoma semilaeve*, with left elytron and hindwing spread. [Redrawn from Essig, 1942.]

hindwings at rest; some flightless with fused elytra; a few species
are wingless.

LEGS typically cursorial; some adapted for swimming, jumping, or
digging; number of tarsal segments variable and taxonomically
useful; larvae: commonly 6 legs, some legless.

ABDOMEN Larvae: some with posterior pair of prolegs (no crochets);
usually a pair of spiracles on each of the first 8 abdominal segments plus 1 or 2 pairs on the thorax.

COMMENTS pronotum = single, conspicuous sclerite; larvae are
often called grubs.

This is the largest order of insects and therefore the largest order
in the animal kingdom. Beetles comprise approximately 40% of all
species of insects.

As a group, beetles are found in nearly as wide a diversity of
habitats as the entire class of insects. A similar statement can be
made relative to their feeding habits. However, surprisingly few are
parasitic. The largest numbers are herbivorous or predatory, while
somewhat fewer are scavengers or feed on fungi. The predatory and
scavenging species are beneficial in that they either destroy other
insects or take part in the breakdown of organic material.

Coleoptera may be divided into four suborders: Archostemmata,
Myxophaga, Adephaga (Fig. 11-35), and Polyphaga (Fig. 11-36).
The first two groups are small and composed of uncommon families.

Most beetles are oviparous, but a few are viviparous and a few
are parthenogenetic. Hypermetamorphosis occurs in some. The pupae are exarate. They often reside in earthen cells or within host
plants. A number of beetles form pupal cocoons, which in some
species are composed of a secretion from the Malpighian tubules or
collapsed peritrophic membrane (see ''Metamorphosis'' in Chapter
5), or they may be protected by the exuviae of the last larval instar.

The beetles rival moths in their economic importance. This order
contains a large number of very destructive pests of agricultural
crops; stored grains, seeds, and grain products; other stored products, such as tobacco, nuts, and chocolate; and shade trees and
shrubs. Several species of beetles serve as vectors of plant disease.

Fig. 11-35
Representative members of the suborder Adephaga. A. Tiger beetle
(Cicindellidae). B. Ground beetle (Carabidae). C. Whirligig beetle
(Gyrinidae). D. Predaceous diving beetle (Dytiscidae). (S to L.)
[Redrawn from Boreal Labs key card.]

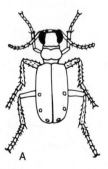

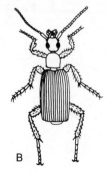

A   B   C   D

One of the best known is the European bark beetle, which transmits the Dutch elm disease. This disease is caused by a fungus, and the spores are introduced into the cambium of a host tree by the boring activities of the beetles. Following are the common names of some examples of pest beetle species: white grub (June beetle larva), wireworm (click beetle larva), Japanese beetle, Colorado potato beetle, boll weevil, corn rootworm, bronze birch borer, alfalfa weevil, cane borer, flat-headed apple borer, powder post beetle, drug store beetle, granary weevil, rice weevil, mealworm, cadelle, confused and red flour beetles, saw-toothed grain beetle, pea weevil, and bean weevil.

Information on these insects may be found in applied entomology texts such as Pfadt (1978) and Metcalf, Flint, and Metcalf (1962).

Order Strepsiptera   (*strepsi*, turning or twisting; *ptera*, wing); stylopids or twisted-wing parasites (Fig. 11-37).
BODY CHARACTERISTICS   minute.
MOUTHPARTS   reduced mandibulate.
EYES and OCELLI   compound eyes present in males and free-living females.
ANTENNAE   males: flabellate; females: lacking.
WINGS   males: mesothoracic, clublike; metathoracic, membranous and fan-shaped; females: lacking.
LEGS   larvae and adult females mostly legless; 2- to 5-segmented tarsi.
ABDOMEN   3 to 5 genital pores in adult females.

This is a very small order of mostly endoparasitic insects, which attack many Hemiptera and Hymenoptera. Males are endoparasitic as larvae but emerge as free-living adults. Females of parasitic species are parasitic throughout their lives and are reduced in structure and legless. Strepsipteran larvae undergo hypermetamorphic development. The first instar *(triungulin)* is free-living and possesses legs. When this larva comes into contact with a host, it burrows through the host's cuticle and becomes endoparasitic. The second and remaining larval instars are vermiform.

Neuropteroids.   Here the neuropteroid group is treated as a single order, Neuroptera, although some authorities view the neuropteran suborders—Megaloptera, Raphidiodea, and Plannipennia—as separate orders—Megaloptera, Raphidioptera, and Neuroptera, respectively. Among the ways neuropteroids differ from the other endopterygotes are details of ovipositor and wing structure (there are characteristically many cross veins in the costal space between the costa and subcosta in neuropteroids). Some authorities classify the coleopteroid orders as neuropteroids.

Order Neuroptera   (*neuro*, nerve; *ptera-* wing); lacewings, dobsonflies, and related forms (Figs. 11-38, 11-39, and 11-40).
BODY CHARACTERISTICS   small to very large; soft-bodied; larvae: campodeiform; pupae: exarate in silken cocoon.
MOUTHPARTS   chewing; larvae: grasping-sucking.
EYES and OCELLI   compound eyes present; dorsal ocelli present or absent.

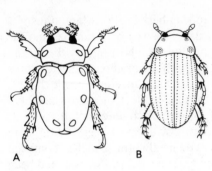
A
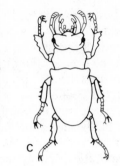
B   C   D
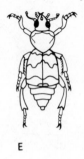
E

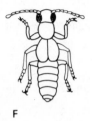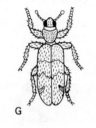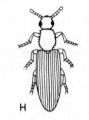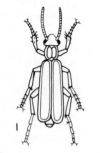
F   G   H   I   J

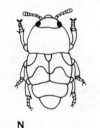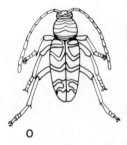
K   L   M   N   O

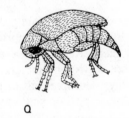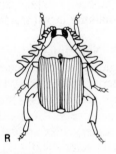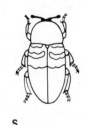
P   Q   R   S

T   U   V

Fig. 11-36 (Opposite)
Representative members of the suborder Polyphaga. A. Scarab beetle
(Scarabaeidae). B. Water scavenger beetle (Hydrophilidae). C. Stag
beetle (Lucanidae). D. Click beetle (Elateridae). E. Carrion beetle
(Silphidae). F. Rove beetle (Staphylinidae). G. Checkered beetle
(Cleridae). H. Cucujid beetle (Cucujidae). I. Blister beetle (Meloidae).
J. Brentid beetle (Brentidae). K. Darkling beetle (Tenebrionidae). L.
Languriid beetle (Languriidae). M. Lady beetle (Coccinellidae). N. Sap
beetle (Nitidulidae). O. Long-horned beetle (Cerambycidae). P. Firefly
(Lampyridae). Q. Tumbling flower beetle (Mordellidae). R. Seed beetle
(Bruchidae). S. Carpet beetle (Dermestidae). T. Snout beetle
(Curculionidae). U. Leaf beetle (Chrysomelidae). V. Metallic wood
borer (Buprestidae). [Redrawn from Boreal Labs key card.]

ANTENNAE usually filiform.

WINGS 2 pairs; similar in size and appearance; usually held rooflike
  over body at rest.

LEGS 5-segmented tarsi.

ABDOMEN cerci lacking; variously modified ovipositor present.

COMMENTS mostly weak fliers.

Neuropterans are found about vegetation and often near bodies of
water since many have aquatic larvae. Adults feed on fluid materials
and various soft-bodied insects. Larvae are carnivorous, capturing
and holding prey in their well-developed mandibles.

Some Neuroptera undergo hypermetamorphosis.

Neuroptera may be divided into three suborders: Megaloptera,
(Fig. 11-39B–E), Raphidiodea (Fig. 11-39A), and Planipennia
(Figs. 11-38 and 11-40). Megaloptera have aquatic larvae with
chewing mouthparts, hindwings basally slightly broader than forewings, with the anal region of the hindwings folding at rest. Members of the Raphidiodea, the snakeflies (Fig. 11-39A) are very
distinct in appearance; the prothorax is elongate, with legs attached
to the posterior end, giving the impression of a very long neck.
Snakefly larvae are terrestrial. Planipennia is the largest neuropteran
suborder. Adults are characterized by having wings of similar shape
and size with a nonfolding anal region on the hindwing. The larvae
are usually terrestrial and have mandibulate sucking mouthparts.

Megaloptera contains two families, Corydalidae (dobsonflies and
fishflies) and Sialidae (alderflies). Dobsonflies (Fig. 11-39B–D) are
comparatively large and their aquatic larvae, or hellgrammites (Fig.
11-39D), are well known to fisherman as fish bait. Fishflies are
smaller than dobsonflies. Alderflies (Fig. 11-39E) are also smaller
than dobsonflies, also have aquatic larvae, and tend to be darkish in
color.

The suborder Planipennia can be grouped into two superfamilies,
Hemerobioidea and Myrmeleontoidea.

Among the families in Hemerobioidea are Mantispidae (mantidflies; Fig. 11-40A), Chrysopidae (common lacewings; Fig. 11-38),
Hemerobiidae (brown lacewings; Fig. 11-40B), and Sisyridae (spongillaflies; Fig. 11-40C). Mantidflies (Fig. 11-40A) bear a superficial
resemblance to snakeflies but have raptorial forelegs attached anteriorly to the elongate prothorax. Their larvae are terrestrial and attack
the spiderlings in the nests of ground spiders. Both brown and green

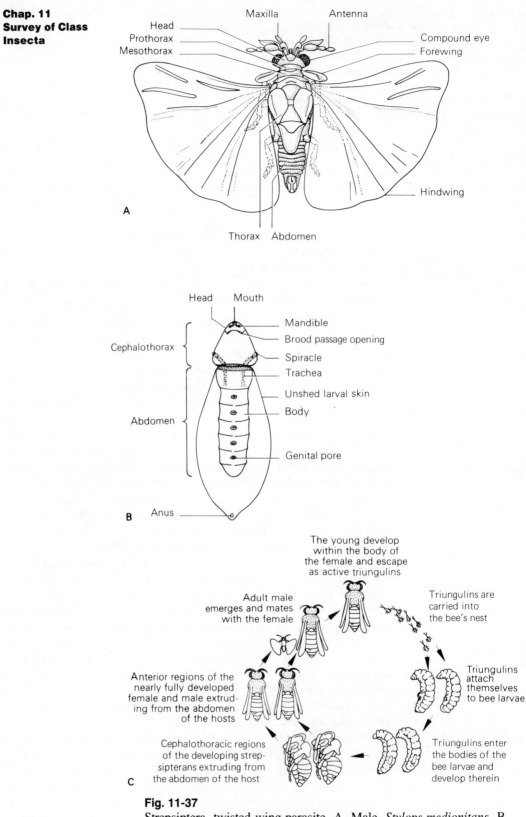

Fig. 11-37
Strepsiptera, twisted-wing parasite. A. Male, *Stylops medionitans*. B. Female. C. Life cycle. (Mi.) [Redrawn from Essig, 1942.]

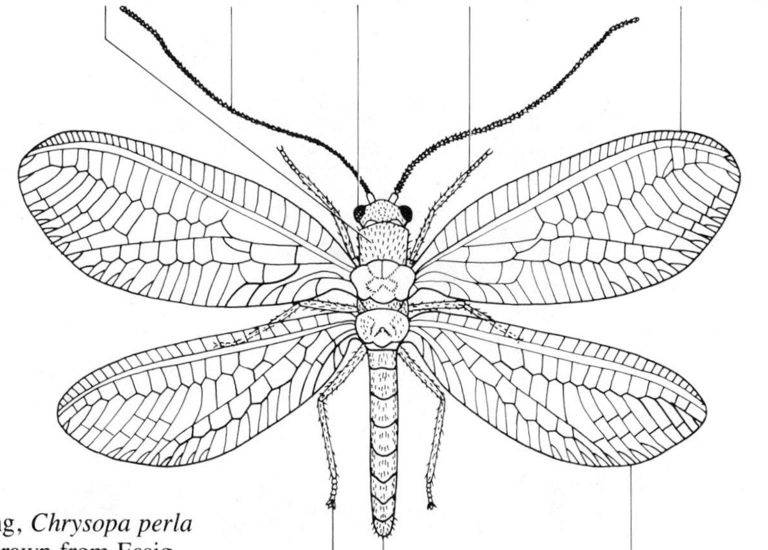

Pronotum   Antenna   Compound eye   Foreleg   Forewing

Fig. 11-38
Neuroptera. Green lacewing, *Chrysopa perla*
(Chrysopidae). (Me.) [Redrawn from Essig,
1942.]

Hindleg   Abdomen   Hindwing

lacewings are terrestrial groups. The larvae of both families live on vegetation and prey on aphids, mites, and other smaller soft-bodied insects. The common lacewings (Fig. 11-38) tend to be greenish in color and often have yellowish metallic-appearing eyes. They are often 2 or 3 cm long, and their eggs are deposited at the ends of stalks that are anchored to leaves. The larvae often carry a layer of debris about on their dorsum, which provides them with camouflage. The brown lacewings (Fig. 11-40B) are usually brownish in color and are smaller than the common lacewings. They do not deposit their eggs at the ends of stalks. Spongillaflies (Fig. 11-40C) are small insects and feed on freshwater sponges during the larval stage.

Myrmeleontoidea contains two families, Myrmeleontidae (antlions) and Ascalaphidae (owlflies). Antlion adults (Fig. 11-40D) superficially resemble damselflies and are similar in size. However, they are weaker fliers, have a very different pattern of wing venation, have long distinctly clubbed antennae, and are soft-bodied. The larvae (Fig. 11-40E) are terrestrial and live at the bottom of funnel-shaped pits in sand. When a potential prey insect or other small arthropod passes close to the pit, the antlion creates a miniature landslide, bringing the prey down within reach of its prehensile, sucking mandibles, and subsequently dines.

Owlflies (Fig. 11-40F) bear a resemblance to dragonflies but differ from them in the same features that the antlion adult differed from the damselflies. In addition, their clubbed antennae are even longer than those of the adult antlions. Their larvae are predatory and are often concealed by a layer of debris on their bodies.

Members of this order have no medical or economic significance. However, because of their predatory habits, many are beneficial.

Panorpoids. The panorpoid orders (Mecoptera, Diptera, Siphonaptera, Trichoptera, and Lepidoptera; Fig. 11-33) are somewhat

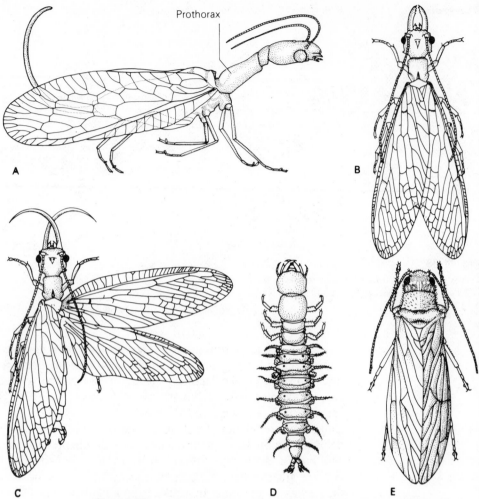

Prothorax

Fig. 11-39
Neuroptera. A. Snakefly, *Agulla bractea* (Raphidiodea; Raphidiidae). B, C, and D. Dobsonfly, *Corydalis cornutus,* female, male, and larva, respectively (Megaloptera; Corydalidae). E. Alderfly, *Sialis mohri* (Megaloptera; Sialidae). (A, Me; B–D, VL; E, Me.) [A redrawn from Wolglum and McGregor, 1958; B–D redrawn from Boreal Labs key card; E redrawn from Frison, 1937.]

closely related to the neuropteroids. Few obvious features characterize this group (Boudreaux, 1979). Included among the defining traits are extreme reduction of the externally apparent parts of the meso- and metasternum and the fact that the media and cubitus of the hindwing arise from a common stem. A fully developed appendicular ovipositor is not present in any member of this group, and specialized, telescoping terminal abdominal segments have come to serve as an ovipositor (see Fig. 2-44E). Each panorpoid order contains at least some members that spin a cocoon from silk secreted by larval labial glands.

Order Mecoptera   (*meco,* long; *ptera,* wing); scorpionflies (Fig. 11-41).

BODY CHARACTERISTICS   small to medium; fragile; larvae: eruciform; pupae: exarate.

MOUTHPARTS chewing; commonly at apex of rostrum formed by
elongation of head capsule; larvae: chewing.

EYES and OCELLI compound eyes present; 0 to 3 dorsal ocelli.

ANTENNAE long and filiform.

Fig. 11-40

Neuroptera, Planipennia. A. Mantispid, *Mantispa brunnea occidentis*
(Mantispidae). B. Brown lacewing, *Sympherobius angustatus*
(Hemerobiidae). C. Spongillafly, *Climacia areolaris* (Sisyridae). D.
Antlion adult, *Hesperoleon abdominalis* (Myrmeleontidae). E. Antlion
larva, *Myrmeleon sp*. F. Owlfly (Ascalaphidae). (A, D and F, L; B and
C, S; E, Me.). [A, B, and F redrawn from Essig, 1942; C and D
redrawn from Froeschner, 1947; E redrawn from Peterson, 1951.]

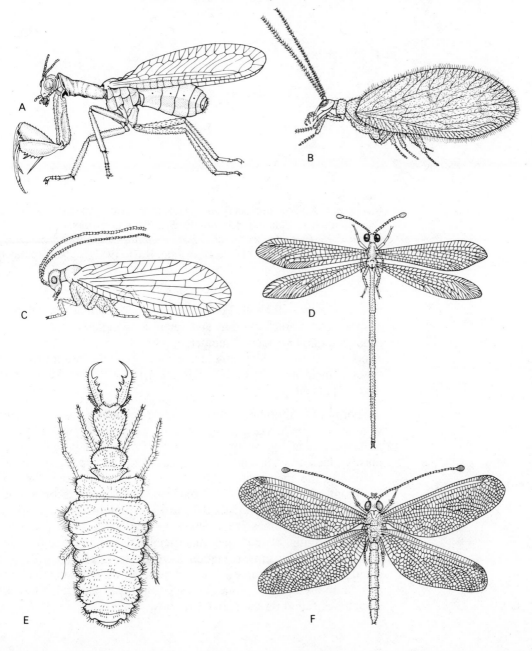

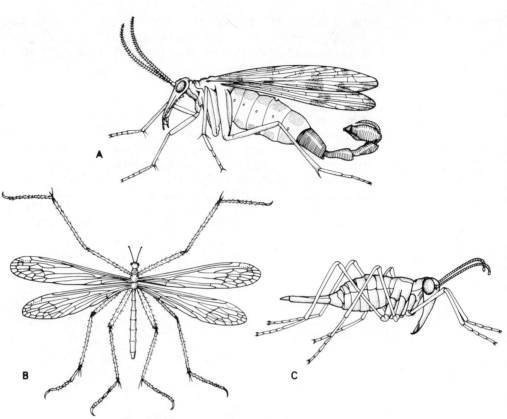

Fig. 11-41
Mecoptera. A. Common scorpionfly (male), *Panorpa chelata*
(Panorpidae). B. Hanging scorpionfly, *Bittacus chlorostigma*
(Bittacidae). C. Snow scorpionfly, *Boreus californicus* (Boreidae). (A,
Me; B, Me to L; C, Mi.) [A redrawn from Ross, 1962; B and C redrawn
from Essig, 1942.]

WINGS   most species with 2 pairs; membranous; commonly with dark
  spots; some brachypterous and apterous (wingless) forms.
LEGS   long and slender; 5-segmented tarsi.
ABDOMEN   in some long and slender and genitalia prominent; last
  segment in male carried upright suggestive of a scorpion; short
  cerci present.

Scorpionfly adults are terrestrial and are found in areas of decaying
vegetation. The larvae are eruciform or scarabeiform with chewing
mouthparts and live in or on the soil. Abdominal prolegs, when
present, lack crochets and are 16 in number. The pupae are exarate.
Among the mecopteran families are Panorpidae, the common
scorpionflies (Fig. 11-41A); Bittacidae, the hanging scorpionflies
(Fig. 11-41B); and Boreidae, the snow scorpionflies (Fig. 11-41C).
Common scorpionflies feed mainly on dead insects and plant materials. Hanging scorpionflies superficially resemble crane flies and
hang suspended from vegetation by their forelegs. They capture
small flies with their raptorial tarsi. Snow scorpionflies inhabit moss
and damp areas under stones and feed on plant material. They are
sometimes found on the surface of snow.

Order Diptera (*di,* two: *ptera,* wing); true flies (Figs. 11-42 through 11-45).

BODY CHARACTERISTICS minute to very large; larvae: apodous (legless), some with well-defined head and thorax and others wormlike (vermiform); pupae: obtect or coarctate.

MOUTHPARTS various modifications of sucking; larvae: various modifications of mandibulate mouthparts.

EYES and OCELLI compound eyes present; most species with 3 ocelli; ocelli lacking in some.

ANTENNAE variable.

WINGS 1 pair; membranous; metathoracic halteres; many apterous species.

LEGS 5-segmented tarsi.

ABDOMEN cerci present or absent.

Diptera is the fourth largest order of insects, and its members are found in a wide diversity of habitats. Some live in biologically adverse environments; for example, *Psilopa petrolei* (Ephydridae) spends its larval stage in crude oil, and larvae of other ephydrid flies are found in the Great Salt Lake. A survey of the feeding habits of the order could well apply to the entire class Insecta. Members of several families are herbivorous as larvae. Some are mycetophagous, "fungus-eaters." Many flies are detritivores and play a vital role in the breakdown of decaying organic material. A large number are predaceous and are to be counted among the beneficial forms. Many feed on the blood of vertebrates or are external or internal parasites.

Fig. 11-42
Diptera. A mosquito, *Anopheles maculipennis* Meigen (female). [Redrawn from James and Harwood, 1969.]

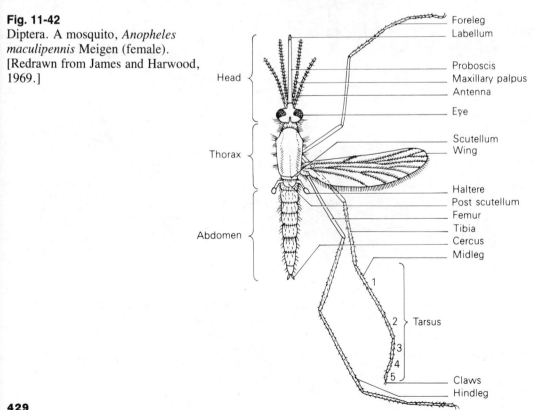

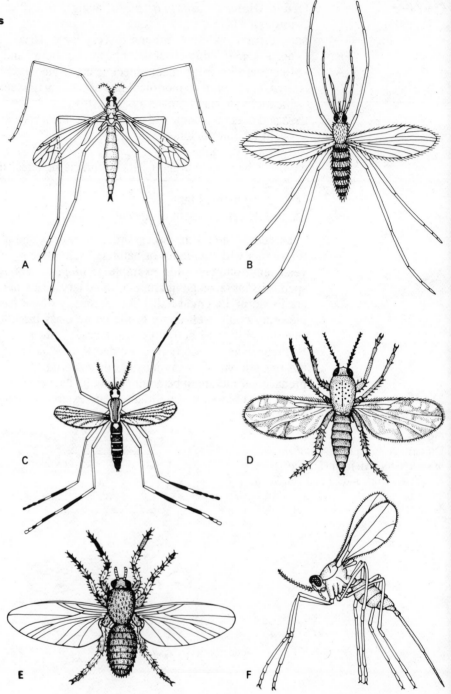

Fig. 11-43
Representative nematocerous Diptera. A. Crane fly (Tipulidae). B. Sand
fly, *Phlebotomus* (Psychodidae). C. Mosquito, *Aedes aegypti*
(Culicidae). D. Biting midge, *Culicoides furens* (Ceratopogonidae). E.
Black fly, *Simulium venustum* (Simuliidae). F. Gall midge
(Cecidomyiidae). (A, Me to L; B, C, and E, S; D and F, Mi.) [A and F
redrawn from Boreal Labs key card; B–E after U.S. Public Health
Service.]

Such associations have resulted in their involvement with human and animal diseases.

Diptera is generally divided into three suborders: Nematocera, Brachycera, and Cyclorrhapha. Members of these suborders are separated on the basis of such features as antennal characteristics, the shape of the exit hole from the pupal case, the nature of the pupa, and larval characteristics.

The Nematocera (Figs. 11-42 and 11-43) are so called because, in the adult stage, they have comparatively long antennae (Nematocera = "thread horn"). The larvae (see Fig. 4-13B) have well-developed heads with mandibles that move in a horizontal plane. The pupae are obtect.

Flies in the suborder Brachycera (Fig. 11-44) have antennae that are shorter than the thorax (Brachycera = "short horn"). The pupae of these flies are exarate. Larvae have weakly developed heads, which are generally retractile, and the mandibles move in the vertical plane. As with Nematocera, Brachycera contains a large number of families. However, many of them are composed of uncommon species.

Members of the suborder Cyclorrhapha (Fig. 11-45) have coarctate pupae. Cyclorrhapha ("circular seam") refers to the circular shape of the opening through which the adult emerges from the puparium (see Fig. 5-24C, D). Near one end of the puparium there is a circular line of weakness along which the cuticle splits at adult emergence, much like a cap popping from a bottle. The split results

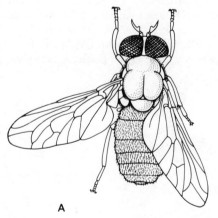
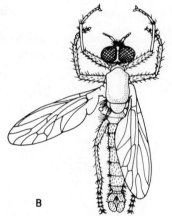
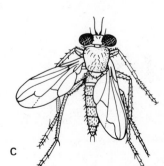

Fig. 11-44
Representative brachycerous Diptera. A. Horse fly (Tabanidae); B. Robber fly (Asilidae). C. Long-legged fly (Dolichopodidae). D. Bee fly (Bombyliidae). (A and B, Me to L; C, S; D, Me.) [Redrawn from Boreal Labs key card.]

from pressure being applied by means of an eversible bladderlike structure, the *ptilinum*, which protrudes from the head after emergence, usually leaving the frontal suture as a remnant. The larvae lack heads and are vermiform (maggots), with mouth hooks that operate in the vertical plane (see Fig. 2-32D). The adults generally have aristate antennae.

Diptera contains more insects of medical and veterinary importance than any other order. They cause serious problems for humans and domestic and wild animals in several ways: (1) as vectors of many serious diseases, (2) as blood-feeders, (3) as general nuisances, and (4) by direct invasion of organs and tissues of humans and animals.

Flies associated with disease transmission typically, but not always, feed on the blood of vertebrates. There are several outstanding examples of fly-borne diseases. Mosquitoes (Nematocera, Culicidae; Figs. 11-42 and 11-43C) are probably responsible for the transmission of more human diseases than any other group of insects. They are the vectors of malaria (caused by four species of protozoa in the genus *Plasmodium*); filariasis (caused by the nematode *Wuchereria bancrofti*, and others); and several viral diseases (yellow fever, dengue, and various types of encephalitis). Two other nematoceran families also include disease-carrying members: Simuliidae (black flies; Fig. 11-43E) and Psychodidae (moth flies and phlebotomus

Fig. 11-45
Representative cyclorrhaphous Diptera. A. Fruit fly (Drosophilidae). B. Picture-winged fly (Otitidae). C. House fly (Muscidae). D. Syrphid fly (Syrphidae). E. Sheep ked, *Melophagus ovinus* (Hippoboscidae). F. Green bottle fly, *Phaenicia sericata*. (A, Mi; B–F, S.) [A–D redrawn from Boreal Labs key card; E and F after U.S. Public Health Service.]

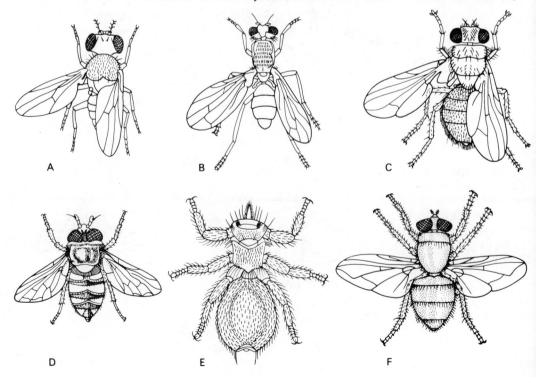

A   B   C

D   E   F

flies; Fig. 11-43B and see also Fig. 8-4). Black flies are vectors of onchocerciasis, a disease caused by a filarial worm, *Onchocerca volvulus*, which induces the development of painful nodules in the skin and often results in blindness. Psychodid flies in the genus *Phlebotomus* transmit the leishmanias, diseases caused by protozoans in the genus *Leishmania*. The tsetse flies, *Glossina* spp. (Cyclorrhapha, Muscidae), serve as vectors of the trypanosomes that cause African sleeping sickness in humans (*Trypanosoma gambiae* and *T. rhodesiense*) and nagana (*T. brucei*), a serious disease of cattle in Africa.

Other biting flies capable of transmitting disease-causing organisms, but more often involved as nuisances through their blood feeding activities, are the biting midges or "no-see-ums" (Ceratopogonidae; Fig. 11-43D); horse flies and deer flies (Tabanidae; Fig. 11-44A); certain snipe flies (Rhagionidae); and the stable fly, horn fly, and other flies in the family Muscidae. An interesting group of biting flies, which usually do not attack humans but are intermittent blood-feeding parasites of sheep, goats, and various wild birds and mammals, are the louse flies and bat flies. The larvae of these flies are retained in the female parent until nearly ready to pupate, at which time they are released. The sheep ked (Fig. 11-45E), *Melophagus ovinus*, is a common wingless ectoparasite of sheep and goats.

Nonbiting flies that pass the larval stage in garbage, excrement, and other decaying organic matter may also be involved in the transmission of disease, generally through the contamination of food. Flies in this category are largely in the families Muscidae (house fly, Fig. 11-45C, and relatives), Calliphoridae (blow flies; Fig. 11-45F), and Sarcophagidae (flesh flies). These flies may transmit various gastrointestinal diseases, both of bacterial (dysentery, typhoid fever, and others) and protozoan (*Entamoeba*) etiology (cholera, yaws, poliomyelitis, and eye infections). These flies also vector various microbes that cause disease in domestic animals, including several helminths (tapeworms; nematodes or roundworms). Chloropidae (fruit flies and relatives) is another family of nonbiting flies, certain species of which transmit diesease—those in the genus *Hippelates*. These are commonly called eye gnats and have been identified as probable vectors of conjunctivitis ("pinkeye"), yaws, and bovine mastitis. These flies are attracted to various body secretions and so tend to cluster around the mouth, eyes, nose, and genital and anal openings as well as sores exuding pus.

Larvae of many flies may invade the organs and tissues of humans and animals, sometimes causing severe injury and even death. Such invasions are referred to as *myiasis*. Invasion may be brought about accidentally (e.g., accidental ingestion of maggots in food) or may be due to adult female flies being attracted to an open wound or body opening and depositing eggs. Several flies (e.g., the screwworm fly, *Cochliomyia hominivorax*) require an animal host in the larval stages. This situation is called *obligate myiasis*. Other examples of obligate myiasis occur in the bot flies in the families Gasterophilidae (horse bot flies) and Oestridae (warble flies and bot flies).

Surprisingly few true flies are pests of agricultural significance. Among those that are, the Hessian fly, *Phytophaga destructor* (family Cecidomyiidae), is outstanding. This fly is the most destructive of the insects that attack wheat in the United States. It was apparently introduced from Europe into the United States during the Revolutionary War in the straw bedding used by Hessian soldiers. One can gain some idea of the kinds of crops attacked by flies from the names of several pest species (e.g., seed corn maggot, apple maggot, cherry fruit fly, black cherry fruit fly, and narcissus bulb fly).

Order Siphonaptera (*siphon*, a tube; *aptera*, wingless); fleas (Figs. 11-46 and 11-47).

BODY CHARACTERISTICS minute; hard-bodied; bilaterally compressed; many posteriorly directed spines; larvae; vermiform; pupae: exarate, in silken cocoons.

MOUTHPARTS piercing-sucking; larvae: chewing.

EYES and OCELLI compound eyes lacking; generally 2 lateral ocelli.

ANTENNAE short; contained within antennal grooves.

WINGS lacking.

LEGS coxae modified for jumping; 5-segmented tarsi.

ABDOMEN male terminalia highly modified with 1 pair of 2-segmented claspers; cerci lacking.

Fleas are ectoparasites of warm-blooded animals (mammals and birds). The bilaterally compressed body and posteriorly directed spines facilitate movement between the hairs or feathers of the host. Although fleas are excellent jumpers, the primary mode of locomotion is walking or running.

Females generally require a blood meal before they are capable of forming eggs. The eggs (Fig. 11-47A) may or may not be deposited on the host, but ultimately they fall to the resting place of the host. The vermiform larvae (Fig. 11-47B) feed on organic detritus and some on partially digested blood in the feces of adults. They are whitish in appearance, quite active, and pupate within silken cocoons (Fig. 11-47C). The pupae (Fig. 11-47C) are exarate. Adults are able to remain quiescent in the cocoon for long periods

Fig. 11-46
Siphonaptera, a flea.
(Mi.) [After U.S.
Public Health Service.]

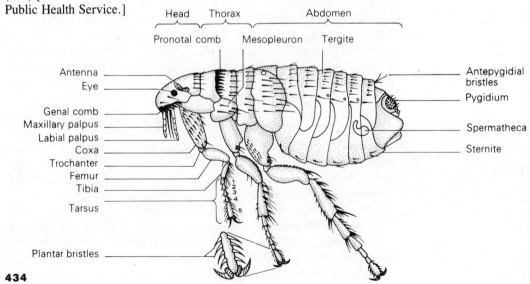

434

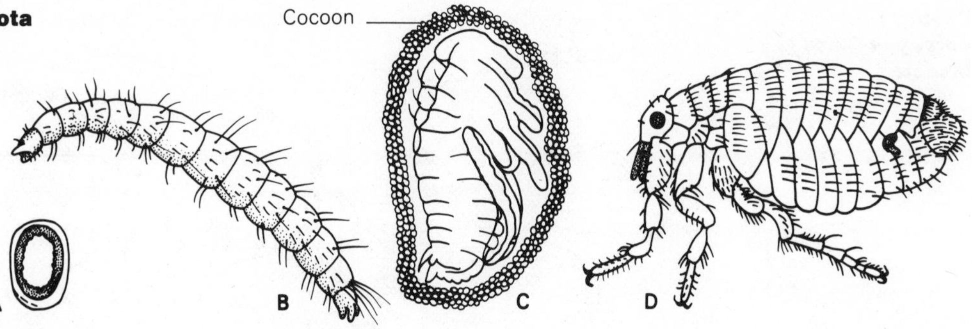

Cocoon

Fig. 11-47
Life stages of a flea.
A. Egg. B. Larva. C.
Pupa. D. Adult. [After
U.S. Public Health
Service.]

of time and exit in response to vibrations of the substrate and other stimuli associated with a potential host. It is common for persons with pets to return from a vacation and be attacked by large numbers of fleas that have been quiescent within their pupal cocoons. This behavior decreases the probability of an adult flea becoming active when no potential hosts are available. In addition, the adults can survive prolonged periods of starvation. The life cycle varies from 3 to several weeks, depending on the species.

Although fleas are highly specialized insects, they show definite affinities with members of the Diptera and Mecoptera.

Fleas are problems from several standpoints. Their blood-feeding is irritating to humans as well as to many wild and domestic animals. Although fleas tend to show host preferences, they will feed on a variety of available hosts. *Pulex irritans*, for example, is often referred to as the human flea, but it has many other hosts too. Other well-known fleas include *Echidnophaga gallinacea*, the sticktight flea of poultry; *Ctenocephalides canis* and *C. felis*, the dog and cat fleas, respectively; and *Xenopsylla cheopis*, the oriental rat flea.

Fleas are involved in the transmission of plague. Plague is a disease of wild rodents; and causative agent, *Yersinia pestis*, is transmitted from rodent to rodent by fleas. Under certain conditions humans may be bitten by infected fleas and develop the symptoms of plague. In regions of high population concentration and poor sanitary conditions, the potential may exist for epidemics of this disease. Throughout recorded history there have been sporadic outbreaks of plague, or the ''Black Death,'' as it has been called at times (Langer, 1964).

One species of flea, *Tunga penetrans*, the chigoe, is purported to burrow into the skin. These fleas do not actually burrow into the skin, but apparently induce tissue changes, which result in their being partially enveloped by host tissues. Only the females ''burrow,'' and they generally attack in the region of the toes and may open the way for secondary infection. In some instances, autoamputation of toes has been attributed to this species. Chigoes are found in the tropical and subtropical regions of North and South America, the West Indies, and Africa.

Order Trichoptera (*tricho*, hair: *ptera*, wing); caddisflies (Figs. 11-48, 11-49, and 11-50).

BODY CHARACTERISTICS small to medium; generally somber colored, usually brownish; larvae; eruciform; pupae: exarate.

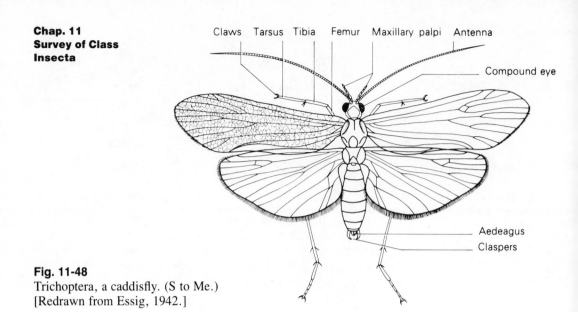

Claws Tarsus Tibia Femur Maxillary palpi Antenna

Compound eye

Aedeagus

Claspers

Fig. 11-48
Trichoptera, a caddisfly. (S to Me.)
[Redrawn from Essig, 1942.]

MOUTHPARTS some nonfeeding in adult stage; adapted for imbibing fluid; mandibles weakly developed or lacking; larvae: chewing.

EYES and OCELLI compound eyes present; 0 or 3 dorsal ocelli.

ANTENNAE range from setaceous to filiform.

WINGS 2 pairs; membranous; hindwings broader than forewings; covered with modified hairlike setae; held rooflike over body at rest.

Caddisflies are aquatic in the egg and larval stages (a few live in brackish water). Adults are terrestrial and may be found some dis-

Fig. 11-49
Caddisfly, *Rhyacophila fenestra*. A. Adult in resting posture. B. Larva. [Redrawn from Ross, 1938.]

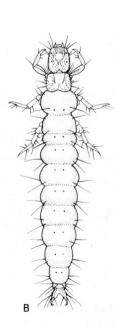

A   B

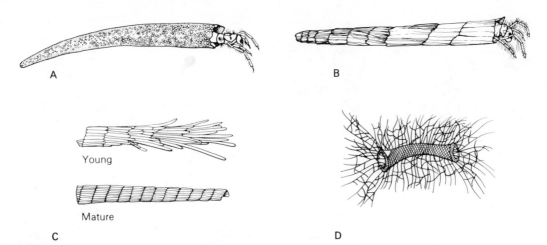

A

B

Young

Mature

C

D

Fig. 11-50
Examples of caddisfly larval cases. A. *Leptocella*. B. *Triaenodes*. C. *Agrypnia*. D. *Polycentropus*. [Redrawn from Cummins, Miller, Smith, and Fox, 1965.]

tance from water. Eggs are deposited in strings or masses near or in the water, on aquatic vegetation, beneath stones, and at other similar locations. Most larvae (Figs. 11-49B and 11-50) construct cases with a silken foundation into which a variety of environmental materials are incorporated. Exactly which environmental materials, together with the method by which they are incorporated, is a family or sometimes a generic characteristic. The larvae that inhabit cases have hooked caudal appendages used to maintain a hold within the cases. Larval and pupal ventilation is by means of abdominal gills. The larvae feed on small aquatic plants and animals, and some are important predators of black flies.

Caddisflies are important as fish food.

Order Lepidoptera (*lepido,* scale; *ptera,* wing); butterflies, moths, and skippers (Figs. 11-51 and 11-52).

BODY CHARACTERISTICS small to very large; usually covered with scales; much diversity in color and color pattern; larvae; eruciform; pupae; obtect.

MOUTHPARTS usually long, sucking proboscis; vestigial in some; larvae: strong mandibles, chewing.

EYES and OCELLI large compound eyes; 0 or 2 dorsal ocelli; larvae: clusters of lateral ocelli (stemmata).

ANTENNAE always prominent; vary in form.

WINGS 2 pairs; membranous; covered with scales; large in proportion to body; few cross veins; a few wingless; forewings larger than hindwings.

LEGS 5-segmented tarsi; larvae: 3 pairs.

ABDOMEN cerci lacking; larvae usually with 5 pairs of prolegs on segments 3 through 6 and 10; apex of each proleg with tiny hooks (crochets).

COMMENTS members in several families have organs of hearing.

Lepidoptera is the second largest order of insects. It can be divided into suborders, Frenatae (or Heteroneura) and Jugatae (or Homoneura) on the basis of wing-coupling mechanisms. In the Frenatae the fore- and hindwings are coupled by a spine or a group of spines, the *frenulum* (see Fig. 2-40B), at the anterior base of the hindwing or by an enlarged area along the anterior base of the hindwing (humeral angle). The hindwings are smaller and have fewer veins than the forewings, hence the alternative name, Heteroneura ("mixed nerves"). The wings of the Jugatae are coupled by the *jugum* (see Fig. 2-40C), a projection from the base of the posterior edge of the forewing. The venation is similar in both fore- and hindwings, hence the alternative name Homoneura ("same nerves"). The vast majority of Lepidoptera are in the suborder Frenatae, the Jugatae being a small group of uncommon insects.

The suborder Frenatae may be further divided into two major groups: the Macrolepidoptera and the Microlepidoptera. The Macrolepidoptera are a diverse group of butterflies (Fig. 11-52A), skippers (Fig. 11-52B), and moths (Fig. 11-52C) of variable size, but usually with a wingspan of more than 2.5 cm. The Microlepidoptera are a large group of moths, most of which are much smaller than Macrolepidoptera, usually with a wingspan of 20 mm or less. Wing venation and other characters are used in further subdivision of these two groups.

Some authors, however, divide the Lepidoptera into the suborders Rhopalocera (butterflies and skippers) and Heterocera (moths). The Rhopalocera are characterized by (1) being diurnally active; (2) holding the wings vertically at rest; (3) having variable antennae, often feathery, but usually not clubbed; and (4) usually having a frenulum.

Lepidoptera includes a large number of families with diverse habits, the treatment of which would be inappropriate in this brief survey.

Many lepidopteran larvae (caterpillars) have repugnatorial glands of various sorts. The lobelike eversible *osmeteria*, located on the dorsum of the prothorax of papilionid (swallowtail butterflies) larvae, are examples of such glands. Caterpillars all have the ability to produce silk. This substance is secreted by large labial silk glands, which open via a common duct into the highly modified labium.

Pupae (Fig. 11-51C) are generally obtect, although some are exarate, and are usually encased in some sort of external protective structure constructed by the last instar larva. This protective structure may, for example, be: (1) a silken cocoon, with or without various kinds of detritus incorporated into it, (2) detritus held together by a sticky secretion that hardens into a foundation matrix, (3) a cell in the soil, or (4) a tentlike structure made by "tying" edges of a leaf together. In many, no protective structure is constructed and they must depend upon various kinds of protective coloration and patterning. Such is the case with most butterflies that attach themselves to twigs or other objects by means of a silken thread and/or a silken patch at the tip of a projection from the posterior part of the body, the *cremaster*. These naked pupae are commonly referred to as *chrysalids*.

Although the vast majority of Lepidoptera are terrestrial, certain members of several families (Arctiidae, tiger and footman moths; Pyralidae, snout moths and relatives; Sphingidae, sphinx moths; and several others) have been found in aquatic environments as larvae and sometimes as adults.

Most Lepidoptera feed on flowering plants and are particularly voracious in the larval stages, most adults being capable of obtaining only fluid meals from flowers and other sources and being unable to

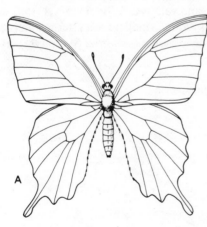

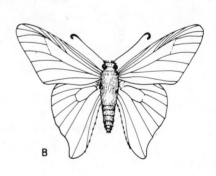

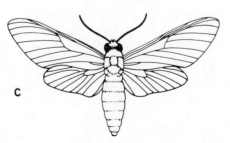

Fig. 11-52
A. Butterfly. B. Skipper. C. Moth. [Redrawn from U.S. Department of Agriculture, 1952.]

masticate plant material. Their plant-feeding habits make several species among the most important agricultural pests. Many attack stored grain and some attack fiber in clothing. Most pest species are moths. The common names of several of the major lepidopteran pests will give you an idea of the various crops they attack and in some instances the part of the plant: European corn borer, corn earworm, beet armyworm, garden webworm, alfalfa webworm, bollworm, imported cabbageworm, oriental fruit moth, cherry fruit worm, grape root borer, strawberry crown moth, and rose budworm. A few lepidopteran species, particularly the silkworm moth, *Bombyx mori* (see Chapter 12), are highly valued insects.

A few species are of medical or veterinary importance (James and Harwood, 1979). For example, certain noctuid moths in Africa feed on the lachrymal secretions of cattle and may be the vectors of the etiological agent of infectious keratitis, a disease that may cause blindness. Some species have been observed to feed in the eyes of humans. Incredibly, a few species of moths in Malaya are blood-feeders (Bänziger, 1971, 1975). The caterpillars of certain members of at least 10 families possess urticating hairs, which may contain a toxin and can cause severe blistering and even blindness if they get into the eyes.

Hymenopteroids.   The single order, Hymenoptera (Fig. 11-53), are a highly specialized group, and there has been much disagreement regarding their evolutionary relationships with other orders.

Order Hymenoptera   (*hymen*, membrane; *ptera*, wing); ants, bees, wasps, and relatives (Fig. 11-53, 11-54, and 11-55).
BODY CHARACTERISTICS   small to very large; larvae: usually legless with distinct head; some (suborder Symphyta) eruciform with legs

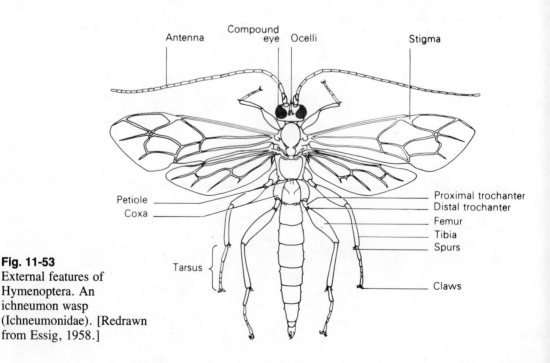

Fig. 11-53
External features of Hymenoptera. An ichneumon wasp (Ichneumonidae). [Redrawn from Essig, 1958.]

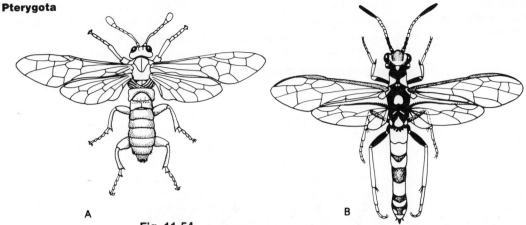

Fig. 11-54
Hymenoptera, Symphyta. A. Sawfly (Tenthredinoidea). B. Raspberry
horntail, *Hartigia cressoni* (Siricoidea). (Me.) [A redrawn from U.S.
Department of Agriculture, 1952; B redrawn from Essig, 1958.]

and prolegs (which lack crochets); pupae: exarate, usually within
a cocoon.

MOUTHPARTS variable; chewing to lapping or sucking.

EYES and OCELLI compound eyes usually well developed; commonly
3 dorsal ocelli, but lacking in some.

ANTENNAE commonly long and filiform or geniculate with elbowed
scape and distal segments clubbed.

WINGS most with 2 pairs of membranous wings; hindwings smaller
than forewings; many wingless.

LEGS usually 5-segmented tarsi.

ABDOMEN in the suborder Apocrita, the first abdominal segment
(*propodeum*) is fused with the metathorax and is constricted posteriorly, forming the *petiole* between the thorax and abdomen;
members of the suborder Symphyta lack a petiole; females with
ovipositor or homolog modified for piercing plant tissues, sawing,
or stinging.

On the basis of complexity and diversity of behavior the Hymenoptera are generally recognized as the most advanced group of
insects.

Hymenoptera can be divided into two suborders, Symphyta (Chal-
astogastra) and Apocrita (Clistogastra). The most obvious characteristic separating these two suborders is the relationship between
the thorax and abdomen. The Symphyta (Fig. 11-54) lack the petiole
that is characteristic of the Apocrita (Fig. 11-55). Other differences
include the following. The Symphyta are less behaviorally sophisticated than the Apocrita, parasitic forms are almost nonexistent, the
ovipositor is usually fitted for sawing or piercing plant tissues, and
the larvae are eruciform. Among the Apocrita are many parasitoid
and predatory forms, and the ovipositor is specialized for piercing
and/or stinging. The larvae are legless, and a distinct head is lacking
in some of the parasitoids.

The habits of Hymenoptera are very diverse and represent nearly
every mode of insectan life. The ants (Formicidae): paper wasps and

relatives (Vespidae); one species of sphecid wasp, *Microstigmus comes* (Sphecidae; Matthews, 1968); and many bees (superfamily Apoidea, e.g., Apidae) have elaborate patterns of social behavior (see Chapter 8). Many of these insects, as well as many other Hymenoptera, are pollen-feeders and are important plant pollinators (see Chapters 9 and 12). A large number of Hymenoptera are par-

Fig. 11-55
Hymenoptera, Apocrita. A. Ichneumon wasp (Ichneumonoidea; Ichneumonidae). B. Chalcid wasp (Chalcidoidea; Chalcidae). C. California gallfly, *Andricus californicus* (Cynipoidea). D. Velvet ant (Scolioidea; Mutillidae). E. Thief ant, *Solenopsis molesta* (Scolioidea; Formicidae). F. Golden polistes, *Polistes aurifer* (Vespoidea; Vespidae). G. Thread waisted wasp (Sphecoidea; Specidae). H. Honey bee, *Apis mellifera* (Apoidea; Apidae). (A, F, and G., Me; B, C, and E, Mi; D, S to Me.) [A, B, G, and H redrawn from U.S. Department of Agriculture, 1952; C and F redrawn from Essig, 1958; D and E after U.S. Public Health Service.]

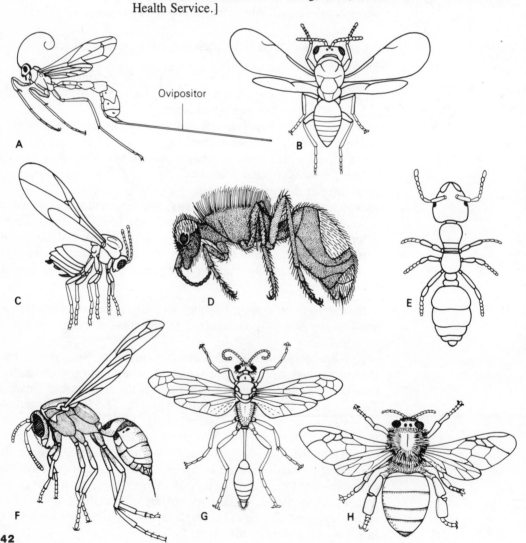

Ovipositor

A   B

C   D   E

F   G   H

asitoids (see Chapters 8 and 9), for example, all members of the superfamilies Ichneumonoidea and Proctotrupoidea and most members of the Chalcidoidea. Members of the superfamilies Sphecoidea and Vespoidea are predatory. Most Symphyta are herbivorous, some being of major economic importance. Some members of the superfamily Cynipoidea induce plants to form galls (see Chapter 9). A few Hymenoptera are found in aquatic habitats (Hagen, 1978).

Hypermetamorphosis is common in certain families. Parthenogenesis is also very common, and parthenogenetic generations may alternate with sexual generations. In some groups parthenogenesis plays a role in sex determination; for example, in the honey bee, *Apis mellifera,* males (drones) develop from unfertilized eggs, whereas queens and workers (all females) develop from fertilized eggs (see Chapter 5). *Polyembryony*, the development of more than one individual from one egg, has been reported in a few parasitoid species. Hymenopteran larvae are usually legless and generally have distinct heads. Some are eruciform and have thoracic legs and abdominal prolegs that lack the tiny hooks or crochets characteristic of lepidopteran larvae. The pupae are exarate and typically within a cocoon or waxen or earthen cell.

The members of this order are certainly more beneficial than harmful, since many are involved in plant pollination; are parasitoids or predatory on other, often harmful, insects; and produce useful products (e.g., honey and beeswax). However, there are several species that are destructive pests. Most of these are sawflies in the families Diprionidae (conifer sawflies), Tenthredinidae (common sawflies), and Cephidae (stem sawflies). The larvae of common sawflies feed on foliage of trees and can be very damaging. The larvae of stem sawflies bore into the stems of grasses and berries, and two species are important pests of wheat. Ants (Formicidae) may be a problem when they invade the household and infest foodstuffs.

Although no hymenopterous insects serve as disease vectors, many are capable of inflicting a painful sting that can lead to serious consequences if the person stung happens to be hypersensitive to the injected venom. Stinging hymenoptera are in the following seven superfamilies (James and Harwood, 1969): (1) Chrysidoidea—cuckoo wasps and relatives; (2) Bethyloidea—bethyloid wasps; (3) Scolioidea—scoliid wasps, velvet ants, and relatives; (4) Formicoidea—ants; (5) Vespoidea—hornets, yellow jackets, spider wasps, and relatives; (6) Sphecoidea—sphecoid wasps; and (7) Apoidea—bees.

Fossil Neopterous Endopterygotes

The Glosselytrodea (see Fig. 10-14) are the only extinct neopterous endopterygote insects that meet Carpenter's criteria (see Chapter 10) for recognition as an order. The pattern of hairs on the wings and appearance of head and thorax are similar to the Neuroptera, but the venation is more specialized and the forewings somewhat hardened.

Selected References

GENERAL

Askew (1971); Bland and Jaques (1978); Borror, DeLong, and Triple-horn (1976); Borror and White (1970); Brues, Melander, and Carpenter (1954); Cheng (1976); Chu (1949); C.S.I.R.O. (1970); Edmondson (1959); Essig (1958); Merritt and Cummins (1978); Metcalf and Metcalf (1928); Peterson (1948, 1951); Pennak (1978); Remington (1975); Richards and Davies (1977); Swain (1957); Swan and Papp (1972); Tietz (1963); Linsenmaier (1972); Tweedie (1974); Usinger (1956); Zim and Cottam (1956).

ORDERS

Apterygota—General: Kevan (1963); Kühnelt (1961); Schaller (1968); Wallwork (1970).

Protura: Ewing (1940); Tuxen (1964).

Collembola: Butcher et al. (1971); Christiansen (1964, 1978); Christiansen and Bellinger (1978); Maynard (1951); Mills (1934); Salmon (1964, 1965); Scott (1961).

Diplura: Remington (1954); Smith (1959, 1960).

Archeognatha: Remington (1954, 1956); Sharov (1966); Smith (1969, 1970).

Thysanura: Remington (1954); Slabaugh (1940); Smith (1969, 1970); Wygodzinsky (1972).

Odonata: Corbet (1963); Gloyd and Wright (1955); Needham and Westfall (1955); Smith and Pritchard (1956); Walker (1953, 1958); Walker and Corbet (1975); Westfall (1978).

Ephemeroptera: Burks (1953); Day (1956); Edmunds (1959a,b, 1978); Edmunds et al. (1976); McCafferty (1975); Needham et al. (1935); Pennak (1978).

*Plecoptera:*Claassen (1931); Frison (1935); Harper (1978); Hitchcock (1974); Hynes (1976, 1977); Jewett (1956); Needham and Claassen (1925); Ricker (1959).

Embioptera: Ross (1940, 1944, 1970).

Phasmida: Blatchley (1920); Helfer (1963); Ragge (1973).

Orthoptera: Blatchley (1920); Dirsh (1975); Helfer (1963); Uvarov (1966, 1977).

Dermaptera: Blatchley (1920); Helfer (1963); Popham (1965).

Grylloblattodea: Gurney (1948); Helfer (1963).

Isoptera: Behnke (1977); Krishna and Weesner (1969, 1970); Lee and Wood (1971); Weesner (1965); Wilson (1971, 1975).

Blattaria: Blatchley (1920) Cornwell (1968); Guthrie and Tindall (1968); Helfer (1963); McKittrick (1964); Ragge (1973); Roth and Willis (1960).

Mantodea: Blatchley (1920); Gurney (1951); Helfer (1963).

Zoraptera: Gurney (1938).

Psocoptera: Chapman (1930); Gurney (1950); Mockford (1951); Mockford and Gurney (1956).

Mallophaga: Askew (1971); Clay (1969); Emerson (1972).

Anoplura: Askew (1971); Clay (1973); Ferris (1951); Harwood and James (1979); Hopkins (1949); Kim and Ludwig (1978); Stojanovich and Pratt (1965).

Thysanoptera: Bailey (1957); Cott (1966); Lewis (1973); Stannard (1957, 1968).

Hemiptera (Heteroptera): Blatchley (1926); Britton (1923); DeCoursey (1971); Ghauri (1973); Herring and Ashlock (1971); Hungerford (1959);

Pterygota

Pennak (1978); Polhemus (1978); Slater and Baranowski (1980); Usinger (1956).

Hemiptera (Homoptera): Britton (1923); Dixon (1973); Kennedy and Stroyan (1959); Metcalf (1945, 1954–1963, 1962–1967).

Coleoptera: Arnett (1980); Blackwelder and Arnett, Jr. (1977); Dillon and Dillon (1972); Evans (1975); Jaques (1951); Leech and Chandler (1956); Leech and Sanderson (1959); USDA/Smithsonian Institution (in prep.).

Strepsiptera: Askew (1971); Bohart (1941); Ulrich (1966).

Neuroptera: Aspöck and Aspöck (1975); Chandler (1956); Evans (1978); Gurney and Parfin (1959).

Mecoptera: Byers (1954, 1963, 1965, 1971); Carpenter (1931a,b).

Diptera: Askew (1971); Cole (1969); Curran (1934); Greenberg (1971); Harwood and James (1979); Merritt and Schlinger (1978); Oldroyd (1964); Smith (1973); Stone et al. (1965); Teskey (1978); Wirth and Stone (1956).

Siphonaptera: Askew (1971); Ewing and Fox (1943); Fox (1940); Harwood and James (1979); Holland (1949, 1964); Hopkins and Rothschild (1953–1971); Hubbard (1947); Lewis (1972–1975); Smit (1973).

Trichoptera: Denning (1956); Pennak (1978); Ross (1944, 1959); Wiggins (1978).

Lepidoptera: Chu (1949); Common (1975); Ehrlich and Ehrlich (1961); Emmel (1975); Holland (1968, 1931); Howe (1975); Klots (1951, 1958); Lewis (1973); Mitchell and Zim (1964); Peterson (1948); Sargent (1976); *The Moths of America North of Mexico* (1971 to present); Tietz (1973); Urquhart (1960); Watson and Whalley (1975).

Hymenoptera: Andrewes (1971); Askew (1971); Bohart and Menke (1976); Clausen (1940); Creighton (1950); Evans (1963); Evans and Eberhard (1970); von Frisch (1967); Hagen (1978); Michener (1974); Mitchell (1960, 1962); Muesebeck et al. (1951–1967); Spradberry (1973); Wilson (1971, 1975); Yarrow (1973).

PART THREE

Applied Aspects
of Entomology

Applied Entomology

This chapter deals with the pragmatic aspects of the science of entomology. This is the point at which the results of both basic and applied research are brought to bear on specific problems. The approach will be first to consider how insects are problems for human beings, then to review the attempted solutions to these problems, and finally to discuss briefly insects that are directly or indirectly beneficial. A number of applied entomology texts are available (see Selected References).

The Problems

Fortunately, the vast majority of insects are directly or indirectly beneficial, but the relatively few that are pests will no doubt continue to tax our ingenuity to its fullest. According to the National Academy of Sciences (N.A.S.) Subcommittee on Insect Pests (1969),

> It is estimated that in the United States 150 to 200 species or complexes of related species frequently cause serious damage. From time to time, 400 to 500 additional species are pests and may cause serious damage. Approximately another 6,000 species of insects are pests at times, but seldom cause severe damage . . ..

These figures represent but a small fraction of the approximately 91,223 described species of insects in North America north of Mexico (Borror, DeLong, and Triplehorn, 1976).

There is nothing, biologically speaking, that clearly defines an insect as a pest. Two similar insects may have almost identical biological patterns and even be in the same family, yet one will be considered a pest, and the other will not, simply because one attacks humans or something valued by humans. Rolston and McCoy (1966) illustrate this point with three chrysomelid beetles.

> The Colorado potato beetle effectively defoliates the potato, and the dock beetle, *Gastrophysa cyanea* Melsheimer, defoliates its host with equal regularity. That the former is a pest and the latter is not reflects the value we place on the potato and our disinterest in dock. We may even encourage the destruction of some plants by insects. St. John's wort, or Klamath weed, has been cleared from thousands of acres of rangeland in western states by a group of insects, but principally by a leaf beetle, *Chrysolina quadrigemina* (Suffrian), introduced specifically to control this weed.

Many factors contribute to the development of pest situations. These may be grouped into three categories: modification of the environment, transportation, and human attitudes and demands.

Human activity has modified the environment, sometimes to advantage, often to disadvantage. One undesirable effect of human activity has been the creation of comparatively simple ecosystems. Generally, the simpler the system of interacting organisms, the less the system's inherent stability and the greater the likelihood of large fluctuations of populations of the component species. Thus it is not difficult to understand why there are a multitude of problems associated with the great monocultures of organisms, particularly plants, which man has developed. These *agroecosystems* are much less complex than the "natural" ecosystems that preceded them and hence are prone to spawn pest situations.

A good example of a vast monoculture is the corn grown yearly in the United States. This crop occupies approximately 85 million acres and represents about one fifth of the total cropland (Burkhardt, 1978). About $900 million worth of damage is caused annually by the pests that compete with us for this $6 billion crop. Another concentration process we have brought about has been the storage of vast quantities of materials that are palatable to certain insects. Among these stored materials are millions of bushels of grains, fresh vegetables, various other food products, fiber, and so on. Such "habitats" furnish optimal conditions for certain species. Also, we have often concentrated ourselves and commonly under other than sanitary conditions. Such concentrations of human populations provide an ideal circumstance for insect-carried diseases, particularly louse-borne typhus and plague.

With the many forms of modern transportation, insects are no longer limited in their dispersal capabilities by natural geographic barriers such as mountains and oceans, but are carried about as hitchhikers.

The introduction of an exotic insect into a new region away from its natural enemies may have disastrous results. For example, the European corn borer is a major pest of corn in the United States, but it is only a minor pest in its original home (Fronk, 1978). Conversely, plants introduced into a new region may prove to be ideal food for indigenous insects, which then constitute a pest problem. For example, the cucumber originated in the East Indies, while one of its major pest insects, the striped cucumber beetle, originated in North America (Fronk, 1978).

Human attitudes and demands may create a pest problem or greatly exaggerate what might, in fact, be a minor one. For instance, even slight evidence of insect damage, may decrease or destroy the economic value of a vegetable or fruit, even though the food value may be unaffected, for the simple and unfortunate reason that the average consumer is very particular about such matters and demands blemish-free produce. This commonly forces producers to employ costly and ecologically undesirable means of insect control. Many insects are considered to be pests not because they attack us or destroy any of our goods, but simply because some of us find them distasteful.

We will examine the conflict between humans and insects in three general categories on the basis of what is attacked by insects.

1. Growing plants,
2. Stored products, household goods, and structural materials,
3. Humans and wild and domestic animals valued by humans.

Examples of many pest insects are given in Chapter 11. The applied entomology text edited by R. E. Pfadt (1978) provides a good introduction to insect control and deals with pests of specific crops, etc. Metcalf and Luckman (1975) deal with fundamental ideas in pest management. The following are interesting and informative essays on the impact of insects on the welfare of humans: Cushing (1957), Pimentel (1975), McKelvey (1975), Ritchie (1979), and Southwood (1977a, b). Jones (1973), Philip and Rozeboom (1973), and Schwerdtfeger (1973) consider various aspects of the history of agricultural, medical-veterinary, and forest entomology, respectively.

Growing Plants

Virtually every growing plant we value is shared, to a greater or lesser extent, with one or usually more species of insect. For example, it is estimated that approximately 400 species of insects infest the apple, and 25 of these are of economic importance (Johansen, 1978). Almost all the damage these insects cause is associated with their feeding activities. Depending on the species and life stage, they remove chunks of tissue by chewing, suck sap from, and /or bore through every part of a plant. Not only are all parts of a plant susceptible to insect attack but all life stages as well. The feeding of insects not only causes the removal of tissue and sap but also may poison a plant by the injection of toxic saliva. As a result of feeding several species of insects are responsible for the introduction, transmission, and dissemination of plant diseases caused by viruses, primarily, but also by bacteria, fungi, and a few protozoans (see "Kinds of Intraspecific Interactions" in Chapter 9). For information on insects and plant disease, see Carter (1973), Pyenson (1977), and Sylvester (1980). Pfadt's (1978) description of the various kinds of injury to small grains (e.g., wheat, barley, and oats) and the guilty insects, illustrates the above points.

> Insects injure small grains in the field from the time the seed is planted until the grain is harvested. Both wireworms and false wireworms feed on the planted seed. In dry soil they may even destroy a crop before rains stimulate germination and growth. After the grain germinates, these pests devour the tender sprouts just as they push out of the seeds.
>
> . . .
>
> Wireworms also kill seedlings by boring into and shredding the underground portion of stems. Wireworms and white grubs feed on roots and sever them from the plant. Cutworms, white grubs, and false wireworms cut off young plants near soil level. In addition, the wounds left by soil pests allow rot pathogens to enter the plant.
>
> . . .

Grasshoppers, Mormon crickets, and armyworms may devour

young plants completely; they may strip the leaves from older plants, feed on maturing heads, or cut through the stems below the heads.

. . .

By injecting toxic secretions while feeding, larvae of Hessian flies retard or kill seedlings and reduce the yields of older plants. Weakened stems of older plants are likely to cause the crop to lodge. Wheat stem sawfly, wheat jointworm, and wheat strawworm bore within culms and obstruct the flow of sap. This damage reduces the number and weight of kernels. Boring insects also cause grain to lodge.

. . .

Insects such as the chinch bug and various aphids impoverish plants by sucking juices from leaves or stems. Moreover, they produce fatal necroses by injecting toxic saliva. Small grain pests may transmit serious plant diseases, such as wheat streak mosaic by the wheat curl mite, striate by the painted leafhopper, barley yellow dwarf by several species of aphids, and aster yellows disease of barley by the aster leafhopper.

. . .

Barley yellow dwarf, a new and widespread virus disease of cereals characterized by leaves rapidly turning light green and yellow, beginning at the tips, is causing much concern in the United States and Canada. Transmission of the disease is solely by aphids. Several species have already been incriminated, including the English grain aphid, birdcherry oat aphid . . . Recent research has demonstrated that three different strains of the virus are present and that they are usually carried by different species of aphids.

Some insects cause damage to growing plants through oviposition. Several kinds of insects have well-developed ovipositors that can penetrate rather hard surfaces like the bark of woody stems. The periodical cicada provides a good example of an insect that can cause considerable oviposition damage. The larvae of these insects spend 13 or 17 years feeding on plant roots and then emerge from the ground for a brief period of time to mate and lay eggs. Mated females deposit their eggs on young stems of shrubs and trees, causing severe damage and often necrosis of the affected area.

Stored Products, Household Goods, and Structural Materials

The wide variety of materials accumulated by man and stored in containers ranging from large bins to small boxes present ideal environmental conditions for certain insects to thrive, essentially free of any natural enemies. The presence of these insects may go undetected for long periods of time, and hence the potential extent of damage is great. Stored grains are particularly susceptible to attack and probably incur the most damage of any stored material. Stored grain pests not only consume grain but also render large quantities useless by contaminating it with fecal material, webbing, odors, shed exoskeletons, and whole or fragmented dead individuals. The activities of these insects may also cause heating of grain. This heating causes moisture-laden warmed air to rise to the surface, where it is cooled, resulting in the condensation of the accumulated moisture on the surface of the grain. This in turn causes caking of the grain and affords an ideal situation for the growth of molds and encourages spoilage (Wilbur and Mills, 1978).

Major stored-grain pests include the rice and granary weevils, the flour moths, and the Angoumois grain moth. The damage done by stored-grain insects is estimated to be 5–10% of the world's production, and much higher in certain areas.

> Destruction of food by stored grain insects is a major factor responsible for the low levels of subsistence in many tropical countries. If these losses could be prevented, it would alleviate much of the food shortages in the famine areas of the world. According to one estimate, 130 million people could have lived for one year on the grain destroyed or contaminated by stored grain insects in 1968.—Wilbur and Mills (1978).

Packaged food materials may be attacked by insects at any point from the processing plant to the consumer's home. What constitutes food for an insect is not necessarily human food. Certain insects, such as the cigarette beetle and drug-store beetle, both members of the Anobiidae (Coleoptera), eat a wide variety of nonhuman food items, including such delicacies as tobacco and several drugs.

Several species of insects have become common cohabitants with humans. These include cockroaches, silverfish and firebrats, and a large variety of ants, beetles, and moths. Stored-products insects also fall into this category. Even when these insects do little or no damage, their presence in a modern home is deemed highly undesirable. Several household items are susceptible to attack, including rugs, furniture, clothing, books, paper, and food.

Any structures made out of wood may be attacked by insects, and extensive and serious damage can result. Structural timbers in houses, buildings, and bridges, and wooden structures such as fences, railroad ties, and telephone poles are all susceptible to attack. Subterranean termites are the major pests involved, although insects such as powder post beetles may also cause trouble.

People and Their Animals

The ways in which insects attack humans and the animals they value are essentially the same and hence will be treated together. The terms *medical* and *veterinary entomology* are commonly used in reference to this area of applied entomology. As might be expected, this category also has strong economic overtones. Animals, wild or domestic, may be killed, weakened, and/or decreased in value as a result of being attacked by insects. Similarly, people may be killed or weakened and literally millions of man-hours of productive labor lost. Insect-borne diseases such as malaria and trypanosomiasis are particularly responsible for the latter. Vast areas of the world have been made uninhabitable or nearly so by the presence of insects and the diseases they carry. For example, the tsetse flies, *Glossina* spp., because they transmit the trypanosomes that cause nagana in cattle and two types of African sleeping sickness in humans, have prevented the development of millions of square miles of tropical Africa (McKelvey, 1973; Nash, 1969).

Insects affect the health of humans and other animals in two ways:

(1) directly as the causative agents of disease and discomfort and (2) indirectly as the transmitters (*vectors*) of disease-causing microorganisms (bacteria, viruses, spirochetes, protozoans, helminths, etc.). Many of the diseases of man and animals transmitted by insects are listed in Chapter 9 (see "Kinds of Interspecific Interactions") and in Chapter 11.

Information on insects of medical and veterinary importance may be found in Busvine (1966, 1975), Greenberg (1971, 1973), Harwood and James (1979), Horsfall (1962), Pfadt (1978), Smith (1973), Snow (1974), and Theiler and Downs (1973).

Insects as the Causative Agents of Disease and Discomfort   Insects cause disease and/or discomfort by various combinations of feeding activities, physical injury, secretions, invasion and infestation, and psychological disturbances.

The most important feeding activity that affects people and animals is the tendency of several species of insects, such as mosquitoes, black flies, other biting flies, bed bugs, and conenose bugs, to suck blood. These insects can make life miserable and may in some instances cause serious illness or even death. Illness due to blood loss in human beings is usually not significant since they can generally defend themselves or escape. However, the pain of being bitten and other complications make blood-feeding insects of major significance. Moreover, blood loss in wild and domestic animals may be considerable, and when the population of the offending insects is large, blood loss coupled with interference with feeding causes death.

> An extreme instance is the oft-quoted census made in Rumania, Bulgaria, and Yugoslavia in the year 1923, where nearly 20,000 domestic animals, horses, cattle, sheep, and goats are said to have been killed, along with many wild animals, by the Golubatz fly, *Simulium columbaschense*—Oldroyd (1964)

Some nonblood-feeding insects are attracted to animals and humans and may feed on sebaceous and lachrymal secretions or perspiration and in this way cause considerable disturbance. Certain species of noctuid and other moths in Africa and Asia have been observed to feed on lachrymal secretions of cattle and sometimes of humans (Harwood and James, 1979). Flies in the genus *Hippelates* (Diptera, Chloropidae) are attracted to lachrymal secretions as well as to mucous and sebaceous secretions, pus, and blood (Harwood and James, 1979).

Physical injury may occur if an insect flies into the eye or ear, or as a result of a defensive reaction of an insect that is carelessly handled, or because of inadvertent contact with an insect. Many insects equipped with powerful jaws or legs may pinch, bite, or jab. For example, many of the predaceous Hemiptera, giant water bugs and backswimmers, may inflict a very painful bite. As with anything that causes a break in the skin of human or animal, secondary infection is always a possibility.

Harmful insect secretions may be divided into two groups: the inherently toxic *venoms* and *the allergens*, whose effect depends on the physiological response of the victim. Introduction of a venom usually follows a predictable course, causing pathological conditions; the response to the introduction of an allergen may vary from no reaction to anaphylactic shock, which may result in death. To become allergic or hypersensitive requires an initial introduction (injection or inhalation) of allergen, at which time there is no response except that antibodies are produced that will couple in some way with the allergen when subsequently introduced and cause the allergic response. A given secretion may act both as a venom and as an allergen (e.g., bee venom usually causes pain and a local reddening of the skin and if introduced into an allergic or hypersensitive individual may cause serious complications).

Harwood and James (1979) describe four ways by which venoms and allergens may be introduced: bite, sting, contact, and active projection. Blood-feeding insects typically inject an amount of saliva into their host before they begin to remove blood. This saliva may act as an allergen, as has been demonstrated with mosquitoes, bed bugs, and others. Thrips, certain phytophagous Hemiptera, and certain predatory Hemiptera have also been observed to bite humans, and their salivary secretions may act as venoms or allergens or both. Several Hymenoptera (i.e., many ants, bees, and wasps) are equipped with stinging apparatuses and venom glands on the posterior part of the abdomen. These structures serve in defense and/or prey capture. See Akre and Davis (1978) for information on venomous wasps. Venoms injected by stinging insects can act as allergens. Direct contact with secretions from the bodies of blister beetles (Coleoptera, Meloidae) can cause blistering of the skin. Some lepidopteran larvae in the families Saturiidae (e.g., the Io moth larva), Lymantriidae (e.g., the brown-tail moth larva), Megalopygidae (e.g., certain flannel-moth larvae), and others possess urticating (stinging) hairs that contain ducts from one or more poison glands. These hairs can cause trouble if contact is made with either a live caterpillar or airborne fragments of dead, dried caterpillars. The effect of urticating hairs is much the same as stinging nettles. Butterfly scales and the dried exuviae and fragmented bodies of dried dead mayflies and other insects may cause allergic reactions in sensitive individuals. Some insects, such as certain ants and predatory Hemiptera and others, can actively project or spray venom/allergen from their bodies. This ability is associated with the defense mechanisms of insects (see Chapter 8). Feingold et al. (1968), Frazier (1969), and Shulman (1967) deal with allergic responses to insects. Bucherl and Buckley (1972), Minton (1974), and Tu (1977) contain information on insect venoms.

There has been much concern during the past decade or so about the northward spread of the very aggressive Africanized honey bees from South America. These bees (*Apis mellifera adansonii*) were imported to Brazil from Africa because of their outstanding honey production. The hope was to cross them with local bees (previously imported from Europe) and produce a well-adapted, highly productive strain. Unfortunately, several African queens escaped in 1957, successfully interbred with wild populations, and began to spread. Since that time there have been problems with large numbers of these very aggressive bees attacking humans and various animals. It is conceivable that they will continue to spread from South America into Central America and finally to the United States. For more information on these bees, see Fletcher (1978), Michener (1975), and the article in the April 1976 issue of *National Geographic* magazine by Gore and Lavies.

Imported fire ants, *Solenopsis* spp., are another example of a major stinging insect problem. Two species, *S. richteri* and *S. invicta,* were imported into the United States many years ago, and they now constitute a serious problem in the southeastern states (see Lofgren, Banks, and Glancey, 1975).

Several species of insects spend all or at least a portion of their existence on humans or other animals. Some of these are *facultative parasites*; that is, they can complete their life cycles without invading or infesting man or animal but can take advantage of such hosts should they become available. Others are *obligate parasites*, requiring a human or animal host in order to survive. Among the larvae of true flies there are several that will invade various cavities or open wounds of animals. This type of invasion is referred to as *myiasis* (Zumpt, 1965). Other animal-infesting insects include the chewing (Mallophaga) and sucking (Anoplura) lice.

Occasionally psychological disturbances are associated with the presence of insects, both in human and in wild and domestic animals. These disturbances may be totally unrelated to whether or not the offending insects are harmless. In humans, Pomeranz (1959) recognizes two categories of psychological disturbances caused by insects.

> 1. *Arthropod phobia (entomophobia* in specific reference to insects)—''the irrational, persistent fear of recurrent arthropod infestation;''
> 2. *Hallucination of arthropod infestation*—''a condition in which the subject imagines he is being molested by small and difficult-to-locate forms which reach and localize on the body despite all sorts of extraordinary preventive measures.''

In wild and domestic animals, the persistent buzzing and biting or oviposition attempts by flying insects may cause considerable behavioral disturbance and interfere severely with grazing. For example, horses become extremely nervous in response to the buzzing and oviposition attempts of adult female bot flies (Diptera, Gasterophilidae). For further information on psychological disturbances associated with insects, see Olkowski and Olkowski (1976).

Solutions

This section reviews the major approaches that have been used in attempting to solve insect problems and some that show promise for the future.

Biological Control

Under natural conditions, insect populations are kept in check by the influences of a multitude of environmental factors. Together these factors bring about the ''natural control'' of a given population.

Among these natural control factors are parasites, predators, pathogenic microbes, and competing species (see Chapter 9). Biological control may be defined as "the action of parasites, predators and pathogens in maintaining another organism's density at a lower average than would occur in their absence" (DeBach, 1964).

Biological control offers certain advantages that many other methods of insect control do not. First, it is inherently safe. There is no danger involved during its application, and no toxic residues contaminate the environment and destroy beneficial animals as well as harmful insects. Second, once started, it promises to provide long-lasting control. Thus, although the initial outlay of money for research and development may be comparatively large, it is economical in the long term.

Recent general treatments of biological control include Coppel and Mertins (1977), DeBach (1974), van den Bosch and Messenger (1973), Huffaker (1971), Huffaker and Messenger (1976), Huffaker, Luck, and Messenger (1977), Swan (1964), and Sweetman (1936a, b). Hagen and Franz (1973) trace the history of the development of biological control.

Parasitoids and Predators.   Insect parasitoids and predators are discussed in Chapter 9. The major parasitoids used in biological control are in the order Hymenoptera (including chalcid, braconid, and ichneumonid wasps) and Diptera (Tachinidae and several other families). More than two thirds of the cases of successful biological control have involved the use of Hymenoptera (DeBach, 1974). Among the predatory insects that have been used in biological control are various Coleoptera (especially Coccinellidae, ladybird beetles), Neuroptera (Chrysopidae, green lacewings; Hemerobiidae, brown lacewings), Hymenoptera (e.g., certain ants), Diptera (e.g., Syrphidae and Asilidae), and certain Hemiptera.

There are three basic ways parasites and predators have been manipulated for use in biological control: introduction, conservation, and augmentation.

The introduction of exotic species with hopes of controlling pest species has been by far the most successful of the three methods. There are two situations in which this method is appropriate: (1) when there are "unoccupied niches in the life system of the pest, which could be filled by an introduced species," and (2) when "a certain niche is occupied by an organism that is inherently inefficient as a regulator and that might be displaced by a more efficient exotic regulator" (N.A.S., 1969). Both situations exist particularly when a given pest species has been accidentally introduced from another area. A large percentage of the pest insect species in the United States fit into this category. Introduced species find themselves in an environment in which they are free to multiply in the absence of their natural enemies, and if they happen to feed on something of value, they can quickly become problems.

Although it has long been known that pest insects are attacked by various natural enemies, the purposeful use of their natural enemies has occurred only in the last couple of centuries. The first written record of the use of beneficial insects describes the protection of date palms in Arabia from phytophagous ant species by the introduction of another species of ant that was predatory on the phytophagous forms (Sweetman, 1936). The first noteworthy example of the application of this type of control in the United States and probably the most significant early demonstration of this method as a valid approach was in California in 1887–88 (N.A.S., 1969). At that time the citrus industry was severely threatened by the cottony-cushion scale, *Icerya purchasi* (Homoptera, Coccoidea), and the available insecticides were failing to control it. Since this pest species was a native of Australia, an appropriate control species was sought there, resulting in the discovery of the vedalia beetle, *Rodolia cardinalis* (a Ladybird beetle; Coleoptera, Coccinellidae). Within a year following the introduction of this predatory beetle, control had been attained and is still in effect.

According to DeBach (1974) the method of introduction of parasitoids and predators worldwide up to 1970 had brought about complete, substantial, or partial control on a permanent basis of at least 120 pest insect species in nine insect orders. Following initial successes many of the same natural enemies were used in other places, adding another 133 cases to the list of successes.

Conservation methods include the application of pesticides at times when they are likely to do the least harm to beneficial species, provision of a favorable habitat for beneficial species, possible reduction of populations of natural enemies of beneficial insects, and similar approaches. Augmentation (or inundation) involves the mass rearing and subsequent dissemination of parasites or predators in order to supplement the actions of the natural populations of these same species. An example of this method is the use of parasitic wasps (e.g., *Trichogramma* spp.), which attack the eggs of several Lepidoptera. Neither conservation nor augmentation has been as effective as the introduction of exotic species.

A fourth possible approach is the artificial selection of highly resistant strains of beneficial insects that could be used in association with insecticides without themselves being harmed. Insecticide resistance has, in fact, been observed to develop in certain beneficial species.

Many vertebrates (e.g., birds, reptiles and amphibians, fish, and some mammals) are predatory on insects, and many are active in the natural control of insect populations. Some vertebrates have been used in biological control, and many more may have potential. A good example is the use of certain cyprinodont fishes, especially *Gambusia affinis*, against mosquito larvae. *Gambusia* has several characteristics that make it valuable in this regard: broad tolerance of salinity and organic pollution, viviparous reproduction, high fecundity, small adult size, surface-feeding habits, ability to penetrate regions where mosquito larvae breed, and ease of transport (Sweetman, 1936a; Bay, 1967). The giant toad *Bufo marinus* has also been used in insect control, for example, against sugar cane white grubs in Puerto Rico (Sweetman, 1936a).

For more information on the use of parasitoids and predators in insect control, see van den Bosch (1975), Ridgway and Vinson (1977), and Stehr (1975).

Pathogenic Microbes. Although man has been long aware of the "natural" control of insect populations by microbes, the first record of the idea of using them for insect control was in the eighteenth century. The study of insect microbiology/pathology did not become a serious topic of study until the early 1900s.

More than 1000 microbes have been reported as insect pathogens. However, there have been only a handful of successful attempts at insect control with microbes. The microbes that may be associated with insects (i.e., bacteria, fungi, viruses, rickettsiae, protozoans, and nematodes) are discussed in Chapter 9.

The use of microbes in control has great promise despite numerous drawbacks. Microbes can often be cultured in large numbers economically, they are generally fairly specific in action and usually not harmful to other plants and animals, and they leave no toxic residues. Disadvantages include dependency on certain weather conditions (e.g., most fungi pathogenic for insects are effective only under warm, moist conditions); difficulties in mass producing certain microbes (e.g., viruses); killing hosts too slowly (e.g., rickettsiae); and potential hazard to other animals (e.g., rickettsiae and certain protozoans). Falcon (1976) discusses the problems involved in the use of viruses in insect control.

An example of the successful use of microbes in insect control is the treatment of soil with spores of the bacterium *Bacillus thuringiensis* to control the Japanese beetle, *Popillia japonica*. This same bacterium has also been used successfully against the alfalfa caterpillar, European corn borer, the gypsy moth, and the pink bollworm. Brown, red, and yellow fungi that attack whiteflies (Homoptera, Aleyrodidae) have been effective against these citrus pests in Florida. There are numerous examples where certain microbes are known to be very effective in natural control but have not yet been developed as artificially applied control agents. Nevertheless, several pathogens (including bacteria, fungi, and viruses) have been developed and are available commercially for insect pest control. Several are listed in DeBach (1974).

For additional information on the use of microbes in insect control, see Briggs (1975), Bulla (1973), Burges and Hussey (1971), Ferron (1978), Ignoffo (1975), Maddox (1975), Maramorosch (1977), N.Y. Academy of Science (1972), Poinar and Thomas (1978), Steinhaus (1963), Tinsley (1977), and Weiser (1969). Cameron (1973) outlines the history of insect pathology.

Competitors.   The use of introduced competitors in insect control is a comparatively new idea and is based on the principle of competitive displacement. This process occurs when two species have essentially identical niches or closely overlapping niches (i.e., they fit into an ecosystem in the same or nearly the same way; see Chapter 9). For competitive displacement to work in insect control, the introduced species would need to have a slight edge over the pest species. Practical applications of competitive displacement have not as yet been extensively developed, but some possibilities are

1. Introduction of a nonbiting species to displace a biting species (e.g., if both competitors shared the same larval niche).
2. Introduction of an imported species that, although sharing the

same food as a pest species, is unable to survive the winter in the area of application, but which during favorable periods is capable of displacing the original species.

3. Introduction of a potential pest species, known to be controllable by biological or other means, that is capable of displacing an otherwise uncontrollable pest species.

Genetic Control

Genetic control involves manipulation of the mechanisms of heredity.

An outstanding example of genetic control is the sterile-male method. Many years ago the idea was conceived that if a large enough percentage of the matings of a given population in the field resulted in no offspring, then over a period of generations the population would decrease (Knipling, 1955). Thus if sexually sterilized males are introduced into or induced in a wild population each generation and if the matings of these sterilized individuals exceed normal matings, the population will decline. If the number of sterile individuals is kept constant (by additional releases) for each generation, the ratio between sterile and normal matings will increase rapidly and the rate of population decline will increase correspondingly.

The earliest application of the sterile-male technique was against the screwworm fly, *Cochliomyia hominivorax,* in the southeastern United States (Knipling, 1959; Baumhover, 1966). During 1958–1959 3.7 billion sterilized (by gamma irradiation) screwworm pupae were reared and released throughout large portions of Florida and Georgia. This resulted in the successful eradication of the fly from this part of the country. Since that time there have been sporadic outbreaks traceable to the movement of infested animals into the territory, but the screwworm has not been a problem since 1959. Baumhover (1966) describes the cost and savings of the screwworm-eradication program:

> During the two-year campaign, 3.7 billion screwworm pupae were produced, and 6.3 million lb. of horsemeat and whale meat were used. Twenty light aircraft were used to release flies over a maximum area of 85,000 square miles, and peak employment, including plant personnel, fly distributors, field inspectors, and clerical and administrative help, totalled 500. However, for a research cost of only $250,000 and an eradication-program cost of $10 million, ranchers in the Southeast have experienced $140 million in savings since inception of the program in 1958.

The screwworm-eradication program has since been expanded and is now active in the southwestern United States and Mexico. A barrier zone of sterilized males is maintained to block the northward movement of flies from overwintering sites in Mexico. Although this program has faced difficulties (Novy, 1978), it constitutes a notable success in insect control. As a result, losses of nearly $2 billion have been prevented (Scruggs, 1978).

Because of the vast numbers of sterilized males that need to be released to ensure successful mating with wild females and effect eradication, the sterile-male method is practical only against insects that occur in relatively small populations as adults. Small populations may occur naturally or may be brought about by the employment of other methods of insect control, insecticides in particular. Further, it is very important that the sterilized flies have the vigor to compete with wild individuals for a mate. Finally, the species involved must be amenable to artificial rearing in huge numbers and at a resonable cost. The sterile-male method is impractical against insects that are very prolific and widespread or against insects that appear in large numbers sporadically and unpredictably (e.g., floodwater mosquitoes) because large numbers of artifically reared individuals would have to be maintained at all times (N.A.S., 1969).

Since the screwworm-eradication campaign, the use of the sterile-male method against other species has been studied and is currently being applied or in the pilot test stage for a number of pest species (Knipling, 1972).

In addition to radiation, several chemicals have been discovered in recent years that induce sterility when ingested by insects. Out of approximately 6000 compounds screened, more than 300 show promise as chemosterilants (N.A.S., 1969). Chemicals produce sterility primarily by (1) causing insects to fail to produce sperm or ova, (2) causing the death of sperm or ova after they have been produced, or (3) producing genetic defects in spermatozoa that prevent zygote development (N.A.S., 1969). The last of these mechanisms for the production of sterility is the most desirable because such sterilized males are generally competitive with the unsterilized males in mating with available females.

Chemosterilants are useful in situations where the sterile-release method would be inappropriate (e.g., against species that occur in very large populations and are difficult or impossible to rear in large numbers in the laboratory). Since the best chemosterilants so far known must be ingested to be effective, they are usually applied with baits. Chemosterilants that are effective by contact might be used in association with luring stimuli such as light and sex attractants. Another possibility is the application of a chemosterilant to breeding places. However, since all the promising chemosterilants are mutagenic agents, they present a serious hazard to other animals, including humans. According to Metcalf and Metcalf (1975), "Their use in pest management cannot be recommended."

Other genetic approaches to insect control include the use of such factors as (World Health Organization, 1967):

1. Sex-ratio distorters.
2. Detrimental genes incorporated into chromosomes that have meiotic drive.
3. Chromosome translocations in heterozygous males.
4. Conditional lethal genes that allow the parents to survive in the laboratory but are lethal to their descendants under field conditions.

For more information on genetic control of insects, see Baumhover (1966), Bŏrkovec (1975), Davidson (1974), Foster et al. (1972), Hoy and McKelvey (1979), Huffaker and Messenger (1976),

Knipling (1972), Metcalf and Metcalf (1975), Pal and Whitten (1974), Richardson (1978), and Smith and von Borstell (1972).

Breeding Insect-Resistant Hosts

Under natural conditions, some individuals of a given species possess characteristics that make them more able than others to cope with various environmental stresses (e.g., insect attack). Individuals possessing such characteristics will tend to be more successful in reproducing and hence in passing the heredity determinants of these characteristics to their progeny. For this reason, over time a population exposed for many generations to a given stress will become composed predominately of members that are able to cope with this stress. This, in a very general way, is the process of natural selection. In a number of instances, artificial selection has been carred out in the laboratory to produce strains or varieties of organisms that are resistant, or more resistant than members of other strains or varieties, to an environmental stress. Another approach has been to encourage the proliferation of some naturally occurring resistant strain or variety.

Resistance may be defined as the heritable ability of an individual, strain, variety, race, and so on (plant or animal) to repel or withstand the effects of some environmental stress or stresses to a greater degree than other individuals or groups. In the context of our discussion insect pests are the "environmental stresses."

Much more emphasis has been placed on studies of insect resistance in plants than in animals. This is due at least in part to greater use and lower costs of plant breeding.

Kogan (1975) describes several categories of plant resistance mechanisms. Among these are phenological asynchrony, induced resistance, resistance factors that influence insect behavior, resistance factors that influence insect metabolism, phenetic resistance and tolerance. These various forms of resistance may occur alone or in combination in a given individual. See "Adaptations Associated with Interspecific Interactions" in Chapter 9 for more information on resistance mechanisms in plants.

Phenology is a term applied to the timing of the various stages in the life cycle of an organism. Various strains of both plants and insects have characteristic phenological patterns throughout their life cycles. The phenological pattern of a successful phytophagous insect is closely synchronized with that of its host plant such that its active feeding stages coincide with the time when the plant provides suitable food. Phenological asynchrony refers to a lack of correspondence between the phenological pattern of a given plant and the insect's life cycle. For example,

> The early varieties of soybeans planted in Illinois mature in the beginning of September. When the second generation of the bean leaf beetle, *Cerotoma trifurcata* (Forster), emerges in soybean fields in the early part of September, most plants are already mature and almost ready to harvest. Thus these beetles cannot harm the crop, since the mature pods are not an adequate food.—(Kogan, 1975)

Various environmental influences may alter the physiology of a plant such that it becomes nutritionally unsuitable as food for certain herbivorous insects; that is, it acquires induced resistance. Thus, for example, application of fertilizer may make a given strain of plant unsuitable for a given pest species by influencing the nitrogen and potassium (both important for insect survival and development) balance (Kogan, 1975).

Plant resistance may be due to an effect or lack of effect a given plant has on pest species relative to stimulating feeding and/or oviposition behavior. Thus, certain varieties of a given crop plant are preferred over other varieties, and nonpreferred varieties may be artifically selected for.

Plants may have an *antibiotic* effect on herbivorous insects by somehow interfering with metabolism. Symptoms induced in insects range from death to developmental malformations to decreased fecundity.

Phenetic resistance refers to the presence of structural characteristics that interfere with insect feeding, oviposition, shelter-seeking, and so on. Several examples may be found in Chapter 9.

Plants have been successfully bred for resistance to more than 25 insect pest species (see Kogan, 1975). Beck (1965), de Wilde and Schoonhoven (1969), Galun et al. (1975), Hanover (1975), Kogan (1975), Maxwell and Jennings (1979), Maxwell et al. (1972), Painter (1958), Pathak (1975), Sondheimer and Simeone (1970), and Van Emden (1966) should be consulted for more information on plant resistance to insects.

Environmental Control

Environmental control procedures involve the removal, destruction, modification, or isolation of materials that might favor the survival of an insect pest by affording food or making a site suitable for breeding and/or dormancy.

Removal and Disposal of Material.   One of the most effective ways of controlling many insect pests is the maintenance of high standards of sanitation. Accumulations of trash, garbage, and untreated sewage provide food and breeding sites for numerous annoying and disease-carrying insects (not to mention rats). Thus these materials must be removed from their source of accumulation and be properly treated and disposed of. Garbage disposals, trash and garbage collection services, sewage collection systems and treatment plants, sanitary land fills, and incineration are among the methods used.

Crop rotation (varying the kind of crop planted from season to season), in addition to benefitting the soil, may also be responsible for the reduction of insects that attack one of the crops being rotated but not the other(s). For example, May beetle larvae (Coleoptera, Scarabaeidae, *Phyllophaga* spp.), which feed on the roots of many crops, do poorly on certain leguminous crops. Hence the use of these crops in rotation has the effect of reducing the larvae (N.A.S., 1969).

Destruction or Modification of Material.   In many instances it is not practical to remove potential insect-breeding material. Under these circumstances the material is destroyed or modified at the site of its accumulation. Burning, turning under, or allowing farm animals access to crop refuse is an excellent way of destroying the insects harbored within. Thus the boll weevil (*Anthonomus grandis*), the European corn borer (*Ostrinia nubilalis*), and others are controlled in part by the destruction of infested plants after harvesting. Destruction or removal of slash following cutting operations in a forest is an appropriate means for controlling numerous insects that attack trees. The borders of fields used for cropping and uncultivated fields may contain plants that serve as alternative hosts for pest species or harbor plant-disease organisms. Cleaning out these borders may aid in the control of certain insects. On the other hand, vegetation on field borders may provide a suitable habitat for beneficial parasitic, predatory, or pollinating insects. Various methods of tillage have been used at appropriate times to kill susceptible stages of insects in the soil. For example, fall plowing is effective against the pupae of the corn earworm (Lepidoptera, Noctuidae, *Heliothis zea*), which overwinters in the soil. Various crop management and treatment practices, such as planting soil-building crops, strip cropping, thinning, topping, preening, and defoliating, have been used at least partly for the purpose of insect control.

Naturally or artificially produced aquatic or semiaquatic habitats can produce severe insect problems, for example, outbreaks of mosquitoes, biting midges, and horse and deer flies. Examples of naturally occurring aquatic or semiaquatic habitats that provide suitable conditions for the breeding of these insects are salt- and freshwater marshes, swamps, and various depressions in which water can collect. Artificial habitat creation includes the construction of water impoundments for a variety of reasons (e.g., flood control, hydroelectric power, water storage, drinking water for animals, recreational purposes, borrow pits), and the development of water distribution systems, particularly for irrigation.

To effect insect control, marshes, swamps, and depressions containing water may be either flooded or drained, depending on a variety of factors. For example, along the east coast of Florida many of the salt marshes, which produce large numbers of salt-marsh mosquitoes, *Aedes taeniorhynchus* and *A. sollicitans*, and biting midges, particularly *Culicoides furens*, have been impounded, a measure that has provided successful control.

Careful management of artificially impounded water and water distribution systems is necessary to avoid insect problems. Practices include (1) proper construction so that water does not stand in any one place longer than necessary (e.g., proper land grading for irrigation purposes, appropriate drain construction); (2) deepening or filling very shallow areas; (3) periodical removal of vegetation within the water-level fluctuations boundaries; and (4) periodic removal of vegetation and accumulated debris from drains to avoid clogging, seepage, and overflow.

Isolation of Breeding Material.   In many instances, it is inappropriate to remove, destroy, or modify potential breeding material and hence the only alternative is to isolate it from potential insect pests. This isolation may be temporal or spatial (Rolston and McCoy, 1966). Temporal isolation involves the timing of any practice (e.g., plant harvesting, shearing, dehorning) such that problems with insect attack will be minimal. A classic example is the sowing of winter wheat on "fly-free" dates to avoid infestation by the Hessian fly, *Mayetiola destructor*. The adult flies that develop on summer wheat live only a few days, and if winter wheat is planted on a fly-free date, all the flies will be dead by the time it comes through the ground. Fly-free dates, of course, vary with latitude. Dehorning and castration of cattle are done during cool periods of the year to avoid the attacks of screwworm flies. A number of procedures may logically be looked upon as providing spatial isolation of breeding materials. These include (1) planting a given crop as far as possible from an earlier infested one of the same kind; (2) the use of trap crops and trap logs, which may divert pests from the main crop; (3) appropriate pasturing of domestic animals to avoid pests [e.g., the late summer use of upland pastures in the western United States to avoid the attacks of tabanid flies, which breed in lowland pastures where there are small streams and abundant trees (N.A.S., 1969)]; and (4) the use of properly designed and constructed storage boxes, bins, rooms, and so on, in which cool, dry conditions are usually ideal for the inhibition of pest species of insects.

Chemical Control

This section deals with those natural or synthetic chemicals that cause directly the death, repulsion, or attraction of insects.

Insecticides.   An *insecticide* may be defined as a chemical or a mixture of chemicals employed to kill insects and related arthropods. The term *pesticide* is more inclusive, referring to insecticides as well as herbicides, rodenticides, and other substances.

Insecticides were apparently used before recorded history.The writings of the Greeks, Romans, and Chinese all allude to the use of various substances, such as sulfur, hellebore (a poisonous herb), and arsenic. Prior to the 1940s the insecticidal value of a number of inorganic chemicals (e.g., arsenic, mercuric chloride, and carbon disulfide) and organic chemicals of botanical origin (e.g., pyrethrum, cube, and nicotine; see "Adaptations Associated with Interspecific Interactions" in Chapter 9) was known and put to extensive use. The discovery of DDT by Paul Müller in Europe in 1939 revolutionized insect control and marked the beginning of the development and application of synthetic organic insecticides. In 1948 Dr. Müller was awarded the Nobel Prize in medicine for his discovery. Since that time hundreds of compounds of varying insecticidal value have been discovered, and thousands of new potential toxicants are being evaluated each year by a detailed screening process. The vast majority (about 90%) of currently used pesticides are of the synthetic organic variety.

Insecticides may be classified in different ways. Some are more useful than others for a given stage in the insect life cycle. Thus there are ovicides, larvicides, and adulticides. A useful classification is based on the primary route of entry: *stomach poisons* act via ingestion and absorption from the alimentary canal; *contact poisons* are absorbed through the cuticle; and *fumigants* enter through the spiracles and tracheal system or body wall in the gaseous state. These categories are by no means completely exclusive (e.g., many contact insecticides also act quite effectively via the oral route). Caswell (1977), Spencer (1973), and *Pesticide Index* (1976) contain useful and up-to-date lists of commercial and experimental insecticides. Melnikov (1971) discusses pesticide chemistry. General references on chemical control include: Cremyln (1978), Metcalf (1975), Fronk (1978), Sheets and Pimentel (1979), Ware (1975, 1978), and Watson and Brown (1977).

Understanding the physiological and biochemical modes of toxic action of insecticides is of great importance in, for example, the development of new, more effective, and more specific insecticides, dealing with the problem of resistance to insecticides; and dealing with pesticide poisoning of humans and other animals. Recent information on this topic may be found in O'Brien (1967, 1978). Matsumura (1975), and Wilkinson (1976).

Stomach poisons are useful against insects with chewing mouthparts, which ingest the toxicants with their food or during grooming These poisons are mostly inorganic compounds and have been used more in the past than they are at present. However, with the development of resistance to synthetic organic insecticides in populations of a number of species, many are coming back into use (e.g., Paris green and sodium fluoride). Paris green has been used effectively against DDT-resistant mosquito larvae. The arsenicals (arsenic trioxide, Paris green, lead arsenate, and so on); compounds of flourine boron, and phosphorus; tartar emetic (an antimony compound); and thallium sulfate are stomach poisons that have been applied against a wide variety of insect pests. Most of these form persistent residues and are applied to the natural food plants of pest species, mixed with an attractive substance (poisoned bait), or applied to a surface where pest insects are likely to contaminate their legs and other body parts later to ingest the poison during grooming.

A group of synthetic organic insecticides called *systemics* are absorbed and distributed to the various tissues of plants or animals rendering them toxic to insects. Systemics are effective against insects that penetrate host tissues with piercing-sucking mouthparts (e.g., sucking lice, thrips, aphids) and ingest blood, sap, and so on as well as against chewing insects. Internal parasites of animals (e.g., cattle grubs, screwworm larvae, and bot fly larvae) are also susceptible to systemic insecticides. Systemics of plants may be applied to the soil, seeds, foliage, and bark of trees or injected into the inner tissues of woody plants. They are then transported throughout the plant by the vascular tissues. Animal systemics are applied as surface sprays, which are absorbed dermally; as boluses (large pills), which animals are induced to swallow; or in food, water, or salt. Fortunately, these insecticides are gradually degraded by enzymes so that eventually the animals can be safely used for milk production or slaughtered for meat (Metcalf, Flint, and Metcalf 1962).

Contact poisons, those absorbed directly through the cuticle, can be divided into the following three groups: (1) in organic, (2) natural organic, and (3) synthetic organic. Inorganic contact insecticides include sulfur, mercury, and copper compounds, which have been used to kill mites and fungi as well as insects. Many have been replaced by synthetic organic compounds. Finely divided silica compounds (e.g., silica aerogel) may also be considered as a part of the inorganic category. However, they are not absorbed into the body, but act by abrading or adsorbing the waterproofing wax layer of the epicuticle (see Chapter 2), resulting in the subsequent desiccation of an insect. These materials have the advantage of being nontoxic to humans and other vertebrates.

Natural organic contact insecticides include the oils and the *botanicals* (see ''Adaptations Associated with Interspecific Interactions'' in Chapter 9). Although oils are used mostly as solvents for insecticides, many are themselves toxic to insects. For example, kerosene and certain other oils have been long used as mosquito larvicides. The botanicals are as a group probably the safest insecticides. They are prepared either by extraction of active ingredients or by being dried and ground into dusts. Pyrethrum and nicotine are probably the best-known botanicals. Pyrethrum is prepared from ground-up *Chrysanthemum* flowers and has been used as an insecticide for hundreds of years. It is widely used in many ''bug bombs'' (canned aerosols) because of its rapid knockdown effect and low mammalian toxicity, two features that make it particularly attractive for household use. Nicotine is prepared from tobacco plants. It is no longer widely used. Other botanicals include allethrin, rotenone, ryania, and sabidilla.

Since the discovery of DDT in 1939, synthetic organic contact insecticides have become the most widely used group. This group includes the organic thiocyanates, dinitrophenols, several sulfur compounds, chlorinated hydrocarbons, organophosphates, and carbamates. The dinitrophenols act to increase oxygen uptake, enhance oxidative metabolism, interfere with the production of ATP, and quickly cause muscle paralysis (N.A.S., 1969). The remaining groups act on the insect nervous system in different ways. For example, the organophosphates, carbamates, and organic sulfur compounds are cholinesterase inhibitors. Cholinesterase is an enzyme that breaks down the neurotransmitter acetylcholine, which is liberated at synapses of neurons in response to an impulse (see Chapter 3). When cholinesterase is blocked by an inhibitor, acetylcholine accumulates, and as a result the insect becomes hyperactive and eventually paralyzed. The synthetic organic insecticides that are probably the most dangerous to humans are certain of the chlorinated hydrocarbons and organophosphates.

Fumigants enter insects via the body wall and tracheal system in the gaseous state. These insecticides are used where the pest insects can be contained in an essentially airtight structure for a period of time. This closed container could range from a small plastic bag to an entire building. Some fumigants, in addition to being toxic to man, are highly flammable (e.g., carbon disulfide) and must be used mixed with nonflammable gases such as carbon dioxide or carbon tetrachloride. Fumigants may be initially in the solid, liquid, or gaseous state. Naphthalene and paradichlorobenzene (PDB) are examples of solids that give off toxic and repellent gases. The latter is the most important, being widely used to protect clothing against clothes moths and insect collections against dermestid beetles. Other fumigants include hydrogen cyanide, methyl bromide, carbon disulfide, and carbon tetrachloride.

The ideal insecticide would be highly specific, killing only the target pest species and leaving beneficial insects unharmed. Unfortunately, to date no such insecticide is known. The systemic insecticides probably come the closest to being specific, in that only insects that feed on the treated plants or animals are directly affected by the poison. However, the parasitoids, parasites, and predators that feed upon the pest species may also be poisoned. Fundamental metabolism is remarkably similar among most organisms; hence most poisons, whatever their particular mode of action, tend to be toxic to a broad spectrum of animals. Whatever degree of specificity a given insecticide may show depends, at least in part, on its mode and ease of entry. For instance, a sucking insect may fail to ingest a lethal dose of a stomach poison on the surface of foliage since it feeds beneath the surface. Insects with very thick cuticles may resist the absorption of some contact poisons. Another factor that may account for some specificity is resistance to a given insecticide. For example, some insects (e.g., the tomato hornworm, *Manduca quinquemaculata*) are unable to detoxify DDT, whereas other species (e.g., the tobacco hornworm, *Manduca sexta; Melanoplus* grasshoppers; the reduviid bug, *Triatoma protracta*) are able to detoxify this insecticide (N.A.S., 1969).

Substances that are used in combination with insecticides are called *adjuvants* since they complement, in some way, insecticidal action. Examples include dusts, granules, solvents, emulsifiers, wetting and spreading agents, adhesives, and synergists. The mixture of an insecticide with one or more adjuvants is called a *formulation*. Dusts are finely divided materials, somewhere in the range 1–40 micrometers in diameter (Fronk, 1978), which are used as inert carriers and diluents for insecticides when dry application is necessary or appropriate. In some instances the insecticides themselves (e.g., sulfur) are ground into dust-sized particles and may be diluted with inert dusts of various types. Many materials have been used as inert carriers and diluents, for example, pyrophyllite (a hydrous aluminum silicate), talc (a hydrous magnesium silicate), and organic substances such as walnut shell flour. Granules are similar to dusts except that they are larger in particle size.

Solvents are used to dissolve insecticides and serve both as diluents and carriers. Most synthetic organic pesticides are insoluble in water, and organic solvents such as fuel oil or kerosene are used. However, insecticides dissolved in an organic solvent can be further diluted with water by forming an emulsion. This requires an emulsifying agent, a substance that makes a finely dispersed mixture of water and an insoluble substance (such as an organic solvent) possible. Insecticide plus organic solvent plus emulsifying agent, an emulsifiable concentrate, is the most common formulation (Fronk,

1978). Wettable powders are similar to emulsifiable concentrates in that they contain a wetting and spreading agent that allows them to be suspended in water and applied as sprays, but they are in dust form rather than dissolved in an organic solvent.

Synergists are substances mixed with insecticides that render the mixture more toxic than would be expected based on a simple additive effect of the toxicities of the individual components. A good example of synergistic action is that between piperonyl butoxide (synergist) and pyrethrum. This is a fortunate relationship since the combination of the two is cheaper than would be an equally effective straight pyrethrum formulation (Rolston and McCoy, 1966). Insecticides may also be mixed with attractive baits (poisoned baits), perfumes to nullify disagreeable odors, and so on. Certain insecticides may be prepared in pill (bolus) form for introduction into maggot-infested alimentary tracts of animals. Insecticides may also be applied in the form of sprays made up of extremely minute particles (droplet size between 0.1 and 50 micrometers in diameter; Pratt and Littig, 1962). These sprays are referred to as *aerosols* and are particularly valuable in space spraying (e.g., outdoors against mosquitoes and other flying insects). Smokes are similar to aerosols both in their properties and application, but they are comprised of much smaller particles of insecticide. They are produced by burning or evaporating insecticides. Smokes are not considered to be particularly effective.

A wide variety of mechanical devices (sprayers, aerosol generators, dusters, etc.) exists for the application of insecticides. Information on such devices may be found in many of the general applied entomology texts listed at the end of this chapter.

Among the advantages afforded by insecticides are the following: (1) they are usually very effective, (2) they generally act within a short period of time, (3) they are effective when applied against large pest populations, and (4) they are readily available for use when needed (N.A.S., 1969). The use of insecticides has played an important role in the development of agriculture as we know it today. Both increased yield per unit land area or animal and improved quality of agricultural products are in large part attributable to the successful use of insecticides. Agricultural yield has shown a 54% increase in the 20 years following the general use of synthetic organic insecticides (N.A.S., 1969). Undoubtedly other factors have played a role in this impressive increase, but application of insecticides is certainly among the most important. Insecticides have also enabled man to bring several dreaded diseases under control or at least reduce their extent.

On the negative side, application of an insecticide can be very hazardous, and direct contact with a highly toxic insecticide can cause severe illness and death. This is why the careful reading and following of instructions on the labels on pesticide containers and constant concern wih safety during their application are so important. Other important considerations are (1) storage of pesticides in well-labeled containers out of reach of children; (2) extreme care in application, including avoidance of spillage on clothing or skin, or inhalation of sprays or dusts; (3) not smoking or eating when working with toxicants; and (4) proper disposal of empty insecticide containers. A pamphlet, *Safe Use of Pesticides,* may be obtained from the American Public Health Association. Caswell (1977) also contains information on pesticide safety.

Also of concern is the presence of residues in food products, plant and animal, that have been treated with pesticides at some point in their production. Every pesticide that is shipped in interstate commerce in the United States must be registered under the Federal Insecticide, Fungicide, and Rodenticide Act. The objective of registration is the protection of the consumer, not only in terms of safety, but also with regard to the value of the products. When an insecticide is registered, the amount of residue (tolerance) to be allowed in various food products is established. Some insecticides have such low mammalian toxicities that they are tolerance exempt, whereas others, which are considered dangerous following an extensive series of tests, are subject to a rigid tolerance. If these tolerance levels are exceeded, various degrees of legal action against the violators result. In addition to the strict controls applied by the federal and state governments, the development of a new insecticide is a rigorous, time-consuming process which increases the probability that an insecticide that reaches the market will be effective and reasonably safe if directions are followed carefully.

Despite the hazards involved, the safety record of pesticide usage in the United States is not as bad as one might suspect:

> In recent years there has been so much publicity about pesticide hazards that much of the general public has developed a distorted impression of the magnitude of the problem. Vital statistics show that pesticides account for only 1 out of every 700 accidental deaths and only 5% of all poison deaths in the United States. As a cause of death, pesticides are far outranked by common drugs and by household agents such as cleaners, polishes, and solvents. For many years, the ubiquitous aspirin tablet has caused about the same number of accidental deaths annually in the United States as all the pesticides combined, and since 1957 aspirin has caused slightly more accidental deaths. As a matter of record, the death rate for all pesticides has remained constant at slightly over one per million of population for the past 25 years.—(N.A.S., 1969)

The toxicity of an insecticide is often expressed in terms of LD_{50} (lethal dosage), which refers to the dosage (milligrams of insecticide per kilogram of body weight of test animal) that is lethal to 50% of a test population. For example, if the LD_{50} dosage for laboratory rats were given to each individual in a population of these rats, one would expect about 50% of the exposed animals to die. The remaining 50% are less susceptible, and larger doses of insecticide would be required to kill them. Theoretically, in a randomly distributed population, the relationship between insecticide (or other toxicant) dosage and percent mortality follows a bell-shaped curve (normal distribution). LD_{50}'s vary depending on the route of entry of the insecticide into a test animal. Thus LD_{50}'s are commonly determined for both the oral and dermal routes.

Environmental pollution with insecticides has become a matter of great concern. Highly residual insecticides can pass well beyond their intended targets and may reduce populations of beneficial insects and wildlife. DDT in particular has been attacked in this regard. Two major factors have contributed to its being a problem: (1) it is highly residual, and (2) it becomes concentrated in a stepwise fashion along a food chain. Woodwell et al. (1969) provide a graphic description of this process of "biological concentration":

> . . . in the food web which the herring gull is a scavenger in Lake Michigan, DDT (DDE and DDD) concentrations in the bottom muds at 33–96 feet averaged 0.014 part per million. In a shrimp *(Pontoporeia affinis)* they were 0.44 ppm, more than ten times higher. Levels increased in fish to the range of a few ppm (3.3 alewife; 4.5 chub; 5.6 whitefish), another tenfold increase, and jumped in the scavenging, omnivorous herring gull to 98.8 ppm, twenty times higher still, and 7000 times as high as in the mud.

Unfortunately, such accumulations of insecticides kill or inhibit reproduction of certain animals, birds in particular. Woodwell et al. (1969) point out that many important predatory birds, including the bald eagle, are in danger of becoming extinct in the United States. Although concern for these animals is justified in itself, of far greater concern is the long-term indirect effects pesticide residues may have by disrupting the intricate balance of ecosystems.

Rachel Carson in her book *Silent Spring* painted a particularly vivid and frightening picture of what could happen. Brown (1978), Haque and Freed (1975), Khan (1977), McEwen and Stephenson (1979),Miller and Berg (1969), Perring and Mellanby (1977), Rudd (1964), van den Bosch (1978); White-Stevens (1977), and Woodwell et al. (1971) discuss various aspects of insecticides in the environment. Johansen (1977) considers the effects of pesticides on pollinating insects.

Another major problem associated with the use of insecticides has been the development of insecticide resistance in strains of several pest species (more on this under "Insect Pest Management").

Repellents. Repellents are substances that are mildly toxic or nontoxic to pests, but prevent damage by causing the pests to make oriented movements away from the source. In very early times, smoke from wood fires was used to keep away biting and annoying insects. As early as 1897, oil of citronella was used as a mosquito repellent (Shepard, 1951). Most of the earlier repellent substances were quite odorous and perhaps somewhat repellent to humans as well as insects. However, many of the more modern, synthetic repellents have little, if any, disagreeable odor.

To be useful as an insect repellent a substance has to be more than merely repellent to insects. It must be nontoxic, nonirritating, and nonallergenic to humans and domestic animals; inoffensive in odor; harmless to fabrics; persistent (not easily removed by perspiration, rubbing, laundering, an so on); and effective against a broad spectrum of pest species. In addition, a good repellent should be cheap and damaging to plastics, painted surfaces, and the like.

The use of good repellents has a number of advantages: (1) They afford individual protection from insects without the necessity for expensive and time-consuming population eradication; (2) they do not damage or kill beneficial animals or plants; and (3) the ones that are available for use are nontoxic to man. On the other hand, disadvantages include the following: (1) repellents are at best a temporary measure (a few hours at most) and tend to evaporate from or rub off skin or clothing due to perspiration and the like; (2) they commonly have an oily feel and may have a somewhat disagreeable odor; (3) they must be applied in comparatively large doses (in the range of 20–40 mg/cm^2 of skin; Smith, 1966); and (4) they may damage certain plastics or painted surfaces.

So far, repellents have been primarily used for the protection of man and animals against attacks from bloodsucking or otherwise annoying insects. The armed forces have over the years had a particular interest in repellents, realizing that protection of troops from disease vectors would greatly increase their military advantage. However, interest has also developed in the use of repellents in plant protection. Metcalf, Flint, and Metcalf (1962) describe three general groups of repellents: those used against (1) crawling insects, (2) the egg laying of insects, and (3) the feeding of insects. Repellents used against crawling insects usually consist of a repellent barrier interposed between an insect whatever material happens to be attractive to it. For example, creosote has been used as a barrier against the migration of chinch bugs, *Blissus leucopterus,* and trichlorobenzene and other repellent insecticidal chemicals to protect buildings from termite invasion. Creosotes derived from coal and wood tar have been used extensively for the protection of wood against termites, powder post beetles, and rot organisms (Shepard, 1951). Creosote and other oils apparently ". . . smother the natural attraction of the insect to its food or oviposition site" (Shepard, 1951). Several chemicals have been found that are reasonably effective in repelling insects from feeding. Washes containing bordeaux, lime and other materials are used to repel leafhoppers and some chewing insects, and inert dusts have been useful on cucurbits to repel cucumber beetles (Metcalf, Flint, and Metcalf, 1962). An ideal repellent for plant protection would be one that would somehow block the natural attractants (kairomones; see Chapter 8) to which pest species respond. Diethyltoluamide, considered to be one of the best general repellents yet discovered, dimethyl phthalate, ethyl hexanediol, dimethyl carbate, and powdered sulfur are examples of repellent chemicals that have been applied to human skin and/or clothing.

When a broad spectrum of effectiveness is desired, a mixture of repellents may be appropriate. For example according to Rolston and McCoy (1966):

The Armed Forces used a repellent containing dimethyl phthalate, ethyl hexanediol, and dimethyl carbate for application to skin, and a repellent containing benzyl benzoate, *n*-butylacettanilide, and 2-butyl-2-ethyl-1,3-propanediol for application to clothing. Both mixtures provided protection against medically important mosquitoes, feas, ticks, and chiggers.

Smokes, smudges, and burning of pyrethum (insecticidal and repellent) are useful repellent measures for outdoors. Chemicals that have been used as repellents against pests of livestock include low concentrations of pyrethrums, butoxypolypropylene glycol, and dibutyl succinate.

In recent years there has been interest in the development of a systemic insect repellent (i.e., one that would be taken orally and following ingestion appear in the skin or skin secretions). Ideally such a systemic would be repellent to a wide variety of arthropods, of low toxicity, and effective for a long period of time (12–24 hours or more after ingestion; Sherman, 1966). There have been repeated failures in the search for a workable systemic repellent, but research is still being conducted along these lines (N.A.S., 1969).

For recent and extensive information on insect repellents, see Metcalf and Metcalf (1975) and Shorey and McKelvey (1977).

Attractants.   Substances that are strongly attractive are known to be involved in a wide variety of insect activities, ranging from feeding, mating, and oviposition to assembling of large groups (see Chapter 8). Attractants (sex pheromones in particular) have great potential in insect control from several stanpoints. For example, they may be of value in conjunction with traps used for sampling insect populations for survey and surveillance purposes, in tracing the movement of marked insects in studies of migration and dispersal, in basic studies of insect behavior, and in insect control. In survey and surveillance, the use of a specific attractant in a trap promotes selective trapping and hence makes the work easier and much more efficient. The design, color, and location of traps are also important factors. The sex pheromone of the gypsy moth, *Porthetria dispar,* is a good example of an attractant that is of value in survey work. Male moths fly into the wind and are attracted to the airborne scent of the nonflying females (Jacobson, 1965). Traps currently used are made of paper and contain a potent synthetic sex pheromone for the gypsy moth, *gyplure*. These traps contain dental-roll wicks impregnated with gyplure and a liner coated with Tanglefoot (a sticky substance in which insects become mired). They are hung on tree limbs. According to Jacobson, "Each year immediately before the male moth flight season (July–August), approximately 50,000 traps are placed in the infested New England area." He describes the potency of gyplure: "a single pound of the attractant, which may be made quite inexpensively, is sufficient to last for more than 300 years if used for survey alone."

There are several applications for attractants in insect control. They can be used to attract insects to traps that alone might be ineffective in reducing some populations. In addition, the traps may contain insecticides, pathogens, or chemosterilants, which further increase the control potential. The use of an attractant with an insecticide avoids the necessity for complete spray coverage and may still be very effective. For example, "The oriental fruit fly, *Dacus dorsalis* Hendel, was eradicated from the Pacific island of Rota by aerial distribution of 5 × 5 cm fiberboard squares saturated with a bait containing an insecticide, naled, and an attractant, methyleugenol" (N.A.S., 1969). Methyleugenol is a strong sex attractant for males.

Attractants may also be used in the perversion of normal insect behavior to accomplish the following:

1. Treatment of weeds and other undesirable plants with insect attractants, feeding stimulants, and oviposition stimulants to create susceptibility to an insect. Materials need to be highly active in order that the treated plants compete with normal host plants in nature.
2. Treatment of host plants to induce greater susceptibility for purposes of luring insects to specifically treated portions of a crop. This method presents greater possibilities than treatment of weeds, because a preference which already exists may be increased with less effort than that needed to create a preferred host from a nonpreferred plant.
3. Use of chemicals to distort sexual activity, diverting the males or females in their search for mates, or confusing their orientation mechanisms—(N.A.S., 1969).

For a given attractant to be useful as described above, the insect behavior with which it is associated must be understood, the attractant must be extracted and chemically identified, and if possible it should be synthesized. Ideally the synthesized attractant or an analogue would be even more attractive to an insect than the original. Another avenue of approach is to test various chemicals for attractiveness. The various sex pheromones that are commercially available in the United States are listed in Caswell (1977). Beroza (1976), Metcalf and Metcalf (1975), Shorey and McKelvey (1977), Rockstein (1978), and Roelofs (1975b, 1978) contain information on the use of chemicals to control insect behavior.

Other Chemical Controls. Additional substances with promise as chemical control agents include *antimetabolites, feeding deterrents,* and *hormones.*

Antimetabolites chemically resemble essential nutrients and interfere with metabolism. They are low in mammalian toxicity and are thus safe to use (e.g., for insect-proofing fabrics). They may be effective against insects that have access only to treated food; however, they have limited value against polyphagous insects (N.A.S., 1969).

Feeding deterrents or antifeeding compounds may be defined as compounds "which will prevent the feeding of pests on a treated material, without necessarily killing or repelling them" (Wright, 1963). Antifeeding compounds have been used for several years in the mothproofing of fabric. However, the use of these compounds in the protection of crops is a fairly new idea. The use of feeding deterrents is still in the experimental stage. They offer considerable specificity since they would affect only the insects that feed on treated plants and would spare the parasites and predators of these pest species. They are also low in mammalian toxicity. Further information on feeding deterrents may be found in Shorey and McKelvey (1977).

Since growth, development, and sexual maturation are largely regulated by hormones (see "Control of Growth and Metamorphosis" in Chapter 5), these substances have potential value in insect control. Ecdysone, juvenile hormone, and various analogues of these compounds have been shown to disrupt the development of an insect if applied at appropriate times and in appropriate doses (Williams, 1970). For example, cyasterone, a substance related to ecdysone, when injected into a diapausing *Cynthia* (moth) pupa in a very low dose (0.2 microgram), stimulates termination of diapause and formation of a normal moth. However, if a high dose (10 micrograms) is injected, the developmental events are accelerated and their sequence disrupted. This results in the premature deposition of cuticle, which literally "locks in" the epidermal tissues before they have completed the developmental changes necessary for survival into the adult stage. Compounds related to ecdysone (i.e., the phytoecdysones) have been found in many different kinds of plants, particularly ferns (see "Adaptations Associated with Interspecific Interactions" in Chapter 9). In addition to lethal effects, there is evidence that certain of the phytoecdysones may function as feeding deterrents. As with ecdysone and its relatives, juvenile hormone can be applied with lethal effects, either by preventing the transformation of the pupa into an adult or by inhibiting the development of eggs. Again, as with ecdysone, several species of plants have been shown to produce compounds that mimic juvenile hormone activity (see Chapter 9). Understanding the chemistry and physiological effects of ecdysone and juvenile hormone and their analogues (i.e., of insect growth regulators or IGRs) may ultimately provide the key to the synthesis of insecticides with extreme specificity. Riddiford and Truman (1978) present recent information on the biochemistry of insect growth regulators. Williams (1967) traces the early developments in the area of the applied use of insect hormones. Menn and Pallos (1975) review the use of IGRs in insect control. See also Menn and Beroza (1972), Riddiford (1972), and Schneiderman (1972).

Mechanical and Physical Control

Mechanical control involves the use of devices that destroy insects directly or act indirectly as barriers, excluders, or collectors. Physical controls include the use of heat, light, electricity, X rays, and so on, to kill insects directly, reduce their reproductive capacity, or to attract them to something that will kill them.

Mechanical Methods. Mechanical methods include the use of simple manual techniques such as handpicking, swatting and crushing, jarring and shaking, and the use of various kinds of barriers, excluders, and traps. Handpicking is effective only when comparatively small numbers of insects are to be dealt with (e.g., the handpicking of bagworms from an ornamental shrub or the handpicking of tomato and tobacco hornworms, extensively used in the past, but now limited to small-garden situations). The fly swatter, the bare hand, or any of a wide variety of implements can be useful against a few insects within a dwelling or on one's person. Jarring and shaking or hand beating of shrubs or trees, especially fruit trees, is sometimes used to remove insects particularly beetles. Sheets, or buckets of kerosene or other material, can be used beneath the plant being shaken to trap the insects that fall into them. Various sorts of collecting devices have been used against insects. These have ranged from a bucket and paddle to horse-drawn "hopper-dozers" and similar machines that were used for grasshopper control in the western United States in the late 1800s (N.A.S., 1969).

Several mechanical means are employed to act as barriers to insect movement. Sticky materials in which insects become hopelessly entangled have been used, for example, in the form of flypaper that traps numerous flying insects. Sticky materials have also been applied in bands about the trunks of trees to protect them from oviposition damage caused by the periodical cicada. Metal collars around tree trunks have also been used for the same purpose and are effective against any nonflying insects that would otherwise attack the branches and foliage. Metal is also used in the construction of shields around the foundation of houses and buildings to prevent attack by subterranean termites. Screens of metal, cloth, fiberglass, or plastic have been used to cover various openings (doors, windows, vents, etc.) to containers and dwellings to allow the passage of air but exclude most insects. Cloth netting (e.g., cheesecloth) is useful in excluding the periodical cicada from young fruit trees and has been used extensively to protect sleeping persons from mosquitoes and other biting insects. Other barrier techniques include the protective packaging of food products, low sheet-metal fences against the flightless Mormon and coulee crickets (Metcalf, Flint, and Metcalf, 1962), plastic sheeting and bags as liners and containers, and the digging of deep furrows around fields being threatened by chinch bug and armyworm attacks (Metcalf, Flint, and Metcalf, 1962).

Traps are used for control, survey, and surveillance purposes. Control traps are usually used in conjunction with some attractive stimulus (e.g., light, food, or sex pheromone) and with some means of killing the insects that enter (e.g., a pesticide or an electrically charged grid). Survey and surveillance traps are used to detect the presence of potential pest species, to evaluate the effectiveness of any control procedures that may have been carried out in a given area, and to monitor levels of various economically or medically important species.

Physical Methods.   Both high and low temperatures have been used to destroy pest insects in a variety of situations. Most insects become inactive at temperatures of about 4°C or below, and many stored products maintained at such temperatures are not damaged, although the insects present would not likely be killed. However, cold can be used to kill drywood termites in furniture in vaults for 4 days at −9°C (N.A.S., 1969). High temperatures are more effective. Few, if any insects can survive prolonged exposure to temperatures much above 60–66°C. High temperatures have been used against insects that infest stored grain, coffee bean, various seeds, citrus fruits, clothing, bedding, furniture, baled fibrous materials, bulbs, soil, and logs. Whether low or high temperatures are used depends in part on the nature of the product to be protected or disinfested.

Insects such as clothes moths and carpet beetles prefer soiled wool garments. Hence laundering or drycleaning not only kills any insects that may be present, but also makes clothing less susceptible to attack.

The burning of crop stubble is highly effective against all life stages of any insects. Electrically charged grids near windows and doors or in association with an attracting light source have been used to kill insects on contact.

Several possible physical means for insect control are still in the experimental stages of development. These include high-frequency electric fields, particularly against stored-grain species; ionizing radiation such as X rays, gamma rays, and high-energy beta particles against insects attacking materials that would not themselves be harmed by the radiation (see also ''Genetic Control'' earlier in this chapter); laser beams, short-duration light used, for example, to induce overwintering species to break diapause and succumb to unfavorable environmental conditions; light reflection from aluminum foil against aphids; and sounds that destroy, repel, attract, or confuse insects.

Regulatory Control

Regulatory control involves the enactment and enforcement of quarantines. Quarantines are designed to prevent the entry of potential pest species, to confine them to as small an area as practicable once introduced, or to prevent them from being exported to other countries. Even if quarantine measures only retard the spread of a given species, the money saved may very well justify the cost. In the definition of regulatory control the phrase ''potential pest'' is especially significant. It should be borne in mind that because a given species is not a major pest in its native land does not mean that it will not be one in another region where few or none of its natural enemies exist.

Prior to the advent of human means of transport, insect dispersal consisted entirely of natural means (migration, etc.). However, when man developed methods of covering long distances, a new means of dispersal became available. The environmental changes that man has brought about have also profoundly influenced the dispersal and distribution of insects. For example, the boll weevil and harlequin bug have extended their ranges from Mexico into the United States because of the extensive irrigation of large areas of the desert that was once an effective natural barrier (Swain, 1952).

Examples of the introduction of insect pests directly attributable to human transport are numerous. A particularly interesting example is the way in which the gypsy moth, an important forest pest, was introduced into the United States. In 1869 an amateur entomologist who was studying silkworms acquired living specimens of the gypsy moth and brought them to Massachusetts (Swan, 1964). Since that time the gypsy moth has spread to several other states in the north-eastern United States and poses a constant threat to other areas of the country where susceptible trees are present.

Apparently the first regulatory control legislation was passed in Germany in 1873. It was intended to prohibit the entry into that country of any materials that might harbor the grape phylloxera, *Phylloxera vitifoliae,* from America (N.A.S., 1969). In the United States, although there was earlier regulatory legislation both at the state and national levels, the first major and effective legislation was passed in 1905. This was the Federal Insect Pest Act, which provided for the regulation of importation and interstate movement of potentially injurious insects. The Plant Quarantine Act of 1912 supplemented and extended earlier legislation and gave the Secretary of Agriculture the authority to enforce laws designed to protect the agriculture of the United States from insect pests and plant diseases by the regulation of importation and interstate movement of potential carrier materials. Additional pertinent legislation has included the Postal Terminal Inspection Act of 1915 and subsequent amendments and several more recent legislative actions, both at state and national levels. Under the authority provided by these acts, trained inspectors at key locations are able to examine materials as they cross the international boundaries into the United States or cross boundaries of regions within the United States that are under quarantine for one or more plant diseases or insect pests. These inspectors have the power to prevent movement of infested or infected materials across these boundaries or to render these materials "safe" by appropriate treatment whether it be fumigation, application of liquid insecticides, dry heat, or otherwise. Also the regulatory control legislations of other countries are respected, and potentially pestiferous materials that are to be moved from the United States require inspection and export certification.

There are currently several quarantines being enforced in the United States and in the past there have been many effective as well as no-so-effective quarantines. The Japanese beetle has been under quarantine regulation since 1919, and as a result many areas of this country are uninfested although they provide favorable environments for this insect (N.A.S., 1969). Current and past control and quarantine programs have confined the gypsy moth to regions of the northeastern United States. Past domestic quarantines, in combination with other kinds of control, have been in part responsible for the virtual elimination of several pest species of insects and ticks—the cattle tick, *Boophilus annulatus;* red tick, *Rhipicephalus evertsi,* from Florida, parlatoria date scale, *Parlatoria blanchardii,* from Arizona and California; Mediterranean fruit fly, *Ceratitis capitata,,* from Florida and Texas; and many others (N.A.S., 1969). Quarantines against the European corn borer, the satin moth, the Asiatic garden beetle, and others were terminated because these insects became widespread in spite of the quarantine and other control measures. In addition, the face fly, *Musca autumnalis,* and the cereal leaf beetle, *Oulema melanopus,* have recently been introduced into the United States.

Insect Pest Management

Prior to the 1940s pest insects were controlled by means of a variety of cultural, mechanical, and biological methods as well as by the use of inorganic and botanical insecticides. During the early 1940s disease-carrying lice, mosquitoes, and other flies were wreaking havoc on military forces and civilian populations in Europe. Spurred by the need for quick, effective control of these insects, an intensive research program was initiated to screen various compounds for insecticidal action (Cushing, 1957). As a result of this program the chlorinated hydrocarbon DDT was found to be a strong insecticide and proved invaluable in quelling the disease vectors mentioned earlier. In addition to its high toxicity to arthropods, DDT was (unlike most other insecticides) highly residual; once applied, it remained stable, and hence effective, for long periods of time. Along with its highly desirable insecticidal properties, DDT turned out to have a very low acute toxicity for laboratory animals and humans. Following the war, DDT proved as effective against numerous agricultural pests as against public health pests. The use of DDT spread rapidly, and many additional synthetic organic insecticides were discovered. These compounds plus DDT made possible tremendous increases in yields of agricultural crops and the development and execution of extensive programs aimed at major arthropod vectors of disease organisms. The notion that entire pest species and the diseases they vectored could be eradicated developed (e.g., "*Aedes aegypti* Eradication Program" and "Malaria Eradication Program"). Eradication programs were impressively successful for some time. In fact, out of 149 countries originally classified as malarious, malaria has been eradicated from 37 and almost eradicated in 16 others. This means that about 40% of the population originally at risk from malaria became free of risk (Luck et al., 1977).

With the success of synthetic organic insecticides, other control measures were deemed less important or even unnecessary and in many instances completely eliminated. Regularly scheduled prophylactic spraying, regardless of actual pest insect levels, became commonplace. Unfortunately, serious problems have arisen from the long-standing overdependence on insecticides.

One of these problems has been the development of resistance in a large number of species. The World Health Organization (Hoskins, 1963) defines the phenomenon of insecticide resistance as "the development of an ability in a strain of insects to tolerate doses of toxicants which would prove lethal to a majority of individuals in a normal population of the same species." In more practical terms, insecticide resistance is "the failure of the customary programs to give practical control, usually even when the dosage was raised and/or frequency of application increased" (Hoskins, 1963). The failure of a given dosage, which previously was effective in control, is due to selection of the individuals in a population that possess the ability to survive this dosage. The traits that enable certain individuals of this normal population to be resistant were present before the introduction of the insecticide; the application of the insecticide gives already resistant individuals a selective (reproductive) advantage relative to the nonresistant ones. Georghiou (1972) and Georghiou and Taylor (1977) discuss the evolution of pesticide resistance.

In the most common situation the resistant insect possesses a detoxification mechanism that enables it to render a particular insecticide or group of insecticides nontoxic. For example, resistant house flies, *Musca domestica* (Diptera, Muscidae), produce an enzyme, DDT dehydrochlorinase, that catalyzes the breakdown of DDT to HCl and nontoxic DDE (N.A.S., 1969). *Culex tarsalis*, a mosquito (Diptera, Culicidae), is able to detoxify the organophosphate insecticide malathion by means of a carboxyesterase enzyme. Strains of some species have been observed to possess a "behavioral" resistance to certain pesticides. According to Hoskins (1963),

> The chief example is a hyperirritability which causes insects to leave a treated surface before picking up a lethal dose. Thus female *Anopheles albimanus* mosquitoes were observed to survive DDT house spraying in certain districts of Panama in much greater numbers than in former years. When confined over treated paper they quickly left the paper and showed signs of restlessness to a greater degree than did mosquitoes from other districts.

Other resistant individuals have been discovered to have cuticles or gut walls that allow much slower penetration of toxicant than the nonresistant individuals. For example, certain strains of German cockroaches (*Blatella germanica*, Blattaria) are resistant to the carbamate Sevin as a result of the slow penetration of the insecticide through the integument. This has been called structural resistance (Hoskins, 1963). Dauterman and Hodgson (1978) and Perry and Agosin (1975) should be consulted for recent information on mechanisms of insecticide resistance.

The first known case of insecticide resistance in the United States was in 1908, when it was found that strains of San Jose scale, *Aspidiotus perniciosus* (Homoptera, Coccoidea), were resistant to lime-sulfur sprays (N.A.S., 1969). In 1946, DDT-resistant house flies were first discovered in Sweden. Prior to that time only 9 species of insects and ticks were known to have developed insecticide resistance (Brown, 1968). Resistance in these cases was, of course, to inorganic and botanical insecticides. Since 1946 insecticide resistance has developed in strains of 364 species of insects and acarines (mites and ticks). Of these, 225 are of public health and veterinary importance, and 139 are pests of stored products or field or forest crops (Georghiou and Taylor, 1977). There has also been a significant increase in the geographic distribution of resistant strains as well as an increase in strains which are resistant to more than one kind of insecticide. Pal (1977) discusses insecticide resistance in vectors of human disease and Vaughn and Leon (1977) consider resistance problems associated with major agricultural crops.

In response to the occurrence of resistance in a given pest species, the dosages of chemical applied have usually been increased (with consequent increases in production costs). When even large doses of a given chemical have failed, alternative insecticides have been substituted. This has worked for a time, but development of resistance to the alternative insecticides has been common.

In addition to the development of resistance, the intensive use of synthetic organic insecticides has tended to kill the natural enemies (especially parasitoids and predators) of pest species, removing or reducing these elements of natural control. This has often allowed the emergence of new pest species, which had previously been held in check by natural enemies, that also became resistant to the insecticide(s) being used.

As the sequence of complete dependence on chemical control, development of resistance in primary pests, and elevation of secondary pests to primary pest levels proceeds, the cost of production of a given crop continually escalates. There are instances of the complete or near complete collapse of a whole agricultural industry, for example, the cotton industry in northeastern Mexico (Luck et al., 1977). Thus, in a number of cases overreliance on chemical control has literally increased the severity of the problem it was intended to ameliorate.

It has become increasingly apparent that chemical control with synthetic organic insecticides is not the panacea it once promised to be. Nevertheless, it is still very important. For instance, there appear to be no immediately available alternatives to chemicals in combating malaria (Luck et al., 1977). Even DDT, which was banned in the United States in 1972, is authorized for use in certain emergency situations, such as the control of rabid bats and control of rodent fleas where plague is endemic (Whittemore, 1977). However, the modern approach is *insect pest management*(IPM).

In the context of insect pest management, problem insects are viewed in an ecological/economic frame of reference. Of prime importance is a thorough understanding of the life system (see Chapter 9) of a pest, including population dynamics and natural enemies. Of equal importance is an understanding of the economic aspects of the situation, especially the level of pest population that actually does economically significant damage (*economic injury level*) and the level of pest population at which control measures should be applied to prevent the pest from reaching the economic injury level (*economic threshold*). The control measures to be applied in a given situation are chosen from *all* available methods (biological, cultural, chemical, etc.), the objective being to keep pests below the economic threshold while avoiding development of resistance, destruction of beneficial nontarget organisms, environmental pollution, and so on.

A good example of the value of the pest management approach is the history of control of cotton pests in Canete Valley, Peru (N.A.S., 1969). Prior to the development of synthetic organic insecticides, the cotton growers of this valley had relied entirely on inorganic insecticides and hand picking of insects. During the late 1940s, they shifted entirely to the use of synthetic organic insecticides. Although these produced good results at first, resistance began to develop, and by 1956 nearly one half the cotton crop was destroyed by insects that survived the chemical treatments. In 1957 an integrated system of pest management was instituted which consisted of the following measures.

1. Reduction in the planting of ratoon cotton (crop produced from plants cut back and allowed to produce new growth from the crown) to less than 25% of the total area.
2. Preparation of soil without irrigation to obtain increased destruction of *Heliothis virescens* pupae.
3. Repopulation of predators and parasites by introduction from other valleys or foreign countries, or by artificial rearing for release.
4. Establishment of mandatory planting and crop-residue destruction dates.
5. Adoption of recommended irrigation schedules.
6. Adherence to the various cultural methods considered to be sound agronomic practice for production of uniformly early crops.
7. Limiting insecticides used to arsenical and botanical materials only, except under very special circumstances, which allowed application of systemic insecticides at dosage rates one fourth to one half of that recommended by the manufacturer to control *Aphis gossypii*.—(N.A.S., 1969)

This program has been highly successful and yields of cotton have increased substantially as a result.

Further information on the philosophy and methods of insect pest management may be found in Apple and Smith (1976), Geier (1966), Klassen (1975), Levins and Wilson (1980), Luckman and Metcalf (1975), McNew (1972), Newsom (1975), Smith and Pimentel (1978), van den Bosch (1978), Watson, Moore, and Ware (1976), and Welch and Croft (1980).

Beneficial Insects

Although many insects are pests to man and often cause serious problems, the vast majority are not pests and in the overall scheme of nature must be considered to be very important animals. This section considers several ways in which insects are directly or indirectly beneficial.

Insect Products

Honey. Honey is a highly nutritive liquid material prepared from flower nectar by several species of bees. It is a sticky, viscous material that ranges from nearly colorless to dark amber. It is composed mainly of water and several sugars (levulose and dextrose in particular) with small amounts of various other substances, including fatty acids, proteins, vitamins, and minerals. The major insect producer of honey is the honey bee, *Apis mellifera*, which has been domesticated and is maintained in artificial hive containers throughout the world. Maintenance of honey bees for the purpose of harvesting honey, wax, and other products is called *apiculture* or *beekeeping*.

Honey has been sought out and collected by humans for thousands of years and has found use mainly as a food material. Meads, beverages similar to wines, are prepared from honey and were possibly among man's earliest alcoholic drinks (White, 1963). Honey is also widely used in the preparation of baked goods, candies, and ice cream. World production of honey totalled 850,569 metric tons in 1972 (Atkins, 1978).

Information on beekeeping and bee products may be found in Adams (1972), Butler (1971), Crane (1963), Grout (1963), More (1976), and Morse (1975). Townsend and Crane (1973) trace the history of apiculture. Morse (1978) contains information on the pests, predators, and diseases of honey bees.

Beeswax. Beeswax is a yellowish white solid waxy material that is secreted by specialized epidermal glands between the abdominal sternites of honey bees. It is used to construct hexagonally cross-sectioned cells (honeycomb) in which bees store honey or rear larvae. Beeswax has long been used for a variety of purposes and was probably the major wax material of ancient times. It was used by man as early as the sixth century. In the past seventy years or so, several other wax or waxlike substances have come into prominence, causing the demand for beeswax to decrease somewhat. Nonetheless, it is still widely used in many cosmetics, nearly smokeless church candles, various pharmaceuticals, some polishes, dental wax, wax museum figures, and several other manufactured materials. One of the most extensive uses to which beeswax is put is in the preparation of comb foundation, which is affixed to the frames of a commercial beehive. This foundation serves to induce the bees to construct honeycomb in the frames, which in turn makes the hive much easier to manage.

Silk. Silk is a product of specialized glands of some lepidopteran larvae, which use it in the construction of pupal cocoons. It is composed of two proteins, fibroin and sericin. Approximately 4000 years ago in China it was discovered that by boiling a cocoon, the filament of silk used to construct it became loose and could be unwound, fortunately in a single strand (Clausen, 1954). The thread produced by winding several of these filaments together could be woven into a soft, lustrous, easily dyed fabric. From that time on the silk industry (*sericulture*) developed and flourished in China, and its methods were closely guarded secrets. Finally, in about A.D. 550 silkworm eggs and the secret techniques for preparing silk were smuggled to Constantinople, and the silk industry began in Europe (Clausen, 1954).

Several species of silkworms have been cultured for the commercial production of silk, but *Bombyx mori* (Bombycidae) is and has been the most widely used. It has become, through generation after generation of careful selection, a totally domesticated insect. Larvae feed on mulberry leaves, and close, continuous attention and great care are required to rear them and harvest silk. Many of the steps in silk production require hours of tedious hand labor, a factor that has allowed the silk industry to flourish in the Orient, particularly China and Japan, where such labor has been fairly cheap. However, silk has been produced in more than twenty countries throughout the world, including the United States (Yokoyama, 1963). In addition to its use in producing fabric, the gummy material that is secreted by the silk glands and drawn out into the filaments by the silkworm is dissected directly from the glands and artificially drawn into thin threads that still find use in surgical stitching and have in the past been used as fishing leaders (Clausen, 1954). In recent years cheaper and in many ways superior fibers have been prepared synthetically and have threatened to destroy the silk industry.

Yokoyama (1973) traces the history of sericulture. Horie and Wantanabe (1980) consider recent advances in sericulture, including topics like nutrition and metabolism, silk protein synthesis, biochemical aspects of virus infections (see Chapter 9), and defense mechanisms against disease.

Lac. Lac is the crude resinous material from which commercial shellac is prepared. It is a glandular secretion of the scale insect *Laccifer lacca* (Homoptera, Coccoidea), which inhabits trees in India and Burma. Tiny immature "crawlers" suck sap from branches of these trees and eventually cover themselves with lac, which serves as a protective shield. Thousands of crawlers feed in close proximity to one another and the lac comes to cover branches almost entirely. When they mature, the females remain wingless and sedentary while winged males emerge and fertilize the females through tiny openings in the lac. When the fertilized eggs hatch, the young crawlers migrate to uninfested areas and commence to feed, beginning the cycle anew. Lac production is encouraged by removing twigs covered with mature females and tieing them to uninfested trees. These twigs are referred to as *brood lac,* and crawlers emerge from them by the thousands and infest their new host tree (Clausen, 1954). Harvesting of lac is done by removing branches covered with lac *(stick lac)* and grinding them. The resultant *seed lac* is washed, dried and bleached in the sun, and, after drying, heated in cloth bags over open charcoal fires. As the lac melts, it is squeezed onto the floor and quickly pressed and stretched into thin sheets, which are then flaked. Shellac is prepared by dissolving this *flake lac* (Metcalf, Flint, and Metcalf, 1962). Lac is a basic ingredient of

> stiffening agents in the toes and soles of shoes, and in felt, fur, and composition hats; shoe polishes; artificial fruits and flowers; lithographic ink; electrical insulation; protective coverings for wood, paper, fabric, wax emulsions, wood fillers, sealing wax and buttons; glazes on confections; coffee bean burnishing; paints; cements and adhesives, shellac varnishes and moldings, photographic products, phonographic records; playing card finishes; dental plates, pyrotechnics; foundry work and hair dyes—(Clausen, 1954)

Glover (1937) describes lac cultivation and production in India.

As with beeswax and silk, lac has been pretty much replaced by synthetic materials. However, with petroleum shortages, this tendency may well be reversed.

Cochineal. Cochineal is also a scale insect product, being secreted by *Coccus cacti*, which lives and feeds on the prickly pear. It is prepared from dried, ground bodies of the insects and in this form is a red pigment that has been used widely, particularly in the past, as a permanent dye, for example, in rouge, for cake decoration, and as a coloring agent for beverages and medicines (Metcalf, Flint, and Metcalf, 1962). The major producers of cochineal are in Mexico, Honduras, and the Canary Islands. Approximately 70,000 insects are required to produce a pound of dye (Bishopp, 1952). In the past dyes have also been prepared from other species of scale insects.

Insect galls. Several kinds of insect galls have been the source of various pigments used for dyeing wool, skin, hair, leather, and so on, and for the production of permanent inks. Tannic acid, a substance widely used in tanning, dyeing, and preparation of inks, is also derived from insect galls. See Askew (1971) and Felt (1940) for information on gall-making insects.

Use of Insects in Medicine

Aside from the many purported medicinal uses recorded in ancient literature and folklore, insects or insect secretions have proved to be some value in the treatment of certain human ailments.

Maggot Therapy. During the last three or four centuries the observation was made on a number of occasions that wounded soldiers on the battlefield fared better if their wounds became infested with the larvae of certain flies (Leclercq, 1969). These larvae appeared to devour necrotic tissue and inhibit infection. In the early 1930s, the use of a strain of *Lucilia sericata* larvae, which attack only necrotic tissues, was advocated for the treatment of osteomyelitis and chronic open sores. Subsequently it was discovered that allantoin, a nitrogenous waste excreted by the maggots, produced the same inhibition of infection as the whole insects. This substance was then produced commercially. In recent years, the antibiotics have replaced maggot therapy and the use of allantoin.

Cantharidin. Cantharidin is derived from the bodies of blister beetles (Coleoptera, Meloidae). The best-known species is *Lytta vesicatoria,* the Spanish fly, found throughout Europe. When taken internally, cantharidin acts as a strong urogenital irritant. For this reason it has been used as an aphrodisiac, e.g., in cattle breeding, and for the treatment of certain urogenital diseases. It is a very dangerous substance and is no longer used in humans.

Miscellaneous. Several other insect substances have been used as medicaments or have promise for such use. Among these are bee venom, which has been used in the treatment of certain forms of arthritis, myalgias, neuralgias and other ailments (Leclercq, 1969; Metcalf, Flint, and Metcalf, 1962). Cochineal is purported to relieve pain associated with whooping cough and neuralgia (Metcalf, Flint, and Metcalf, 1962). The Aleppo gall or gallnut formed on several species of oaks in Asia and Europe has been used as a "tonic, astringent, and antidote for certain poisons" (Bishopp, 1952). According to Leclercq (1969), there is evidence that antibiotic substances are present in the hemolymph of, or are secreted by, certain insect species. An ancient technique still applied by primitive peoples in various parts of the world is the use of insects such as ants and carabid beetles, which have well-developed mandibles, in the suturing of a wound (Leclerq, 1969). The insect is induced to bite a wound so that the two edges are brought together and is then decapitated; the head remains in the clamped position and provides a suture.

Insects in Biological Research

Insects make ideal organisms for fundamental biological research. They are usually easy to collect and rear in large numbers, are small in size and can be easily manipulated within a small space, have comparatively short generation times, and as a group display a diversity in form and function nearly unapproached by other groups of animals. *Drosphila melanogaster* has been widely utilized as a research species in genetics. A large amount of experimental-developmental endocrine work has been carried out on the bug *Rhodnius prolixus* (Hemiptera, Reduviidae). Among the many other genera that appear often in biological literature are Tribolium (Coleoptera, Tenebrionidae), *Calliphora* (Diptera, Calliphoridae), *Musca* (Diptera, Musicidae), *Periplaneta* (Blattaria), and *Apis* (Hymenoptera, Apidae), *Aedes* (Diptera, Culicidae). See Chapter 1 for more information on the significance of insects in biological research.

No student should study entomology (or, for that matter, general biology) without reading Dethier's delightful, enlightening book *To Know a Fly* (1962).

Pollination by Insects

In addition to countless wild flowering plants (see Chapter 9), domesticated plants pollinated by insects include vegetables such as tomatoes, peas, beans, and onions; most fruit crops; and field crops such as alfalfa, red and white clover, and tobacco. Insects involved in the pollination of flowering plants number in the thousands of species and are found mainly among the Hymenoptera, Diptera, Lepidoptera, and Coleoptera, although members of some smaller orders (e.g., Thysanoptera) may also be involved. Several Hymenoptera, in particular the bees (Apoidea), are the most important pollinators in commercial crops. In the past, wild, native pollinating insects were sufficient in number to accomplish the pollination of our food crops, but with the intensive land use, clean cultivation, and excessive application of insecticides characteristic of modern agriculture, the populations of many of these species have been so reduced that they can no longer maintain an adequate level of pollination. According to Bohart (1952), "An estimated 80 percent of the insect pollination of our commercial crops is performed by honey bees." Thus honey bees are no doubt our most valuable pollinators and have the advantage that they can be easily cultured and moved about at will. In fact, "It has been claimed that the value of bees in the pollination of crops is 10 to 20 times the value of the honey and wax which they produce'' (Lovell, 1963). Pimentel (1975) provides a graphic description of the pollinating abilities of honey bees.

A single honey bee may visit and pollinate 1,000 blossoms in a single day. In New York state, which has about 3 million domestic honeybee colonies, each with about 10,000 worker bees, honeybees could visit 30 trillion blossoms in a day. Wild bees pollinate a number at least equal to that. Thus on a bright, sunny day 60 trillion blossoms may be pollinated by the bees in New York, a task impossible for man to accomplish today.

Additional information on pollinating insects can be found in Bohart (1972), Free (1970), Martin and McGregor (1973), McGregor (1976) and Nørgaard Holm (1976).

Insects Consumed and as Consumers

As Human Food. Insects actually have a high nutritional value, being quite rich in protein and lipids, and may therefore be a very important supplement to the diets of otherwise vegetarian peoples (Leclercq, 1969; Taylor, 1975). In Africa today, meals are sometimes supplemented with insects as a source of protein, which may partially stave off the protein-deficiency disease of children, kwashiorkor (Leclercq, 1969). The Australian aborigines realize that grubs are an essential part of their diet, even though they do not realize that they constitute their sole source of protein (Clausen, 1954). In less acute food situations, various insects have for centuries been valued as delicacies and have even become articles of food commerce. For example, ''gusanos''—fried caterpillars, earthworms and beetle grubs found in agave plants—are exported from Mexico and can be purchased in gourmet shops in the United States (Clausen, 1954). Other insects that have been or are still prepared in some way or eaten raw by various peoples in the world include termites, silkworm pupae, migratory locusts, ants, caterpillars, diving beetles, cicadas, and eggs of water boatmen.

See Taylor (1975) for a discussion of insects as human food and Taylor and Carter (1976) for some recipes.

As Food for Wildlife. Aside from being eaten by a wide variety of arthropods, including their own kind, insects are major food organisms for many kinds of wildlife, in particular birds and fishes (see Chapter 9). Insects probably comprise more than half the food consumed by the more than 1,400 species and subspecies of birds in North America (Swan, 1964). Metcalf, Flint, and Metcalf (1962) cite the work of Forbes, who concluded that two fifths of the food of adult freshwater fishes is insects, the most important of which are bloodworms, mayfly larvae, and caddisfly larvae.

Insects as Consumers. The magnitude of the roles played by insects as plant eaters, scavengers, predators, and parasites in the total picture of nature is inestimable. They are a part of nearly every terrestrial and freshwater food web. Our understanding, all too incomplete at times, has enabled us in some instances to manipulate populations of insects to control other insects and, even, undesirable weeds (e.g., klamath weed and prickly pear). The food habits of insects are discussed in Chapter 8.

Insects, Esthetics, Philosophy, and Blatant Anthropomorphism

Beyond the rigors of science, one cannot help but be in awe of insects as organisms worthy of admiration and respect. Along with pathogenic microbes, they are formidable adversaries. But, on the other hand, they often appeal to our emotions in a positive way. Many display great beauty and grace (see, for example, the marvelous photographs in Sandved and Brewer, 1976, and in Dalton, 1975). Their beauty has been celebrated in art forms ranging from oil paintings to jewelry to postage stamps. Some have even been elevated to gods (e.g., the dung-rolling scarab beetles in ancient Egypt).

If not inspiring our esthetic senses, some insects entertain and astound us with the bizarre: devouring a mate during copulation (praying mantids), having strangely placed penises (dragonflies and damselflies), flitting around releasing disagreeable odors when nervous about being eaten, flying against burning light bulbs or into fires, rolling sundry pieces of fecal material into little balls, or using a "jet propulsion" system to escape from predators the way dragonfly larvae do.

Further, the organizational skills of insects are without equal. Intricately run termite, ant, bee, and wasp societies put our ever-present bureaucrats to shame.

Public awareness of the fascinating world of insects has increased in the past few years. Insects are becoming popular attractions at zoos and museums, for example, the Cincinnati zoo and the National Museum of Natural History in Washington, D.C, which has an exhibit of living insects. Not too long ago, *Time* Magazine (July 21, 1976), which usually reserves its cover for notable human beings of the moment, featured a frontal view of a cicada killer wasp.

Insects have been "known" to be philosophers, like the cockroach "Archie" who befriended a journalist in the 1930s. This was fortunate, because that journalist, Don Marquis, conveyed Archie's outlook on life to us (see *Archie and Mehitabel*, 1960, and *Archie's Life of Mehitabel*, 1966):

insects have their own point of view about civilization a man thinks he amounts to a great deal but to a flea or a mosquito a human being is merely something good to eat—Marquis (1960)

Archie typed his messages to Mr. Marquis by butting the keys, but was unable to reach the shift key and therefore typed everything in lower case.

Selected References

GENERAL

Anderson and Kaya (1976); Baker (1972); Davidson and Lyon (1979); Davidson and Peairs (1966); Ebeling (1975); Fletcher (1978); Frankie and

Beneficial Insects Koehler (1976, 1978); Harwood and James (1979); Johnson and Lyon (1979); National Academy of Science (1969, 1972); Pfadt (1978); Pyenson and Barke (1977); Roberts (1978); Sweetman (1965); van Emden (1974); Watson et al. (1976); Wilson et al. (1977).

INSECTS AND HUMAN WELFARE
   Cushing (1957); Jones (1973); McKelvey (1975); Philip and Rozeboom (1973); Pimentel (1975); Ritchie (1979); Schwerdtfeger (1973); Southwood (1977a, b).

INSECTS AND PLANT DISEASE
   Carter (1973); Pyenson (1977); Sylvester (1980).

MEDICAL/VETERINARY ENTOMOLOGY
   Busvine (1966, 1975); Greenberg (1971, 1973); Harwood and James (1979); Horsfall (1962); Pfadt (1978); Smith (1973); Snow (1974); Theiler and Downs (1973).

BIOLOGICAL CONTROL
   General: Coppel and Mertins (1977); Debach (1974); van den Bosch and Messenger (1973); Huffaker (1971); Huffaker and Messenger (1976); Huffaker et al. (1977); Swan (1964); Sweetman (1936 a,b).
   Parasitoids and Predators: van den Bosch (1975); Ridgway and Vinson (1977); Stehr (1975).
   Pathogens: Briggs (1975); Bulla (1973); Burges and Hussey (1971); Cameron (1973); Falcon (1976); Ferron (1978); Ignoffo (1975); Maddox (1975); Maramorosch (1977); N.Y. Academy of Science (1972); Poinar and Thomas (1978); Steinhaus (1963); Tinsley (1977); Weiser (1969).

GENETIC CONTROL
   Baumhover (1966); Bŏrkovec (1975); Davidson (1974); Foster et al. (1972); Hoy and McKelvey (1979); Huffaker and Messenger (1976); Knipling (1972); Metcalf and Metcalf (1975); Pal and Whitten (1974); Richardson (1978); Smith and von Borstell (1972).

PLANT RESISTANCE TO INSECTS
   Beck (1965); de Wilde and Schoonhoven (1969); Gallun et al. (1975); Hanover (1975); Kogan(1975); Maxwell and Jennings (1979); Maxwell et al. (1972); Painter (1958); Pathak (1975); Sondheimer and Simeone (1970); van Emden (1966).

CHEMICAL CONTROL
   General: Cremyln (1978); Fronk (1978); Metcalf (1975); Sheets and Pimentel (1979); Ware (1975, 1978); Watson and Brown (1977).
   Modes of Action: O'Brien (1967, 1978); Matsumura (1975); Wilkinson (1976).
   Pesticide Safety: American Public Health Association (1967); Caswell (1977).
   Pesticides in the Environment: Brown (1978); Haque and Freed (1975); Johansen (1977); Khan (1977); McEwen and Stephenson (1979); Miller and Berg (1969); Perring and Mellanby (1977); Rudd (1964); van den Bosch (1978); White-Stevens (1977); Woodwell et al. (1971).
   Sex Pheromones: Beroza (1976); Caswell (1977); Metcalf and Metcalf (1975); Shorey and McKelvey (1977); Rockstein (1978); Roelofs (1975b, 1978) (See also Chapter 8).
   Insect Growth Regulators: Menn and Beroza (1972); Menn and Pallos (1975); Riddiford (1972); Riddiford and Truman (1978); Schneiderman (1972); Williams (1967).

Insect Resistance to Insecticides: Brown (1968); Dauterman and Hodgson (1978); Georghiou (1972); Georghiou and Taylor (1977); Hoskins (1963); Pal (1977); Perry and Agosin (1975); Vaughn and Leon (1977).

INSECT PEST MANAGEMENT

Apple and Smith (1976); Geier (1966); Klassen (1975); Levins and Wilson (1980); Luck et al. (1977); Luckman and Metcalf (1975); McNew (1972); Newsom (1975); Smith and Pimentel (1978); van den Bosch (1978); Watson et al. (1976); Welch and Croft (1980).

BENEFICIAL INSECTS

General: Bishopp (1952); Clausen (1954); Metcalf et al. (1962).

Honey Bees and Products: Adams (1972); Butler (1971); Crane (1963); Grout (1963); More (1976); Morse (1975, 1978); Townsend and Crane (1973).

Silk: Horie and Wantanabe (1980); Yokoyama (1973).

Lac: Glover (1937).

Insects in Medicine: Leclercq (1969); Taylor (1975).

Insects as Pollinators: Bohart (1972); Free (1970); Martin and McGregor (1973); McGregor (1976); Nørgaard Holm (1966).

Insects as Human Food: Clausen (1954); Leclercq (1969); Taylor (1975); Taylor and Carter (1976).

References Cited

Adams, John F. 1972. *Beekeeping—The Gentle Craft.* New York, Avon Books.

Adiyodi, K. G., and R. G. Adiyodi. 1974. Comparative physiology of reproduction in arthropods. Adv. Comp. Physiol. Biochem., 5:37–107.

Agrell, I. P. S., and A. M. Lundquist. 1973. Physiological and biochemical changes during insect development. In *The Physiology of Insecta,* 2nd ed., Vol. I, pp. 159–247, M. Rockstein, ed. New York, Academic Press.

Akre, R. D., and H. G. Davis. 1978. Biology and pest status of venomous wasps. Ann. Rev. Entomol., 23:215–38.

Alcock, J. 1975. *Animal Behavior.* Sunderland, MA, Sinauer Associates.

Alexander, R. D. 1967. Acoustical communication in arthropods. Ann. Rev. Entomol., 12:495–596.

————. 1975. Natural selection and specialized chorusing behavior. In *Insects, Science and Society,* pp. 35–77, D. Pimentel, ed. New York, Academic Press.

———— and D. J. Borror. 1956. The songs of insects, a phonograph record. Ithaca, NY, Cornell University Press.

———— and W. L. Brown, Jr. 1963. Mating behavior and the origin of insect wings. University of Michigan, Occasional Papers Mus. Zool. No. 628.

———— and T. E. Moore. 1962. The evolutionary relationships of 17-year and 13-year cicadas, and three new species (Homoptera, Cicadidae, Magicicada). University of Michigan, Misc. Publ. Mus. Zool. No. 121.

Alloway, T. M. 1972. Learning and memory in insects. Ann. Rev. Entomol., 17:43–56.

————. 1973. Learning in insects except Apoidea. In *Invertebrate Learning,* Vol. 2, pp.131–71, W. C. Corning, J. A. Dyal, and A. O. D. Willows, eds. New York, Plenum.

American Public Health Association. 1967. *Safe Use of Pesticides,* prepared by the Subcommittee on Pesticides. New York, American Public Health Association, Inc. 92 pp.

Anderson, D. T. 1973. *Embryology and Phylogeny in Annelids and Arthropods.* New York, Pergamon.

————. 1979. Embryos, fate maps and the phylogeny of Arthropods. In *Arthropod Phylogeny,* pp. 59–105, A. P. Gupta, ed. New York, Van Nostrand Reinhold.

Anderson, J. F. and H. K. Kaya, eds. 1976. *Perspectives in Forest Entomology.* New York, Academic Press.

Anderson, S. O. 1976. Cuticular enzymes and sclerotization in insects. In *The Insect Integument,* pp. 121–44, H. P. Hepburn, ed. New York, Elsevier Scientific.

————. 1979. Biochemistry of insect cuticle. Ann. Rev. Entomol., 24:29–61.

Andrewartha, H. G. 1971. *Introduction to the Study of Animal Populations,* 2nd ed. Chicago, University of Chicago Press.

———— and L. C. Birch. 1954. *The Distribution and Abundance of Animals.* Chicago, University of Chicago Press.

———— and ————. 1973. The history of insect ecology. In *History of Entomology,* pp. 229–66, R. F. Smith, T. E. Mittler, and C. N. Smith, eds. Palo Alto, CA, Annual Reviews.

————, P. M. Miethke, and A. Wells. 1974. Induction of diapause in the pupa of *Phalaenoides glycinae* by a hormone from the subesophageal ganglion. J. Insect Physiol., 20:679–701.

Andrews, C. 1971. *The Lives of Wasps and Bees.* New York, American Elsevier.

Apple, J. L., and R. F. Smith. 1976. *Integrated Pest Management.* New York, Plenum.

Arnett, R. H., Jr. 1970. *Entomological Information Storage and Retrieval.* Baltimore, The Bio-Rand Foundation.

————. 1980. *The Beetles of the United States.* 2nd ed. Marlton, NJ, World Natural History.

References Cited

Arnold, J. W. 1974. The hemocytes of insects. In *The Physiology of Insecta,* 2nd ed., Vol. V, pp. 202–54, M. Rockstein, ed. New York, Academic Press.

Asahina, E. 1966. Freezing and frost resistance in insects. In *Cryobiology*, H. T. Meryman, ed. London, Academic Press.

Ashburner, M. 1970. Function and structure of polytene chromosomes during insect development. Adv. Insect Physiol., 7:1–95.

———. 1972. Puffing patterns in *Drosophila melanogaster* and related species. In *Developmental Studies on Giant Chromosomes,* pp. 101–51, N. Beerman, ed. New York, Springer-Verlag.

Askew, R. R. 1971. *Parasitic Insects.* New York, American Elsevier.

Aspöck, H., and A. Aspöck. 1975. The present state of knowledge on the Raphidioptera of America (Insecta, Neuroptoidea). Pol. Pismo Entomol., 45:537–46.

Atkins, M. D. 1978. *Insects in Perspective.* New York, Macmillan.

———. 1980. *Introduction to Insect Behavior.* New York, Macmillan.

Baccetti, B. 1972. Insect sperm cells. Adv. Insect Physiol., 9:316–84.

Bailey, E. 1975. Biochemistry of insect flight Part 2—fuel supply. In *Insect Biochemistry and Function,* D. J. Candy and B. A. Kilby, eds., New York, Wiley.

Bailey, S. F. 1957. The thrips of California, Part I: suborder Terebrantia. Bull. Calif. Insect Surv., 4:143–220.

Baker, E. W., T. M. Evans, D. J. Gould, W. B. Hull, and H. L. Keegan. 1956. *A Manual of Parasitic Mites of Medical or Economic Importance.* New York, National Pest Control Association.

Baker, H. G., and I. Baker. 1975. Studies of nectar-constitution and pollinator-plant coevolution. In *Animal and Plant Coevolution,* pp. 100–40, L. E. Gilbert and P. H. Raven, eds. Austin, University of Texas Press.

——— and P. D. Hurd, Jr. 1968. Intrafloral ecology. Ann. Rev. Entomol., 13:385–409.

Baker, W. L. 1972. *Eastern Forest Insects.* USDA Misc. Pub. No. 1175. Washington, D.C., U.S. Government Printing Office, 642 pp.

Baldus, K. 1926. Experimentelle Untersuchungen über die Entfernungslokalisation der Libellen (*Aeschna cyanea*). Z. vergl. Physiol., 3:375–505.

Banziger, H. 1971. Bloodsucking moths of Malaya. Fauna 1:5–16.

———. 1975. Skin-piercing bloodsucking moths I: ecological and ethological studies on *Calpe eustrigota* (Lepidoptera, Noctuidae). Acta Trop., 32:125–44.

Barash, D. P. 1977. *Sociobiology and Behavior.* New York, Elsevier North-Holland.

Bartell, R. J. 1977. Behavioral responses of Lepidoptera to pheromones. In *Chemical Control of Insect Behavior,* pp. 201–13, H. H. Shorey and J. J. McKelvey, Jr., eds. New York, Wiley.

Barth, R. 1937. Muskulatur und Bewegungsart der Raupen. Zool. Jahrb., 62:507–66.

Barth, R. H., and L. J. Lester. 1973. Neuro-hormonal control of sexual behavior in insects. Ann. Rev. Entomol., 18:455–72.

Barton Browne, L. 1974. *The Experimental Analysis of Insect Behavior.* New York, Springer-Verlag.

———. 1975. Regulatory mechanisms in insect feeding. Adv. Insect Physiol., 11:1–116.

Bastock, M. 1956. A gene mutation which changes a behavior pattern. Evolution, 10:421–39.

Bates, H. W. 1862. Contributions to the insect fauna of Amazon Valley. Trans. Linn. Soc. (London), 23:495–566.

Batra, L. R. ed. 1979. *Insect-Fungus Symbiosis—Nutrition, Mutualism, and Commensalism.* New York, Wiley (Halstead Press).

Batra, S. W. T., and L. R. Batra. 1967. The fungus gardens of insects. Sci. Amer., 217(5):112–20.

Baumhover, A. H. 1966. Eradication of the screwworm fly. J. Amer. Med. Assoc., 196(3):150–58.

Baust, J. G., and R. E. Morrissey. 1977. Strategies of low temperature adaptation. Proc. XV Int. Congr. Entomol., pp. 173–84.

Bay, E. C. 1967. Mosquito control by fish: a present day appraisal. WHO Chronicle, 21(10):415–23.

Beach, R. 1979. Mosquitoes: biting behavior inhibited by ecdysone. Science, 205:829–31.

References Cited

Beament, J. W. L. 1958. The effect of temperature on the waterproofing mechanism of an insect. J. Exp. Biol., 35:494–519.

Beck, S. D. 1965. Resistance of plants to insects. Ann. Rev. Entomol., 10:207–32.

———. 1968. *Insect Photoperiodism*. New York, Academic Press.

Beermann, W., ed. 1977. *Developmental Studies on Giant Chromosomes*. Berlin, Springer-Verlag. 227 pp.

Behnke, F. L. 1977. *A Natural History of Termites*. New York, Scribner.

Bennet-Clark, H. C. 1962. Active control of the mechanical properties of insect endocuticle. J. Insect Physiol., 8:627–33.

———. 1970. The mechanism and efficiency of sound production in mole crickets. J. Exp. Biol., 52:619–52.

———. 1976. Energy storage in jumping insects. In *The Insect Integument*, pp. 421–43, H. R. Hepburn, ed. New York, Elsevier Scientific.

——— and A. W. Ewing. 1970. The love song of the fruit fly. Sci. Amer., 223:84–92 (July).

——— and E. C. A. Lucey. 1967. The jump of the flea: a study of the energetics and a model of the mechanism. J. Exp. Biol., 47(1):59–76.

Bentley, D. R. 1971. Genetic control of an insect neuronal network. Science, 174:1139–41.

———. 1975. Single gene cricket mutations: effects on behavior, sensilla, sensory neurons, and identified inter-neurons. Science, 187:760–64.

——— and R. Hoy. 1972. Genetic control of the neuronal network generating cricket (*Telogryllus gryllus*) song patterns. Anim. Behav., 20:478–92.

——— and ———. 1975. The neurobiology of cricket song. Sci. Amer., 231:34–44.

Benzer, S. 1973. Genetic dissection of behavior. Sci. Amer., 229:24–37.

Bergerard, J. 1972. Environmental and physiological control of sex determination and differentiation. Ann. Rev. Entomol., 17:57–74.

Bergstrom, J. 1979. Morphology of fossil arthropods as a guide to phylogenetic relationships. In *Arthropod Phylogeny*, pp. 3–56, A. P. Gupta, ed. New York, Van Nostrand Reinhold.

Bernhard, C. G., ed. 1966. *The Functional Organization of the Compound Eye*. Oxford, Pergamon.

Beroza, M., ed. 1970. *Chemicals Controlling Insect Behavior*. New York, Academic Press.

———. 1976. *Pest Management with Insect Sex Attractants and Other Behavior-Controlling Chemicals*. Washington, DC, American Chemical Society.

Berridge, M. J. 1970. A structural analysis of intestinal absorption. In *Insect Ultrastructure*, pp. 135–51, A. C. Neville, ed. Roy. Entomol. Soc. (London), Symp. 5.

Birch, M. C., ed. 1974. *Pheromones*. Amsterdam, North-Holland.

Birukow, G. 1966. Orientation behavior in insects and factors which influence it. In *Insect Behavior*, pp. 2–12, P. T. Haskell, ed. Roy. Entomol. Soc. (London), Symp. 3.

Bishop, J. A., and L. M. Cook. 1975. Moths, melanism and clean air. Sci. Amer., 232:90–99.

Bishopp, F. C. 1952. Insects as helpers. In *Yearbook of Agriculture*, pp. 79–87. Washington, DC, U. S. Department of Agriculture.

Blackwelder, R. E. 1967. *Taxonomy: A Text and Reference Book*. New York, Wiley.

———. and R. H. Arnett, Jr. 1977. *Checklist of the Beetles of Canada, United States, Mexico, Central America, and the West Indies*. Marlton, NJ, World Natural History.

Bland, R. C., and H. E. Jaques. 1978. *How to Know the Insects,* 3rd ed. Dubuque, IA, Wm. C. Brown.

Blatchley, W. S. 1920. *Orthoptera of Northeastern America*. Indianapolis, Nature Publishing Company.

———. 1926. *Heteroptera or True Bugs of Eastern North America, with Especial Reference to the Faunas of Indiana and Florida*. Indianapolis, Nature Publishing Company.

Blest, A. D. 1957. The function of eyespot patterns in the Lepidoptera. Behaviour, 11:209–56.

Blum, M. S. 1969. Alarm pheromones. Ann. Rev. Entomol., 14:57–80.

References Cited

————. 1978. Biochemical defenses of insects. In *Biochemistry of Insects,* pp. 466–513, M. Rockstein, ed. New York, Academic Press.

———— and J. M. Brand. 1972. Social insect pheromones: their chemistry and function. Amer. Zool., 12:5553–76.

Bodenstein, D., ed. 1971. *Milestones in Developmental Physiology of Insects.* New York, Appleton-Century-Crofts.

Bohart, G. E. 1952. Pollination by native insects. In *Yearbook of Agriculture,* pp. 107–21. Washington, DC, U. S. Department of Agriculture.

————. 1972. Management of wild bees for the pollination of crops. Ann. Rev. Entomol., 17:287–312.

Bohart, R. M. 1941. A revision of the Strepsiptera with special reference to the species of North America. Univ. Calif. Pub. Entomol., 7:91–160.

———— and A. S. Menke. 1976. *Sphecid Wasps of the World.* Berkeley, University of California Press.

Bonhag, P. F. 1958. Ovarian structure and vitellogenesis in insects. Ann. Rev. Entomol., 3:136–60.

Borden, J. H. 1977. Behavioral responses of Coleoptera to pheromones, allomones and kairomones. In *Chemical Control of Insect Behavior,* pp. 169–98, H. H. Shorey and J. J. McKelvey, Jr., eds. New York, Wiley.

Bŏrkovec, A. G. 1975. Control of insects by sexual sterilization. In *Insecticides of the Future,* pp. 61–69, M. Jacobson, ed. New York, Marcel Dekker.

Bornemissza, G. F. 1964. Sex attractant of male scorpionflies. Nature, 203:786–87.

Borror, D. J. 1960. *Dictionary of Word Roots and Combining Forms.* Palo Alto, CA, Mayfield Publishing, Company.

————, D. M. DeLong, and C. A. Triplehorn. 1976. *An Introduction to the Study of Insects,* 4th ed. New York, Holt, Rinehart & Winston.

———— and R. E. White. 1970. *A Field Guide to the Insects of America North of Mexico.* Boston, Houghton Mifflin.

Bosch, R. van den. 1978. *The Pesticide Conspiracy.* Garden City, NY, Doubleday.

———— and P. S. Messenger. 1973. *Biological Control.* New York, Intext Educational Publications.

————. 1975. Biological control of insects by predators and parasites. In *Insecticides of the Future,* pp. 5–21, M. Jacobson, ed. New York, Marcel Dekker.

Boudreaux, H. B. 1979. *Arthropod Phylogeny with Special Reference to Insects.* New York, Wiley.

Boughey, A. S. 1971. *Fundamental Ecology.* San Francisco, Intext Educational Publications.

Bowers, W. S., T. O. Ohta, J. S. Cleere, and P. A. Marsella. 1976. Discovery of insect antijuvenile hormones in plants. Science, 193:542–47.

Brady, J. 1974. The physiology of insect circadian rhythms. Adv. Insect Physiol., 10:1–115.

Brammer, J. D., and R. H. White. 1969. Vitamin A deficiency: effect on mosquito eye ultrastructure. Science, 163:821–23.

Breland, O. P., C. D. Eddleman, and J. J. Biesele. 1968. Studies of insect spermatozoa I. Entomol. News, 79(8):197–216.

Briggs, J. D. 1975. *Biological Regulation of Vectors—The Saprophytic and Aerobic Bacteria and Fungi.* DHEW Publication No. (NIH) 77–1180.

Britton, W. E., ed. 1923. *The Hemiptera or Sucking Insects of Connecticut.* Connecticut State Geol. Nat. Hist. Surv. Bull. 34.

Brooks, M. A. 1963. Symbiosis and aposymbiosis in Arthropods. Symp. Soc. Gen. Microbiol., 13:200–31.

Brooks, W. M. 1974. Protozoan infections. In *Insect Diseases,* pp. 237–300, G. E. Cantwell, ed. New York, Marcel Dekker.

Brower, L. P. 1969. Ecological chemistry. Sci. Amer., 220(2):22–29.

————. 1977. Monarch migration. Nat. Hist., 86(6):40–53.

———— and J. V. Z. Brower. 1964. Birds, butterflies and plant poisons: A study in ecological chemistry. Zoologica, 49:137–59.

———— and S. C. Glazier. 1975. Localization of heart poisons in the monarch butterfly. Science, 188:19–25.

———— and W. N. Ryerson, L. L. Coppinger, and S. C. Glazier. 1968. Ecological chemistry and the palatability spectrum. Science, 161:1349–51.

Brown A. W. A. 1968. Insecticide resistance comes of age. Bull. Entomol. Soc. Amer., 14(1):3–9.

References Cited

————. 1978. *The Ecology of Pesticides*. New York, Wiley–Interscience.

Brown, J. L. 1975. *The Evolution of Behavior*. New York, Norton.

Brown, W. L., Jr., T. Eisner, and R. H. Whittaker. 1970. Allomones and kairomones: transpecific chemical messages. BioScience, 20:21–22.

Brues, C. T. 1946. *Insect Dietary*. Cambridge, MA, Harvard University Press.

————, A. L. Melander, and F. M. Carpenter. 1954. *Classification of Insects*. Bull. Mus. Comp. Zool.; Harvard University, 108.

Bücherl, W., and E. E. Buckley, eds. 1972. *Venomous Animals and Their Venoms*, Vol. 3, *Venomous Invertebrates*. New York, Academic Press. 560 pp.

Buchner, P. 1965. *Endosymbiosis of Animals with Plant Micro-organisms*. New York, Interscience.

Buck, J. B. 1953. Physical properties and chemical composition of insect blood. In *Insect Physiology*, pp. 147–90, K. Roeder, ed. New York, Wiley.

————and E. Buck. 1968. Mechanisms of rhythmic synchronous flashing of fireflies. Science, 159:1319–28.

———— and————. 1976. Synchronous fireflies. Sci. Amer., 234:74–85.

Bulla, L. A., ed. 1973. Regulation of insect populations by microorganisms, Ann. N.Y. Acad. Sci., vol. 217.

Bunning, E. 1967. *The Physiological Clock*, rev. 2nd ed. New York, Springer-Verlag.

Burges, H. D., and N. W. Hussey, eds. 1971. *Microbial Control of Insects and Mites*. New York, Academic Press.

Burkhardt, C. C. 1978. Insect pests of corn. In *Fundamentals of Applied Entomology*, 3rd ed., pp. 303–34, R. E. Pfadt, ed. New York, Macmillan.

Burks, B. D. 1953. The mayflies, or Ephemeroptera, of Illinois. Illinois Nat. Hist. Surv. Bull., 26:1–216.

Burns, M. D. 1974. Structure and physiology of the locust femoral chordotonal organ. J. Insect Physiol., 20:1319–39.

Bursell, E. 1970. *An Introduction to Insect Physiology*. New York, Academic Press.

————. 1974a. Environmental Aspects—Temperature. In *The Physiology of Insecta*, 2nd ed., Vol I, pp. 2–41, M. Rockstein, ed. New York, Academic Press.

————. 1974b. Environmental Aspects—Humidity. In *The Physiology of Insecta*, 2nd ed. Vol. I, pp. 44–84, M. Rockstein, ed. New York, Academic Press.

Busvine, J. R. 1966. *Insects and Hygiene*. London, Methuen.

————.1975.*Arthropod Vectors of Disease*. London, Edward Arnold.

Butcher, J. W., R. Snider, and R. J. Snider. 1971. Biology of endaphic Collembola and Acarina. Ann. Rev. Entomol., 6:249–88.

Butler, C. G. 1963. The honey bee colony—life history. In *The Hive and the Honey Bee*, ch. 3, R. A. Grout, ed. Hamilton, IL, Dadant & Sons.

————. 1971. *The World of the Honeybee*. London, Collins. 226 pp.

Byers, G. W. 1954. Notes on North American Mecoptera. Ann. Entomol. Soc. Amer., 47:484–510.

————. 1963. The life history of *Panorpa nuptialis* (Mecoptera: Panorpidae). Ann. Entomol. Soc. Amer., 56:142–49.

————. 1965. Families and genera of Mecoptera. Proc. XII Int. Congr. Entomol., 1964:123.

————. 1971. Ecological distribution and structural adaptation in the classification of Mecoptera. Proc. XIII Int. Congr. Entomol., 1968, I:486.

Callahan, P. S. 1972. *The Evolution of Insects*. New York, Holiday House.

Calvert, W. H., L. E. Hedrick, and L. P. Brower. 1979. Mortality of the monarch butterfly (*Danaus plexippus* L.): avian predation at five overwintering sites in Mexico. Science, 204:847–51.

Cameron, J. W. MacBain. 1973. Insect pathology. In *History of Entomology*, pp. 267–84, R. F. Smith, T. E. Mittler, and C. N. Smith, eds. Palo Alto, CA, Annual Reviews.

Cantwell, G. E., ed. 1974. *Insect Diseases*, Vols. 1–2. New York, Marcel Dekker.

Carlson, S. D., and Che Chi. 1979. The functional morphology of the insect photoreceptor. Ann. Rev. Entomol., 24:379–416.

Carpenter, F. M. 1931a. The biology of the Mecoptera. Psyche, 38(2):41–55.

————. 1931b. Revision of Nearctic Mecoptera. Bull. Mus. Comp. Zool., Harvard University, 72:205–77.

———— 1953. The geological history and evolution of insects. Amer. Sci., 41:256–70.

References Cited

————. 1973. Geological history and evolution of insects, In *Syllabus—Introductory Entomology*, pp. 77–88, V. J. Tipton, ed. Provo, UT, Brigham Young University Press.

————. 1977. Geological history and evolution of the insects. Proc. XV Int. Congr. Entomol., pp. 63–70.

Carson, R. L. 1962. *Silent Spring*. Boston, Houghton Mifflin.

Carter, W. 1973. *Insects in Relation to Plant Disease*, 2nd ed. New York, Wiley.

Carthy, J. D. 1958. *An Introduction to the Behavior of Invertebrates*. London, Allen & Unwin.

————. 1965. *The Behavior of Arthropods*. San Francisco, Freeman.

Caswell, R. L., ed. 1977. *Pesticide Handbook—Entomology*. College Park, MD, The Entomological Society of America.

Caudell, A. N. 1918. Zoraptera not an apterous order. Proc. Entomol. Soc. Wash., 22:84–97.

Caveney, S. 1969. Muscle attachment related to cuticle architecture in apterygota. J. Cell Sci., 4:541–59.

Chamberlin, W. J. 1952. *Entomological Nomenclature and Literature*, 3rd ed. Dubuque, IA, Wm. C. Brown.

Chandler, H. P. 1956. Megaloptera. In *Aquatic Insects of California*, pp. 229–33, R. L. Usinger, ed. Berkeley, University of California Press.

Chapman, P. J. 1930. Corrodentia of the United States of America. I. Suborder Isotecnomera. J. N.Y. Entomol. Soc., 38:219–90, 319–403.

Chapman, R. F. 1971. *The Insects—Structure and Function*, 2nd ed. New York, American Elsevier.

Chauvin, R. 1967. *The World of an Insect*. London, World University Library; New York, McGraw-Hill.

Chen, P. S. 1971. *Biochemical Aspects of Insect Development. Monographs in Developmental Biology*, Vol. 3. New York, S. Karger.

————. 1978. Protein synthesis in relation to cellular activation and deactivation. In *Biochemistry of Insects*, pp. 145–203, M. Rockstein, ed. New York, Academic Press.

Cheng, L., ed. 1976. *Marine Insects*. Amsterdam, North-Holland.

China, W. E. 1962. South American Peloridiidae (Hemiptera-Homoptera: Coleorrhyncha). Trans. Royal Entomol. Soc. Lond., 114:131–61.

Christiansen, K. 1964. Bionomics of Collembola. Ann. Rev. Entomol., 9:147–78.

———— 1978. Aquatic collembola. In *An Introduction to the Aquatic Insects of North America*, pp. 51–55, R. W. Merritt and K. W. Cummins, eds. Dubuque, IA, Kendall/Hunt Publishing Company.

———— and P. Bellinger. 1978. *The Collembola of North America North of the Rio Grande*. Los Angeles, Entomological Reprint Specialists.

Chu, H. F. 1949. *How to Know the Immature Insects*. Dubuque, IA, Wm. C. Brown.

Cisne, J. L. 1974. Trilobites and the origin of arthropods. Science, 186:13–18.

Claassen, P. W. 1931. *Plecoptera Nymphs of America (North of Mexico)*. Thomas Say Foundation Publ. 3. College Park, MD, The Entomological Society of America.

Clark, A. M., and M. Rockstein. 1964. Aging in insects. In *The Physiology of Insecta*, Vol. 1, pp.227–81, M. Rockstein, ed. New York, Academic Press.

Clark, L. R., P. W. Geier, R. D. Hughes, and R. F. Morris. 1967. *The Ecology of Insect Populations in Theory and Practice*. London, Methuen.

Clarke, K. U. 1973. *The Biology of the Arthropods*. New York, American Elsevier.

Clausen, C. P. 1940. *Entomophagous Insects*. New York, McGraw-Hill.

————. 1976. Phoresy among entomophagous insects. Ann. Rev. Entomol., 21:343–68.

Clausen, L. W. 1954. *Insect Fact and Folklore*. New York, Macmillan (Collier Books).

Clay, T. 1969. A key to the genera of the Menoponidae (Amblycera: Mallophaga: Insecta). Bull. Brit. Mus. (Nat. Hist.), Entomol., 24:1–26.

————. 1973. Phthiraptera (lice). In *Insects and Other Arthropods of Medical Importance*, pp. 395–97, K. G. Smith, ed. London, British Museum (Natural History).

Clements, A. N. 1963. *The Physiology of Mosquitoes*. New York, Macmillan.

References Cited

Cleveland, L. R., S. R. Hall, E. P. Saunders, and J. Collier, 1934. The wood-feeding roach, *Cryptocercus,* its protozoa, and the symbiosis between protozoa and roach. Mem. Amer. Acad. Sci., 17:185–342.

Cloudsley-Thompson, J. L. 1975. Adaptations of arthropods to arid environments. Ann. Rev. Entomol., 20:261–83.

———. 1976. *Insects and History.* London, Weidenfeld & Nicolson.

Cochran, D. G. 1975. Excretion in insects. In *Insect Biochemistry and Function,* pp. 179–281, D. J. Candy and B. A. Kilby, eds. New York, Wiley (Halsted Press).

Cole, F. R. 1969. *The Flies of Western North America.* Berkeley, University of California Press.

Collier, B. D., G. W. Cox, A. W. Johnson, and P. C. Miller. 1973. *Dynamic Ecology.* Englewood Cliffs, NJ, Prentice-Hall.

Common, I. F. B. 1975. Evolution and classification of the Lepidoptera. In *The Insects of Australia,* pp. 765–866. Canberra, C.S.I.R.O.

Comstock, J. H. 1918. *The Wings of Insects.* Ithaca, NY, Cornell University Press.

———. 1940. *An Introduction to Entomology,* 9th ed. Ithaca, NY, Cornell University Press.

Coppel, H. C., and J. W. Mertins. 1977. *Biological Insect Pest Suppression.* New York, Springer-Verlag.

Corbet, P. S. 1963. *A Biology of Dragonflies.* Chicago, Quadrangle Books.

———. 1966. The role of rhythms in insect behavior. In *Insect Behavior,* pp.13–28, P. T. Haskell, ed. Roy. Entomol. Soc. (London), Symp. 3.

Cornwell, P. B. 1968. *The Cockroach,* Vol. 1. London, Hutchinson.

Cott, H. B. 1957. *Adaptive Coloration in Animals.* London, Methuen.

Cott, H. E. 1966. *Systematics of Suborder Tubulifera (Thysanoptera) of California.* Berkeley, University of California Press.

Cottrell, C. B. 1964. Insect ecdysis with particular emphasis on cuticular hardening and darkening. Adv. Insect Physiol., 2:175–218.

Counce, S. J. 1961. The analysis of insect embryogenesis. Ann. Rev. Entomol., 6:295–312.

———. 1973. The causal analysis of insect embryogenesis. In *Developmental Systems: Insects,* Vol. 2, S. J. Counce and C. H. Waddington, eds. New York, Academic Press.

——— and C. H. Waddington, eds. 1972, 1973. *Developmental Systems: Insects,* Vols. 1, 2. New York, Academic Press.

Crabtree, B., and E. A. Newsholme. 1975. Comparative aspects of fuel utilization and metabolism by insects. In *Insect Muscle,* pp. 405–500, P. N. R. Usherwood, ed. New York, Academic Press.

Crane, E. 1963. The world's beekeeping—past and present. In *The Hive and the Honey Bee,* rev. ed., pp.1–18, R. A. Grout, ed. Hamilton, IL, Dadant & Sons.

Creighton, W. S. 1950. *The Ants of North America.* Harvard University, Mus. Comp. Zool. Bull. 104.

Cremyln, R. 1978. *Pesticides—Preparation and Mode of Action.* New York, Wiley–Interscience.

Crossley, A. C. 1975. The cytophysiology of insect blood. Adv. Insect Physiol., 11:117–221.

C.S.I.R.O. 1970. *The Insects of Australia.* Melbourne, Melbourne University Press.

Cummins, K. W., L. D. Miller, N. A. Smith, and R. M. Fox. 1965. *Experimental Entomology.* New York, Van Nostrand Reinhold.

Curran, C. H. 1934. *The Families and Genera of North American Diptera.* Woodhaven, NY, H. Tripp.

Cushing, E. C. 1957. *History of Entomology in World War II.* Smithsonian Inst. Publ. 4294.

Dadd, R. H. 1970. Digestion in insects. In *Chemical Zoology,* Vol. 5, Arthropoda, Part A, pp. 117–45, M. Florkin and B. T. Scheer, eds. New York, Academic Press.

———. 1973 Insect nutrition: current developments and metabolic implications. Ann. Rev. Entomol., 18:381–420.

Dalton, S. 1975. *Borne on the Wind. The Extraordinary World of Insects in Flight.* New York, Dutton.

References Cited

Daly, H. V., J. T. Doyen, and P. R. Ehrlich. 1978. *Introduction to Insect Biology and Diversity*. New York, McGraw-Hill.

Danilevskii, A. S. 1965. *Photoperiodism and Seasonal Development of Insects*. Edinburgh, Oliver & Boyd. 283 pp.

Dauterman, W. C., and E. Hodgson. 1978. Detoxification mechanisms in insects. In *Insect Biochemistry*, pp. 541–77, M. Rockstein, ed. New York, Academic Press.

Davey, K. G., 1964. The control of visceral muscles in insects. Adv. Insect Physiol., 2:219–45.

———. 1965. *Reproduction in the Insects*. Edinburgh, Oliver & Boyd.

——— and J. E. Treherne. 1963. Studies on crop function in the cockroach (*Periplaneta americana* L.) I–III. J. Exp. Biol., 40:763–73, 775–80; 41:513–24.

David, W. A. L. 1975. The status of viruses pathogenic for insects and mites. Ann. Rev. Entomol., 20:97–117.

Davidson, G. 1974. *Genetic Control of Insect Pests*. New York, Academic Press.

Davidson, R., and W. F. Lyon. 1979. *Insect Pests of Farm, Garden and Orchard*, 7th ed. New York, Wiley.

——— and L. M. Peairs. 1966. *Insect Pests of Farm, Garden and Orchard*, 6th ed. New York, Wiley.

Dawkins, R. 1976. *The Selfish Gene*. Oxford, Oxford University Press.

Day, M. F. 1954. The mechanism of food distribution to the midgut or diverticula in the mosquito. Australian J. Biol. Sci., 7:515–24.

Day, W. C. 1956. Ephemeroptera. In *Aquatic Insects of California*, pp. 79–105, R. L. Usinger, ed. Berkeley, University of California Press.

DeBach, P. ed. 1964. *Biological Control of Insect Pests and Weeds*. London, Chapman & Hall.

———. 1974. *Biological Control by Natural Enemies*. London, Cambridge University Press.

DeCoursey, R. M. 1971. Keys to the families and subfamilies of the nymphs of North American Hemiptera-Heteroptera. Proc. Entomol. Soc. Wash., 73:413–28.

Denning, D. G. 1956. Trichoptera. In *Aquatic Insects of California*, pp. 237–70, R. L. Usinger, ed. Berkeley, University of California Press.

Dethier, V. G. 1953. Chemoreception. In *Insect Physiology*, pp. 544–76, K. Roeder, ed. New York, Wiley.

———. 1955. The physiology and histology of the contact chemoreceptors of the blowfly. Quart. Rev. Biol., 30:348–71.

———. 1962. *To Know a Fly*. San Francisco, Holden-Day.

———. 1963. *The Physiology of Insect Senses*. New York, Wiley.

———. 1966. Feeding behavior. In *Insect Behavior*, pp. 46–58, P. T. Haskell, ed. Roy. Entomol. Soc. (London), Symp. 3.

———. 1970 Chemical interaction between plants and insects. In *Chemical Ecology*, pp. 83–102, E. Sondheimer and J. B. Simeone, eds. New York, Academic Press.

———. 1976. *The Hungry Fly. A Physiological Study of the Behavior Associated with Feeding*. Cambridge, MA, Harvard University Press.

——— and E. Stellar. 1970. *Animal Behavior*, 3rd ed. Englewood Cliffs, NJ, Prentice-Hall.

de Wilde, J. and A. de Loof. 1973a. Reproduction. In *The Physiology of Insecta*, 2nd ed., Vol. I, pp. 12–95, M. Rockstein, ed. New York, Academic Press.

——— and ———. 1973b. Reproduction—endocrine control. In *The Physiology of Insecta*, 2nd ed., Vol. I, pp. 97–157, M. Rockstein, ed. New York, Academic Press.

——— and L. M. Schoonhaven, eds. 1969. *Insect and Host Plant*. Amsterdam, North-Holland. (Reprinted from Entomol. Exp. Appl., 12:471–810.)

Dillon, E. S., and L. S. Dillon. 1972. *A Manual of Common Beetles of Eastern North America* (2 vols.). New York, Dover (reprint of 1961 edition).

Dingle, H. 1972. Migration strategies of insects. Science, 175:1327–35.

———, ed. 1978a. *Evolution of Insect Migration and Diapause*. New York, Springer-Verlag.

———. 1978b. Migration and diapause in tropical, temperate, and island milkweed bugs. In *Evolution of Insect Migration and Diapause*, pp. 254–76. New York, Springer-Verlag.

References Cited

Dirsh, V. M. 1975. *Classification of the Acridomorphoid Insects*. Faringdon, England, E. W. Classey. 184 pp.

Dixon, A. F. G. 1973. *Biology of Aphids*. The Institute of Biology's Studies in Biology No. 44. London, Edward Arnold.

Doane, W. W., 1973. Role of hormones in insect development. In *Developmental Systems: Insects,* Vol. 2. New York, Academic Press.

Dobzhansky, T., F. J. Ayala, G. L. Stebbins, and J. W. Valentine. 1977. *Evolution*. San Francisco, Freeman.

Downes, J. A. 1965. Adaptations of insects in the Arctic. Ann. Rev. Entomol., 10:257–74.

Duffey, S. S. 1977. Arthropod allomones: Chemical effronteries and antagonists. Proc. XV Int. Congr. Entomol., pp. 323–94.

———. 1980. Sequestration of plant natural products by insects. Ann. Rev. Entomol. 25:447–77.

Dunn, J. A. 1960. The natural enemies of the lettuce root aphid, *Pemphigus bursarius* (L.). Entomol. Res. Bull., 51:271–78.

Dunning, D. C., and K. D. Roeder. 1965. Moth sounds and the insect-catching behavior of bats. Science, 147:173–74.

DuPorte, E. M. 1961. *Manual of Insect Morphology*. New York, Van Nostrand Reinhold.

Dyar, H. G. 1890. The number of moults of Lepidopterous larvae. Pysche, 5:420–22.

Eaton, J. L. 1971. Insect photoreceptor: An internal ocellus is present in sphinx moths. Science, 173:822–23.

Ebeling, W. 1974. Permeability of insect cuticle. In *The Physiology of Insecta,* 2nd ed., Vol VI, pp. 271–343, M. Rockstein, ed. New York, Academic Press.

———. 1975. *Urban Entomology*. Berkeley, University of California, Div. Agr. Sci.

———. 1976. Insect integument: a vulnerable organ system. In *The Insect Integument,* pp. 383–400, H. R. Hepburn, ed. New York, Elsevier Scientific.

Ebling, J., and K. C. Highnam. 1970. *Chemical Communication*, Studies in Biology, No. 19. New York, Crane Russak.

Edmonson, W. T., ed. 1959. *Freshwater Biology*. New York, Wiley.

Edmunds, G. F., Jr. 1959a. Ephemeroptera. In *Freshwater Biology,* pp. 908–16, W. T. Edmonson, ed. New York, Wiley.

———. 1959b. Biogeography and evolution of the Ephemeroptera. Ann. Rev. Entomol. 17:21–42.

———. 1978. Ephemeroptera. In *An Introduction to the Aquatic Insects of North America,* pp. 57–80, R. W. Merritt and K. W. Cummins, eds. Dubuque, IA, Kendall/Hunt Publishing Co.

———, S. L. Jensen, and L. Berner. 1976. *The Mayflies of North and Central America*. Minneapolis, University of Minnesota Press. 330 pp.

Ehrlich, P. R., and A. H. Ehrlich. 1961. *How to Know the Butterflies*. Dubuque, IA, Wm. C. Brown.

——— and P. H. Raven. 1964. Butterflies and plants: a study in coevolution. Evolution, 18:586–608.

Ehrman, L., and P. A. Parsons. 1976. *The Genetics of Behavior*. Sunderland, MA, Sinauer Associates. 390 pp.

Eibl-Eibesfeldt, I. 1970. *Ethology. The Biology of Behavior*. New York, Holt, Rinehart, and Winston. 530 pp.

Eisenstein, E. M. 1972. Learning and memory in isolated insect ganglia. Adv. Insect Physiol., 9:111–81.

——— and M. J. Cohen. 1966. Learning in an isolated insect ganglion. Anim. Behav., 13:104–108.

Eisner, T. 1965. Defensive spray of a phasmid insect. Science, 148:966–68.

——— 1970. Chemical defense against predation in arthropods. In *Chemical Ecology,* pp. 157–217, E. Sondheimer and J. B. Simeone, eds. New York, Academic Press.

———, K. Hicks, M. Eisner, and D. S. Robson. 1978. ''Wolf-in-sheep's-clothing'' strategy of a predaceous insect larva. Science, 199:790–93.

——— and Y. C. Meinwald. 1965. Defensive secretion of a caterpillar. Science, 150:1733–35.

References Cited

—— and ——. 1966. Defensive secretions of arthropods. Science, 153:1341–50

——, R. E. Silberglied, D. Aneshansley, J. E. Carrel, and H. C. Howland. 1969. Ultraviolet video-viewing: the television camera as an insect eye. Science, 166:1172–74.

——, E. van Tassell, and J. E. Carrel. 1967. Defensive use of a "fecal shield" by a beetle larva. Science, 158:1471–73.

—— and E. O. Wilson, eds. 1953–1977. *The Insects,* Readings from Scientific American. San Francisco, Freeman. 334 pp.

Elder, H. Y. 1975. Muscle structure. In *Insect Muscle,* pp. 1–74, P. N. R. Usherwood, ed. New York, Academic Press.

Elzinga, R. J. 1978. *Fundamentals of Entomology.* Englewood Cliffs, NJ, Prentice-Hall. 325 pp.

Emden, H. F. van. 1966. Plant insect relationships and pest control. World Rev. Pest Control, 5:115–23.

——. 1972. *Insect/Plant Relationships.* Roy. Entomol. Soc. (London), Symp. 6. 215 pp.

——. 1974. *Pest Control and Its Ecology.* London, Edward Arnold. 60 pp.

Emerson, K. C. 1972. Checklist of the Mallophaga of North America (North of Mexico). Part 1, Suborder Ischnocera, 200 pp. Part 2, Suborder Amblycera, 118 pp. Part 3, Mammal Host List, 28 pp. Part 4, Bird Host List. Dugway, UT, Desert Test Center, Proving Ground.

Emmel, T. C. 1975. *Butterflies: Their World, Their Life Cycle, Their Behavior.* New York, Knopf. 260 pp.

Engelmann, F. 1968. Endocrine control of reproduction in insects. Ann. Rev. Entomol. 13:1–26.

—— 1970. *The Physiology of Insect Reproduction.* New York, Pergamon.

Erber, J. 1975. The dynamics of learning in the honeybee. J. Comp. Physiol., 99(3):231–55.

Essig, E. O. 1931. *A History of Entomology.* New York, Macmillan.

——. 1942. *College Entomology.* New York, Macmillan.

——. 1958. *Insects and Mites of Western North America.* New York, Macmillan. 1050 pp.

Etkin, W., and L. I. Gilbert, eds. 1968. *Metamorphosis: A Problem in Developmental Biology.* Amsterdam, North-Holland.

Evans, E. D. 1978. Megaloptera and aquatic Neuroptera. In *An Introduction to the Aquatic Insects of North America,* pp. 133–45, R. W. Merritt and K. W. Cummins, eds. Dubuque, IA, Kendall/Hunt Publishing Co.

Evans, G. 1975. *The Life of Beetles.* London, Allen and Unwin. 232 pp.

Evans, H. E. 1957. *Studies on the Comparative Ethology of Digger Wasps of the Genus* Bembix. Ithaca, NY, Comstock.

——. 1963. *Wasp Farm.* Garden City, NY, Natural History Press. 178 pp.

—— and M. T. W. Eberhard. 1970. *The Wasps.* Ann Arbor, University of Michigan Press.

Ewing, A. W. 1977. Communication in Diptera. In *How Animals Communicate,* pp. 403–17, T. A. Sebeok, ed. Bloomington, Indiana University Press.

—— and A. Manning. 1967. The evolution and genetics of insect behavior. Ann. Rev. Entomol., 12:471–94.

Ewing, H. E. 1940. The Protura of North America. Ann. Entomol. Soc. Amer., 33:495–551.

—— and I. Fox. 1943. *The Fleas of North America.* U. S. Department of Agriculture, Misc. Publ. 500.

Faegri, K., and L. van der Pijl. 1978. *The Principles of Pollination Ecology,* 3rd ed. New York, Pergamon. 242 pp.

Falcon, L. A. 1976. Problems associated with the use of arthropod viruses in pest control. Ann. Rev. Entomol., 21:305–24.

Fallon, A. M., and H. H. Hagedorn. 1972. Synthesis of vitellogenin by the fat body in *Aedes aegypti:* the effect of injected ecdysone. Amer. Zool., 12:697.

——, ——, and G. R. Wyatt. 1974. Activation of vitellogenin synthesis in the mosquito *Aedes aegypti* by ecdysone. J. Insect Physiol., 20:1815–23.

Faust, R. M. 1974. Bacterial diseases. In *Insect Diseases,* Vol. 1, pp. 87–183, G. E. Cantwell, ed. New York, Marcel Dekker.

Feingold, B. F., E. Benjamini, and D. Michaeli. 1968. The allergic responses to insect bites. Ann. Rev. Entomol., 13:137–58.

Felt, E. P. 1940. *Plant Galls and Gall Makers.* Ithaca, NY, Comstock. 364 pp.

References Cited

Ferris, G. F. 1928. *The Principles of Systematic Entomology*. Stanford, CA, Stanford University Press.

————. 1951. *The Sucking Lice*. Pacific Coast Entomol. Soc. Mem. 1. 320 pp.

Ferron, P. 1978. Biological control of insect pests by entomogenous fungi. Ann. Rev. Entomol. 23:409–42.

Flanagan, T. R., and H. H. Hagedorn. 1977. Vitellogenin synthesis in the mosquito: the role of juvenile hormone in the development of responsiveness to ecdysone. Physiol. Entomol. 2:173–78.

Fletcher, B. S. 1977. Behavioral responses of Diptera to pheromones, allomones and kairomones. In *Chemical Control of Insect Behavior*, pp. 129–48, H. H. Shorey and J. J. McKelvey Jr., eds. New York, Wiley.

Fletcher, D. J. C. 1978. The african bee, *Apis mellifera adansonii*, in Africa. Ann. Rev. Entomol. 23:151–71.

Fletcher, W. W. 1978. *The Pest War*. New York, Wiley (Halsted Press).

Florkin, M., and C. Jeuniaux. 1974. Hemolymph: composition. In *The Physiology of Insecta*, 2nd ed., Vol. V, pp. 256–307, M. Rockstein, ed. New York, Academic Press.

Flower, J. W. 1964. On the origin of flight in insects. J. Insect Physiol., 10:81–88.

Folsom, J. W., and R. A. Wardle. 1934. *Entomology, with Special Reference to Its Ecological Aspects*, 4th ed. Philadelphia, Blakiston.

Foote, R. H. 1977. *Thesaurus of Entomology*. College Park, MD, The Entomological Society of America, 188 pp.

Foster, G. G., M. J. Whitten, T. Prout, and R. Gill. 1972. Chromosome rearrangements for the control of insect pests. Science, 176:875–80.

Fox, I. 1940. *Fleas of Eastern United States*. New York, Hafner (1968 reprint). 191 pp.

Fox, R. M., and J. W. Fox. 1964. *Introduction to Comparative Entomology*. New York, Van Nostrand Reinhold.

Frankel, G. S. 1932. Untersuchungen über die Koordination von Reflexen und automatisch-nervösen Rhythmen bei Insekten III. Z. vergl. Physiol., 16:394–460.

———— and C. Hsiao. 1965. Bursicon, a hormone which mediates tanning of the cuticle in the adult fly and other insects. J. Insect Physiol., 11:513–56.

———— and D. L. Gunn. 1961. *The Orientation of Animals*. New York, Dover (reprint of 1940 Oxford edition).

Frankie, G. W., and C. S. Koehler, eds. 1976. *Perspectives in Urban Entomology*. Symposium held at the XV Int. Congr. Entomol. (Washington, DC). New York, Academic Press. 417 pp.

Frazier, C. A. 1969. *Insect Allergy: Allergic and Toxic Reactions to Insects and Other Arthropods*. St. Louis, MO, Warren H. Green. 493 pp.

Free, J. B. 1970. *Insect Pollination of Crops*. New York, Academic Press. 544 pp.

————. 1977. *The Social Organization of Honeybees*. London, Edward Arnold. 68 pp.

Friend, W. G., and J. J. B. Smith. 1977. Factors affecting feeding by bloodsucking insects. Ann. Rev. Entomol., 22:309–31.

Frings, H., and M. Frings. 1977. *Animal Communication*, 2nd ed., rev. Norman, University of Oklahoma Press.

Frisch, K. von. 1950. *Bees, Their Vision, Chemical Senses, and Language*. Ithaca, NY, Cornell University Press.

————. 1967. *The Dance, Language, and Orientation of Bees* (trans. by L. E. Chadwick). Cambridge, Harvard University Press.

————. 1971. *Bees—Their Vision, Chemical Senses, and Language*, rev. ed. Ithaca, NY, Cornell University Press.

Frison, T. H. 1935. *The Stoneflies, or Plecoptera, of Illinois*. Ill. Nat. Hist. Surv. Bull., 20:281–471.

————. 1937. Descriptions of Plecoptera. Illinois Natural History Survey, Bull. 21(3):78–99.

Froeschner, R. C. 1947. Notes and keys to the Neuroptera of Missouri. Ann. Entomol. Soc. Amer., 40(1):123–36.

Fronk, W. D. 1978. Chemical control. In *Fundamentals of Applied Entomology*, 3rd ed., pp. 209–40, R. E. Pfadt, ed. New York, Macmillan.

Frost, S. W. 1959. *Insect Life and Insect Natural History*, 2nd ed. New York, Dover.

Fukuda, S., and S. Takeuchi. 1967. Diapause factor-producing cells in the sub-

References Cited

esophageal ganglion of the silkworm, *Bombyx mori* L. Proc. Japan Acad., 43:51–56.

Fuller, J. L., and W. R. Thompson. 1960. *Behavior Genetics*. New York, Wiley.

Furneaux, P. J. S., and A. L. Mackay. 1976. The composition, structure and formation of the chorion and the vitelline membrane of the insect egg-shell. In *The Insect Integument*, pp. 157–76, H. R. Hepburn, ed. New York, Elsevier Scientific.

Futuyma, D. J. 1979. *Evolutionary Biology*. Sunderland, MA, Sinauer Associates. 565 pp.

Fuzeau-Braesch, S. 1972. Pigments and colour changes. Ann. Rev. Entomol., 17:403–24.

Gallun, R. L., K. J. Starks, and W. D. Guthrie. 1975. Plant resistance to insects attacking cereals. Ann. Rev. Entomol. 20:337–57.

Galun, R. 1977a. The physiology of hematophagous insect/animal host relationships. Proc. XV Int. Congr. Entomol., pp. 257–65.

———. 1977b. Responses of bloodsucking arthropods to vertebrate hosts. In *Chemical Control of Insect Behavior—Theory and Application*, pp. 103–15, H. H. Shorey and J. J. McKelvey, eds. New York, Wiley.

Ganesalingam, W. K. 1974. Mechanism of discrimination between parasitized and unparasitized hosts by *Venturia canescens* (Hymenoptera: Ichneumonidae). Entomol. Exp. Appl., 17:36–44.

Garman, P. A. 1927. The Odonata of Connecticut. Conn. State Geol. and Nat. Hist. Survey, Bull. 39.

Gehring, W. J., and R. Nöthiger. 1973. The imaginal discs of *Drosophila*. In *Developmental Systems: Insects*, Vol. 2, pp. 212–90, S. J. Counce and C. H. Waddington, eds. New York, Academic Press.

Geier, P. W. 1966. Management of insect pests. Ann. Rev. Entomol., 11:471–90.

Gelperin, A. 1966. Control of crop emptying in the blow fly. J. Insect Physiol., 12:331–45.

———. 1971. Regulation of feeding. Ann. Rev. Entomol., 16:365–78.

Georghiou, G. P. 1972. The evolution of resistance to pesticides. Ann. Rev. Ecol. Syst., 3:133–68.

——— and C. E. Taylor. 1977. Pesticide resistance as an evolutionary phenomenon. Proc. XV Int. Congr. Entomol., pp. 759–85.

Gerber, G. H. 1970. Evolution of the methods of spermatophore formation in pterygote insects. Can. Entomol., 102:358–62.

Ghauri, M. S. K. 1973. Hemiptera (bugs). In *Insects and Other Arthropods of Medical Importance*, pp. 373–93, K. G. V. Smith, ed. London, British Museum (National History).

Gibbs, A. J., ed. 1973. *Viruses and Invertebrates*. New York, American Elsevier.

Gilbert, L. E. 1971. Butterfly-plant coevolution: Has *Passiflora adenopoda* won the selectional race with Heliconiine butterflies? Science, 172:585–86.

———, and P. H. Raven, eds. 1975. *Coevolution of Animals and Plants*. Austin, University of Texas Press. 275 pp.

Gilbert, L. I. 1967. Lipid metabolism and function in insects. Adv. Insect Physiol., 4:69–211.

———, ed. 1976. *The Juvenile Hormones*. New York, Plenum Press.

——— and D. S. King. 1973. Physiology of growth and development: endocrine aspects. In *The Physiology of Insecta*, 2nd ed., Vol. I, pp. 249–370, M. Rockstein, ed. New York, Academic Press.

Gillet, J. D., and V. B. Wigglesworth. 1932. The climbing organ of an insect, *Rhodnius prolixus* (Hemiptera: Reduviidae). Proc. Roy. Soc. (London), Ser. B, 111:364–76.

Gilmour, D. 1965. *The Metabolism of Insects*. San Francisco, Freeman.

Glover, P. M. 1937. *Lac Cultivation in India*. Nankum, Ranchi, Indian Lac Institute. 147 pp.

Gloyd, L. K., and M. Wright, Jr. 1955. Odonata. In *Fresh-water Biology*, pp. 917–40, W. T. Edmondson, ed. New York, Wiley.

Gochnauer, T. A. 1963. Diseases and enemies of the honey bee. In *The Hive and the Honey Bee*, pp. 477–511, R. A. Grout, ed. Hamilton, IL, Dadant & Sons.

Goetsch, W. 1957. *The Ants*. Ann Arbor, University of Michigan Press. 173 pp.

Goldsmith, T. H., and G. D. Bernard. 1974. The visual system of insects. In *The Physiology of Insecta*, 2nd ed., Vol. II, pp. 165–272, M. Rockstein, ed. New York, Academic Press.

References Cited

Gooding, R. H. 1972. Digestive processes of haematophagous insects I. A literature review. Quaestiones Entomologicae, 8:5–60.

Goodman, L. J. 1970. The structure and function of the insect dorsal ocellus. Adv. Insect Physiol., 7:97–195.

Graham, S. A., and F. B. Knight. 1965. *Principles of Forest Entomology,* 4th ed. New York, McGraw-Hill.

Greenberg, B. 1971. *Flies and Disease,* Vol. I, *Ecology, Classification, and Biotic Associations.* Princeton, NJ, Princeton University Press.

―――. 1973. *Flies and Disease,* Vol. II, *Biology and Disease Transmission.* Princeton, NJ, Princeton University Press, 447 pp.

Grégoire, Ch. 1974. Hemocyte coagulation. In *The Physiology of Insecta,* 2nd ed., Vol. V, pp. 309–60, M. Rockstein, ed. New York, Academic Press.

Griffin, D. R. 1959. *Echoes of Bats and Men.* Garden City, NY, Doubleday (Anchor Books).

Grosch, D. S. 1974. Environmental aspects: radiation. In *The Physiology of Insecta,* 2nd ed., Vol. II, pp. 85–126, M. Rockstein, ed. New York, Academic Press.

Grout, R. A. 1963. The production and uses of beeswax. In *The Hive and the Honey Bee,* pp. 425–36, R. A. Grout, ed. Hamilton, IL, Dadant & Sons.

Gupta, A. P. 1969. Studies of the blood of Meloidae (Coleoptera) I. The haemocytes of *Epicauta cinerea* (Forster), and a synonymy of haemocyte terminologies. Cytologia, 34(2):311–44.

―――. 1976. *Entomology in the U.S.A.* College Park, MD, The Entomological Society of America.

―――, ed. 1979. *Arthropod Phylogeny.* New York, Van Nostrand Reinhold. 762 pp.

Gurney, A. B. 1938. A synopsis of the Order Zoraptera, with notes on the biology of *Zorotypus hubbardi* Caudell. Proc. Entomol. Soc. Wash., 40:57–87.

―――. 1948. The taxonomy and distribution of the Grylloblattidae. Proc. Entomol. Soc. Wash., 50:86–102.

―――. 1950. Corrodentia. In *Pest Control Technology, Entomological Section,* pp. 129–63. New York, National Pest Control Association.

―――. 1951. Praying mantids of the United States: native and introduced. Smithsonian Inst. Rept., 1950:339–62.

Gurney, A. B., and S. Parfin. 1959. Neuroptera. In *Freshwater Biology,* pp. 973–80, W. T. Edmundson, ed. New York, Wiley.

Guthrie, D. M., and A. R. Tindall. 1968. *The Biology of the Cockroach.* London, Edward Arnold. 408 pp.

Hackman, R. H. 1974. Chemistry of the insect cuticle. In *The Physiology of Insecta,* Vol. VI, pp. 215–70, M. Rockstein, ed. New York, Academic Press.

―――. 1976. The interactions of cuticular proteins and some comments on their adaptation to function. In *The Insect Integument,* pp. 107–20, H. R. Hepburn, ed. New York, Elsevier Scientific.

――― and M. Goldberg. 1975. Peripatus: its affinities and its cuticle. Science, 190:582–83.

Hagedorn, H. H. 1974. The control of vitellogenesis in the mosquito, *Aedes aegypti.* Am. Zool., 14:1207–17.

――― and J. G. Kunkel. 1979. Vitellogenin and vitellin in insects. Ann. Rev. Entomol., 24:475–505.

Hagen, H. R. 1951. *Embryology of the Viviparous Insects.* New York, Ronald Press.

Hagen, K. S. 1978. Aquatic Hymenoptera. In *An Introduction to the Aquatic Insects of North America,* pp. 233–43, R. W. Merritt and K. W. Cummins, eds. Dubuque, IA, Kendall/Hunt Publishing Co.

――― and J. M. Franz. 1973. A history of biological control. In *History of Entomology,* pp. 433–76, R. F. Smith, T. E. Mittler, and C. N. Smith, eds. Palo Alto, CA, Annual Reviews.

Hailman, J. P. 1977. Communication by reflected light. In *How Animals Communicate,* pp. 184–210, T. A. Sebeok, ed. Bloomington, Indiana University Press.

Hamilton, K. G. A. 1971. The insect wing. Part I. Origin and development of wings from notal lobes. J. Kansas Entomol. Soc., 44:421–33.

―――. 1972. The insect wing. Part II. Vein homology and the archetypal insect wing. J. Kansas Entomol. Soc., 45:54–58.

References Cited

Hamilton, W. D. 1964. The genetical theory of social behavior, I and II. J. Theoret. Biol., 7(1):1–52.

———. 1972. Altruism and related phenomena, mainly in social insects. Ann. Rev. Ecol. Syst., 3:193–232.

Hammack, G. M. 1970. *The Serial Literature of Entomology—A Descriptive Study.* College Park, MD, The Entomological Society of America.

Handlirsch, A. 1908. Die fossilen Insekten und die Phylogenie der rezenten Formen. *Ein Handbuch für Paleontologen und Zoologen* (2 vols). Leipzig, Engelman. 1430 pp.

Hanover, J. W. 1975. Physiology of tree resistance to insects. Ann. Rev. Entomol., 20:75–95.

Haque, R., and V. J. Freed, eds. 1975. *Environmental Dynamics of Pesticides.* New York, Plenum. 387 pp.

Harcourt, D. G. and E. J. Leroux. 1967. Population regulation in insects and man. Amer. Scientist, 55(4):400–15.

Harper, P. P. 1978. Plecoptera. In *An Introduction to the Aquatic Insects of North America,* pp. 105–18, R. W. Merritt and K. W. Cummins, eds. Dubuque, IA, Kendall/Hunt Publishing Co.

Harwood, R. F., and M. T. James. 1979. *Entomology in Human and Animal Health,* 7th ed. New York, Macmillan. 548 pp.

Haskell, P. T. 1960. Stridulation and associated behavior in certain Orthoptera. 3. The influence of the gonads. Anim. Behav., 8:76–81.

———. 1961. *Insect Sounds.* Chicago, Quadrangle.

———. 1974. Sound production. In *The Physiology of Insecta,* 2nd ed., pp. 354–410, M. Rockstein, ed. New York, Academic Press.

Hassell, M. P. 1978. *The Dynamics of Arthropod Predator-Prey Systems.* Monog. Pop. Biol. 13. Princeton, NJ, Princeton University Press. 237 pp.

Hebard, M. 1934. The Dermaptera and Orthoptera of Illinois, Ill. Nat. Hist. Survey, Bull. 20(3).

Heinrich, B. 1973. The energetics of the bumblebee. Sci. Amer., 228:96–102.

———. 1974. Thermoregulation in endothermic insects. Science, 185:747–56.

———, and G. A. Bartholomew. 1972. Temperature control in flying moths. Sci. Amer., 226:70–77.

——— and P. H. Raven. 1972. Energetics and pollination ecology. Science, 176:597–602.

Helfer, J. R. 1963. *How to Know the Grasshoppers, Cockroaches, and Their Allies.* Dubuque, IA, Wm. C. Brown. 353 pp.

Hennig, W. 1965. Phylogenetic systematics. Ann. Rev. Entomol., 10:97–116.

———. 1966. *Phylogenetic Systematics.* Chicago, University of Chicago Press.

———. 1969. *Die Stammesgeschichte der Insekten.* Frankfurt-am-Main, Dramer. 436 pp.

Hepburn, H. R., ed. 1976. *The Insect Integument.* New York, Elsevier Scientific. 571 pp.

Hepburn, H. R., and I. Joffe. 1976. On the material properties of insect exoskeletons. In *The Insect Integument,* pp. 207–35, H. R. Hepburn, ed. New York, Elsevier Scientific.

Hermann, H. R. 1979. *Social Insects,* Vol. I. New York, Academic Press. 437 pp.

Herms, W. B., and M. T. James. 1961. *Medical Entomology,* 5th ed. New York, Macmillan.

Herried, C. F., II. 1977. *Biology.* New York, Macmillan. 884 pp.

Herring, J. L., and P. D. Ashlock. 1971. A key to the nymphs of the families of Hemiptera (Heteroptera) of America North of Mexico. Fla. Entomol., 54:207–12.

Hertz, M. 1929. Die Organization des optischen Feldes bei der Biene I. Z. vergl. Physiol., 8:693–748.

Heslop-Harrison, Y. 1978. Carnivorous plants. Sci. Amer., 238:104–15.

Hess, W. N. 1917. The chordotonal organs and pleural discs of cerambycid larvae. Ann. Entomol. Soc. Amer., 10:63–78.

Highnam, K. C., ed. 1964. *Insect Reproduction.* Roy. Entomol. Soc. (London), Symp. 3.

——— and L. Hill. 1980. *The Comparative Endocrinology of the Invertebrates,* 2nd ed. New York, Elsevier Scientific.

Hinde, R. A. 1969. *Animal Behavior: A Synthesis of Ethology and Comparative*

Psychology, 2nd ed. New York, McGraw-Hill.

Hinton, H. E. 1946a. A new classification of insect pupae. Proc. Zool. Soc. (London), 116:282–328.

———. 1946b. Concealed phases in the metamorphosis of insects. Nature (London), 157:552–53.

———. 1948. On the origin and function of the pupal stage. Trans. Roy. Entomol. Soc. (London), 99:395–409.

———. 1951. A new Chironomid from Africa, the larva of which can be dehydrated without injury. Proc. Zool. Soc. (London), 121:371–80.

———. 1958. Concealed phases in the metamorphosis of insects. Sci. Progr. (London), 46:260–75.

———. 1960. Cryptobiosis in the larva of *Polypedilum vanderplanki* Hint. (Chironomidae). J. Insect Physiol., 5:286–300.

———. 1963a. The origin of flight in insects. Proc. Roy. Entomol. Soc. (London) [C], 28:24–25.

———. 1963b. The origin and function of the pupal stage. Proc. Roy. Entomol. Soc. (London) [A], 50:96–113.

———. 1964. Sperm transfer in insects and the evolution of haemocoelic insemination. In *Insect Reproduction,* pp. 95–107, K. C. Highnam, ed. Roy. Entomol. Soc. (London), Symp. 3

———. 1968. Spiracular gills. Adv. Insect Physiol., 5:65–162.

———. 1969. Respiratory systems of insect egg shells. Ann. Rev. Entomol., 14:343–68.

———. 1971. Some neglected phases in metamorphosis. Proc. Roy. Entomol. Soc. (London) [C], 35:55–64.

———. 1973. Neglected phases in metamorphosis: a reply to V. B. Wigglesworth. J. Entomol. [A], 48:57–68.

———. 1974. Accessory functions of seminal fluid. J. Med. Entomol., 11:19–25.

———. 1976a. Notes on neglected phases in metamorphosis, and a reply to J. M. Whitten. Ann. Entomol. Soc. Amer., 69(3):560–66.

———. 1976b. Recent work on physical colours of insect cuticle. In *The Insect Integument,* pp. 475–96, H. R. Hepburn, ed. New York, Elsevier Scientific.

———. 1977. Enabling mechanisms. Proc. XV Int. Congr. Entomol., pp. 71–83.

———. 1979. *Biology of Insect Eggs* (3 vols). New York, Pergamon.

Hirsch, J., ed. 1967. *Behavior—Genetic Analysis.* New York, McGraw-Hill.

Hirsch, J., and L. Erlenmeyer-Kimling. 1961. Sign of taxis as a property of the genotype. Science, 134:835–36.

——— and ———. 1962. Studies in experimental behavioral genetics. IV. Chromosome analyses for geotaxis. J. Comp. Physiol. Psychol. 55:732–39.

Hitchcock, S. W. 1974. *The Plecoptera or Stoneflies of Connecticut.* Guide to the Insects of Connecticut VII, Bull. 107. 262 pp.

Hocking, B. 1971. Blood-sucking behavior of terrestrial arthropods. Ann. Rev. Entomol. 16:1–26.

Hodgson, E. S. 1958. Chemoreception in arthropods. Ann. Rev. Entomol. 3:19–36.

———. 1965. The chemical senses and changing viewpoints in sensory physiology. Viewpoints in Biol., 4:83–124.

———. 1974. Chemoreception. In *The Physiology of Insecta,* Vol. II, pp. 127–64, M. Rockstein, ed. New York, Academic Press.

Hoeglund, G., K. Hamdorf, H. Langer, R. Paulsen, and J. Schwemer. 1973. The photopigments in an insect retina. In *Biochemistry and Physiology of Visual Pigments,* H. Langer, ed. New York, Springer-Verlag.

Holland, G. P. 1949. The Siphonaptera of Canada. Can. Dept. Agr. Publ. 817, Tech. Bull. 70. 306 pp.

———. 1964. Evolution, classification, and host-relationships of Siphonaptera. Ann. Rev. Entomol., 9:123–46.

Holland, W. J. 1968. *The Moth Book.* New York, Dover (reprint of 1903 Doubleday edition).

———. 1931. *The Butterfly Book,* rev. ed. Garden City, NY, Doubleday.

Hölldobler, B. 1977. Communication in social Hymenoptera. In *How Animals Communicate,* pp. 418–71, T. A. Sebeok, ed. Bloomington, Indiana University Press.

Hopkins, G. H. E. 1949. The host-associations of the lice of mammals. Proc.

Zool. Soc. (London), 119:387–604.

———— and M. Rothschild. 1953–1971. *An Illustrated Catalog of the Rothschild Collection of Fleas (Siphonaptera) in the British Museum* (5 vols.). London, British Museum (Natural History).

Horie, Y., and H. Wantanabe. 1980. Recent advances in sericulture. Ann. Rev. Entomol., 25:49–71.

Horn, D. J. 1976. *Biology of Insects*. Philadelphia, Saunders. 439 pp.

Horridge, G. A. 1962a. Learning of leg position by the ventral nerve cord of headless insects. Proc. Roy. Soc. (London) [B], 157:33–52.

————. 1962b. Learning of leg position by headless insects. Nature (London), 193:697–98.

————. 1975. *The Compound Eye and Vision of Insects*. Oxford, Clarendon Press. 595 pp.

————. 1977. The compound eye of insects. Sci. Amer., 237:108–20.

Horsfall, F. L., and I. Tamm, eds. 1965. *Viral and Rickettsial Infections of Man*, 4th ed. Philadelphia, Lippincott. 1282 pp.

Horsfall, W. R. 1962. *Medical Entomology—Arthropods and Human Disease*. New York, Ronald Press.

Hoskins, W. M. 1963. Resistance to insecticides. Int. Rev. Trop. Med., 2:119–74.

Hotta, Y., and S. Benzer. 1972. Mapping of behavior in *Drosophila* mosaics. Nature (London), 240:527–35.

Houk, E. J., and G. W. Griffiths. 1980. Intracellular symbiotes of the Homoptera. Ann. Rev. Entomol., 25:161–87.

House, H. L. 1969. Effects of different proportions of nutrients on insects. Entomol. Exp. Appl., 12:659–69.

————. 1974a. Nutrition. In *The Physiology of Insecta*, 2nd ed., Vol. V, pp. 1–62, M. Rockstein, ed. New York, Academic Press.

————. 1974b. Digestion. In *The Physiology of Insecta*, 2nd ed., Vol. V, pp. 63–117, M. Rockstein, ed. New York, Academic Press.

Howard, L. O. 1930. *A History of Applied Entomology*. Smithsonian Inst. Misc. Coll. 84.

Howe, W. H. 1975. *The Butterflies of North America*. Garden City, NY, Doubleday. 633 pp.

Howse, P. E. 1975. Brain structure and behavior in insects. Ann. Rev. Entomol., 20:359–79.

Hoy, M. A., and J. J. McKelvey, Jr. 1979. *Genetics in Relation to Insect Management*. New York, The Rockefeller Foundation.

Hoy, R. R. 1974. Genetic control of acoustic behavior in crickets. Amer. Zool., 14:1067–80.

———— and R. C. Paul. 1972. Genetic control of song specificity in crickets. Science, 180:82–83.

Hoyle, G. 1970. Cellular mechanisms underlying behavior—neuroethology. Adv. Insect Physiol., 7:349–444.

————. 1974. Neural control of skeletal muscle. In *The Physiology of Insecta*, 2nd ed., Vol. IV, pp. 176–269, M. Rockstein, ed. New York, Academic Press.

————. 1975. The neural control of skeletal muscles. In *Insect Muscle*, pp. 501–43, P. N. R. Usherwood, ed. New York, Academic Press.

Hubbard, C. A. 1947. *Fleas of Western North America*. New York, Hafner (1968 reprint). 533 pp.

Huber, F. 1974. Neural integration (central nervous system). In *The Physiology of Insecta*, 2nd ed., Vol. IV, pp. 4–100, M. Rockstein, ed. New York, Academic Press.

Huffaker, C. B. 1971. *Biological Control*. New York, Plenum.

————, R. F. Luck, and P. S. Messenger. 1977. The ecological basis of biological control. Proc. XV Int. Congr. Entomol., pp. 560–86.

———— and P. S. Messenger, 1976 *Theory and Practice of Biological Control*. New York, Academic Press. 788 pp.

Hughes, G. M., and P. J. Mill. 1974. Locomotion: terrestrial. In *The Physiology of Insecta*, 2nd ed., Vol. III, pp. 335–79, M. Rockstein, ed. New York, Academic Press.

Hull, W. B., and G. C. Odland, eds. 1979. Directory of North American Entomologists and Acarologists. College Park, MD, The Entomological Society of America.

References Cited

Hungerford, H. B. 1959. Hemiptera. In *Freshwater Biology*, pp. 959–72, W. T. Edmondson, ed. New York, Wiley.

Huxley, H. E. 1965. The contraction of muscle. In *The Living Cell*, Readings from Scientific American, pp. 279–289. San Francisco, Freeman.

Hynes, H. B. N. 1976. Biology of Plecoptera. Ann. Rev. Entomol., 21:135–53.

———. 1977. A key to the adults and nymphs of stoneflies (Plecoptera). Sci. Pubs. Freshwater Biol. Assn., No. 14. 90 pp.

Ignoffo, C. M. 1975. Entomopathogens as insecticides. In *Insecticides of the Future*, pp. 23–40, M. Jacobson, ed. New York, Marcel Dekker.

Ilan, J., J. Ilan, and N. G. Patel. 1972. Regulation of messenger RNA translation mediated by juvenile hormone. In *Insect Juvenile Hormones—Chemistry and Action*, pp. 43–68, J. J. Menn and M. Beroza, eds. New York, Academic Press.

Ishay, J. 1977. Acoustical communication in wasp colonies (Vespinae). Proc. XV Int. Congr. Entomol., pp. 406–35.

Ivanova-Kasas, O. M. 1972. Polyembryony in insects. In *Developmental Systems: Insects*, Vol. 1, pp. 243–71, S. J. Counce and C. H. Waddington, eds. New York, Academic Press.

Jacobson, M. 1965. *Insect Sex Attractants*. New York, Wiley–Interscience.

———. 1972. *Insect Sex Pheromones*. New York, Academic Press. 382 pp.

———. 1974. Insect pheromones. In *The Physiology of Insecta*, 2nd ed., Vol. III, pp. 229–76, M. Rockstein, ed. New York, Academic Press.

Jander, R. 1963. Insect orientation. Ann. Rev. Entomol. 8:94–114.

———. 1975. Ecological aspects of spatial orientation. Ann. Rev. Ecol. Syst. 6:171–88.

Janzen, D. H. 1967. Interaction of the bull's-horn acacia (*Acacia cornigera* L.) with an ant inhabitant (*Pseudomyrmex ferruginea* F. Smith) in eastern Mexico. Univ. of Kansas Sci. Bull., 47(6):315–558.

———. 1977. Why are there so many species of insects? Proc. XV Int. Congr. Entomol., pp. 84–94.

Jaques, H. E. 1951. *How to Know the Beetles*. Dubuque, IA, Wm. C. Brown.

Jeannel, R. 1949. Classification et phylogénie des insectes. In *Traité, de Zoologie*, 9:1–110, P. P. Grassé, ed. Paris, Masson.

Jenkin, P. M., and H. E. Hinton, 1966. Apolysis in arthropod molting cycles. Nature, 211(5051):871.

Jermy, T., ed. 1974. *The Host-plant in Relation to Insect Behavior and Reproduction*. New York, Plenum. 322 pp.

Jewett, S. G., Jr. 1956. Plecoptera. In *Aquatic Insects of California*, pp. 155–81, R. L. Usinger, ed. Berkeley, University of California Press.

Johannsen, O. A., and F. H. Butt. 1941. *Embryology of Insects and Myriapods*. New York, McGraw-Hill.

Johansen, C. A. 1977. Pesticides and pollinators. Ann. Rev. Entomol., 22:177–92.

Johnson, C. G. 1963. The aerial migration of insects. Sci. Amer., 209–132–38.

———. 1966. A functional system of adaptive dispersal by flight. Ann. Rev. Entomol., 11:233–60.

———. 1969. *Migration and Dispersal of Insects by Flight*. London, Methuen.

———. 1974. Insect migration: aspects of its physiology. In *The Physiology of Insecta*, 2nd ed., Vol. III, pp. 279–334, M. Rockstein, ed. New York, Academic Press.

———. 1976. *Insect Migration*, Carolina Biology Reader, No. 84. Burlington, NC, Carolina Biological Supply Co.

Johnson, W. T., and H. H. Lyon. 1979. *Insects That Feed on Trees and Shrubs—An Illustrated Practical Guide*. Ithaca, NY, Cornell University Press.

Jones, D. P. 1973. Agricultural entomology. In *History of Entomology*, pp. 307–32, R. F. Smith, T. E. Mittler, and C. N. Smith, eds. Palo Alto, CA, Annual Reviews.

Jones, J. C. 1964. The circulatory system of insects. In *The Physiology of Insecta*, Vol. III, pp. 1–107, M. Rockstein, ed. New York, Academic Press.

———. 1968. The sexual life of a mosquito. In *The Insects*, pp. 71–78, T. Eisner and E. O. Wilson, eds. San Francisco, Freeman.

———. 1977. *The Circulatory System of Insects*. Springfield, IL, Charles C Thomas.

———. 1978. The feeding behavior of mosquitoes. Sci. Amer., 238:138–48.

Kafatos, F. C. 1972. The cocoonase zymogen cells of silk moths: a model of

References Cited

terminal cell differentiation for specific protein synthesis. Cur. Top. Dev. Biol., 7:125–91.

———— A. M. Tartakoff, and J. H. Law, 1967. Cocoonase. I. Preliminary characterization of a proteolytic enzyme from silkmoths. J. Biol. Chem., 242:1477–87.

Kamp, J. W. 1973. Numerical classification of the orthopteroids, with special reference to the Grylloblattodea. Can. Entomol., 105:1235–49.

Karlson, P., and A. Butenandt. 1959. Pheromones (ectohormones) in insects. Ann. Rev. Entomol., 4:39–58.

Karlson, P., and C. E. Sekeris. 1976. Control of tyrosine metabolism and cuticle sclerotization by ecdysone. In *The Insect Integument*, pp. 145–56, H. R. Hepburn, ed. New York, Elsevier Scientific.

Kellogg, F. E. 1970. Water vapour and carbon dioxide receptors in *Aedes aegypti*. J. Insect Physiol., 16:99–108.

Kenchington, W. 1976. Adaptations of insect peritrophic membranes to form cocoon fabrics. In *The Insect Integument*, pp. 497–513, H. R. Hepburn, ed. New York, Elsevier Scientific.

Kennedy, J. S., ed. 1961. *Insect Polymorphism*, Roy. Entomol. Soc. (London), Symp.1.

————. 1975. Insect dispersal. In *Insects, Science and Society*, pp. 103–19, D. Pimentel, ed. New York, Academic Press.

————. 1977. Olfactory responses to distant plants and other odor sources. In *Chemical Control of Insect Behavior—Theory and Application*, pp. 67–91, H. H. Shorey and J. J. McKelvey, Jr., eds. New York, Wiley.

Kennedy, J. S., and H. L. G. Stroyan. 1959. Biology of aphids. Ann. Rev. Entomol., 4:139–60.

Kettlewell, H. B. D. 1959. Darwin's missing evidence. Sci. Amer. 200:48–53.

————. 1961. The phenomenon of industrial melanism in Lepidoptera. Ann. Rev. Entomol., 6:245–62.

————. 1973. *The Evolution of Melanism. The Study of a Recurring Necessity. With Special Reference to Industrial Melanism in the Lepidoptera*. New York, Clarendon. 424 pp.

Kevan, D. K. McE. 1962. *Soil Animals*. New York, Philosophical Library. 237 pp.

Khan, M. A. Q., ed. 1977. *Pesticides in Aquatic Environments*. Symposium held at XV Int. Congr. Entomol. New York, Plenum. 257 pp.

Kim, K. C., and H. W. Ludwig. 1978. The family classification of the Anoplura. Syst. Entomol., 3(3):249–84.

King, R. C. 1974. Symposium on reproduction of arthropods of medical and veterinary importance. I. Insect gametogenesis. J. Med. Entomol., 11(1):1–7.

Kirk, D. 1975. *Biology Today*. New York, Random House.

Klassen, W. 1975. Pest management: organization and resources for implementation. In *Insects, Science and Society*, pp. 227–56, D. Pimentel, ed. New York, Academic Press.

Klopfer, P. H., and J. P. Hailman. 1967. *An Introduction to Animal Behavior. Ethology's First Century*. Englewood Cliffs, NJ, Prentice-Hall. 297 pp.

Klots, A. B. 1951. *A Field Guide to the Butterflies*. Boston, Houghton Mifflin.

————. 1958. *The World of Butterflies and Moths*. New York, McGraw-Hill. 207 pp.

Knipling, E. F. 1955. Possibilities of insect control or eradication through the use of sexually sterile males. J. Econ. Entomol., 48:459–62.

————. 1959. Screwworm eradication: concepts and research leading to the sterile male method. Smithsonian Inst. Publ. 4365.

————. 1972. Sterilization and other genetic techniques. In *Pest Control Strategies for the Future*, pp. 272–87. Washington, DC, National Academy of Sciences.

Knudsen, J. W. 1966. *Biological Techniques*. New York, Harper & Row.

Koch, A. 1967. Insects and their endosymbionts. In *Symbiosis*, Vol. 2, pp. 1–106, S. M. Henry, ed. New York, Academic Press.

Kogan, M. 1975. Plant resistance in pest management. In *Introduction to Insect Pest Management*, pp. 103–46, R. L. Metcalf and W. H. Luckman, eds. New York, Wiley.

————. 1977. The role of chemical factors in insect/plant relationships. Proc. XV Int. Congr. Entomol., pp. 211–27.

References Cited

Kormondy, E. J. 1976. *Concepts of Ecology,* 2nd ed. Englewood Cliffs, NJ, Prentice-Hall.

Krebs, C. J. 1972. *Ecology: The Experimental Analysis of Distribution and Abundance.* New York, Harper & Row. 694 pp.

Kring, J. B. 1972. Flight behavior of aphids. Ann. Rev. Entomol., 17:461–92.

Krishna, K., and F. M. Weesner, eds. 1969. *Biology of Termites,* Vol. 1. New York, Academic Press. 598 pp.

———— and ————. 1970. *Biology of Termites,* Vol. 2. New York, Academic Press. 643 pp.

Kristensen, N. P. 1975. The phylogeny of hexapod "orders." A critical review of recent accounts. Z. Zool. Syst. Evol.-forsch., 13:1–44.

Kroeger, H. 1968. Gene activities during insect metamorphosis and their control by hormones. In *Metamorphosis: A Problem in Developmental Biology,* pp. 185–219, W. Etkin and L. I. Gilbert, eds. Amsterdam, North-Holland.

Kühnelt, W. 1961. *Soil Biology with Special Reference to the Animal Kingdom.* London, Faber and Faber. 397 pp.

Kullenberg, B. 1961. Studies in *Ophrys* pollination. Zool. Bidr. Upps., 34:1–340.

Langer, W. L. 1964. The black death. Sci. Amer., 210:114–21.

Law, J. H., and F. E. Regnier. 1971. Pheromones. Ann. Rev. Biochem., 40:533–48.

Lawrence, P. A. 1970. Polarity and patterns in the post-embryonic development of insects. Adv. Insect Physiol., 7:197–266.

————. 1973. The development of spatial patterns in the integument of insects. In *Developmental Systems: Insects,* Vol. 2, pp. 157–209, S. J. Counce and C. H. Waddington, eds. New York, Academic Press.

Leclercq, M. L. 1969. *Entomological Parasitology.* New York, Pergamon.

Lee, K. E., and T. G. Wood. 1971. *Termites and Soils.* New York, Academic Press. 251 pp.

Leech, H. B., and H. P. Chandler. 1956. Aquatic Coleoptera. In *Aquatic Insects of California,* pp. 293–371, R. L. Usinger, ed. Berkeley, University of California Press.

———— and M. W. Sanderson. 1959. Coleoptera. In *Fresh-water Biology,* pp. 981–1023, W. T. E. Edmondson, ed. New York, Wiley.

Lees, A. D. 1955. *The Physiology of Diapause in Arthropods.* Cambridge, Cambridge, University Press.

Leftwich, A. W. 1976. *A Dictionary of Entomology.* New York, Crane Russack. 360 pp.

Leopold, R. A. 1976. The role of male accessory glands in insects. Ann. Rev. Entomol., 21:199–221.

Levin, D. A. 1973. The role of trichomes in plant defense. Quart. Rev. Biol., 48:3–15.

Levins, R., and M. Wilson. 1980. Ecological theory and pest management. Ann. Rev. Entomol., 25:287–308.

Lewis, C. T. 1970. Structure and function in some external receptors. In *Insect Ultrastructure,* pp. 59–76, A. C. Neville, ed. Roy. Entomol. Soc. (London), Symp. 5.

Lewis, R. E. 1972–1975. Notes on the geographic distribution and host preferences in the order Siphonaptera. Parts 1–6. J. Med. Entomol., 9:511–20; 10:255–60; 11:147–67, 403–13, 525–40, 658–76.

Lewis, T. 1973. *Thrips—Their Biology, Ecology and Economic Importance.* New York, Academic Press. 350 pp.

Lewis, T., and L. R. Taylor. 1967. *Introduction to Experimental Ecology.* New York, Academic Press. 401 pp.

L'Hélias, C. 1970. Chemical aspects of growth and development in insects. In *Chemical Zoology,* Vol. V, Part A, Arthropoda, pp. 343–400, M. Florkin and B. J. Sheer, eds.. New York, Academic Press.

Lindauer, M. 1965. Social behavior and mutual communication. In *The Physiology of Insecta,* Vol II, pp. 124–87, M. Rockstein, ed. New York, Academic Press.

————. 1974. Social behavior and mutual communication. In *The Physiology of Insecta,* 2nd ed., Vol. III, pp. 149–228, M. Rockstein, ed. New York, Academic Press.

Lindroth, C. H. 1973. Systematics specializes between Fabricius and Darwin:

References Cited

1800–1859. In *History of Entomology*, pp. 119–54, R. F. Smith, T. E. Mittler, and C. N. Smith, eds. Palo Alto, CA, Annual Reviews.

Linsenmaier, W. 1972. *Insects of the World*. Translated from German by L. E. Chadwick. New York, McGraw-Hill. 392 pp.

Linsley, E. G., and R. L. Usinger. 1961. Taxonomy. In *The Encyclopedia of the Biological Sciences*, P. Gray, ed. New York, Van Nostrand Reinhold.

Little, V. A. 1972. *General and Applied Entomology*. New York, Harper & Row. 527 pp.

Lloyd, J. E. 1966. Studies on the flash communication system in *Photinus* fireflies. Misc. Publ. Mus. Zool., Univ. Michigan, No. 130. 95 pp.

———. 1971. Bioluminescent communication in insects. Ann. Rev. Entomol., 16:97–122.

———. 1977. Bioluminescence and communication. In *How Animals Communicate*, pp. 164–83, T. A. Sebeok, ed. Bloomington, Indiana University Press.

Locke, M. 1965. Permeability of the insect cuticle to water and lipids. Science, 147:295–98.

———. 1973. Body wall. In *Syllabus: Introductory Entomology*, pp. 165–82, V. J. Tipton, ed. Provo, UT, Brigham Young University Press.

———. 1974. The structure and formation of the integument in insects. In *The Physiology of Insecta*, 2nd ed., pp. 124–213, M. Rockstein, ed. New York, Academic Press.

Lofgren, C. S., W. A. Banks, and B. M. Glancey. 1975. Biology and control of imported fire ants. Ann. Rev. Entomol., 20:1–30.

Lovell, H. B. 1963. The honey bee as a pollinating agent. In *The Hive and the Honey Bee*, pp. 463–76, R. A. Grout, ed. Hamilton, IL, Dadant & Sons.

Luck, R. F., R. van den Bosch, and R. Garcia. 1977. Chemical insect control— a troubled pest management strategy. BioScience, 27(9):606–11.

Luckmann, W. H., and R. L. Metcalf. 1975. The pest-management concept. In *Introduction to Insect Pest Management*, pp. 3–35, R. L. Metcalf and W. H. Luckman, eds. New York, Wiley.

Lunt, G. G. 1975. Synaptic transmission in insects. In *Insect Biochemistry and Function*, D. J. Candy and B. A. Kilby, eds. New York, Wiley (Halsted Press).

Lüscher, M. 1961a. Social control of polymorphism in termites. In *Insect Polymorphism*, J. S. Kennedy, ed. Roy. Entomol. Soc. (London), Symp. 1, pp. 57–67).

———. 1961b. Air-conditioned termite nests. Sci. Amer., 205:138–45.

———, ed. 1976. *Phase and Caste Determination in Insects*. Elmsford, NY, Pergamon.

———, and A. Springhetti. 1963. Functions of the corpora allata in the development of termites. Proc. XVI Int. Congr. Zool., Washington, 4:244–50.

MacArthur, R. H. 1972. *Geographical Ecology: Patterns in the Distribution of Species*. New York, Harper & Row.

Mackerras, I. M. 1970. Evolution and classification of insects. In *The Insects of Australia*, pp. 152–67, C.S.I.R.O., Canberra, Melbourne, Melbourne University Press.

Maddrell, S. H. P. 1971. The mechanisms of insect excretory systems. Adv. Insect Physiol., 8:200–324.

Maddox, J. V. 1975. Use of diseases in pest management. In *Introduction to Insect Pest Management*, pp. 189–233, R. L. Metcalf and W. H. Luckmann, eds. New York, Wiley.

Madelin, M. F. 1966. Fungal parasites of insects. Ann. Rev. Entomol., 11:423–48.

Mahowald, A. P. 1972. Oogenesis. In *Developmental Systems: Insects*, Vol. I, pp. 1–47, S. J. Counce and C. H. Waddington, eds. New York, Academic Press.

Mallis, A. 1971. *American Entomologists*. New Brunswick, NJ, Rutgers University Press. 549 pp.

Mangan, A. 1934. *La Locomotion chez les Animaux, I: le Volume des Insectes*. Paris, Hermann & Cie.

Manning, A. 1961. Effects of artificial selection for mating speed in *Drosophila melanogaster*. Anim. Behav., 9:82–92.

———. 1966. Sexual behavior. In *Insect Behavior*, pp. 59–68, P. T. Haskell, ed. Roy. Entomol. Soc. (London), Symp. 3.

References Cited

————. 1967. Genes and the evolution of insect behavior. In *Behavior—Genetic Analysis*, pp. 44–60, J. Hirsch, ed. New York, McGraw-Hill.

Mansingh, A. 1971. Physiological classification of dormancies in insects. Can. Entomol., 103:983–1009.

Manton, S. M. 1964. Mandibular mechanisms and the evolution of arthropods. Phil. Trans. Roy. Soc. (London), B. Biol. Sci., 247:1–183.

————. 1970. Arthropoda: Introduction. In *Chemical Zoology*, Vol. 5, pp. 1–34, M. Florkin and B. J. Sheer, eds. New York, Academic Press.

————. 1972. The evolution of arthropodan locomotory mechanisms. Part 10: Locomotory habits, morphology and evolution of the hexapod classes. Zool. J. Linn. Soc., 51:203–400.

————. 1977. *The Arthropoda. Habits, Functional Morphology and Evolution.* London, Oxford University Press.

————. 1979. Functional morphology and the evolution of the hexapod classes. In *Arthropod Phylogeny*, pp. 387–465, A. P. Gupta, ed. New York, Van Nostrand Reinhold.

Maramorosch, K., ed. 1977. *The Atlas of Insect and Plant Viruses.* New York, Academic Press. 512 pp.

———— and K. F. Harris. 1979. *Leafhopper Vectors and Plant Disease Agents.* New York, Academic Press. 654 pp.

Markl, H. 1962. Schweresinnesorgane bei Ameisen und anderen Hymenopteren. Z. vergl. Physiol., 44:475–569.

————. 1974. Insect behavior: functions and mechanisms. In *The Physiology of Insecta*, 2nd ed., Vol. III, pp. 3–148, M. Rockstein, ed. New York, Academic Press.

———— and M. Lindauer. 1965. Physiology of insect behavior. In *The Physiology of Insecta*, Vol. II, pp. 3–122, M. Rockstein, ed. New York, Academic Press.

Marler, P., and W. J. Hamilton III. 1966. *Mechanisms of Animal Behavior.* New York, Wiley.

Marquis, D. 1960. *Archy and Mehitabel.* Garden City, NY, Doubleday (Dolphin Books).

————. 1966. *Archy's Life of Mehitabel.* Garden City, NY, Doubleday (Dolphin Books).

Martin, E. C., and S. E. McGregor. 1973. Changing trends in insect pollination of commercial crops. Ann. Rev. Entomol., 18:207–26.

Martin, M. M. 1970. The biochemical basis of the fungus–attine ant symbiosis. Science, 169:16–20.

Maruyama, K. 1974. The biochemistry of the contractile elements of insect muscle. In *The Physiology of Insecta*, 2nd ed., Vol. IV, pp. 237–69, M. Rockstein, ed. New York, Academic Press.

Masek Fialla, K. 1941. Die Korpertemperatur Poikilothermes Thiere im Abhangigkeit vom Kleinklima. Z. wiss. Zool., 154:170–247.

Matheson, R. 1944. *Handbook of the Mosquitos of North America*, 2nd ed. Ithaca, NY, Cornell University Press.

Matsuda, R. 1965. Morphology and evolution of the insect head. Mem. Amer. Entomol. Inst. 4. 334 pp.

————. 1970. Morphology and evolution of the insect thorax. Entomol. Soc. Canada, Mem. 76.

————. 1976. *Morphology and Evolution of the Insect Abdomen.* Elmsford, NY, Pergamon.

Matsumura, F. 1975. *Toxicology of Insecticides*, New York, Plenum. 504 pp.

Matthews, R. W. 1968. *Microstigmus comes:* Sociality in a sphecid wasp. Science, 160:787–88.

———— and J. R. Matthews. 1978. *Insect Behavior*, New York, Wiley. 507 pp.

Maxwell, F. G., J. N. Jenkins, and W. L. Parrott. 1972. Resistance of plants to insects. Adv. Agron., 24:187–265.

———— and P. R. Jennings. 1979. *Breeding Plants Resistant to Insects.* New York, Wiley.

May, M. L. 1979. Insect thermoregulation. Ann. Rev. Entomol., 24:313–49.

May, R. M., ed. 1976. *Theoretical Ecology. Principles and Applications.* Oxford, Blackwell Scientific Publications.

Maynard, E. A. 1951. *A Monograph of the Collembola or Springtail Insects of*

New York State. Ithaca, NY, Comstock. 339 pp.

Mayr, E. 1963. *Animal Species and Evolution*. Cambridge, MA, Harvard University Press.

———. 1969. *Principles of Systematic Zoology*. New York, McGraw-Hill Book Company.

———. 1970. *Populations, Species and Evolution*. Cambridge, MA, Belknap Press/Harvard University Press. 453 pp.

——— E. G. Linsley, and R. L. Usinger. 1953. *Methods and Principles of Systematic Zoology*. New York, McGraw-Hill.

Mazokhin-Porschnyakov, G. A. 1969. *Insect Vision*. New York, Plenum. 306 pp.

McCafferty, W. P. 1975. The burrowing mayflies (Ephemeroptera: Ephemeroidea) of the United States. Trans. Amer. Entomol. Soc., 101:447–504.

McCann, F. V. 1970. Physiology of insect hearts. Ann. Rev. Entomol., 15:173–200.

McClearn, G. E., and J. C. DeFries. 1973. *Introduction to Behavior Genetics*. San Francisco, Freeman.

McConnell, E., and A. G. Richards. 1955. How fast can a cockroach run? Bull. Brooklyn Entomol. Soc., 50:36–43.

McDonald, T. J. 1975. Neuromuscular pharmacology of insects. Ann. Rev. Entomol., 20:151–66.

McElroy, W. D., and H. H. Seliger. 1962. Biological luminescence. Sci. Amer. 207:76–89.

———, ———, and M. DeLuca. 1974. Insect bioluminescence. In *The Physiology of Insecta*, 2nd ed., Vol. II, pp. 411–60, M. Rockstein, ed. New York, Academic Press.

McEwen, F. L., and G. R. Stephenson. 1979. *The Use and Significance of Pesticides in the Environment*. New York, Wiley.

McGregor, S. E. 1976. *Insect Pollination of Cultivated Crop Plants*. USDA Agr. Handbook No. 496. 411 pp.

McIver, S. B. 1975. Structure of cuticular mechanoreceptors of arthropods. Ann. Rev. Entomol., 20:381–98.

McKelvey, J. J., Jr. 1973. *Man Against Tsetse—Struggle for Africa*. Ithaca, NY, Cornell University Press.

———. 1975. Insects and human welfare. In *Insects, Science and Society*, pp. 13–24, D. Pimentel, ed. New York, Academic Press.

McKittrick, F. A. 1964. Evolutionary Studies of Cockroaches. Cornell Univ. Agr. Exp. Sta. Mem. 389. 197 pp.

McNew, G. L. 1972. Concept of pest management. In *Pest Control Strategies for the Future*, pp. 119–33. Washington, DC, National Academy of Sciences.

Meglitsch, R. A. 1967. *Invertebrate Zoology*. London, Oxford University Press.

Melnikov, N. N. 1971. *Chemistry of Pesticides*. New York, Springer-Verlag. 480 pp.

Menn, J. J., and M. Beroza, eds. 1972. *Insect Juvenile Hormones—Chemistry and Action*. New York, Academic Press.

——— and F. M. Pallos. 1975. Development of morphogenetic agents in insect control. In *Insecticides of the Future*, pp. 71–88, M. Jacobson, ed. New York, Marcel Dekker.

Menzel, R., and J. Erber. 1978. Learning and memory in bees. Sci. Amer., 239:102–10.

Merritt, R. W., and K. W. Cummins, eds. 1978. *An Introduction to the Aquatic Insects of North America*. Dubuque, IA, Kendall/Hunt Publishing Company. 512 pp.

——— and E. I. Schlinger. 1978. Aquatic Diptera, Part 2, Adults of aquatic Diptera. In *An Introduction to the Aquatic Insects of North America*, pp. 259–83, R. W. Merritt and K. W. Cummins, eds. Dubuque, IA, Kendall/Hunt Publishing Company.

Messenger, P. S. 1975. Parasites, predators, and population dynamics. In *Insects, Science and Society*, pp. 201–23, D. Pimentel, ed. New York, Academic Press.

Metcalf, C. L., W. P. Flint, and R. L. Metcalf. 1962. *Destructive and Useful Insects*. New York, McGraw-Hill.

Metcalf, R. L. 1975. Insecticides in pest management. In *Introduction to Insect Pest Management*, pp. 235–73, R. L. Metcalf and W. H. Luckmann, eds. New York, Wiley.

——— and W. H. Luckmann, eds. 1975. *Introduction to Insect Pest Management*. New York, Wiley. 587 pp.

References Cited

—— and R. A. Metcalf. 1975. Attractants, repellents, and genetic control in pest management. In *Introduction to Insect Pest Management,* pp. 275–306, R. L. Metcalf and W. H. Luckmann, eds. New York, Wiley.

Metcalf, Z. P. 1945. *A Bibliography of the Homoptera (Auchenorrhyncha),* Vols. 1 and 2. Raleigh, NC, North Carolina State College, Dept. of Zoology and Entomology).

——. 1954–1963. *General Catalogue of the Homoptera.* Raleigh, NC, University of North Carolina at Raleigh.

——. 1962–1967. *General Catalogue of the Homoptera.* USDA Agr. Res. Service.

—— and C. L. Metcalf. 1928. *A Key to the Principal Orders and Families of Insects.* Published by the authors.

Michener, C. D. 1974. *The Social Behavior of Bees. A Comparative Study.* Cambridge, MA, Belknap Press/Harvard University Press. 404 pp.

——. 1975. The Brazilian bee problem. Ann. Rev. Entomol., 20:399–416.

Milburn, N. S., and K. D. Roeder. 1962. Control of efferent activity in the cockroach terminal abdominal ganglion by extracts of the corpora cardiaca. Gen. Comp. Endocrinol. 2:70–76.

——, E. A. Weiant, and K. D. Roeder. 1960. The release of efferent nerve activity in the roach, *Periplaneta americana,* by extracts of the corpus cardiacum. Biol. Bull. Mar. Biol. Lab., Woods Hole, 118:111–19.

Miles, P. W. 1972. The saliva of Hemiptera. Adv. Insect Physiol., 9:183–256.

Mill, P. J. 1974. Respiration: aquatic insects. In *The Physiology of Insecta,* 2nd ed., Vol. VI, pp. 403–67, M. Rockstein, ed. New York, Academic Press.

——, ed. 1976. *Structure and Function of Proprioceptors in the Invertebrates.* New York, Wiley (Halsted Press). 686 pp.

Miller, L. A. 1970. Structure of the green lacewing tympanal organ (*Chrysopa carnea,* Neuroptera). J. Morphol., 131:359–82.

Miller, M. W., and G. C. Berg, eds. 1969. *Chemical Fallout.* Springfield, IL, Charles C Thomas.

Miller, P. L. 1974a. Respiration—aerial gas transport. In *The Physiology of Insecta,* 2nd ed., Vol VI, pp. 345–402, M. Rockstein, ed. New York, Academic Press.

——. 1974b. The neural basis of behavior. In *Insect Neurobiology,* pp. 359–430, J. E. Treherne, ed. New York, American Elsevier.

Miller, T. A. 1974. Electrophysiology of the insect heart. In *The Physiology of Insecta,* 2nd ed., Vol. V, pp. 169–200, M. Rockstein, ed., New York, Academic Press.

——. 1975a. Insect visceral muscle. In *Insect Muscle,* pp. 545–606, P. N. R. Usherwood, ed. New York, Academic Press.

——. 1975b. Neurosecretion and the control of visceral organs in insects. Ann. Rev. Entomol., 20:133–149.

Miller, W. H., C. D. Bernard, and J. L. Allen. 1968. The optics of insect compound eyes. Science, 162:760–67.

Mills, H. B. 1934. *A Monograph of the Collembola of Iowa.* Ames, IA, Iowa State College Press, Monog. 3. 143 pp.

Minton, S. A. 1974. *Venom Diseases.* Springfield, IL, Charles C Thomas. 235 pp.

Mitchell, R. T., and H. S. Zim. 1964. *Butterflies and Moths.* New York, Golden Press. 160 pp.

Mitchell, T. B. 1960 & 1962. *Bees of the Eastern United States.* Vol. 1 (1960), Tech. Bull. 141, 538 pp. Vol. 2 (1962), Tech. Bull. 152, 557 pp. North Carolina Agr. Exp. Sta.

Mockford, E. L. 1951. The Psocoptera of Indiana. Proc. Indiana Acad. Sci., 60:192–204.

—— and A. B. Gurney. 1956. A review of the psocids, or book-lice and bark-lice of Texas (Psocoptera). J. Wash. Acad. Sci., 46:353–68.

Montagner, H. 1977. Learning and communication. Proc. XV Int. Congr. Entomol., pp. 397–99.

More, D. 1976. *The Bee Book.* New York, Universe Books. 143 pp.

Morgan, E. D., and C. F. Poole. 1976. The extraction and determination of ecdysones in arthropods. Adv. Insect Physiol., 12:17–62.

Morse, R. A. 1975. *Bees and Beekeeping.* Ithaca, NY, Cornell University Press. 320 pp.

——. 1978. *Honey Bee Pests, Predators and Diseases.* Ithaca, NY, Cornell

University Press. 430 pp.

Moths of America North of Mexico. 1971 to present (issued as fasicles for each family). London, E. Classey and R.B.D. Publishers.

Muesebeck, C. F. W., K. V. Krombein, H. K. Townes et al. 1951–1967. Hymenoptera of America North of Mexico; Synoptic Catalogue. USDA Agr. Monog. 2, 1951, 1420 pp. First Supplement to Monog. 2, 1958, 305 pp. Second Supplement to Monog. 2, 1967, 584 pp.

Müller, H. J. 1970. Formen der Dormanz bei Insekten. Nova Acta Leopoldina, 35:1–27.

Müller, J. F. T. 1879. *Ituna* and *Thyridia:* a remarkable case of mimicry in butterflies. Trans. Entomol. Soc., (Proc.) 1879:20–28.

Nachtigall, W. A. 1960. Über Kinematik, Dynamik und Energetik des Schwimmens einheimischen Dytisciden. Z. vergl. Physiol., 43:48–118.

———. 1963. Zur Lokomotionsmechanik schwimmender Dipterenlarven. Z. vergl. Physiol., 46:449–66.

———. 1974a. Locomotion: mechanics and hydrodynamics of swimming in aquatic insects. In *The Physiology of Insecta,* 2nd ed., Vol. III, pp. 381–432, M. Rockstein, ed. New York, Academic Press.

———. 1974b. *Insects in Flight.* New York, McGraw-Hill. 153 pp.

Naisse, J. 1966. Controle endocrine de la différenciation sexuelle chez l'insecte *Lampyris noctiluca.* I–III, Arch. Biol., Liège, 77:139–201. Gen. Comp. Endocrinol., 7:85–104, 105–110.

———. 1969. Role des neurohormones dans la différenciation sexuelle de *Lampyris noctiluca.* J. Insect Physiol., 15:877–892.

N.A.S. 1969. *Insect-Pest Management and Control,* Subcommittee on Insect Pests. Washington, DC, National Academy of Sciences. 508 pp.

———. 1972. *Pest Control Strategies for the Future.* Washington, DC, National Academy of Sciences, 376 pp.

Nash, T. A. M. 1969. *Africa's Bane—The Tsetse Fly.* London, Collins.

Needham, A. E. 1978. Insect biochromes: their chemistry and role. In *Biochemistry of Insects,* pp. 233–305, M. Rockstein, ed. New York, Academic Press.

Needham, J. G., and P. W. Claassen. 1925. *A Monograph of the Plecoptera or Stoneflies of America North of Mexico.* Thomas Say Foundation Publ. 2. College Park, MD, The Entomological Society of America. 397 pp.

——— (chm.), P. S. Galtsoff, F. E. Lutz, and P. S. Welch. 1937. *Culture Methods for Invertebrate Animals.* New York, Dover. 590 pp.

———, J. R. Traver, and Y. C. Hsu. 1935. *The Biology of Mayflies, with a Systematic Account of North American Species.* Ithaca, NY, Comstock. 759 pp.

——— and M. J. Westfall. 1955. *A Manual of the Dragonflies of North America (Anisoptera).* Berkeley, University of California Press. 615 pp.

Nelson, W. A., J. E. Keirans, J. F. Bell, and C. M. Clifford. 1975. Host-ectoparasite relationships. J. Med. Entomol., 12(2):143–66.

Neville, A. C. 1970. Cuticle ultrastructure in relation to the whole insect. In *Insect Ultrastructure,* pp. 1–16, A. C. Neville, ed. Roy. Entomol. Soc. (London), Symp. 5.

———. 1975. *Biology of the Arthropod Cuticle.* New York, Springer-Verlag. 448 pp.

Newsom, L. D. 1975. Pest management: concept to practice. In *Insects, Science and Society,* pp. 257–77, D. Pimentel, ed. New York, Academic Press.

New York Academy of Sciences. 1972. *Regulation of Insect Populations by Microorganisms.* Conference on Regulation of Insect Populations by Microorganisms.

Nienhaus, F., and R. A. Sikora. 1979. Mycoplasmas, spiroplasmas, and rickettsialike organisms as plant pathogens. Ann. Rev. Phytopathol., 17:37–58.

Noirot, C., and A. Quennedy. 1974. Fine structure of insect epidermal glands. Ann. Rev. Entomol., 19:61–80.

Nørgaard Holm, S. 1966. The utilization and management of bumble bees for red clover and alfalfa seed production. Ann. Rev. Entomol., 11:155–82.

Novák, V. J. A. 1975. *Insect Hormones.* New York, Wiley (Halsted Press).

Novy, J. E. 1978. Operation of a screwworm eradication program. In *The Screwworm Problem—Evolution of Resistance to Biological Control,* pp. 19–36, R. H. Richardson, ed. Austin, University of Texas Press.

Nowock, J., and L. I. Gilbert. 1967. *In vitro* analysis of factors regulating the juvenile hormone titer of insects. In *Invertebrate Tissue Culture,* pp. 203–12, E. Kurstak and K. Maramorosch, eds. New York, Academic Press.

O'Brien, R. D. 1967. *Insecticides Action and Metabolism.* New York, Academic Press. 332 pp.

———. 1978. The biochemistry of toxic action of insecticides. In *Biochemistry of Insects,* pp. 515–39, M. Rockstein, ed. New York, Academic Press.

Odum, E. P. 1971. *Fundamentals of Ecology,* 3rd ed. Philadelphia, Saunders. 574 pp.

——— and H. T. Odum. 1959. *Fundamentals of Ecology,* 2nd ed. Philadelphia, Saunders.

Oldroyd, H. 1964. *The Natural History of Flies.* New York, Norton.

Olkowski, H., and W. Olkowski. 1976. Entomophobia in the urban ecosystem; some observations and suggestions. Bull. Entomol. Soc. Amer., 22(3):313–17.

Osborn, H. 1937. *Fragments of Entomological History.* Columbus, OH, published by the author.

Oster, G. F., and E. O. Wilson. 1978. *Caste Ecology in the Social Insects.* Princeton, NJ, Princeton University Press.

Otte, D. 1974. Effects and functions in the evolution of signalling systems. Ann. Rev. Ecol. Syst., 5:385–417.

———. 1977. Communication in Orthoptera. In *How Animals Communicate,* pp. 334–61, T. A. Sebeok, ed. Bloomington, Indiana University Press.

Packard, A. S. 1898. *A Textbook of Entomology.* New York, Macmillan.

Painter, R. H. 1958. Resistance of plants to insects. Ann. Rev. Entomol., 3:267–90.

Pal, R. 1977. Problems of insecticide resistance in insect vectors of human disease. Proc. XV Int. Congr. Entomol., pp. 800–11.

——— and M. J. Whitten, eds. 1974. *The Use of Genetics in Insect Control.* New York, Elsevier Scientific. 241 pp.

Pan, M. L., W. J. Bell, and W. H. Telfer. 1969. Vitellogenic blood protein synthesis by insect fat body. Science, 165:393–94.

Park, T. 1962. Beetles, competition and populations. Science, 138:1369–79.

Parnas, I., and D. Dagan. 1971. Functional oragnization of giant axons in the central nervous systems of insects: new aspects. Adv. Insect Physiol., 8:95–144.

Pasteels, J. M. 1977. Evolutionary aspects in chemical ecology and chemical communication. Proc. XV Int. Congr. Entomol., pp. 281–93.

Pathak, M. D. 1975. Utilization of insect–plant interactions in pest control. In *Insects, Science and Society,* pp. 121–48, D. Pimentel, ed. New York, Academic Press.

Patton, R. L. 1963. *Introductory Insect Physiology.* Philadelphia, Saunders.

Pennak, R. W. 1978. *Fresh-water Invertebrates of the United States.* New York, Ronald Press. 769 pp.

Percival, M. S. 1965. *Floral Biology.* Oxford, Pergamon. 243 pp.

Perring, F. H., and K. Mellanby, eds. 1977. *Ecological Effects of Pesticides.* Linnaean Society Symposium Series, No. 5. New York, Academic Press. 204 pp.

Perry, A. S., and M. Agosin. 1975. The physiology of insecticide resistance by insects. In *The Physiology of Insecta,* 2nd ed., Vol. VI, pp. 3–121, M. Rockstein, ed. New York, Academic Press.

Pesticide Index, 5th ed. 1976. College Park, MD, Entomol. Soc. Amer.

Peters, W. 1976. Investigations on the peritrophic membranes of Diptera. In *The Insect Integument,* pp. 515–43, H. R. Hepburn, ed. New York, Elsevier Scientific.

Peterson, A. 1948. *Larvae of Insects,* Part I, Lepidoptera and Plant Infesting Hymenoptera. Columbus, OH, published by the author.

———. 1951. *Larvae of Insects,* Part II, Coleoptera, Diptera, Neuroptera, Siphonaptera, Mecoptera, Trichoptera. Columbus, OH, published by the author.

———. 1959. *Entomological Techniques.* Ann Arbor, MI, Edwards Brothers.

Pfadt, R. E., ed. 1978. *Fundamentals of Applied Entomology,* 3rd ed. New York, Macmillan.

Philip, C. B., and L. E. Rozeboom. 1973. Medico-veterinary entomology: a generation of progress. In *History of Entomology,* pp. 333–60, R. F. Smith, T. E. Mittler, and C. N. Smith, eds. Palo Alto, CA, Annual Reviews.

References Cited

Phillips, D. M. 1970. Insect sperm: their structure and morphogenesis. J. Cell Biol., 44:243–77.

Pichon, Y. 1974. The pharmacology of the insect nervous system. In *The Physiology of Insecta*, 2nd ed., Vol. IV, pp. 102–74, M. Rockstein, ed. New York, Academic Press.

Pillemer, E. A., and W. M. Tingey. 1976. Hooked trichomes: a physical plant barrier to a major agricultural pest. Science, 193:482–84.

Pimentel, D. 1975. Introduction. In *Insects, Science and Society*, D. Pimentel, ed. New York, Academic Press.

Platt, R. B., and J. F. Griffiths. 1964. *Environmental Measurement and Interpretation*. New York, Reinhold. 235 pp.

Poinar, G. O., Jr. 1972. Nematodes as facultative parasites of insects. Ann. Rev. Entomol., 17:103–22.

———. 1975. *Entomogenous Nematodes*. Leiden, Netherlands, Brill. 317 pp.

——— and G. M. Thomas. 1978. *Diagnostic Manual for the Identification of Insect Pathogens*. New York, Plenum. 210 pp.

Polhemus, J. T. 1978. Aquatic and semiaquatic Hemiptera. In *An Introduction to the Aquatic Insects of North America*, pp. 119–31, R. W. Merritt and K. W. Cummins, eds. Dubuque, IA, Kendall/Hunt Publishing Co.

Pomeranz, C. 1959. Arthropods and psychic disturbances. Bull. Entomol. Soc. Amer., 5:65–67.

Popham, E. S. 1965. A key to Dermaptera subfamilies. Entomologist, 98:126–36.

Portmann, A. 1959. *Animal Camouflage*. Ann Arbor Science Library, University of Michigan Press. 111 pp.

Powell, J. A., and R. A. Mackie. 1966. *Biological Interrelationships of Moths and Yucca whipplei (Lepidoptera: Gelechiidae, Blastobasidae, Prodoxidae)*. Berkeley, University of California Press.

Pratt, H. D., and K. S. Littig. 1962. *Insecticides for the Control of Insects of Public Health Importance*. U. S. Public Health Service, Publ. 772.

Price, G. M. 1972. Protein and nucleic acid metabolism in insect fat body. Biol. Rev., 48:333–75.

Price, P. W. 1975. *Insect Ecology*. New York, Wiley. 514 pp.

Pringle, J. W. S. 1957. *Insect Flight*. Cambridge, Cambridge University Press.

———. 1968. Comparative physiology of the flight motor. Adv. Insect Physiol., 5:163–228.

———. 1974. Locomotion: flight. In *The Physiology of Insecta*, 2nd ed., Vol. III, pp. 433–76, M. Rockstein, ed. New York, Academic Press.

———. 1975. *Insect Flight*. Oxford Biology Reader, No. 52. London, Oxford University Press. 16 pp.

Proctor, M., and P. Yeo. 1972. *The Pollination of Flowers*. New York, Taplinger. 418 pp.

Pyenson, L. L., and H. E. Barke. 1977. *Fundamentals of Entomology and Plant Pathology*. Westport, AVI Publishing. 344 pp.

Rabe, W. 1953. Beitrage zum Orientierungsproblem der Wasserwanzen. Z. vergl. Physiol., 35:300–25.

Ragge, D. R. 1973. Dictyoptera (cockroaches & mantises). In *Insects and Other Arthropods of Medical Importance*, pp. 399–403, K. G. V. Smith, ed. London, British Museum (Natural History).

Rainey, R. C., ed. 1976. *Insect Flight*. Roy. Entomol. Soc. (London,) Symp. 7. London, Blackwell Scientific.

Rathcke, B. J., and R. W. Poole. 1975. Coevolutionary race continues: butterfly larval adaptation to plant trichomes. Science, 187:175–76.

Raup, D. M., and S. M. Stanley. 1978. *Principles of Paleontology*. San Francisco, Freeman.

Rees, H. H. 1977. *Insect Biochemistry*. New York, Wiley (Halsted Press).

Reichstein, T., J. von Euw, J. A. Parsons, and M. Rothschild. 1968. Heart poisons in the monarch butterfly. Science, 161:861–66.

Reik, E. F. 1970. Fossil history. In *The Insects of Australia*, pp. 168–86, C.S.I.R.O., Canberra. Melbourne, Melbourne University Press.

Remington, C. L. 1954. The "Apterygota." In *A Century of Progress in the Natural Sciences, 1853–1953*, pp. 495–505, E. L. Kessel, ed. San Francisco, California Academy of Sciences.

References Cited

————. 1956. The suprageneric classification of the order Thysanura (Insecta). Ann. Entomol. Soc. Amer., 47:277–86.

Remington, J. E. 1975. *Insects of the World*. New York, Ridge Press. 159 pp.

Rempel, J. G. 1975. The evolution of the insect head: the endless dispute. Quaestiones Entomologicae, 11:7–25.

Rettenmeyer, C. W. 1970. Insect mimicry. Ann. Rev. Entomol., 15:43–74.

Richard, G. 1973. The historical development of Nineteenth and Twentieth Century studies on the behavior of insects. In *History of Entomology*, pp. 477–502, R. F. Smith, T. E. Mittler, and C. N. Smith, eds. Palo Alto, CA, Annual Reviews.

Richards, A. G. 1951. *The Integument of Arthropods*. Minneapolis, University of Minnesota Press.

————. 1973. Anatomy and morphology. In *History of Entomology*, pp. 185–202, R. F. Smith, T. E. Mittler, and C. N. Smith, eds. Palo Alto, CA, Annual Reviews.

————. 1978. The chemistry of insect cuticle. 1978. In *Biochemistry of Insects*, pp. 205-32, M. Rockstein, ed. New York, Academic Press.

———— and P. A. Richards. 1977. The peritrophic membranes of insects. Ann. Rev. Entomol., 22:219–40.

Richards, O. W. 1953. *The Social Insects*. New York, Harper & Row. 219 pp.

————. 1961. An introduction to the study of polymorphism in insects. In *Insect Polymorphism*, pp. 2–10, J. S. Kennedy, ed. Roy. Entomol. Soc. (London), Symp. 1.

———— and R. G. Davies. 1977. *Imm's General Textbook of Entomology*, 10th ed., Vol. 1, Structure, Physiology and Development, 418 pp. Vol. 2, Classification and Biology, 1354 pp. New York, Wiley (Halsted Press).

———— and ————. 1978. *Imm's Outlines of Entomology*, 6th ed. New York, Wiley. 254 pp.

Richardson, R. H. ed. 1978. *The Screwworm Problem—Evolution of Resistance to Biological Control*. Austin, University of Texas Press.

Ricker, W. E. 1959. Plecoptera. In *Fresh-water Biology*, pp. 941–57, W. T. Edmondson, ed. New York, Wiley.

Ricklefs, R. E. 1973. *Ecology*. Newton, MA, Chiron Press.

Riddiford, L. M. 1972. Juvenile hormone and insect embryonic development: its potential role as an ovicide. In *Insect Juvenile Hormones—Chemistry and Action*, pp. 95–111, J. J. Menn and M. Beroza, eds. New York, Academic Press.

———— and J. W. Truman. 1974. Hormones and insect behavior. In *Experimental Analysis of Insect Behavior*, pp. 286–96, L. Barton Browne, ed. New York, Springer-Verlag.

———— and ————. 1978. Biochemistry of insect hormones and insect growth regulators. In *Biochemistry of Insects*, pp. 307–57, M. Rockstein, ed. New York, Academic Press.

Ridgway, R. L., and S. B. Vinson, eds. 1977. *Biological Control by Augmentation of Natural Enemies*. New York, Plenum. 480 pp.

Riley, C. V. 1888. The habits of *Thalessa* and *Tremex*. U.S. Div. Ent. Ins. Life, 1:168–79.

Ritchie, C. I. A. 1979. *Insects, The Creeping Conquerors*. New York, Elsevier/Nelson Books. 139 pp.

Roberts, D. A. 1978. *Fundamentals of Plant-Pest Control*. San Francisco, Freeman. 242 pp.

Rockstein, M., ed. 1973–1974. *The Physiology of Insecta*, Vol. I, 1973; Vols. II–VI, 1974. New York, Academic Press.

————, ed. 1978. *Biochemistry of Insects*. New York, Academic Press. 649 pp.

———— and J. Miquel. 1973. Aging in insects. In *The Physiology of Insecta*, 2nd ed., Vol. I, pp. 371–478, M. Rockstein, ed. New York, Academic Press.

Roeder, K. D. 1953. *Insect Physiology*. New York, Wiley.

————. 1959. A physiological approach to the relation between prey and predator, In *Studies in Invertebrate Morphology*, pp. 287–306. Smithsonian Institution, Misc. Coll. 157.

————. 1963. *Nerve Cells and Insect Behavior*. Cambridge, MA, Harvard University Press.

————. 1965. Moths and ultrasound. Sci. Amer. 212:94–102.

————. 1966. Auditory system of noctuid moths. Science, 154:1515–21.

References Cited

————. 1967. *Nerve Cells and Insect Behavior,* rev. ed. Cambridge, MA, Harvard University Press. 238 pp.

————. 1970. Episodes in insect brains. Amer. Sci., 58:378–89.

————. 1972. Acoustic and mechanical sensitivity of the distal lobe of the pilifer in choerocampine hawkmoths. J. Insect Physiol., 18:1249–64.

————. 1974. Some neuronal mechanisms of simple behavior. Adv. Study Behav., 5:2–46.

Roelofs, W. L. 1975a. Insect Communication—Chemical. In *Insects, Science and Society,* pp. 79–99, D. Pimentel, ed. New York, Academic Press.

————. 1975b. Manipulating sex pheromones for insect suppression. In *Insecticides of the Future,* pp. 41–59, M. Jacobson, ed. New York, Marcel Dekker.

————. 1978. Chemical control of insects by pheromones. In *Biochemistry of Insects,* pp. 419–64, M. Rockstein, ed. New York, Academic Press.

———— and R. T. Cardé. 1977. Responses of Lepidoptera to synthetic sex pheromone chemicals and their analogues. Ann. Rev. Entomol., 22:377–405.

Rohdendorf, B. B. 1969. Palaontologie. In *Handbuch der Zoologie,* 4(2), Lfg. 9:1–27, J. G. Helmcke, D. Starck, and H. Wermuth, eds. Berlin, Walter de Gruyter.

————. 1973. The history of paleoentomology. In *History of Entomology,* pp. 155–70, R. F. Smith, T. E. Mittler, and C. N. Smith, eds. Palo Alto, CA, Annual Reviews.

Rolston, L. H., and C. E. McCoy. 1966. *Introduction to Applied Entomology.* New York, Ronald Press.

Rosenthal, G. A., and D. H. Janzen, eds. 1979. *Herbivores—Their Interaction with Secondary Plant Metabolites.* New York, Academic Press. 744 pp.

Ross, E. S. 1940. A revision of the Embioptera of North America. Ann. Entomol. Soc. Amer., 33:629–76.

————. 1944. A revision of the Embioptera, or webspinners, of the new world. Proc. U.S. Natl. Mus., 94:401–504.

————. 1970. Biosystematics of the Embioptera. Ann. Rev. Entomol., 15:157–72.

Ross, H. H. 1938. Descriptions of nearctic caddisflies. Ill. Nat. Hist. Survey, Bull. 21(4).

————. 1944. The caddisflies or Trichoptera of Illinois. Ill. Nat. Hist. Survey, Bull. 23(1):1–326.

————. 1955. The evolution of the insect orders. Entomol. News, 66(8):197–208.

————. 1959. Trichoptera. In *Fresh-water Biology,* pp. 1024–49, W. T. Edmondson, ed. New York, Wiley.

————. 1962. *How to Collect and Preserve Insects.* Ill. Nat. Hist. Surv., Circ. 39.

————. 1965. *A Textbook of Entomology,* 3rd ed. New York, Wiley. 539 pp.

————. 1973. Evolution and phylogeny. In *History of Entomology,* pp. 171–84, R. F. Smith, T. E. Miller, and C. N. Smith, eds. Palo Alto, CA, Annual Reviews.

————. 1974. *Biological Systematics.* Reading, MA, Addison-Wesley. 345 pp.

Roth, L. M., and T. Eisner. 1962. Chemical defenses of arthropods. Ann. Rev. Entomol., 7:107–36.

———— and E. R. Willis. 1957. *The Medical and Veterinary Importance of Cockroaches.* Smithsonian Inst. Misc. Coll., 134(10).

———— and ————. 1960. *The Biotic Associations of Cockroaches.* Smithsonian Inst. Misc. Coll., 141. 470 pp.

Rothenbuhler, W. C. 1964. Behavior genetics of nest cleaning in honey bees IV. Response of F_1 and backcross generations to disease-killed brood. Amer. Zool., 4:111–23.

————. 1967. Genetic and evolutionary considerations of social behavior of honey bees and some related insects. In *Behavior—Genetic Analysis,* pp. 61–106, J. Hirsch, ed. New York, McGraw-Hill.

Rothschild, M. 1965a. Fleas. Sci. Amer., 213:44–50.

————. 1965b The rabbit flea and hormones. Endeavor, 24:162–68.

————. 1972. Secondary plant substances and warning colouration in insects. In *Insect/Plant Relationships,* pp. 59–83, H. F. van Emden, ed. Roy. Entomol. Soc. (London), Symp. 6. London, Blackwell Scientific.

———— and B. Ford. 1966. Hormones of the vertebrate host controlling ovarian regression and copulation of the rabbit flea. Nature, 211:261–66.

References Cited

————— and B. Ford. 1966. Hormones of the vertebrate host controlling ovarian regression and copulation of the rabbit flea. Nature, 211:261–66.

————— and —————. 1972. Breeding cycle of the flea *Cediopsylla simplex* is controlled by breeding cycle of host. Science, 178:625–26.

—————, Y. Schlein, K. Parker, C. Neville, and S. Sternberg. 1973. The flying leap of the flea. Sci. Amer., 229:92–100.

Rowell, H. F. 1976. The cells of the insect neurosecretory system: constancy, variability and the concept of the unique identifiable neuron. Adv. Insect Physiol., 12:63–123.

Ruck, P. 1964. Retinal structures and photoreception. Ann. Rev. Entomol., 9:83–102.

Rudall, K. M., and W. Kenchington. 1971. Arthropod silks: the problem of fibrous proteins in animal tissues. Ann. Rev. Entomol., 16:73–96.

Rudd, R. L. 1964. *Pesticides and the Living Landscape.* Madison, University of Wisconsin Press. 320 pp.

Ruse, M. 1979. *Sociobiology: Sense or Nonsense?* Dordrecht, Netherlands, Reidel.

Sacktor, B. 1970. Regulation of intermediary metabolism, with special reference to the control mechanisms in insect flight muscles. Adv. Insect Physiol., 7:267–347.

—————. 1974. Biological oxidations and energetics in insect mitochondria. In *The Physiology of Insecta,* 2nd ed., Vol. IV, pp. 272–353, M. Rockstein, ed. New York, Academic Press.

—————. 1975. Biochemistry of insect flights, Part 1—utilization of fuels by muscle. In *Insect Biochemistry and Function,* D. J. Candy and B. A. Kilby, eds. New York, Wiley (Halsted Press).

Salmon, J. T. 1964–1965. An Index to the Collembola. Bull. Roy. Soc. New Zealand, No. 7, Vols. 1–3. 651 pp.

Salt, R. W. 1959. Role of glycerol in the cold-hardening of *Bracon cephi* (Gahan). Can. J. Zool., 37:59–69.

—————. 1961. Principles of insect cold-hardiness. Ann. Rev. Entomol., 6:55–74.

Sander, K. 1976. Specification of the basic body pattern in insect embryogenesis. Adv. Insect Physiol., 12:125–238.

Sandved, K. B., and J. Brewer. 1976. *Butterflies.* New York, N. Abrams. 176 pp.

Sargent, T. D. 1969. Legion of Night: The Underwing Moths. Amherst, MA, University of Massachusetts Press. 222 pp.

Saunders, D. S. 1974. Circadian rhythms and photoperiodism in insects. In *The Physiology of Insecta,* 2nd ed., Vol. II, pp. 461–533, M. Rockstein, ed. New York, Academic Press.

—————. 1976a. *Insect Clocks.* New York, Pergamon.

—————. 1976b. The biological clock of insects. Sci. Amer., 234:114–21.

Schaefer, G. W. 1976. Radar observations of insect flight. In *Insect Flight,* pp. 157–97, R. C. Rainey, ed. Roy. Entomol. Soc. (London), Symp. 7. London, Blackwell Scientific.

Schaller, F. 1968. *Soil Animals.* Ann Arbor, University of Michigan Press.

—————. 1971. Indirect sperm transfer by soil arthropods. Ann. Rev. Entomol., 16:407–46.

Schneider, D. 1964. Insect antennae. Ann. Rev. Entomol., 9:103–22.

—————. 1969. Insect olfaction: deciphering system for chemical messages. Science, 163:1031–37.

—————. 1974. The sex-attractant receptor of moths. Sci. Amer., 231:28–35.

—————. 1975. Pheromone communication in moths and butterflies. In *Sensory Physiology and Behavior,* pp. 173–93, R. Galun, P. Hillman, I. Parnas, and R. Werman, eds. New York, Plenum.

Schneider, F. 1962. Dispersal and migration. Ann. Rev. Entomol., 7:223–42.

Schneiderman, H. A. 1972. Insect hormones and insect control. In *Insect Juvenile Hormones—Chemistry and Action,* pp. 3–27, J. J. Menn and M. Beroza, eds. New York, Academic Press.

————— and L. I. Gilbert. 1964. Control of growth and form in insects. Science, 143:325–33.

Schneirla, T. C. 1953. Insect behavior in relation to its setting. In *Insect Physiology,* pp. 685–722, K. D. Roeder, ed. New York, Wiley.

—————. 1971. *Army Ants—A Study in Social Organization.* San Francisco, Freeman.

References Cited

Schön, A. 1911. Bau und Entwicklung des tibialen Chordotonalorgane bei der Honigbiene und bei Ameisen. Zool. Jahrb. Anat., 31:439–72.

Schoonhoven, L. M. 1972a. Plant recognition by lepidopterous larvae. In *Insect/Plant Relationships,* H. F. van Emden, ed. Roy. Entomol. Soc. (London), Symp. 6. London, Blackwell Scientific.

———. 1972b. Secondary plant substances and insects. In *Structural and Functional Aspects of Phytochemistry,* pp. 197–224, V. C. Runeckler and T. C. Tso, eds. New York, Academic Press.

———. 1977. Insect chemosensory responses to plant and animal hosts. In *Chemical Control of Insect Behavior,* pp. 7–14, H. H. Shorey and J. J. McKelvey, Jr., eds. New York, Wiley.

Schwabe, J. 1906. Beiträge zur Morphologie und Histologie der tympanalen Sinnesapparate der Orthopteren. Zoologica, Stuttgart.

Schwartzkopff, J. 1974. Mechanoreception. In *The Physiology of Insecta,* 2nd ed., Vol. II, pp. 273–352, M. Rockstein, ed. New York, Academic Press.

Schwerdtfeger, F. 1973. Forest entomology. In *History of Entomology,* pp. 361–86, R. F. Smith, T. E. Mittler, and C. N. Smith, eds. Palo Alto, CA, Annual Reviews.

Scientific American. 1978. Evolution. Vol 239, No. 3.

Scott, H. G. 1961. Collembola: pictorial keys to the Nearctic genera. Ann. Entomol. Soc. Amer., 54:104–13.

Scruggs, C. G. 1978. The origin of the screwworm control program. In *The Screwworm Problem—Evolution of Resistance to Biological Control,* pp. 11–18, R. H. Richardson, ed. Austin, University of Texas Press.

Scudder, G. G. E. 1971. Comparative morphology of insect genitalia. Ann. Rev. Entomol., 16:379–406.

———. 1973. Recent advances in the higher systematics and phylogenetic concepts in entomology. Can. Entomol., 105:1251–63.

Seabrook, W. D. 1977. Insect chemosensory responses to other insects. In *Chemical Control of Insect Behavior,* pp. 15–43, H. H. Shorey and J. J. McKelvey, Jr. eds. New York, Wiley.

———. 1978. Neurobiological contributions to understanding insect pheromone systems. Ann. Rev. Entomol., 23:471–85.

Sebeok, T. A., ed. 1968. *Animal Communication. Techniques of Study and Results of Research.* Bloomington, Indiana University Press. 686 pp.

———, ed. 1977. *How Animals Communicate.* Bloomington, Indiana University Press. 1128 pp.

Sharov, A. G. 1966. *Basic Arthropodan Stock with Special Reference to Insects.* New York, Pergamon.

Sheets, T. J., and D. Pimentel, eds. 1979. *Pesticides: Contemporary Roles in Agriculture, Health and the Environment.* Clifton, NJ, The Humana Press. 186 pp.

Shepard, H. H. 1951. *The Chemistry and Action of Insecticides.* New York, McGraw-Hill.

Sheppard, P. M. 1961. Recent genetical work on polymorphic mimetic Papilios. In *Insect Polymorphism,* pp. 20–29, J. S. Kennedy, ed. Roy. Entomol. Soc. (London), Symp. 1.

Sherman, J. L. 1966. Development of a systemic insect repellent. J. Amer. Med. Assoc., 196(3):166–68.

Shorey, H. H. 1973. Behavioral responses to insect pheromones. Ann. Rev. Entomol., 18:349–80.

———. 1976. *Animal Communication by Pheromones.* New York, Academic Press.

———. 1977a. Pheromones. In *How Animals Communicate,* pp. 137–63, T. A. Sebeok, ed. Bloomington, Indiana University Press.

———. 1977b. The adaptiveness of pheromone communication. Proc. XV Int. Congr. Entomol., pp. 294–307.

——— and J. J. McKelvey, Jr., eds. 1977. *Chemical Control of Insect Behavior—Theory and Application.* New York, Wiley.

Shulman, S. 1967. Allergic responses to insects. Ann. Rev. Entomol., 12:323–46.

Silberglied, R. E. 1977. Communication in the Lepidoptera. In *How Animals Communicate,* pp. 362–402, T. A. Sebeok, ed. Bloomington, Indiana University Press.

References Cited

Simpson, G. G. 1961. *Principles of Animal Taxonomy*. New York, Columbia University Press.

Siverly, R. E. 1962. *Rearing Insects in Schools*. Dubuque, IA, Wm. C. Brown.

Skaife, S. H. 1961. *Dwellers in Darkness*. New York, Doubleday.

Slabaugh, R. E. 1940. A new thysanuran, and a key to the domestic species of Lepismatidae (Thysanura) found in the United States. Entomol. News, 5(4):95–98.

Sláma, K. 1969. Plants as a source of materials with insect hormone activity. Entomol. Exp. Appl., 12:721–28.

———, M. Romanŭk, and F. Sŏrm. 1974. *Insect Hormones and Bioanalogues*. New York, Springer-Verlag.

——— and C. M. Williams. 1966. The juvenile hormone, V. The sensitivity of the bug, *Pyrrhocoris apterus*, to a hormonally active factor in American paperpulp. Biol. Bull., 130:235–46.

Slater, J. A., and R. M. Baranowski. 1978. *How to Know the True Bugs (Hemiptera-Heteroptera)*. Dubuque, IA, Wm. C. Brown.

Slifer, E. H. 1970. The structure of arthropod chemoreceptors. Ann. Rev. Entomol., 15:121–42.

Smart, J. 1959. Notes on the mesothoracic musculature of Diptera. In *Studies in Invertebrate Morphology*, pp. 331–64. Smithsonian Institution, Misc. Coll. 157.

——— and N. F. Hughes. 1972. The insect and the plant: progressive palaeoecological integration. In *Insect/Plant Relationships*, pp. 143–55, H. F. van Emden, ed. Roy Entomol. Soc. (London), Symp. 6. London, Blackwell Scientific.

Smit, F. G. A. M. 1973. Siphonaptera (fleas). In *Insects and Other Arthropods of Medical Importance*, pp. 325–71, K. G. V. Smith, ed. London, British Museum (Natural History).

Smith, C. N. 1966. Personal protection from bloodsucking arthropods. J. Amer. Med. Assoc., 196(3):146–49.

Smith, D. S. 1968. *Insect Cells: Their Structure and Function*. Edinburgh, Oliver & Boyd.

———. 1972. *Muscle*. New York, Academic Press. 60 pp.

——— and J. E. Treherne. 1963. Functional aspects of the organization of the insect nervous system. Adv. Insect Physiol., I:401–484.

Smith, E. H., and D. Pimentel. 1978. *Pest Control Strategies*. New York, Academic Press. 342 pp.

Smith, E. L. 1969. Evolutionary morphology of external insect genitalia, I. Origin and relationships to other appendages. Ann. Entomol. Soc. Amer., 62(5):1051–79.

———. 1970. Biology and structure of some California bristletails and silverfish (Apterygota: Microcoryphia, Thysanura). Pan-Pacific Entomol., 46:212–25.

Smith, K. G. V., ed. 1973. *Insects and Other Arthropods of Medical Importance*. London, British Museum (Natural History). 561 pp.

Smith, K. M. 1967. *Insect Virology*. New York, Academic Press. 270 pp.

———. 1977. *Virus-Insect Relationships*. New York, Longmans. 291 pp.

Smith, L. M. 1959–1960. Series publications on Diplura: Ann. Entomol. Soc. Amer., 52:363–68 (1959); 53:137–43, 575–83 (1960).

Smith, R. C. 1958. *Guide to the Literature of the Zoological Sciences*, 5th ed. Minneapolis, Burgess.

——— and R. H. Painter. 1966. *Guide to the Literature of the Zoological Sciences*, 7th ed. Minneapolis, Burgess.

——— and W. M. Reid. 1972. *Guide to the Literature of the Life Sciences*, 8th ed. Minneapolis, Burgess.

Smith, R. F., T. E. Mittler, and C. N. Smith, eds. 1973. *History of Entomology*. Palo Alto, CA, Annual Reviews. 517 pp.

——— and A. E. Pritchard. 1956. Odonata. In *Aquatic Insects of California*, pp. 106–53, R. L. Usinger, ed. Berkeley, University of California Press.

Smith, R. H., and R. C. von Borstel. 1972. Genetic control of insect populations. Science, 178:1164–74.

Smith, W. J. 1977. *The Behavior of Communicating—An Ethological Approach*. Cambridge, MA, Harvard University Press.

Snodgrass, R. E. 1925. *Anatomy and Physiology of the Honey Bee*. New York, McGraw-Hill.

———. 1935. *Principles of Insect Morphology*. New York, McGraw-Hill.

———. 1950. *Comparative Studies on the Jaws of Mandibulate Arthropods*. Smithsonian Inst. Misc. Coll., 116(1). Publ. 4018.

References Cited

————. 1952. *A Textbook of Arthropod Anatomy*. Ithaca, NY, Comstock.

————. 1954. The dragonfly larva. Smithsonian Institution, Misc. Coll. 123(2).

————. 1956. *The Anatomy of the Honeybee*. Ithaca, NY, Cornell University Press.

————. 1957. A revised interpretation of the external reproductive organs of male insects. Smithsonian Inst. Misc. Coll., 135(6):1–60.

————. 1958. *Evolution of Arthropod Mechanisms*. Smithsonian Inst. Misc. Coll., 138(2):1–77.

————. 1959. *The Anatomical Life of the Mosquito*. Smithsonian Institution, Misc. Coll. 138(8).

————. 1960. *Facts and Theories Concerning the Insect Head*. Smithsonian Inst. Misc. Coll., 142(1):1–61.

————. 1961. *The Caterpillar and the Butterfly*. Smithsonian Institution, Misc. Coll., 143(6).

————. 1963a. *A Contribution Toward an Encyclopedia of Insect Anatomy*. Smithsonian Inst. Misc. Coll., 146(2):1–48.

————. 1963b. The anatomy of the honey bee. In *The Hive and the Honey Bee*, pp. 141–90, R. A. Grout, ed. Hamilton, IL, Dadant and Sons, Inc.

Snow, K. R. 1974. *Insects and Disease*. New York, Wiley. 208 pp.

Sokal, R. R., and P. A. A. Sneath. 1963. *Principles of Numerical Taxonomy*. San Francisco, Freeman.

Solomon, M. E. 1969. *Population Dynamics*. The Institute of Biology's Studies in Biology, no. 18. London, Edward Arnold.

Sondheimer, E., and J. B. Simeone, eds. 1970. *Chemical Ecology*. New York, Academic Press. 336 pp.

Southwood, T. R. E. 1966. *Ecological Methods with Particular Reference to the Study of Insect Populations*. London, Methuen.

————. 1975. The dynamics of insect populations. In *Insects, Science and Society*, pp. 151–99, D. Pimentel, ed. New York, Academic Press.

————. 1977a. Entomology and mankind. Proc. XV Int. Congr. Entomol., pp. 36–51.

————. 1977b. *Entomology and mankind*. Amer. Sci. 65:30–39.

————. 1979. *Ecological Methods with Particular Reference to the Study of Insect Populations*, 2nd ed. New York, Wiley (Halsted Press).

Spencer, E. Y. 1973. *Guide to the Chemicals Used in Crop Protection*, 6th ed. Publ. 1093, Information Canada, Ottawa, Ontario, Canada. 542 pp.

Spradbery, J. P. 1973. *Wasps. An Account of the Biology and Natural History of Social and Solitary Wasps*. Seattle, University of Washington Press. 416 pp.

Staedler, E. 1977. Sensory aspects of insect plant interactions. Proc. XV Int. Congr. Entomol., pp. 228–48.

Stannard, L., Jr. 1957. The phylogeny and classification of the suborder Tubulifera (Thysanoptera). Ill. Biol. Monog. No. 25. 200 pp.

————. 1968. The Thrips, or Thysanoptera, of Illinois. Ill. Nat. Hist. Surv. Bull., 29:215–552.

Steele, J. E. 1976. Horomonal control of metabolism in insects. Adv. Insect Physiol., 12:239–323.

Stehr, F. W. 1975. Parasitoids and predators in pest management. In *Introduction to Insect Pest Management*, pp. 147–88, R. L. Metcalf and W. H. Luckmann, eds. New York, Wiley.

Steinhaus, E. A. 1949. *Principles of Insect Pathology*. New York, McGraw-Hill.

————, ed. 1963. *Insect Pathology, An Advanced Treatise*. New York, Academic Press. Vol. I, 661 pp, Vol. II, 689 pp.

Stellwaag, F. 1916. Wie steuern die Insekten wahrend des Fluges. Biol. Zentrbl., 36:30–44.

Stiles, K. A., R. W. Hegner, and R. Boolootian. 1969. *College Zoology*, 8th ed. New York, Macmillan.

Stobbart, R. H., and J. Shaw. 1974. Salt and water balance; excretion. In *The Physiology of Insecta*, 2nd ed., Vol. V, pp. 362–446, M. Rockstein, ed. New York, Academic Press.

Stoffolano, J. G., Jr. 1974. Control of feeding and drinking in diapausing insects. In *Experimental Analysis of Insect Behavior*, pp. 32–47, L. Barton Browne, ed. New York, Springer-Verlag.

References Cited

Stojanovich, C. J., and H. D. Pratt. 1965. *Key to Anoplura of North America*. Atlanta, GA, U. S. Public Health Service, Center for Disease Control. 24 pp.

Stone, A., C. W. Sabrosky, W. W. Wirth, R. H. Foote, and J. R. Coulson, eds. 1965. A Catalog of the Diptera of America, North of Mexico. USDA Agr. Handbook 276. 1969 pp.

Strausfeld, N. 1976. *Atlas of an Insect Brain*. New York, Springer-Verlag. 214 pp.

Stuart, A. M. 1977. Some aspects of communication in termites. Proc. XV Int. Congr. Entomol., pp. 400–405.

Sturm, H. 1956. Die Paarung beim Silberfischen, *Lepisma saccharina*. Z. Tierpsychol., 13:1–2.

Sudd, J. H. 1967. *An Introduction to the Behavior of Ants*. New York, St. Martin's Press. 200 pp.

Swain, R. B. 1952. How insects gain entry. In *Insects—The Yearbook of Agriculture,* pp. 350–55. Washington, DC, U.S. Dept. of Agriculture.

———. 1957. *The Insect Guide*. Garden City, NY, Doubleday.

Swan, L. A. 1964. *Beneficial Insects*. New York, Harper & Row.

——— and C. S. Papp. 1972. *The Common Insects of North America*. New York, Harper & Row. 750 pp.

Swartz, R. 1974. *Carnivorous Plants*. New York, Avon Books. 128 pp.

Sweetman, H. L. 1936a. *The Biological Control of Insects*. Ithaca, NY, Comstock.

———. 1936b. *The Principles of Biological Control*. Dubuque, IA, Wm. C. Brown.

———. 1965. *Recognition of Structural Pests and Their Damage*. Dubuque, IA, Wm. C. Brown. 384 pp.

Sylvester, E. S. 1980. Circulative and propagative virus transmission by aphids. Ann. Rev. Entomol., 25:257–86.

Tauber, M. J., and C. A. Tauber. 1976. Insect seasonality: diapause maintenance, termination, and postdiapause development. Ann. Rev. Entomol., 21:81–107.

Taylor, R. L. 1975. *Butterflies in My Stomach—Or: Insects in Human Nutrition*. Santa Barbara, CA, Woodbridge Press. 224 pp.

——— and B. J. Carter. 1976. *Entertaining with Insects*. Santa Barbara, CA, Woodbridge Press. 160 pp.

Telfer, W. H. 1965. The mechanism and control of yolk formation. Ann. Rev. Entomol., 10:161–84.

———. 1975. Development and physiology of the oocyte-nurse cell syncytium. Adv. Insect Physiol., 11:223–319.

——— and D. S. Smith. 1970. Aspects of egg formation. In *Insect Ultrastructure,* pp. 117–34, A. C. Neville, ed. Roy. Entomol. Soc. (London), Symp. 5.

Teskey, H. J. 1978. Aquatic Diptera; Part one. Larvae of aquatic Diptera. In *An Introduction to the Aquatic Insects of North America*, pp. 245–57, R. W. Merritt and K. W. Cummins, eds., Dubuque, IA, Kendall/Hunt Publishing Company.

Thiele, H. U. 1973. Remarks about Mansingh's and Müller's classification of dormancies in insects. Can. Entomol., 105:925–28.

Theiler, M., and W. G. Downs. 1973. *Arthropod Borne Viruses of Vertebrates*. New Haven, CT, Yale University Press. 578 pp.

Thomson, J. A. 1976. Major patterns of gene activity during development in holometabolous insects. Adv. Insect Physiol., 11:321–98.

Thorpe, W. H. 1950. A note on detour experiments with *Ammophila pubescens* Curt. Behavior, 12:257–63.

———. 1963. *Learning and Instinct in Animals,* 2nd ed. London, Methuen.

Tiegs, O. W., and S. M. Manton. 1958. The evolution of the Arthropoda. Biol. Rev., 33:255–337.

Tietz, H. M. 1963. *Illustrated Keys to the Families of North American Insects*. Minneapolis, Burgess.

———. 1973. *An index to the Described Life Histories, Early Stages, and Hosts of the Macrolepidoptera of the Continental United States and Canada*. Hampton, Engl., Classey. 1042 pp.

Tinsley, T. W. 1977. Viruses and the biological control of insect pests. BioScience, 27(10):659–61.

Torre-Bueno, J. R. de la. 1937. *A Glossary of Entomology,* Brooklyn Entomol. Soc. 336 pp.

References Cited

Townsend, G. F., and E. Crane. 1973. History of apiculture. In *History of Entomology,* pp. 387–406, R. F. Smith, T. E. Mittler, and C. N. Smith, eds. Palo Alto, CA, Annual Reviews.

Treherne, J. E. 1967. Gut absorption. Ann. Rev. Entomol., 12:43–58.

———, ed. 1974. *Insect Neurobiology.* New York, American Elsevier. 450 pp.

Trivers, R. L., and H. Hare. 1976. Haplodiploidy and the evolution of social insects. Science, 191:249–63.

Truman, J. W. 1971–1973. Physiology of insect ecdysis I–III. J. Exp. Biol, 54:805–14; Biol. Bull. Mar. Biol. Lab, Woods Hole, 144:200–11; J. Exp. Biol., 58:821–29.

———. 1973. How moths "turn on": a study of the action of hormones on the nervous system. Amer. Sci., 61:700–706.

——— and L. M. Riddiford. 1970. Neuroendocrine control of ecdysis in silkmoths. Science, 167:1624–26.

——— and ———. 1974. Hormonal mechanisms underlying insect behavior. Adv. Insect Physiol., 10:297–352.

Tu, A. T. 1977. *Venoms: Chemistry and Molecular Biology.* New York, Wiley. 560 pp.

Tulloch, G. S. 1960. Torre-Bueno's Glossary of Entomology—Supplement A. Brooklyn Entomol. Soc. 36 pp.

Tuxen, S. L. 1964. *The Protura. A Revision of the Species of the World with Keys for Determination.* Paris, Hermann. 360 pp.

———. 1970. *Taxonomist's Glossary of Genitalia in Insects,* 2nd ed. New York, Stechert-Hafner Service Agency, Inc.

———. 1973. Entomology systematizes and describes: 1700–1815. In *History of Entomology,* pp. 95–118, R. F. Smith, T. E. Mittler, and C. N. Smith, eds. Palo Alto, CA, Annual Reviews.

Tweedie, M. 1974. *Atlas of Insects.* New York, John Day. 128 pp.

Ulrich, W. 1966. Evolution and classification of the Strepsiptera. Proc. 1st Int. Congr. Parasitol., 1:609–11.

Urquhart, F. A. 1960. *The Monarch Butterfly.* Toronto, University of Toronto Press. 361 pp.

———. 1976. Found at last: the monarch's winter home. Nat. Geographic, 150:161–73.

USDA/Smithsonian Institution. *A Catalog of the Coleoptera of America North of Mexico* (in prep.) Washington, DC.

U.S. Department of Agriculture. 1952. *Insects — The Yearbook of Agriculture.* Washington, DC, U.S. Department of Agriculture.

Ursprung, H., and R. Nöthiger. 1972. *The Biology of Imaginal Discs.* New York, Springer-Verlag.

Usherwood, P. N. R. 1974. Nerve-muscle transmission. In *Insect Neurobiology,* pp. 245–305, J. E. Treherne, ed. New York, American Elsevier.

——— ed. 1975. *Insect Muscle.* New York, Academic Press. 621 pp.

Usinger, R. L., ed. 1956. *Aquatic Insects of California, with Keys to North American Genera and California Species.* Berkeley, University of California Press. 508 pp.

Uvarov, B. 1966. *Grasshoppers and Locusts. A Handbook of General Acridology.* Vol. 1. Anatomy, physiology, development, phase polymorphism, introduction to taxonomy. Cambridge, Cambridge University Press. 481 pp.

———. 1977. *A Handbook of General Acridology.* Vol. 2. Behavior, ecology, biogeography, population dynamics. London, Centre for Overseas Pest Research. 613 pp.

Varley, G. C., G. R. Gradwell, and M. P. Hassell. 1973. *Insect Population Ecology: An Analytical Approach.* Oxford, Blackwell Scientific Publications. 212 pp.

Vaughn, J. L. 1974. Viruses and rickettsial diseases. In *Insect Diseases,* Vol. 1, pp. 49–85, G. E. Cantwell, ed. New York, Marcel Dekker.

Vaughan, M. A., and G. Leon. 1977. Pesticide management on a major crop with severe resistance problems. Proc. XV Int. Congr. Entomol., pp. 812–24.

Villee, C. A., and V. G. Dethier. 1976. *Biological Principles and Processes.* Philadelphia, Saunders. 999 pp.

Vinson, S. B. 1976. Host selection by insect parasitoids. Ann. Rev. Entomol., 21:109–33.

References Cited

———— and G. F. Iwantsch. 1980. Host suitability for insect parasitoids. Ann. Rev. Entomol., 25:397–419.

Vogel, R. 1921. Zur Kenntnis des Baues und der Funkton des Stachels und des Vorderdarmes der Kleiderlaus. Zool. Jahrb., Anat., 42:229–58.

Volpe, E. P. 1977. *Understanding Evolution,* 3rd ed. Dubuque, IA, Wm. C. Brown. 232 pp.

Waddington, C. H. 1973. The morphogenesis of patterns in *Drosophila.* In *Developmental Systems: Insects,* Vol. 2, pp. 499–535, S. J. Counce and C. H. Waddington, eds. New York, Academic Press.

Walker, E. M. 1953. *The Odonata of Canada and Alaska,* Vol. 1, General, the Zygoptera—Damselflies. Toronto, University of Toronto Press. 292 pp.

————. 1958. *The Odonata of Canada and Alaska,* Vol. 2, The Anisoptera—Four Families. Toronto, University of Toronto Press. 318 pp.

———— and P. S. Corbet. 1975. *The Odonata of Canada and Alaska,* Vol. 3, The Anisoptera—Three Families. Toronto, University of Toronto Press. 307 pp.

Wallwork, J. A. 1970. *Ecology of Soil Animals.* New York, McGraw-Hill. 283 pp.

Warburton, F. E. 1967. The purposes of classification. Syst. Zool., 16(3):241–45.

Ware, G. W. 1975. *Pesticides, an Autotutorial Approach.* San Fracisco, Freeman. 191 pp.

————. 1978. *The Pesticide Book.* San Francisco, Freeman. 197 pp.

Watson, A., and P. E. S. Whalley. 1975. *The Dictionary of Butterflies and Moths in Colour.* London, Michael Joseph. 296 pp.

Watson, D. L., and A. W. A. Brown, eds. 1977. *Pesticide Management and Insecticide Resistance.* New York, Academic Press. 656 pp.

Watson, T. F., L. Moore, and G. W. Ware. 1976. *Practical Insect Pest Management, a Self-Instruction Manual.* San Francisco, Freeman. 196 pp.

Waterhouse, D. F. 1957. Digestion in insects. Ann. Rev. Entomol., 2:1–18.

Way, M. J. 1963. Mutualism between ants and honeydew-producing Homoptera. Ann. Rev. Entomol., 8:307–44.

Weaver, N. 1978a. Chemical control of behavior—intraspecific. In *Biochemistry of Insects,* pp. 360–89, M. Rockstein, ed. New York, Academic Press.

————. 1978b. Chemical control of behavior—interspecific. In *Biochemistry of Insects,* pp. 392–418, M. Rockstein, ed. New York, Academic Press.

Weber, N. A. 1966. Fungus-growing ants. Science, 153(3736):587–609.

————. 1972. *Gardening Ants—The Attines.* Philadelphia, American Philosophical Society. 146 pp.

Weesner, F. M. 1965. *The Termites of the United States—A Handbook.* Elizabeth, NJ, National Pest Control Association. 70 pp.

Wehner, R. 1971. *Information Processing in the Visual Systems of Arthropods.* New York, Springer-Verlag. 334 pp.

————. 1976. Polarized-light navigation by insects. Sci. Amer., 235:106–15.

Weiser, J. 1969. *An Atlas of Insect Diseases.* Shannon, Irish University Press. 292 pp.

Weis-Fogh, T. 1971. Flying insects and gravity. In *Gravity and the Organism,* pp. 177–84, S. A. Gordon and M. J. Cohen, eds. Chicago, University of Chicago Press.

Welch, H. E. 1965. Entomophilic nematodes. Ann. Rev. Entomol., 10:275–302.

Welch, S. M., and B. A. Croft. 1980. *The Design of Biological Monitoring Systems for Pest Management.* New York, Wiley.

Wells, P. H. 1973. Honey bees. In *Invertebrate Learning,* Vol. 2, pp. 173–85, W. C. Corning, J. A. Dyal, and A. O. D. Willows, eds. New York, Plenum.

Wendler, G. 1971. Gravity orientation in insects: the role of different mechanoreceptors. In *Gravity and the Organism,* pp. 195–201, S. A. Gordon and M. J. Cohen, eds. Chicago, University of Chicago Press.

West, L. S. 1951. *The Housefly: Its Natural History, Medical Importance, and Control.* Ithaca, NY, Comstock Publishing Company, Inc.

Westfall, M. J. 1978. Odonata. In *An Introduction to the Aquatic Insects of North America,* pp. 81–98, R. W. Merritt and K. W. Cummins, eds. Dubuque, IA, Kendall/Hunt Publishing Company.

Weygoldt, P. 1979. Significance of later embryonic stages and head development in arthropod phylogeny. In *Arthropod Phylogeny,* pp. 107–35, A. P. Gupta, ed. New York, Van Nostrand Reinhold.

Wharton, G. W., and A. G. Richards. 1978. Water vapor exchange kinetics in insects and acarines. Ann. Rev. Entomol., 23:309–28.

Wheeler, W. M. 1910. *Ants*. New York, Columbia University Press. 663 pp.

Whitcomb, R. F. 1973. Diversity of procaryotic plant pathogens. Proc. N. Central Branch, Entomol. Soc. Amer., 28:38–60.

————— and R. E. Davis. 1970. Mycoplasma and phytarboviruse as plant pathogens persistently transmitted by insects. Ann. Rev. Entomol., 15:405–64.

—————, M. Shapiro, and R. R. Granados. 1974. Insect defense mechanisms against microorganisms and parasitoids. In *The Physiology of Insecta,* 2nd ed., Vol. V, pp. 448–536, M. Rockstein, ed. New York, Academic Press.

White, J. W. 1963. Honey. In The *Hive and the Honey Bee,* pp. 369–406, R. A. Grout, ed. Hamilton, IL, Dadant & Sons.

White, M. J. D. 1964. Cytogenetic mechanisms in insect reproduction. In *Insect Reproduction,* pp. 1–12, K. C. Highnam, ed. Roy. Entomol. Soc. (London), Symp. 2.

White-Stevens, R., ed. 1977. *Pesticides in the Environment,* Vol. 3. New York, Marcel Dekker. 366 pp.

Whittaker, R. H. 1975. *Communities and Ecosystems.* New York, Macmillan.

Whittaker, R. H., and P. P. Feeny. 1971. Allelochemics: chemical interactions between species. Science, 171:757–70.

Whittemore, F. W. 1977. The evolution of pesticides and the philosophy of evolution. Proc. XV Int. Congr. Entomol., pp. 714–18.

Whitten, J. M. 1968. Metamorphic changes in insects. In *Metamorphosis—A Problem in Developmental Biology,* pp. 43–105, W. Etkin and L. I. Gilbert, eds. Amsterdam, North-Holland.

—————. 1972. Comparative anatomy of the tracheal system. Ann. Rev. Entomol., 17:373–402.

—————. 1976. Definition of insect instars in terms of "apolysis" or "ecdysis." Ann. Entomol. Soc. Amer., 69(3):556–59.

Wickler, W. 1968. *Mimicry in Plants and Animals.* New York, McGraw-Hill. 255 pp.

Wiebes, J. T. 1979. Co-evolution of figs and their insect pollinators. Ann. Rev. Ecol. Syst., 10:1–12.

Wiggins, G. B. 1978. *Larvae of the North American Caddisfly Genera (Trichoptera).* Toronto, University of Toronto Press. 410 pp.

Wigglesworth, V. B. 1930. The formation of the peritrophic membrane in insects with special reference to the larvae of mosquitoes. Quart. J. Microscop. Sci., 73:593–616.

—————. 1941. The sensory physiology of the human louse, *Pediculus humanus* corporis DeGeer (Anoplura). Parasitol., 33:67–109.

—————. 1954. *The Physiology of Insect Metamorphosis.* Cambridge, Cambridge University Press.

—————. 1959. *The Control of Growth and Form: A Study of the Epidermal Cell in an Insect.* Ithaca, NY, Cornell University Press.

—————. 1964a. *The Life of Insects.* London, Weidenfeld & Nicolson. 360 pp.

—————. 1964b. The hormonal regulation of growth and reproduction in insects. Adv. Insect Physiol., 2:247–336.

—————. 1965. *The Principles of Insect Physiology,* 6th ed. London, Methuen.

—————. 1970. *Insect Hormones.* Edinburgh, Oliver & Boyd.

—————. 1972. *The Principles of Insect Physiology,* 7th ed. London, Chapman & Hall.

—————. 1973a. The significance of "apolysis" in the moulting of insects. J. Entomol. [A], 47(2):141–49.

—————. 1973b. Evolution of insect wings and flight. Nature, 246:127–29.

—————. 1976a. *Insects and the Life of Man.* New York, Wiley. 217 pp.

—————. 1976b. The distribution of lipid in the cuticle of *Rhodnius.* In *The Insect Integument.* pp. 89–106, H. R. Hepburn, ed. New York, Elsevier Scientific.

—————. 1976c. The evolution of insect flight. In *Insect Flight,* pp. 255–69, R. C. Rainey, ed. Roy. Entomol. Soc. (London), Symp. 7. London, Blackwell Scientific.

Wilbur, D. A., and R. B. Mills. 1978. Stored grain insects. In *Fundamentals of Applied Entomology,* 3rd ed., pp. 573–603, R. E. Pfadt, ed. New York, Macmillan.

References Cited

Wilkinson, C. F., ed. 1976. *Insecticide Biochemistry and Physiology.* New York, Plenum. 769 pp.

Wille, A. 1960. The phylogeny and relationships between the insect orders. Rev. Biol. Trop., 8:93–123.

Williams, C. M. 1946–1953. Physiology of insect diapause. I–IV. Biol. Bull., 90:234–43; 93:90–98; 94:60–65; 103:120–38.

———. 1967. Third-generation pesticides. Sci. Amer., 217:13–17.

———. 1970. Hormonal interactions between plants and insects. In *Chemical Ecology,* pp. 103–32, E. Sondheimer and J. B. Simeone, eds. New York, Academic Press.

Williams, C. M., and F. C. Kafatos. 1972. Theoretical aspects of the action of juvenile hormone. In *Insect Juvenile Hormones—Chemistry and Action,* pp. 29–41, J. J. Menn and M. Beroza, eds. New York, Academic Press.

Willis, J. H. 1974. Morphogenetic action of insect hormones. Ann. Rev. Entomol., 19:97–115.

Wilson, D. M. 1966. Insect walking. Ann. Rev. Entomol., 11:103–22.

———. 1968. The nervous control of insect flight and related behavior. Adv. Insect Physiol., 5:289–334.

———. 1971. Stabilizing mechanisms in insect flight. In *Gravity and the Organism,* pp. 169–76, S. A. Gordon and M. J. Cohen, eds. Chicago, University of Chicago Press.

———. 1972. Genetic and sensory mechanisms for locomotion and orientation in animals. Amer. Sci., 60:358–65.

Wilson, E. O. 1963. Pheromones. Sci. Amer., 208:100–14.

———. 1965. Chemical communication in the social insects. Science, 149:1064–71.

———. 1970. Chemical communication within animal species. In *Chemical Ecology,* pp. 133–55, E. Sondheimer and J. B. Simeone, eds. New York, Academic Press.

———. 1971. *The Insect Societies.* Cambridge, MA, Belknap Press/Harvard University Press. 548 pp.

———. 1975. *Sociobiology—The New Synthesis.* Cambridge, MA, Belknap Press/Harvard University Press.

Wilson, E. O., and W. H. Bossert. 1971. *A Primer of Population Biology.* Stamford, CT, Sinauer Associates. 192 pp.

Wilson, H. F., and M. H. Doner. 1937. *The Historical Development of Insect Classification.* St. Louis, J. S. Swift. 133 pp.

Wilson, M. C., D. B. Broersma, and A. V. Provonsha. 1977. *Fundamentals of Applied Entomology.* Prospect Heights, IL, Waveland Press. 166 pp.

Windley, B. F. 1977. *The Evolving Continents.* New York, Wiley.

Wirth, W. W., and A. Stone. 1956. Aquatic Diptera. In *Aquatic Insects of California,* pp. 372–482, R. L. Usinger, ed. Berkeley, University of California Press.

Wolglum, R. S., and F. A. McGregor. 1958. Observations on the life history and morphology of *Agulla bractea* Carpenter (Neuroptera: Raphidiodea: Raphidiidae). Ann. Entomol. Soc. Amer., 51:129–41.

Woodwell, G. M., P. P. Craig, and H. A. Johnson. 1971. DDT in the biosphere: where does it go? Science, 174:1101–107.

Woodwell, G. M., W. M. Malcolm, and R. H. Whittaker. 1969. A-bombs, bug bombs, and us. In *The Subversive Science . . . Essays Toward an Ecology of Man,* pp. 230–41, P. Shepard and D. McKinley, eds. Boston, Houghton Mifflin.

Wooton, R. J. 1976. The fossil record and insect flight. In *Insect Flight,* pp. 235–54, R. C. Rainey, ed. Roy. Entomol. Soc. (London), Symp. 7. London, Blackwell Scientific.

World Health Organization. 1967. The genetic control of insects. WHO Chronicle, 21(12):517–24.

Wright, D. P., Jr. 1963. Anti-feeding compounds for insect control. In *New Approaches to Pest Control and Eradication,* Advances in Chemistry Series 41. Washington, DC, American Chemical Society.

Wyatt, G. R. 1968. Biochemistry of insect metamorphosis. In *Metamorphosis—A Problem in Developmental Biology,* pp. 143–84, W. Etkin and L. I. Gilbert, eds. Amsterdam, North-Holland.

Wygodzinsky, P. 1972. *A Revision of the Silverfish (Lepismatidae, Thysanura) of the United States and the Caribbean Area.* Amer. Mus. Novit., No. 2481. 26 pp.

References Cited

Yamaoka, K., M. Hoshino, and T. Hirao. 1971. Role of sensory hairs on the anal papillae in oviposition behavior of *Bombyx mori*. J. Insect Physiol., 17:897–911.

Yarrow, I. H. H. 1973. Hymenoptera (bees, wasps, ants, etc.). In *Insects and Other Arthropods of Medical Improtance*, pp. 409–11, K. G. V. Smith, ed. London, British Museum (Natural History).

Yeager, J. F. 1938. Mechanographic method of recording insect cardiac activity, with reference to effect of nicotine on isolated heart preparation of *Periplaneta americana*. J. Agr. Res., 56:267–306.

Yeo, P. F. 1973. Floral allurements for pollinating insects. In *Insect/Plant Relationships*, pp. 51–57, H. F. van Emden, ed. Roy. Entomol. Soc. (London), Symp. 6.

Yokoyama, T. 1973. The history of sericultural science in relation to industry. In *History of Entomology*, pp. 267–84, R. F. Smith, T. E. Mittler, and C. N. Smith, eds. Palo Alto, CA, Annual Reviews.

Zacharuk, R. Y. 1976. Structural changes of the cuticle associated with moulting. In *The Insect Integument*, pp. 299–321, H. R. Hepburn, ed. New York, Elsevier Scientific.

Zim, H. S., and C. Cottam. 1956. *Insects: A Guide to Familiar American Insects*. Racine, WI, Golden Press. 160 pp.

Zumpt, F. 1965. *Myiasis in Man and Animals in the Old World*. London, Butterworth.

Glossary

a-   without; lacking.

ab-   away from; off.

acetylcholine   a kind of neurotransmitter substance.

acetylcholine esterase   an enzyme that breaks down acetylcholine.

acinar   resembling a cluster.

acinus   resembling a single grape in a cluster.

acquired active immunity   immunity induced by injection of toxoids or killed or attenuated microorganisms.

acron   anterior, preoral body cap of a segmented animal, also called the prostomium (Figs. 10-12 and 10-13).

active ventilation   ventilation aided by movements such as pumping.

ad-   next to; at.

adecticous   without functional mandibles (movable mouthparts) used for escape from a cocoon or pupal cell.

adipocytes   cells composing fat body (Fig. 4-9B).

adjuvants   substances (e.g., dusts, solvents, emulsifiers, adhesives) that are combined with insecticides.

aedeagus   *see* penis.

aeropyles   channels in insect eggs by which the gas-filled layer communicates with the outside of the egg.

aerosols   sprays made of extremely small droplets, 0.1–50 micrometers in diameter.

afferent neuron   *see* sensory neuron.

aging   all changes in structure and function that occur from the beginning to the termination of life.

agroecosystem   a relatively simple ecosystem created by man for agricultural purposes, in which the major producer, the crop, is a single plant species.

alary muscles   segmented, fibromuscular structures associated with the heart and having a role in the functioning of that organ (Fig. 4-6).

alate   winged.

alinotum   wing-bearing portion of the tergum (Figs. 2-18 and 2-19).

all-, allo-   other; different.

allergens   insect secretions whose effect depends on the physiological response of the victim.

allometric growth   *see* heterogonic growth.

allomone   chemical emitted by an individual of one species and received by one of another species that induces a response favorable to the emitter.

amnion   inner membrane around the developing embryo (Figs. 5-15, 5-19, and 5-20).

an-   without.

anamorphosis   the addition of abdominal segments at the time of molting.

androconia   specialized scales and associated sex-pheromone-producing glands among the wing scales of some male Lepidoptera.

anlage   precursor; the first recognizable beginning of a developing organ in an embryo.

Anoplura, order   (*anopl,* unarmed; *ura,* tall) sucking lice (Fig. 11-25).

ante-   before; in front of.

antecosta   an internal ridge, associated with the tergum and sternum of an abdominal segment (Fig. 2-11).

antennae   elongate sensory appendages found on the insect head and usually bearing olfactory and tactile receptors (Fig. 2-12).

antennal lobes   lobes in the insect brain that receive both sensory and motor axons from the antennae (Fig. 3-6).

anterior intestine   tubular part of the hindgut between the Malpighian tubules and the rectum (Figs. 4-1 and 4-2).

anthoxanthins   whitish or yellow pigments found in insects.

anthraquinones   red and orange pigments common in scale insects.

anthropomorphism   interpreting behavior patterns observed in animals in terms of human characteristics.

anti-   against.

antibiotic   substance that prevents, inhibits, or destroys life.

antimetabolites   chemicals resembling essential nutrients that interfere with metabolism.

anus   posterior opening of the alimentary system (Figs. 4-1 and 4-2).

aorta   anterior subdivision of the dorsal vessel, lacking ostia and alary muscles (Figs. 4-6 and 4-7).

aphins   red, orange, and yellow pigments found only in aphids.

apical   distal.

apiculture   beekeeping.

apneustic   the closed type of ventilatory system, without functional spiracles (Fig. 4-13C).

apodeme   an internal inflection of the integument to which muscles attach (Fig. 2-10).

apolysis   the separation of the old cuticle from the epidermis during molting.

apophysis   spinelike inflection (apodeme) of the insect integument to which muscles attach (Fig. 2-10).

aposematic coloration   warning coloration such as that of dangerous or poisonous animals.

appetitive behavior   behavior that increases the probability of exposure to a releasing stimulus.

apterous   wingless.

Apterygota   primitively wingless insects that do not undergo metamorphosis.

arboviruses   arthropod-borne viruses.

arch-   primitive; original.

archedictyon   irregular network of veins found between the principal wing veins in many fossil insect groups; similar networks in the wings of contemporary forms (e.g., dragonflies) are probably remnants of the archedictyon.

Archeognatha, order   (*archeo,* ancient; *gnath,* jaw) jumping bristletails (Fig. 11-6A).

arolium   bulbous, lobelike, structure found between the tarsal calws as in Orthoptera (Fig. 2-38B).

arthropod phobia   the persistent fear of infestation by arthropods.

arthropodins   a group of soluble proteins found in the insect cuticle.

association neurons   neurons found within ganglia, which may synapse with one another or with sensory or motor neurons (Fig. 3-1D); also called internuncial neurons.

asynchronous muscles   *see* resonating muscles.

athrocytes   cells of mesodermal origin that absorb particles of colloidal size from the hemolymph, e.g., pericardial cells (Fig. 4-9A).

atrium   spiracular chamber housing the opening to a trachea (Fig. 4-12B, C).

auto-   self; same.

autotomy   ability to amputate a limb spontaneously, as in stick insects.

axon   a fiber of a nerve cell that transmits impulses from the cell body (Fig. 3-1A–C).

basal   proximal.

basalar sclerite   small sclerite anterior to the pleural wing process (Figs. 2-18 and 7-9).

basement membrane   the noncellular innermost layer of the insect integument (Fig. 2-1A).

bi-   two.

bio-   life; living.

bioluminescence   light production by living organisms.

binomen   the two parts, genus and epithet, that comprise an organism's scientific name.

bipolar neuron   neuron with a cell body bearing an axon and a single, branched or unbranched, dendrite (Fig. 3-1B).

blasto-   embryo.

blastocoel an internal cavity surrounded by blastoderm.

blastoderm the first distinct layer of cells to form in morphogenesis (Fig. 5-14D).

blastokinesis all displacements, rotations, or revolutions of the embryo in the egg.

blastomeres cells composing the blastoderm.

Blattaria, order (*blatta,* cockroach) cockroaches (Fig. 11-20).

bombycol female sex pheromone of the silkworm moth.

botanical insecticides secondary plant substances having a toxic, repellent, deterrent, or hormonal effect on insects.

brachypterous having very short wings.

brain the anterior composite ganglion located dorsal to the foregut in the insect head (Figs. 3-3, 3-5, and 3-6).

brain hormone the hormone that stimulates secretory activity of the prothoracic gland; also called prothoraciotropic hormone and ecdysiotropin.

bursa copulatrix a saclike vagina (Fig. 5-7).

bursicon neurosecretory hormone that stimulates sclerotization of the cuticle following ecdysis; also called tanning hormone.

caryo- nucleus.

cardenolides secondary plant substances in milkweed plants that may render milkweed-feeding insects unpalatable to vertebrate predators.

cardo proximal segment of the maxilla (Fig. 2-15).

carnivore zoophagous or animal-eating animal.

carotenoids insect pigments of red and yellow, derived from plant tissues.

castes morphologically and functionally different types within a colony of social insects as in some Hymenoptera and Isoptera (Figs. 11-18 and 11-19).

caudal ganglion the posteriormost abdominal ganglion; innervates the genitalia (Fig. 3-3).

caudal sympathetic system the part of the visceral nervous system associated with the posterior segments of the abdomen.

cavernicolous found in caves.

cell a functional unit of living matter bounded by a membrane and containing one or more nuclei or nuclear materials; a space on an insect wing delineated by various combinations of longitudinal and cross veins and the wing margins.

cement layer outermost layer of the epicuticle (Fig. 2-1C).

cervical sclerites lateral plates found on the membranous region connecting the head with the thorax (Figs. 2-12 and 2-13C).

cervix membranous region connecting the head with the thorax (Figs. 2-12 and 2-13C).

chaetotaxy the arrangement of setae in constant patterns, useful in systematics.

chelicerae the anteriormost pair of preoral appendages found in spiders, ticks, scorpions, and relatives (Fig. 10-4).

chemoreception the perception of atoms and molecules as stimuli.

chitin a high-molecular-weight polymer of *N*-acetyl-*D*-glucosamine that resembles cellulose and constitutes one fourth to one half of the dry weight of the exo- and endocuticle (Fig. 2-3).

chordotonal organ *see* scolopophorus organ.

chorion eggshell secreted by the ovarian follicle cells (Fig. 5-10).

chrom-, -chrome color; pigment.

chrysalid a butterfly pupa that lacks a protective cover.

cibarium the portion of the preoral cavity between the hypopharynx and the labrum (Figs. 2-16 and 4-1).

circadian rhythms behavorial or physiological events that occur at approximately 24-hour intervals.

cleavage the divisions of the zygote and subsequent daughter cells that lead to the formation of the blastoderm (Fig. 5-14B, C).

clypeus lobelike structure located on the anterior part of the insect head just beneath the epistomal suture; hinged with the labrum in many insects (Figs. 2-13, 2-15, and 2-16).

co- with; together.

coarctate pupa a type of pupa typical of cyclorrhaphous Diptera, in which the next-to-last larval skin forms a hard barrel-shaped case called a puparium. The puparium splits at emergence, and the adult fly pushes away the top using an eversible head sac called the ptilinum (Fig. 5-24C, D).

cocoonase an enzyme that digests the silken cocoon in saturniid and bombyliid moths.

coel- cavity; chamber.

Coleoptera, order (*coleo,* sheath; *ptera,* wing) beetles (Figs. 11-34, 11-35, and 11-36.)

Collembola, order (*coll,* glue; *embol,* a wedge) springtails (Fig. 11-5).

colleterial glands glands that secrete adhesive materials.

collophore a nongenital structure found on the venter of the first abdominal segment in Collembola (Fig. 11-5).

colon posterior region of the anterior intestine of the hindgut.

com- together.

commensalism a symbiotic association in which one individual benefits and the other is neither helped nor harmed.

communication the influence of signals from one organism on the behavior and/or physiology of another organism.

compound eyes located on either side of the insect head, compound eyes are composed of a number of individual units called ommatidia (Figs. 2-12, 2-13, 6-11, and 6-12).

concave veins longitudinal veins in troughs formed by wings folded in a pleated fashion.

conjunctivae soft membranous areas separating sclerites in the insect skeleton.

convex veins longitudinal veins located on crests resulting from wings folded in a pleated fashion.

corbiculum specialized structure on the legs of honey bees, used to store pollen for transport to the hive; also called a pollen basket (Fig. 2-37).

corneal lens the outermost part of an ommatidium, derived from cuticle (Fig. 6-11).

cornicles a pair of lobelike projections on the posterior dorsum of the abdomen in aphids (Fig. 11-32A, B).

corpora allata endocrine glands that secrete juvenile hormone (Figs. 3-5B, 3-7, and 5-26).

corpora cardiaca neurohaemal organs where brain hormone is stored and released (Figs. 3-7 and 5-26).

corpora pedunculata cell masses in the insect brain that contain association neurons and are considered to be the location of the "higher centers" controlling the most complex behavior; also called mushroom bodies (Fig. 3-6).

coxa the basal segment of the leg (Figs. 2-21 and 2-36).

cremaster a structure, on the end of the abdomen, bearing small hooks by which some lepidopterous pupae attach to twigs.

crochets minute hooks found on the prolegs of lepidopterous larvae (Fig. 2-42C).

crop enlarged posterior region of the esophagus that functions as a site for temporary food storage (Fig. 4-1).

cross veins transverse supporting structures in the insect wing that connect the longitudinal veins.

crypsis combination of form, color, and pattern that facilitates "hiding" from predators and/or parasites.

crypts packets of cells projecting through the muscle layers of the midgut and containing regenerative cells that replace degenerating gut epithelial cells (Fig. 4-4C).

cursorial legs legs adapted for running and walking (Fig. 2-36A).

cuticle the noncellular outermost layer of the insect integument, secreted by the epidermis (Fig. 2-1).

cuticular gills *see* spiracular gills.

cuticular phase the period of time between the first appearance of new cuticle beneath the old and ecdysis.

cuticulin layer a layer of the epicuticle, probably lipoprotein, that covers the entire integumental surface, including tracheoles and gland ducts (Fig. 2-1C).

-cyte, cyto- cell.

decticous having functional mandibles used for escape from a cocoon or pupal case.

dendrites tiny branching processes associated with or near the cell body of a neuron (Fig. 3-1).

-derm skin; covering.

dermal glands modified epidermal cells on the outer surface of the insect body that secrete irritants, poisons, wax, scents, silk, etc. (Figs. 2-1A and 3-9).

Dermaptera, order (*derm*, skin; *ptera*, wing) earwigs (Fig. 11-16).

deutocerebrum the lobe of the insect brain that receives both sensory and motor fibers from the antennae (Fig. 3-5).

di- two.

diastasis a period of rest between successive heartbeats.

diastole relaxation phase of the heartbeat cycle (Fig. 4-8).

dicondylic joint a joint consisting of two condyles and restricting movement to a single plane (Fig. 7-1B).

Diplura, order (*dipl*, two; *ura*, tail) campodeids and japygids (Fig. 11-4B).

Diptera, order (*di*, two; *ptera*, wings) true flies (Figs. 11-42 through 11-45).

disseminated nephrocytes cells that are similar to pericardial cells and located in the perivisceral sinus.

distal distant from the point of attachment.

diverticula blind sacs that form as evaginations from the alimentary canal, for example, the esophageal diverticula of mosquitoes (Fig. 4-2).

dorsal diaphragm fibromuscular septum that encloses the alary muscles associated with the heart and divides the hemocoel into the pericardial and perivisceral sinuses (Figs. 4-6 and 4-7); also called pericardial septum.

dorsal ocelli simple eyes usually accompanying compound eyes and found dorsally on the insect head (Figs. 2-12, 2-13A, C, and 6-16B).

dorsal vessel principal conducting tube of the circulatory system, extending the length of the dorsal midline (Figs. 4-6 and 4-7).

ecdysial cleavage lines preformed lines of weakness along which the shed cuticle of the cranium splits during ecdysis; cuticle in the region of these lines lacks the exocuticle (Fig. 2-8).

ecdysiotropin *see* brain hormone.

ecdysis shedding of the exuvia.

ecdysone a hormone that initiates the growth and molting activity of epidermal cells.

eclosion the emergence of an immature stage from the egg; sometimes used to refer to pupal–adult ecdysis.

eclosion hormone a hormone, synthesized by the median neurosecretory cells, stored in and released from the corpora cardiaca, and controlled by a circadian clock, that influences behavior associated with pupal–adult ecdysis.

economic injury level the lowest density of pest population that does economically significant damage.

economic threshold the density of pest population at which control measures should be applied to prevent the pest from reaching the economic injury level.

ectad outward relative to the midline of the body.

ecto- outside; external.

ectognathous having mouthparts not pulled into the head.

efferent neurons *see* motor neurons.

ejaculatory duct a duct formed by the joining of vas deferens from each testis and involved in the propulsion of semen (Fig. 5-1A).

elytra hardened forewings that protect membranous hindwings; characteristic of beetles (Fig. 2-39C).

Embioptera, order (*embio*, lively; *ptera*, wing) webspinners (Fig. 11-12).

encapsulation a protective mechanism in which large numbers of hemocytes become layered around a foreign entity, such as a parasitic worm, that has invaded the hemocoel.

endocrine glands ductless glands whose products are released directly into the insect hemolymph.

endocuticle the inner, unstabilized layer of the cuticle, which may or may not be pigmented, that is digested during molting (Figs. 2-1 and 2-8).

endo within

endoderm innermost germ layer.

endogenous arising from within.

endopterygote   describes insects that undergo complete metamorphosis; immature instars dissimilar to adults and usually adapted to different environmental conditions; *syn.* holometabolous.

ento   within.

entad   inward relative to the midline of the body.

entognathous   having mouthparts somewhat pulled into the head.

entomophobia   irrational and persistent fear of insect infestation or attack.

Ephemeroptera, order   (*ephemero*, for a day; *ptera*, wings) mayflies (Fig. 11-10).

epicuticle   unpigmented outer layer of the insect cuticle, 0.05–4.0 micrometers thick (Fig. 2-1).

epidermis   the cellular layer of the insect integument (Fig. 2-1A, B).

epimeron   posterior portion of the thoracic pleuron (Fig. 2-18B).

episternum   anterior portion of a thoracic pleuron (Fig. 2-18B).

esophagus   conducting tube of the foregut immediately posterior to the pharynx (Figs. 4-1, and 4-2).

exarate pupa   pupa with free appendages, usually not covered by a cocoon (Fig. 5-24A).

exocrine glands   glands that discharge their products through apertures or ducts into the external world or into lumens of various viscera (Fig. 3-9).

exocuticle   stabilized layer of cuticle between the endo- and epicuticlar layers; resistant to molting fluid and often pigmented (Figs. 2-1 A, B and 2-8A, B).

exogenous   arising from without.

exopterygote   describes insects having wings that become externally apparent during the immature instars (immatures resemble adults) i.e., undergo simple metamorphosis; *syn.* hemimetabolous.

exoskeleton   a skeleton located outside the body, as in arthropods.

extant   presently existing form.

extrinsic muscles   muscles that originate on the inner surface of the body wall of the head, thorax, or abdomen and insert into an appendage (Fig. 3-15).

exuvia   the undigested exo- and epicuticle shed during ecdysis.

exuvial membrane   a thin, homogeneous membrane that may appear in the exuvial space during molting (Fig. 2-7B).

exuvial phase   period of time between apolysis and the beginning of new cuticle secretion.

exuvial space   the space between the epidermis (and later the new cuticle) and the old cuticle that forms as the old cuticle is digested during molting (Fig. 2-7B, C).

facultative parasite   parasite that can complete its life cycle without invading a host, but can also take advantage of a host if one is availble.

feeding deterrent   a chemical that prevents feeding of pests on treated material.

femur   long leg segment between the trochanter and tibia (Fig. 2-21).

fenestrae   openings in the dorsal diaphragm through which hemolymph can pass.

fiber tracts   groups of axons within a ganglion that run parallel to one another.

flagellum   long, apical part of the antenna, commonly composed of several subsegments (Fig. 2-14).

flavines   greenish yellow pigments found in insects.

foramen magnum   posterior opening of the head capsule through which internal structures (alimentary canal, dorsal nerve cord, ventral nerve cord, etc.) pass to the thorax (Fig. 2-13B).

foregut   anterior region of the insect alimentary canal, also called the stomodaeum (Fig. 4-1).

fossorial legs   forelegs bearing sclerotized claws for digging (Fig. 2-36B).

frenulum   spinelike wing-coupling mechanism characteristic of several Lepidoptera (Fig. 2-40B).

frons   frontal region delimited by the frontal sutures (Fig. 2-13A, C).

frontal ganglion   ganglion of the stomatogastric system, located on the dorsal midline of the foregut just anterior to the brain (Figs. 3-3 and 3-5).

fumigants   insecticides that enter through the spiracles and tracheal system in gaseous form.

furcal pits   external manifestations of the sternal apophyses.

furcula   fork-shaped structure on the abdomen of the springtail that engages the tenaculum and is used to propel the insect through the air (Figs. 2-42B and 11-5).

galea tonguelike, distal lobe attached to the stipes of the maxilla (Fig. 2-15).

galls abnormal growths on plants caused by insect larvae feeding on the plant tissue (Fig. 9-14).

ganglion a group of neurons (Figs. 3-1D and 3-2 through 3-7).

gaster collective term for the hymenopteran abdominal segments posterior to the petiole.

gastric caeca a group of diverticula commonly found in the midgut (Fig. 4-1).

gastrulation formation of the mesoderm and endoderm (Fig. 5-16).

-gen, -geny producing; generation.

gena lateral "cheek" region of the head delimited by the compound eye, subgenal suture, and occipital suture (Fig. 2-13A, C).

germ band thickened area of columnar cells of the blastoderm that develops into the embryo (Fig. 5-14E, F).

germ cells cells that differentiate into gametes, eggs, and sperm (Fig. 5-14C–E).

glossae inner, distal lobes found on the prementum of the labium (Fig. 2-15).

gonopods appendages composing the ovipositor.

gonopore external reproductive opening.

Grylloblattodea, order (*gryll,* cricket; *blatta,* cockroach) rock crawlers or ice bugs (Fig. 11-17).

gustation contact chemoreception; "taste."

gynandromorphs insects displaying some typically female body parts and some typically male body parts (Fig. 5-13).

hair plate cluster of tiny trichoid sensilla, located in appressed or overlapping areas of the body, that have a proprioceptive function (Fig. 6-4).

hallucination of arthropod infestation a condition in which a person imagines he or she is being molested by small, difficult-to-locate forms, which localize on the body despite extraordinary preventive measures.

halteres hindwings modified to club-shaped structures important in flight stability (Fig. 2-39E).

harmonic growth a growth pattern in which all parts of the insect body, and the body as a whole, increase by the same ratio during each molt; also called isogonic growth.

haustellate having mouthparts adapted for sucking activities (Figs. 2-30 and 2-31).

heart posterior division of the dorsal vessel; associated with alary muscles and contains ostia (Figs. 4-6 and 4-7).

hematophagous blood-feeding.

hemelytra partly hardened forewings with membranous distal portions, as in Hemiptera (Figs. 2-39D and 11-27A).

hemimetabolous *see* exopterygote.

Hemiptera, order (*hemi,* one half; *ptera,* wing) true bugs, aphids, leafhoppers, and relatives (Figs. 11-27 through 11-32).

hemocoel body cavity of the insect in which the viscera are located and the hemolymph circulates.

hemocoelic insemination *see* traumatic insemination.

hemocytes insect blood cells (Fig. 4-5).

hemocytopoietic organs organs that produce blood cells.

hemogram a count of the total number of hemocytes and an estimate of the numbers of different types of hemocytes in an insect at a given time.

hemostasis inhibition of hemolymph loss; coagulation.

hetero- other; different.

heterogametic possessing the X0 or XY chromosome configuration.

heterogonic growth growth patterns characterized by some parts of the insect body developing at different rates than others; also called allometric growth.

hist- tissue.

hindgut posterior region of the insect alimentary canal; also called proctodaeum (Fig. 4-1).

holoblastic cleavage method of blastoderm formation characteristic of Collembola and certain parasitic wasps and associated with eggs that contain little yolk; the entire zygote divides (compare with meroblastic cleavage).

holometabolous   *see* endopterygote.

holopneustic   pertaining to open ventilatory system that consists of two lateral rows of 10 spiracles each (Fig. 4-13A).

homeo-, *homo-*   like; similar.

homogametic   possessing two X chromosomes.

hooked trichomes   minute, sharp, hooked hairs found on certain plants that provide defense against insect attack.

hydro-   water; fluid; hydrogen.

hydrophilic   having water-attractant characteristics; wettable.

hydrophobic   having water-repellent characteristics; nonwettable.

hydrostatic balance   maintenance of body position in the water by aquatic insects.

hygroreception   perception of moisture in the air.

Hymenoptera, order   (*hymen*, membrane; *ptera*, wing) ants, bees, wasps, and relatives (Figs. 11-53, 11-54 and 11-55).

hyper-   over; above; excessive.

hypermetamorphosis   a form of holometabolous development in which one or more larval instars is distinctly different from the others (Fig. 5-22E).

hypo-   under; lower.

hypognathous   mouthparts hung vertically from the head capsule (Fig. 2-33A).

imaginal buds   masses of cells that persist in an undifferentiated state throughout the larval instars and become adult tissues; also called imaginal discs (Fig. 5-25).

imago   adult insect.

indirect wing muscles   muscles that contribute to insect flight but are not directly attached to the insect wings (Figs. 3-12 and 7-9B).

Insecta   the taxon made up of insects; means literally ''in-cut,'' describing the segmented appearance of insects.

insecticide   a chemical employed to kill insects.

insectoverdins   green pigments found in insects

insect pest management   a modern approach to dealing with pest insects; control measures chosen from all available methods and applied on the basis of carefully evaluated ecological/economic parameters.

insemination   transfer of semen from the male reproductive system to the female reproductive system.

insertion   movable attachment area of skeletal muscles.

in situ   in a natural or orignal position.

instar   each developmental stage of an insect's life cycle.

integument   general body covering.

inter-   between.

internuncial neurons   see association neurons.

intersegmental muscles   the principal longitudinal muscles running between successive antecostae in insects with secondary segmentation (Fig. 2-11B, C).

intima   chitinous lining of the fore- and hindguts and tracheae, which is continuous with the cuticle of the integument (Figs. 4-1 and 4-11A).

intra-   within.

iso-   equal; uniform.

isogonic growth   *see* harmonic growth.

Isoptera, order   (*iso*, equal; *ptera*, wings) termites (Figs. 11-18 and 11-19).

Johnston's organ   sensory structure located in the antennal pedicel (Fig. 6-5).

jugum   lobelike wing-coupling mechanism characteristic of some Lepidoptera (Fig. 2-40C).

juvabione   juvenile hormone mimic from balsam fir.

juvenile hormone   a hormone that promotes larval development and inhibits development of adult characteristics.

kairomone   a chemical that is adaptively favorable to the receiving organism.

kin-, *kino-*   motion; activity.

labial glands   glands associated with the labial segment, commonly having salivary functions.

labial palpi   a pair of segmented appendages associated with the labium (Fig. 2-15).

labium   the posterior mouthpart or "lower lip" (Fig. 2-15).

labrum   the anterior mouthpart or "upper lip" (Fig. 2-15).

lac   a natural body secretion of the scale insect *Laccifer lacca* (Hemiptera, Homoptera) that is used to make shellac, shoe polish, paint, adhesives, and many other products.

lacinia   distal, mesal lobe of the stipes of a maxilla; bears teeth on its inner edge (Fig. 2-15).

larvae   immature instars of insects; usage commonly limited to imamtures of insects with complete metamorphosis.

larviparity   a form of viviparity in which the female gives "birth" to larvae instead of depositing eggs.

lateral occelli   *see* stemmata.

Lepidoptera, order   (*lepido,* scale; *ptera,* wing) butterflies, moths, skippers (Figs. 11-51 and 11-52).

macro-   large.

macrotrichia   *see* setae.

Mallophaga, order   (*mallo,* wool; *phaga,* eat) chewing lice (Fig. 11-24).

Malpighian tubules   elongate excretory structures of the digestive tract in most insects; located at the anterior end of the hindgut (Figs. 4-1 and 4-18).

mandibles   a pair of highly sclerotized, unsegmented "jaws" located between the labrum and maxillae (Figs. 2-13A, C and 2-15).

mandibulate   having primitive, usually chewing, mouthparts.

Mantodea, order   (*mantid,* soothsayer) praying mantids (Fig. 11-21).

maxillae   paired, segmented, secondary jaws that aid in holding and chewing foods; located between the mandibles and the labium (Figs. 2-13B, C and 2-15).

mechanoreception   the perception of stimuli as various mechanical changes.

Mecoptera, order   (*meco,* long; *ptera,* wing) scorpionflies (Fig. 11-41).

median neurosecretory cells   cells in the pars intercerebralis region of the brain that secrete hormones, including brain and eclosion hormones (Figs. 3-6, 3-7, 3-10, and 3-11).

mega-   large; female.

melanins   insect pigments of yellow, black, or brown.

melanization   darkening.

mentum   distal portion of the postlabium (Fig. 2-15).

meroblastic cleavage   method of blastoderm formation characteristic of most insects and associated with eggs containing a large amount of yolk; the fusion nucleus and associated cytoplasm proliferate mitotically, and the daughter nuclei migrate to the periphery of the egg and form the blastoderm (Fig. 5-14B, C).

meso-   middle.

mesenteron   *see* midgut.

mesocuticle   a procuticular layer lying between the exo- and endocuticle.

mesothorax   middle segment of the thorax (Fig. 2-12).

meta-   later; latter; changing.

metameres   essentially identical primary body segments found in embryos and thought to be characteristic of the primitive arthropod (Fig. 10-12); also called somites.

metamorphosis   the developmental process by which a first instar immature stage is transformed into the adult.

micro-   small.

microtrichia   minute fixed hairs that lack the basal articulation characteristic of setae.

micropyle   channel through which spermatozoa enter an egg (Fig. 5-14A).

microvilli   numerous, fingerlike processes that serve to increase an absorptive surface area, for example, in midgut epithelial cells (Fig. 4-3).

midgut   middle region of the insect alimentary canal; site of most digestion and absorption of food; also called mesenteron (Fig. 4-1).

molting fluid   a substance secreted by the epidermal cells; contains chitinase and protease and digests old endocuticle.

molting hormone   ecdysone; hormone that stimulates molting.

monocondylic joint   joint containing a single condyle, allowing considerable movement (Fig. 7-1A).

-morph, morpho-   form.

morphogenesis   all events occuring between the formation of the zygote and the emergence of a sexually mature adult.

motor neurons   unipolar neurons that stimulate muscles; also called efferent neurons.

mouthparts   organs of ingestion (Figs. 2-12 and 2-15).

multipolar   neurons with an axon and several branched dendrites (Fig. 3-1C).

multiterminal innervation   descriptive of insect motor axons that branch out and form terminals at several regularly spaced points along a muscle fiber (Fig. 3-21).

muscle fiber   basic structural unit of a muscle (Fig. 3-18).

muscle unit   10–20 muscle fibers organized as a discrete morphological entity and enveloped by a tracheolated membrane (Fig. 3-21).

mushroom bodies   *see* corpora pedunculata.

mutualism   intimate relationships between two or more kinds of animals or between plants and animals that are advantageous to both.

mycetocytes   cells specialized to house microbes essential to the normal growth and development of certain insects.

mycetomes   tissues composed of mycetocytes and associated with the gut, fat body, or gonads.

mycetophagous   "fungus-eating."

myiasis   a condition caused by the invasion of open wounds or cavities by certain fly larvae (maggots).

myo-   muscle.

myrmecophilous   "ant-loving"; refers to organisms, other than ants, that interact with ants symbiotically.

naiad   immature instar of an aquatic hemimetabolous insect.

natatorial legs   legs adapted for swimming (Fig. 2-36E).

neo-   new.

nephr-   kidney.

nerve   a bundle of axons that provides connections among ganglia and other parts of the nervous system.

neurohaemal organs   structures that serve as storage and release sites for neurosecretory material such as corpora cardiaca.

neuron   a basic functional unit of the nervous system; an elongated, excitable cell that carries information in the form of electrical charges (Fig. 3-1).

neuropile   central region of a ganglion, consisting of intermingling, synapsing axons encapsulated by glial cell processes (Fig. 3-11).

Neuroptera, order   (*neuro*, nerve; *ptera*, wing) lacewings, dobsonflies, and relatives (Figs. 11-38, 11-39, and 11-40).

neurosecretory cells   modified neurons that secrete hormones.

nicotine   secondary plant substance in the common tobacco plant that is toxic to insects.

nidi   groups of regenerative cells located at the bases of midgut epithelial cells (Fig. 4-4B).

nonresonating muscles   muscles that require a nervous impulse for each contraction (compare with resonating muscles); also called synchronous muscles.

notum   dorsal region of a thoracic segment (Figs. 2-17, 2-18A, 2-19, and 2-20).

nurse cells   nutritive cells in polytrophic and telotrophic ovarioles; also called trophocytes (Fig. 5-8A, B, trophic tissue).

nymphs   immature instars of ametabolous and hemimetabolous insects.

obligate parasites   organisms that require a host to survive.

obtect pupa   pupa with wings, antennae, mouthparts, and legs glued close to the body; commonly covered by a cocoon (Fig. 5-24B).

olfaction   distance chemoreception.

ommatidia   individual functional units of the insect compound eye (Fig. 6-11B–D).

ommochromes   red, yellow, and brown pigments commonly found as insect eye pigments.

oo- (oö-)   egg.

oocytes   germ cells in females (Fig. 5-8).

oosome   specialized region of the egg associated with differentiation of germ cells in the embryo (Fig. 5-14A, B).

ootheca "egg purse"; several eggs contained in a common capsule.

opisthognathus mouthparts mouthparts directed ventroposteriorly relative to the head capsule (Fig. 2-33C).

origin stationary attachment area of skeletal muscle.

Orthoptera, order (*ortho-*, straight; *ptera*, wing) grasshoppers, crickets, and relatives (Figs. 11-14 and 11-15).

osmeterium a usually Y-shaped eversible glandular structure located behind the head in swallowtail butterfly larvae; secretes a repugnatorial substance.

ostia valvular openings in the heart (Figs. 4-6 and 4-7).

ov-, ovi- egg.

ovaries paired gonads in the female reproductive system (Fig. 5-6A).

ovarioles egg-producing units of the ovaries (Figs. 5-6 and 5-8).

ovipositor egg-laying apparatus (Figs. 2-12 and 2-44).

ovoviviparity a condition in which a fully developed egg with a chorion hatches within the parent, and a larva is "born" instead of an egg deposited.

ovulation the movement of an oocyte from the ovariole.

pacemaker a region from which waves of contraction in the heartbeat cycle are propagated.

para- alongside of.

parabiosis joining the hemolymph circulation of one individual with that of another.

paraglossae two outer lobes found on the prementum of the labium (Fig. 2-15).

parasitism an intimate relationship in which one individual (the parasite) benefits at the expense of another (the host).

parasitoids insects that resemble both parasite and predator; they begin life as small parasites, but may outgrow and eventually devour the host.

pars intercerebralis cell mass, located in the middorsal region of the brain above the protocerebral bridge, that contains the median neurosecretory cells (Figs. 3-6 and 3-7).

parthenogenesis production of individuals from unfertilized eggs; asexual reproduction.

passive suction ventilation sucking in of air through nearly closed or fluttering spiracles as a result of negative pressure in the tracheae.

passive ventilation simple diffusion of gases into the tracheae and tracheoles without the aid of pumping or other movements.

paurometabolous undergoing a gradual metamorphosis, passing through immature and adult instars in essentially the same environment.

pedicel second antennal segment, between the scape and the flagellum (Fig. 2-14); the constricted region between the thorax and abdomen of Hymenoptera, Apocrita (Fig. 2-41A), which is also called petiole.

pediculosis infestation with lice.

pedogenesis reproduction by immature insects, for example, the beetle *Micromalthus debilis*.

penis male copulatory organ; also called aedeagus (Fig. 5-1).

peri- surrounding.

pericardial surrounding the heart.

pericardial cells large, multinucleate athrocytes found in association with the heart, alary muscles, and surrounding connective tissue (Fig. 4-9A).

pericardial septum *see* dorsal diaphragm.

pericardial sinus body cavity region above the dorsal diaphragm (Fig. 4-6).

perikaryon cell body of a neuron; also called the soma or cell body (Fig. 3-1A–C).

periodicity recurrence of particular behavior(s) in a regular pattern.

peripheral nervous system all the nerves emanating from the ganglia of the central and visceral nervous systems.

periproct posterior somite of a segmented animal, in which the anus is located; also called the telson (Fig. 10-12A).

peristaltic contractions rhythmic contractions of longitudinal and circular muscles surrounding or composing a tubular organ such as the alimentary canal.

peritrophic membrane a chitinous, noncellular membrane that surrounds the food bolus in most insects (Fig. 4-1).

perivisceral sinus body cavity region below the dorsal diaphragm and above the ventral diaphragm (if present) (Fig. 4-6).

pesticide   chemical employed to kill insects, weeds, rodents, and other pests.

petiole   the constriction following the first abdominal segment of Hymenoptera, Apocrita; also called pedicel (Fig. 2-41A).

phagostimulant   a stimulus that induces feeding.

pharate   period of time between apolysis and ecdysis, while the insect is still within the confines of the old cuticle.

pharynx   foregut structure into which the true mouth opens; often a well-developed pump (Fig. 4-2).

Phasmida, order   (*phasm*, phantom) stick and leaf insects (Fig. 11-13).

pheromone   chemical produced by one individual that influences the behavior of another individual of the same species.

-phore   carrier.

phoresy   a commensalistic relationship in which one kind of animal attaches to another and thereby gains a mode of transportation.

photo-   light.

photoreception   reception of stimuli in the form of light.

phragma   platelike inflections of the terga (nota) that provide surface area for the attachment of the dorsal longitudinal wing muscles (Figs. 2-19 and 7-9A).

-phyll   leaf.

-phyte, phyto-   plant.

phytoecdysones   compounds found in certain plants that mimic the action of ecdysone.

plasm-, plasmo-, -plasm   something molded or modeled.

plasma   fluid portion of insect blood.

plastron   a thin layer of gas held permanently in place by fine hydrofuge hairs or other fine cuticular networks; acts as a permanent "physical gill."

Plecoptera, order   (*pleco*, twist; *ptera*, wing) stoneflies (Fig. 11-11).

pleural apophysis   armlike projection of the pleural ridge (Fig. 2-17).

pleural membrane   the unsclerotized cuticle in the pleural region of the abdomen that separates the terga and sterna (Fig. 2-24).

pleural wing process   ventral articulation of a wing (Figs. 2-18B and 7-9).

pleuron   lateral region of the thorax (Figs. 2-17 and 2-18B).

pleurosternal sutures   sutures dividing the sternum from the laterosternites (Fig. 2-18C).

pollen basket   *see* corbiculum.

poly-   many.

polyembryony   mitotic division of an egg resulting in the development of several embryos from a single egg; found among several groups of parasitic Hymenoptera.

polymorphism   many forms, for example one or both sexes of a single species may occur in two or more clearly distinct forms.

polyneural innervation   the situation in which insect skeletal muscle units receive more than one motor axon (Fig. 3-21).

pore canals   tiny tubes connecting the cellular and cuticular layers of the insect integument (Figs. 2-1A, C and 2-2B).

posterior notal wing process   articular point for a wing (Fig. 2-18A).

postnotum   portion of a thoracic notum that bears the internally inflected phragma (Figs. 2-18A, B and 2-19).

predator   a free-living organism that feeds on other, commonly smaller, living organisms (prey); the prey is killed and eaten.

pretarsus   distal leg segment, usually bearing claws (Fig. 2-21).

primary segmentation   the arrangement of body units such that intersegmental grooves form the attachment points for longitudinal muscles (Fig. 2-11A).

pro-   before.

proctodael feeding   ingestion by one individual of material from the anus of another individual.

proctodaeum   *see* hindgut.

procuticle   endo- and exocuticular layers together.

prognathous mouthparts   anteriorly directed mouthparts (Fig. 2-33B).

prolegs   bilateral abdominal legs characteristic of lepidopteran and many hymenopteran larvae (Fig. 2-42C).

propodeum   first abdominal segment of Hymenoptera-Apocrita; it is completely associated with the thorax and anterior to the petiole (Fig. 2-41A).

prot-, proto- first; primary.

prostomium *see* acron.

protein epicuticle the innermost layer of the epicuticle (Fig. 2-1C, inner epicuticle).

prothoracic gland gland that secretes ecdysone (Fig. 5-26).

prothoracic gland hormone see ecdysone.

prothoraciotropic hormone see brain hormone.

prothorax anterior segment of the thorax (Fig. 2-12).

protocerebrum dorsal lobe of the insect brain; contains optic lobes and other major cell masses and regions of neuropile (Fig. 3-5).

Protura, order (*prot,* first; *ura,* tail) proturans (fig. 11-4A).

proventriculus a structure posterior to the crop that often bears sclerotized teeth or spines; the stomodael valve in Diptera (Fig. 4-1).

proximal adjacent to or nearest to the point of attachment.

pseudo- false; temporary,

Psocoptera, order (*psoco,* rub small; *ptera,* wing) book lice and relatives (Fig. 11-23).

pterins red, yellow, and white pigments commonly found as insect eye pigments.

ptero- wing.

pterothorax collective term describing the wing-bearing segments.

Pterygota taxon composed of winged or secondarily wingless insects.

ptilinum an eversible bladder in the head of coarctate pupae, used to force open the puparial case.

pulvillus padlike structure on the lower surface of tarsi in some Orthoptera or in association with each pretarsal claw in Diptera (Fig. 2-38A).

puparium hardened cuticle of the next-to-last larval instar, which encloses a pupa (Fig. 5-24C, D).

pyloric valve valvular structure at the anterior extremity of the insect hindgut (Fig. 4-1).

pyrethrum a kind of botanical insecticide extracted from *Chrysanthemum* flowers.

raptorial legs forelegs modified for grabbing and holding prey (Fig. 2-36C).

rectal papillae pads projecting into the lumen of the rectum that play an important role in the excretory system.

rectum muscular, enlarged, posteriormost region of the intestine (Figs. 4-1 and 4-2).

resilin a cuticular protein with rubberlike properties; sometimes found in pure form in skeletal articulations.

resonating muscles muscles capable of undergoing several successive contractions when stimulated by a single nerve impluse; usually associated with insect flight and sound production; also called asynchronous muscles.

respiration intracellular oxidative breakdown of carbohydrate to carbon dioxide and water.

salivarium the cavity between the hypopharynx and the labium into which the common salivary duct opens (Fig. 4-1).

saltatorial legs legs adapted for jumping (Fig. 2-36D).

scape basal segment of an insect antenna (Fig. 2-14).

sclerites the hardened plates of the insect skeleton (Fig. 2-10).

sclerotins stabilized proteins responsible for the hard, horny character of the cuticle.

sclerotization hardening

scolopidium *see* scolopophorous organ.

scolopophorus organ typically a bundle of sensilla consisting of one or more specialized, bipolar neurons (scolopidia; also called scolopophores) stretched between two internal integumental surfaces; often has a proprioceptive function; also called chordotonal organ (Figs. 6-3 and 6-5).

scutellum posterior plate of the alinotum (Fig. 2-18A, B).

scutum relatively large plate on the alinotum anterior to the scutoscutellar suture (Fig. 2-18A, B).

secondary plant substances chemicals, found in plants, that have no apparent metabolic role and that often influence the behavior and/or physiology of herbivorous insects.

secondary segmentation characterized by membranous areas between adjacent segments slightly anterior to the attachment points of the longitudinal muscles; typical of adult and many larval insects (Fig. 2-11B, C).

segmental muscles the principal longitudinal muscles running between the intersegmental constrictions associated with primary segmentation (Fig. 2-11A).

semen spermatozoa plus various glandular secretions.

seminal vesicle dilated region of the vas deferens in which spermatozoa are stored (Fig. 5-1A).

senescence changes in structure and function that decrease an individual's capacity for survival and lead to death.

sensilla specialized structures that collect information from the external and internal environments and transmit this information to the central nervous system (Figs. 6-1A–D and 6-2).

sensory neuron usually bipolar neuron with cell body located peripherally in the insect and dendrites associated with sensory structures (Fig. 3-1D); also called afferent neuron.

sericulture the silk industry.

serosa outer membrane around the yolk, amnion, and embryo (Fig. 5-15).

sesamin a mixed-function oxidase inhibitor produced by the *Chrysanthemum* plant.

setae macrotrichia; unicellular, cuticular processes (Fig. 2-9).

sex ratio proportion of males to females.

sexual dimorphism differences between the sexes.

Siphonaptera, order (*siphon*, a tube; *aptera*, wingless) fleas (Figs. 11-46 and 11-47).

skeletal muscles muscles attached at both ends to the integument (Figs. 3-15, 3-16 and 3-17).

skeleton supportive shell.

-soma, somat-, -some body.

soma *see* perikaryon.

somites *see* metameres.

specific epithet second part of a species name, indicating a particular species.

spermatheca outpocketing of the vagina in which spermatozoa are stored prior to fertilization (Fig. 5-6A).

spermatocytes diploid cells that undergo meiotic division and become spermatozoa.

spermatogenesis development of male reproductive cells (spermatozoa) in the testicular follicles (Fig. 5-2).

spermatophore spermatozoa encased in a gelatinous packet that is introduced into the female reproductive tract or deposited on the substrate and taken up by the female (Fig. 5-5).

spermatozoa male reproductive cells (Figs. 5-2, 5-3, and 5-4).

spinneret silk-spinning apparatus of lepidopteran larvae (Fig. 2-32E) and others.

spiracles small openings through tracheae that communicate with the external environment (Figs. 2-12, 4-10, and 4-12).

spiracular gills in aquatic insects with closed ventilatory systems, filamentous outgrowths of thin cuticle that open directly into the tracheae; also called cuticular gills (Fig. 4-16C).

stadium interval of time passed in an instar.

stemmata simple eyes found on the sites of the insect head when compound eyes are lacking (Fig. 6-16A); also called lateral ocelli.

sternal apophyses internal projections arising from the eusternum (Fig. 2-17).

sternites individual sclerites forming the sternum of a thoracic or abdominal segment (Figs. 2-18C and 2-24).

sternum ventral region of a thoracic or abdominal segment (Figs. 2-18C and 2-24).

stipes comparatively large part of a maxilla, borne on the cardo, that bears the galea and lacinia (Fig. 2-15).

stomatogastric system the part of the visceral nervous system associated with the brain, aorta, and foregut; also called the stomodael system (Figs. 3-5 and 3-7).

stomodael valve a structure formed by an intussusception of fore- and midgut tissues (Fig. 4-1).

stomodaeum   *see* foregut.

Strepsiptera, order   (*strepsi,* turning or twisting; *ptera,* wing) stylopids or twisted-wing parasites (Fig. 11-37).

stretch receptors   mechanoreceptors composed of multipolar neurons associated with muscles, the alimentary canal, and other viscera; produce a nervous impulse when associated tissue undergoes a change in length.

stridulation   production of sound by means of a frictional mechanism.

stylets   swordlike or needlelike modifications of the generalized mouthpart structures (Fig. 2-30).

styli   simple, bilateral appendages borne on the abdominal segments (Fig. 2-42A).

subalar sclerite   small sclerite posterior to the pleural wing process.

subcuticle   newly secreted endocuticle in which the microfibers are not yet oriented.

subesophageal ganglion   the ganglion, ventral to the foregut, that innervates sense organs and muscles associated with mouthparts, salivary glands, and the neck region.

subgenual organs   groups of chordotonal sensilla located in the basal portion of the tibial leg segment (Fig. 6-8).

subimago   winged ''subadult'' that undergoes one molt and becomes the adult; unique to mayflies (Ephemeroptera).

submentum   proximal portion of the postlabium (Fig. 2-15).

sutures   external grooves in the insect skeleton (Fig. 2-10); also called sulci.

sym-, syn-   together.

synapse   region of close association between terminal arborizations and dendrites (Fig. 3-1D).

synaptic cleft   space between the terminal arborizations of one neuron and the dendrites of the next.

synchronous muscles   *see* nonresonating muscles.

synergist   chemical that, when mixed with an insecticide, renders the mixture more toxic than the simple sum of the toxicities of the individual components.

systemic   a synthetic organic insecticide that, when absorbed by plants or ingested and absorbed by animals, makes them toxic to insects.

systole   contraction phase of the heartbeat cycle (Fig. 4-8).

taenidia   helical fold in the cuticular lining of tracheae and tracheoles (Fig. 4-11A, B).

tagmata   the three major regions of the insect body: head, thorax, and abdomen.

tanning hormone   *see* bursicon.

tarsomeres   subdivisions of the tarsus.

tarsus   distal, segmented part of the insect leg attached to the tibia (Fig. 2-21).

teleology   ascribing purposiveness to natural phenomena.

telson   *see* periproct.

tenaculum   abdominal structure found on the venter of Collembola; serves as a ''catch'' for the furcula prior to ''springing'' (Figs. 2-42B and 11-5A).

tenent hairs   tiny hairs on the pulvilli and tarsal pads that allow some insects to cling to smooth surfaces (Fig. 7-3).

teneral   term describing newly emerged, pale, soft-bodied individuals prior to the completion of melanization and sclerotization.

tentorial pits   small, externally apparent mouths of the invaginations that form the anterior and posterior tentorial arms (Fig. 2-13).

tentorium   an internal framework in the insect head that provides rigidity and areas for muscle attachment (Fig. 2-13D).

tergites   sclerites that form the tergum.

tergum   dorsal region of the thorax and abdomen (Fig. 2-24); usually called notum when referring to the thorax.

terminal aborizations   branching processes found at the end of an axon (Fig. 3-1).

testes   in the male reproductive system, the bilateral, paired structures housing the germ cells (Fig. 5-1).

testicular follicles   sperm tubes composing the testes (Fig. 5-1).

thermoreception   the reception of stimuli as heat.

thoracic ganglia   three ganglia located posterior to the subesophageal ganglion and innervating the three thoracic segments (Fig. 3-3).

Thysanoptera, order (*thysano,* a fringe; *ptera,* wing) thrips (Fig. 11-26).

Thysanura, order (*thysan,* fringe; *ura,* tail) silverfish, firebrats, and relatives (Fig. 11-6B, C).

tibia long leg segment located between the femur and the tarsus (Fig. 2-21).

tonofibrillae fibrillar structures (bundles of microtubules, Fig. 3-14) of skeletal muscle that anchor muscles to the integument (Fig. 3-13B).

tormogen cells cell that forms the socket and membrane around a seta (Fig. 2-9).

tracheae a system of branching tubules through which gaseous exchange is accomplished in insects (Fig. 4-10 and 4-11).

tracheal end cell end of a trachea; typically branches into several tracheoles (Fig. 4-11C).

tracheal gills integumental evaginations, covered by thin cuticle and supplied with tracheae and tracheoles; found in many aquatic (Fig. 4-16A, B) and some endoparasitic insects.

transovarial via the egg, for example, transovarial transmission of symbiotic microbes from one generation to the next.

transpiration loss of water through the integument or tracheae by evaporation.

traumatic insemination ejaculation of semen directly into the hemocoel; characteristic of bed bugs and close relatives; also called hemocoelic insemination.

trichogen cell cell that forms a seta (Fig. 2-9).

Trichoptera, order (*tricho,* hair; *ptera,* wing) caddisflies (Figs. 11-48, 11-49 and 11-50).

tritocerebrum ventral lobe of the insect brain (Figs. 3-5 and 3-6).

trochanter typically small segment of the insect leg that articulates with the coxa and usually forms an immovable attachment with the femur (Fig. 2-21).

trophocytes *see* nurse cells.

true mouth the opening at the base of the hypopharynx, within the preoral (cibarial) cavity formed by the mouthparts (Figs. 2-16 and 4-1).

tympanic organs auditory organs composed of a thin integumental area (tympanic membrane) and a group of chordontonal sensilla attached to the entad surfaces (Figs. 2-37C, and 6-6).

Tyndall blue a blue color produced by colloidal-sized granules in the epidermal cells of some dragonflies.

urate cells cells in fat body; may store nitrogenous waste.

ungue pretarsal claw (Fig. 2-38).

unipolar neuron neuron with a single stalk from the cell body connecting with the axon and a collateral (Fig. 3-1A).

vagina external opening of the female reproductive system (Fig. 5-6A).

valvifers proximal structures of the ovipositor that bear the valvulae (Fig. 2-24).

valvulae distal structures of the ovipositor borne on the valvifers (Fig. 2-24).

vas deferens a duct, into which the vasa efferentia empty, that connects each testis to a seminal vesicle (Fig. 5-1).

vasa efferentia tiny ducts leading from each testicular follicle to a common lateral duct, the vas deferens (Fig. 5-1B).

vector carrier; for example, *Anopheles* mosquitoes serve as vectors for malaria parasites, carrying the parasites from one human host to another.

venom inherently toxic insect (or other animal) secretion, harmful to man and animals.

venter the ventral region.

ventilation transport of oxygen to and carbon dioxide from the insect body tissues.

ventriculus the usually somewhat enlarged region of the midgut that serves as the insect's stomach (Fig. 4-1).

vermiform resembling a worm.

vertex the dorsal portion of the cranium bisected by the coronal suture (Fig. 2-13A).

visceral muscles muscles responsible for movements of internal organs; usually attached to other visceral muscles.

vitellarium the region of an ovariole in which oocytes undergo vitellogenesis (Fig. 5-6B).

vitellin yolk protein.

vitelline membrane   cell membrane encasing the egg (Fig. 5-10).

vitellogenesis   the process of deposition of nutrients in oocytes; yolk deposition.

vitellogenin   protein in the fat body that is transferred to oocytes during vitellogenesis.

vitellophage   yolk cells considered responsible for the initial digestion of yolk (Fig. 5-14C–F)

wax filaments   tiny canals, continuous with pore canals, that penetrate the protein epicuticle and cuticulin layers and are thought to be involved in transporting wax molecules to the epicuticle; also called wax canals (Fig. 2-1C).

wax layer   an epicuticular lipid layer between the cuticulin layer and cement layer that contributes to the permeability characteristics of the cuticle (Fig. 2-1C).

wing flexion   the folding of the wings posteriorly over the abdomen.

wing pads   externally apparent developing wings of hemimetabolous insects.

wing veins   longitudinal and transverse supportive framework of the wings (Fig. 2-23).

yolk   nutrients (carbohydrates, proteins, and lipids) found in mature insect eggs (Figs. 5-10 and 5-14); also called deutoplasm.

Zoraptera, order   (*zor,* pure; *ptera,* wing) zorapterans (Fig. 11-22).

zygote   fertilized egg.

Author Index

Subject Index